ELECTRONIC STRUCTURE OF MATERIALS

ELECTRONIC STRUCTURE OF MATERIALS

RAJENDRA PRASAD

CRC Press is an imprint of the
Taylor & Francis Group, an **informa** business
A CHAPMAN & HALL BOOK

Taylor & Francis
Taylor & Francis Group
6000 Broken Sound Parkway NW, Suite 300
Boca Raton, FL 33487-2742

Printed on acid-free paper
Version Date: 20130424

International Standard Book Number-13: 978-1-4665-0468-4 (Hardback)

Library of Congress Cataloging-in-Publication Data

Prasad, Rajendra.
Electronic structure of materials / Rajendra Prasad.
pages cm
Includes bibliographical references and index.
ISBN 978-1-4665-0468-4 (hardback)
1. Electronic structure. 2. Materials--Electric properties. 3. Electronic apparatus and appliances--Materials. I. Title.

QC176.8.E4P74 2013
530.4'11--dc23 2013014259

Visit the Taylor & Francis Web site at
http://www.taylorandfrancis.com

and the CRC Press Web site at
http://www.crcpress.com

Dedicated to my teacher Professor S. K. Joshi
who introduced me to this subject.

Contents

Preface xiii
Author xvii
Symbols xix
Abbreviations xxi

1. Introduction 1
Further Reading 8

2. Quantum Description of Materials 11
2.1 Introduction 11
2.2 Born–Oppenheimer Approximation 12
2.3 Hartree Method 18
2.3.1 Interpretation of ε_i 23
2.4 Hartree–Fock (H–F) Method 24
2.4.1 Interpretation of ε_i: Koopmans' Theorem 30
2.5 Configuration Interaction (CI) Method 30
2.6 Application of Hartree Method to Homogeneous Electron Gas (HEG) 30
2.7 Application of H–F Method to HEG 33
2.8 Beyond the H–F Theory for HEG 37
2.8.1 Correlation Energy 37
Exercises 38
Further Reading 39

3. Density Functional Theory 41
3.1 Introduction 41
3.2 Thomas–Fermi Theory 43
3.3 Screening: An Application of Thomas–Fermi Theory 46
3.4 Hohenberg–Kohn Theorems 49
3.5 Derivation of Kohn–Sham (KS) Equations 52
3.6 Local Density Approximation (LDA) 55
3.7 Comparison of the DFT with the Hartree and H–F Theories 57
3.8 Comments on the KS Eigenvalues and KS Orbitals 57
3.9 Extensions to Magnetic Systems 59
3.10 Performance of the LDA/LSDA 60
3.11 Beyond LDA 61
3.11.1 Generalized Gradient Approximations (GGAs) 61
3.11.2 LDA + U Method 62
3.11.3 Self-Interaction Correction (SIC) Method 63
3.11.4 GW Method 63

3.12 Time-Dependent Density Functional Theory (TDDFT) 64
Exercises 65
Further Reading 66

4. Energy Band Theory 67
4.1 Introduction 67
4.2 Crystal Potential 68
4.3 Bloch's Theorem 69
4.4 Brillouin Zone (BZ) 73
4.5 Spin–Orbit Interaction 76
4.6 Symmetry 80
4.7 Inversion Symmetry, Time Reversal, and Kramers' Theorem 82
4.8 Band Structure and Fermi Surface 84
4.9 Density of States, Local Density of States, and Projected Density of States 87
4.10 Charge Density 90
4.11 Brillouin Zone Integration 92
Exercises 92
Further Reading 94

5. Methods of Electronic Structure Calculations I 95
5.1 Introduction 95
5.2 Empty Lattice Approximation 95
5.3 Nearly Free Electron (NFE) Model 98
5.4 Plane Wave Expansion Method 103
5.5 Tight-Binding Method 104
5.6 Hubbard Model 110
5.7 Wannier Functions 111
5.8 Orthogonalized Plane Wave (OPW) Method 112
5.9 Pseudopotential Method 114
Exercises 121
Further Reading 123

6. Methods of Electronic Structure Calculations II 125
6.1 Introduction 125
6.2 Scattering Approach to Pseudopotential 127
6.3 Construction of First-Principles Atomic Pseudopotentials 133
6.4 Secular Equation 137
6.5 Calculation of the Total Energy 142
6.6 Ultrasoft Pseudopotential and Projector-Augmented Wave Method 144
6.7 Energy Cutoff and *k*-Point Convergence 145
6.8 Nonperiodic Systems and Supercells 146
Exercises 149
Further Reading 150

7. Methods of Electronic Structure Calculations III 151
7.1 Introduction 151
7.2 Green's Function 151
7.3 Perturbation Theory Using Green's Function 159
7.4 Free Electron Green's Function in Three Dimensions 163
7.5 Korringa–Kohn–Rostoker (KKR) Method 166
7.6 Linear Muffin-Tin Orbital (LMTO) Method 170
7.7 Augmented Plane Wave (APW) Method 172
7.8 Linear Augmented Plane Wave (LAPW) Method 174
7.9 Linear Scaling Methods 176
Exercises 177
Further Reading 178

8. Disordered Alloys 179
8.1 Introduction 179
8.2 Short- and Long-Range Order 180
8.3 An Impurity in an Ordered Solid 181
8.4 Disordered Alloy: General Theory 184
8.5 Application to the Single Band Tight-Binding Model of Disordered Alloy 194
8.6 Muffin-Tin Model: KKR-CPA 196
8.7 Application of the KKR-CPA: Some Examples 202
8.7.1 Density of States 202
8.7.2 Complex Energy Bands 203
8.7.3 Fermi Surface 205
8.8 Beyond CPA 206
Exercises 207
Further Reading 208

9. First-Principles Molecular Dynamics 209
9.1 Introduction 209
9.2 Classical MD 210
9.3 Calculation of Physical Properties 212
9.4 First-Principles MD: Born–Oppenheimer Molecular Dynamics (BOMD) 214
9.5 First-Principles MD: Car–Parrinello Molecular Dynamics (CPMD) 215
9.6 Comparison of the BOMD and CPMD 220
9.7 Method of Steepest Descent 220
9.8 Simulated Annealing 221
9.9 Hellmann–Feynman Theorem 223
9.10 Calculation of Forces 225
9.11 Applications of the First-Principles MD 230
Exercises 230
Further Reading 231

10. Materials Design Using Electronic Structure Tools ... 233
10.1 Introduction ... 233
10.2 Structure–Property Relationship ... 234
10.3 First-Principles Approaches and Their Limitations ... 234
10.4 Problem of Length and Time Scales: Multiscale Approach ... 235
10.5 Applications of the First-Principles Methods to Materials Design ... 237

11. Amorphous Materials ... 241
11.1 Introduction ... 241
11.2 Pair Correlation and Radial Distribution Functions ... 242
11.3 Structural Modeling ... 243
11.4 Anderson Localization ... 245
11.5 Structural Modeling of Amorphous Silicon and Hydrogenated Amorphous Silicon ... 248
Exercises ... 252
Further Reading ... 253

12. Atomic Clusters and Nanowires ... 255
12.1 Introduction ... 255
12.2 Jellium Model of Atomic Clusters ... 257
12.3 First-Principles Calculations of Atomic Clusters ... 259
12.3.1 Ground-State Structures of Silicon and Hydrogenated Silicon Clusters ... 260
12.3.2 Photoabsorption Spectra ... 266
12.3.3 Carbon Clusters ... 268
12.4 Nanowires ... 270
12.4.1 Peierls Distortion ... 271
12.4.2 Jellium Model of Nanowire ... 272
12.4.3 First-Principles Calculations ... 277
Exercises ... 278
Further Reading ... 278

13. Surfaces, Interfaces, and Superlattices ... 281
13.1 Introduction ... 281
13.2 Geometry of Surfaces ... 282
13.3 Surface Electronic Structure ... 283
13.3.1 Surface States ... 284
13.3.2 First-Principles Calculations of Surface States ... 286
13.4 Surface Relaxation and Reconstruction ... 288
13.5 Interfaces ... 290
13.5.1 Band Offsets in Heterojunctions ... 290
13.6 Superlattices ... 292
Exercises ... 295
Further Reading ... 296

14. Graphene and Nanotubes ... 297
14.1 Introduction ... 297
14.2 Graphene ... 297
14.2.1 Structure and Bands ... 297
14.2.2 Dirac Fermions, Pseudospin, and Chirality ... 304
14.3 Carbon Nanotubes ... 307
Exercises ... 314
Further Reading ... 315

15. Quantum Hall Effects and Topological Insulators ... 317
15.1 Introduction ... 317
15.2 Classical Hall Effect ... 317
15.3 Landau Levels ... 321
15.4 Integer and Fractional Quantum Hall Effects (IQHE and FQHE) ... 324
15.5 Quantum Spin Hall Effect (QSHE) ... 328
15.6 Topological Insulators ... 330
Exercises ... 337

16. Ferroelectric and Multiferroic Materials ... 339
16.1 Introduction ... 339
16.2 Polarization ... 340
16.3 Born Effective Charge ... 345
16.4 Ferroelectric Materials ... 346
16.5 Multiferroic Materials ... 351
Exercises ... 354
Further Reading ... 355

17. High-Temperature Superconductors ... 357
17.1 Introduction ... 357
17.2 Cuprates ... 358
17.3 Iron-Based Superconductors ... 365
Exercises ... 367
Further Reading ... 368

18. Spintronic Materials ... 369
18.1 Introduction ... 369
18.2 Magnetic Multilayers ... 370
18.3 Half-Metallic Ferromagnets ... 375
18.4 Dilute Magnetic Semiconductors ... 380
Exercises ... 383
Further Reading ... 383

19. Battery Materials ... 385
19.1 Introduction ... 385

19.2 $LiMnO_2$ 387
19.3 $LiMn_2O_4$ 395
Exercises 398

20. Materials in Extreme Environments 399
20.1 Introduction 399
20.2 Materials at High Pressures 400
20.3 Materials at High Temperatures 403
Exercises 407
Further Reading 407

Appendix A: Electronic Structure Codes 409

Appendix B: List of Projects 411

Appendix C: Atomic Units 413

Appendix D: Functional, Functional Derivative, and Functional Minimization 415

Appendix E: Orthonormalization of Orbitals in the Car–Parrinello Method 417

Appendix F: Sigma (σ) and Pi (π) Bonds 421

Appendix G: *sp*, sp^2, and sp^3 Hybrids 423

References 425

Index 443

Preface

Recently there has been a move in various universities and Indian Institutes of Technology (IITs) to introduce research in the undergraduate curriculum (BS, BTech, and MSc) so that undergraduates can also experience the excitement of the current research. The biggest problem one faces in implementing such a scheme is the shortage of textbooks that are written at the undergraduate level and at the same time that cover the topics of current research. This book on the electronic structure of materials is a step in this direction. It is aimed at the advanced undergraduate and graduate students who want to gain some understanding of electronic structure methods or want to use these tools in their research.

Electronic structure plays a fundamental role in determining the properties of materials such as the cohesive energy, equilibrium crystal structure, phase transitions, transport properties, magnetism, ferroelectricity, optical properties, and so on. The last two decades have seen an intense activity in the field of electronic structure of materials and the field has been gaining importance day by day. Several factors have contributed to this:

1. The theory of electronic structure has advanced to a level where it is possible to obtain good quantitative agreement with experiments without using any adjustable parameters. Thus, the theory has acquired some predictive power.
2. There has been an intense search for new materials for technological applications such as materials for computer memory, spintronic devices, quantum computations, rechargeable batteries, and so on. There has been a realization that electronic structure studies can greatly help such a search. The new emerging area of nanomaterials has given further impetus to this field.
3. With the advent of fast computers, there has been a tremendous increase in computing power. Due to the development of fast algorithms and the easy availability of codes, complicated electronic structure calculations with large unit cells can be done quite efficiently. With these developments, it is now possible to design a material with desired properties in a computer.

Realizing the importance of this field, a course on this subject was offered at IIT Kanpur for MSc and PhD students. With increasing importance of materials, many other IITs and universities are likely to start such a course. The existing textbooks in the field are either too advanced for MSc students or too elementary to bring them up to date with current research. This book, which is a compilation of lecture notes given in this course, is an attempt to bridge this gap. These notes were circulated among students and their feedback was taken. In addition to pointing out typographical mistakes, they gave some important suggestions that were incorporated in these notes. Since the students attending this course were not only from physics but also from other disciplines such as materials science, chemical engineering, and chemistry, I have only assumed some basic knowledge of quantum mechanics, statistical mechanics, and condensed matter physics. Thus, in all the chapters, I start the subject at a very basic level, slowly come to an advanced level, and then refer the students to more recent work. The idea is to get the students interested in the subject and prepare them adequately so that they can follow the more advanced material that has been referred to. For this reason, I have given a longish list of references at the end. Some important references are given at the end of each chapter as Further Reading.

In this book, we take a microscopic view of materials as composed of interacting electrons and nuclei and aim at explaining all the properties of materials in terms of basic quantities of electrons and nuclei such as electronic charge, mass, and atomic number. This is called the first-principles approach, which does not have any adjustable parameters and is based on quantum mechanics. The book has been divided into two parts. The first part (Chapters 1 through 10) is concerned with the fundamentals and methods of electronic structure and the second part (Chapters 11 through 20) deals with the applications of these methods. In the second part, some selected examples have been given to illustrate the applications of these methods to different materials. The materials chosen include crystalline solids, disordered substitutional alloys, amorphous solids, nanoclusters, nanowires, graphene, topological insulators, battery materials, spintronic materials, materials under extreme conditions, and so on.

Chapter 1 gives a historical introduction and an overview of the electronic structure field. Chapter 2 explains quantum description of matter in terms of electrons and nuclei and sets the stage for the rest of the book. In Chapter 2, we try to solve the many-body problem of interacting electrons and nuclei by using the Born–Oppenheimer approximation. After discussing the Born–Oppenheimer approximation, we focus on the electronic problem by keeping the nuclei fixed. Then Hartree and Hartree–Fock methods are discussed in detail with application to the jellium model.

Chapter 3 is devoted to the discussion of the density functional theory (DFT), which provides the foundation for the first-principles calculations discussed in the subsequent chapters. We start with the Thomas–Fermi theory and then discuss basic theorems of the DFT and derivation of the Kohn–Sham (KS) equations. We then discuss the approximations like LDA (local density approximation), GGA (generalized gradient approximation), LDA + U, GW, and so on. We also discuss very briefly the time-dependent DFT.

In Chapter 4, we discuss basic energy band theory and in Chapters 5 through 7, various methods of electronic structure calculations such as pseudopotential, the KKR (Korringa–Kohn–Rostoker) method, APW (augmented plane wave) methods, and so on are explained. In Chapter 7, we introduce Green's function, which we use extensively in Chapter 8. In Chapter 8, we focus on disordered alloys and discuss approximations such as VCA (virtual crystal approximation), ATA (average *t*-matrix approximation), CPA (coherent potential approximation), and KKR-CPA (Korringa–Kohn–Rostoker-coherent potential approximation). We also discuss very briefly the attempts to go beyond CPA. In Chapter 9, we discuss first-principles molecular dynamics (MD). We start with the discussion of classical MD, and then discuss Born–Oppenheimer MD (BOMD) and Car–Parrinello MD (CPMD), simulated annealing, the Hellman–Feynman theorem, and calculation of forces. Chapter 10 discusses some general principles associated with materials design.

The second part of the book, from Chapters 11 through 20, discusses some applications of electronic structure. In Chapter 11, we discuss amorphous semiconductors and Anderson localization. It turns out that the first-principles approach is ideally suited for studying low-dimensional systems and nanomaterials, which are covered in most of the remaining chapters. In Chapters 12 through 14, we discuss low-dimensional systems. Chapter 12 is devoted to atomic clusters and nanowires, Chapter 13 to surfaces, interfaces, and multilayers and Chapter 14 to graphene and nanotubes. Chapter 15 discusses quantum Hall effects and recently discovered topological insulators. In Chapters 16 through 20, we discuss ferroelectrics and multiferroics, high T_c materials, spintronic materials, battery materials, and materials under extreme conditions.

The exercises given at the end of chapters form an integral part of the course. Although many exercises are analytic in nature, in many exercises the student is asked to use the computer and draw figures. By doing so, the student is expected to get a sense of numbers and visualize the physical picture associated with the problem. There are excellent software packages available for visualization that can be used to see, for example, the structure of a solid, charge density, and Fermi surface. Also, there are many electronic structure packages that are quite user-friendly and are listed in Appendix A. Some of these packages are free of cost and can run on a laptop PC. The students are encouraged to use one of these packages for exercises and projects. It is only through practice that one can learn how to do electronic structure calculations.

The book contains much more material than can be covered in a semester. Generally, I have been able to cover the first 10 chapters with a few applications in one semester. Depending on the level of the students and the course, some applications can be discussed immediately after the theory has been covered. For example, the chapter on graphene and carbon nanotubes can be discussed immediately after the discussion of the tight-binding approximation. Similarly, the chapters on surfaces and interfaces can be discussed after Chapter 6. Most applications given in Chapters 11 through 20 were given as projects. In this project, each student was supposed to use one of the methods of electronic structure. At the end of the semester, they were asked to submit a report and give a presentation. A list of some projects is given in Appendix B. It was expected that by the end of the course, the students could calculate the electronic structure and atomic arrangement of a given material.

In this book I have used SI notation in theoretical derivations, except in a few cases, where I have used atomic units (a.u.). In Appendix C, I have discussed a.u. and their conversion to SI units. In general, all the electronic calculations are done in a.u. However, for convenience, in this book I have expressed all the energies in electron volts and all lengths in angstroms.

I have tried to use the same notation for a physical quantity throughout the book. At times this has been difficult and I had to invent a new notation, not commonly used in the literature. For example, for electron mass m_e is used instead of m, which is used for magnetic quantum number. Therefore, a list of common notations and abbreviations is given at the beginning of the book.

I have been very fortunate to have many great friends and colleagues. Over the years, discussions with them have given me a wider perspective and better understanding of the field, for which I am grateful to them. It is not possible to name all of them, but I am specially grateful to Professors Arun Bansil, Roy Benedek, Deepak Kumar, Abhijit Mookerjee, Sushil Auluck, Deepak Gupta, Manoj Harbola, V. Ravishankar, Ashish Garg, Ramesh Budhani, and Satish Agarwal. I am specially thankful to my PhD students Amritendu Roy, Bahadur Singh, Divya, Himanshu Pandey, and Uday Bhanu Pramanik, who went through the notes, pointed out mistakes, and also helped in drawing many figures. Pictures of scientists and some figures have been taken from Wikipedia, which are thankfully acknowledged. Some figures have been reproduced from journals and I thank the authors and publishers for giving me the permission to reproduce them in the book.

This book could not have been written without the encouragement, understanding, and support of my wife Mamta and my daughters Vibha and Anu. The book was essentially written on the time that belonged to them.

Although the book has been proofread many times, but there could still be some errors. I request the readers to bring them to my notice by sending an e-mail to rprasad@iitk.ac.in. The corrections to these errors will be put on the website of the book: http://home.iitk.ac.in/~rprasad/Electronic_Structure.shtml

Rajendra Prasad
IIT Kanpur

Author

Rajendra Prasad is a professor of physics at the Indian Institute of Technology (IIT) Kanpur. He obtained his PhD in physics from the University of Roorkee (now renamed as IIT Roorkee) in 1976. He did his postdoctoral work at Northeastern University, Boston. After 3 years of postdoctoral work, he joined the faculty of Northeastern University, Boston. In 1983, he joined the Argonne National Laboratory, Lemont, Illinois, as assistant physicist and in 1984, he was appointed at IIT Kanpur as a faculty member. He also spent a year at Abdus Salam International Centre of Physics, Trieste during 1993–1994. Aside from brief periods which he spent at Northeastern University, Boston and Abdus Salam International Centre of Physics, Trieste, he has remained at IIT Kanpur where he has taught many undergraduate and graduate level courses and has guided several PhD students. His research work, which spans over four decades, is mainly in the field of electronic structure of materials and covers a wide spectrum including electronic structure of metals, disordered alloys, atomic clusters, transition metal oxides, ferroelectrics, multiferroics, and topological insulators. He is a fellow of the National Academy of Sciences, India.

Symbols

Ω	Volume of the system
Ω_{cell}	Volume of the unit cell
E	Energy of many-particle system
ε	One-electron energy
ε_F	Fermi energy
$\vec{k}$	Bloch wave vector
k_F	Fermi wave vector
H	Hamiltonian
$\vec{r}$	Electronic coordinate
$\{\vec{r}_i\}$	$\vec{r}_1, \vec{r}_2, \ldots, \vec{r}_n$
$\vec{R}$	Ionic coordinate
m_e	Mass of electron
M	Mass of ion
T	Kinetic energy
$V(\vec{r})$	Potential energy
W	Total pseudopotential
w	Atomic pseudopotential
N_{cell}	Number of unit cells in the solid
$\rho(\varepsilon)$	Density of states
$n(\vec{r})$	Electron density at point $\vec{r}$
$\rho(\vec{r})$	Charge density at point $\vec{r}$
$\vec{P}$	Polarization

Abbreviations

2D-ACAR	Two-dimensional angular correlation of positron annihilation radiation
APW	Augmented plane wave
ARPES	Angle-resolved photoemission spectroscopy
BOMD	Born–Oppenheimer molecular dynamics
CCPA	Cluster coherent potential approximation
CI	Configuration interaction
CPA	Coherent potential approximation
CPMD	Car–Parrinello molecular dynamics
DFT	Density functional theory
dHvA	de Haas–van Alphen
DOS	Density of states
GGA	Generalized gradient approximation
H–F	Hartree–Fock
KKR	Korringa–Kohn–Rostoker
KKR-CPA	Korringa–Kohn–Rostoker coherent potential approximation
LDA	Local density approximation
LDOS	Local density of states
LSDA	Local spin density approximation
MD	Molecular dynamics
PAW	Projected augmented wave
PDOS	Projected density of states
QHE	Quantum Hall effect
STM	Scanning tunneling microscopy
TDDFT	Time-dependent density functional theory

1

Introduction

Materials have played an important role in the evolution of human civilization. This is apparent from the fact that various stages in the history of mankind such as Stone Age, Iron Age, Bronze Age, and now, the Silicon Age have been named after materials. Materials based on silicon are integral parts of many devices and have revolutionized the way we live today. Thus, the progress of our civilization is largely dependent on new materials that exhibit new and exotic properties. Examples of such materials are high T_c superconductors, multiferroic materials, topological insulators, battery materials, and so on. Electronic structure calculation, which is the subject of the book, can play an important role in the discovery of many new materials. In fact, many materials such as topological insulators were first predicted by theoretical calculations.

In this book, we embark on an ambitious journey aimed at understanding and predicting properties of materials from first principles. Such an approach does not involve any adjustable parameter. The only input to the calculation is electronic charge, electron mass, atomic numbers, and masses of the atoms of the material. Materials are made of atoms, which, in turn, are made of electrons and nuclei. Therefore, all the properties of materials can be attributed to the complex behavior of electrons and nuclei interacting with each other. Nuclei are massive compared to electrons and can be handled using classical mechanics except in the case of hydrogen and helium. In solids, the nuclei vibrate about their mean positions. In contrast, electrons do not obey classical mechanics and are governed by quantum mechanics. The behavior of electrons is mainly responsible for all physical and chemical properties of materials. A study of the behavior of electrons in materials and how they are distributed in real space and in different energy levels forms the main subject of electronic structure theory. The electronic structure determines the cohesive energy, which, in turn, determines the structure of the material. The transport, optical, magnetic, and superconducting properties are also governed by the electronic structure. Thus, the electronic structure plays a central role in understanding diverse properties of materials. With the recent advances that have taken place in the electronic structure calculations, now, it is possible to design a new material with desired properties.

Attempts to understand properties of metals based on electronic behavior started as early as 1900 with Drude (1900) and Lorentz (1909). They assumed that metals contain free electrons that move in a uniform positive background provided by the ions. This model, known as the jellium model or

FIGURE 1.1
(a) Paul Drude (1863–1906) played a leading role in developing the free electron model. (b) H. A. Lorentz (1853–1928) further developed Drude's model by including Maxwell distribution function. He is better known for Lorentz transformation, Lorentz contraction, and Lorentz force. He was awarded the Nobel Prize in 1902.

free electron model, completely neglects electron–ion and electron–electron interactions and the electrons are assumed to move as independent particles. This approximation is known as independent particle or one-electron approximation. Drude and Lorentz (Figure 1.1a and b) applied the kinetic theory of gases to this model and were able to understand some properties such as electrical conductivity but could not explain the specific heat of metals at low temperatures. Sommerfeld (1928) (Figure 1.2) modified Drude's model by assuming that electrons follow the Pauli-exclusion principle or Fermi–Dirac statistics. This enabled him to explain specific heat at low temperatures and the Wiedemann–Franz law but failed in explaining many other properties. The question why some solids were insulators was not addressed by the model at all. Further progress was made by including interaction between electrons and ions that were assumed to sit on a periodic lattice. Bloch (1928) provided the first breakthrough by giving Bloch's theorem that puts a constraint on the form of electronic wave function and defines a quantum number $\vec{k}$ as a result of the translational symmetry. The Schrödinger equation for an electron moving in a weak periodic potential could be solved by using the nearly free approximation to obtain energy eigenvalues $\varepsilon_{\vec{k}}$ as a function of $\vec{k}$, known as band structure. One important consequence of the periodic electron–ion potential was that electrons are allowed only in certain energy

FIGURE 1.2
A. Sommerfeld (1868–1951) included the Fermi–Dirac distribution in the free electron model. He is also known for many other contributions to quantum mechanics.

ranges called energy bands and are forbidden in certain energy ranges called energy band gaps. This explained why certain materials were insulators and others were metals (Wilson 1931a,b). If the number of electrons in a material is such that the occupied bands are completely filled, it is an insulator; otherwise, it is a metal. If the band gap is small, say <2 eV, it is a semiconductor. This is because it could have a small conductivity at room temperature due to thermal excitation of electrons across the band gap.

A proper treatment of electron–electron interaction in materials has remained a challenge in the electronic structure theory. The simplest scheme was due to Hartree (1928) who assumed electrons to be independent and included only the average electron–electron interaction in his scheme. One drawback of Hartree's method is that the many-body wave function of electrons is symmetric under the exchange of two electrons although it should be antisymmetric. This drawback was taken care of in the Hartree–Fock method (Fock 1930) by assuming a determinantal wave function. One new feature of the Hartree–Fock method is the appearance of the exchange interaction in the Hartree–Fock equation. The exchange interaction appears only between two parallel spin electrons and has no analog in the classical mechanics. One drawback of the Hartree–Fock method is that it neglects the correlations between the opposite spin electrons although it includes correlations between the parallel spin electrons. When applied to the free electron model, it gives pathological results as the density of states becomes singular near the Fermi energy. The method is computationally very demanding for systems containing a large number of electrons as the wave function has at least $3N$ degrees of freedom where N is the number of electrons.

A big breakthrough in this field occurred in 1964 when a new approach known as the density functional theory (DFT) was proposed by Hohenberg and Kohn (1964) to handle electron–electron interaction. In DFT, the ground-state density of electrons, and not the many-body wave function, plays the basic role in the theory. Since the density has only 3 degrees of freedom, in contrast to $3N$ for the many-body wave function, this approach turns out to be computationally inexpensive in comparison to the wave function-based approaches. The theory is based on two theorems. The first theorem states that the external potential is a functional of the ground-state density and establishes density as the basic variable. The second theorem states that the total energy is minimum for the correct ground-state density and provides a variational principle for determining the total energy. Using Kohn and Sham (1965) ansaltz, it was possible to map the many-body problem of interacting electrons onto a system of noninteracting particles and derive a one-electron equation known as the Kohn–Sham equation, which has a similar structure as Hartree's equation. However, it involves an unknown exchange-correlation functional for which some approximation has to be used. The simplest approximation is called the local density approximation (LDA) that works reasonably well in most cases. However, it fails in many cases such as giving the correct ground state of Fe and generally overestimates binding energy and underestimates the bond lengths and lattice constants. Other approximations such as GGA have been developed, which, to some extent, tend to rectify these problems (Perdew 1991). However, both LDA and GGA underestimate the band gaps in semiconductors and insulators. There is much research activity aimed at finding a good exchange-correlation functional for the DFT calculations.

There exist many methods to solve the Kohn–Sham equation to obtain the band structure of a periodic solid. One of the oldest methods is the augmented plane wave (APW) method given by Slater (1937) that is still one of the most accurate methods to obtain energy bands. The method uses the muffin-tin form of the electron–ion potential, that is, the potential around an ion is assumed to be spherically symmetric within a sphere of radius, r_m, called the muffin-tin radius and is constant outside the sphere up to the cell boundary. Another method that uses the muffin-tin form of the potential is the Korringa–Kohn–Rostoker (KKR) method (Korringa 1947, Kohn and Rostoker 1954) that uses a Green's function formulation. Both APW and KKR methods are quite expensive in terms of computer time. Andersen (1975) realized that these methods could be speeded up by using a linear approximation and developed the linear augmented plane wave (LAPW) and linear muffin-tin orbital (LMTO) methods. Now, it is possible to get rid of the muffin-tin approximation and use full potential (FP). The methods that use FP form are known as FP-LAPW and FP-LMTO. The tight-binding (TB) method or the linear combination of atomic orbitals (LCAO) method that uses tightly bound atomic orbitals as bases has also been developed to obtain the band structure. The LMTO method can be cast in the TB form that is then called the TB-LMTO method. A new approach known as the *N*th order muffin-tin orbital (NMTO) method has been recently developed. However, it should be noted that all these band structure methods should give almost the same band structure and it is a matter of convenience which method to choose. Once the band structure is known, other quantities such as the Fermi surface and density of states can be calculated. The experimental energy band structures for a number of materials obtained by the angle-resolved photoemission spectroscopy (ARPES) are reasonably in good agreement with theoretical band structures. Experimental Fermi surfaces obtained by the de Haas–van Alphen (dHvA) method are also in good agreement with theoretical predictions for a number of metals.

It turns out that the band structure of an alkali metal such as K, Na, etc. is very much similar to a nearly free electron band structure. This was puzzling since the electron–ion potential in Na or K near the ions should be quite strong. So, the question was why the band structure of the alkali metals was almost free electron like. This remained a mystery until the pseudopotential concept was developed (Hellmann 1935, Antoncik 1959, Phillips and Kleinman 1959). A pseudopotential is a weak potential but gives the same energy bands obtained by using the FP. Although the pseudopotential method explained the mystery, it remained a semiempirical method for a long time. Things changed in 1979, when Hamann et al. (1979) showed how to construct a first-principles pseudopotential. Later, better pseudopotentials such as Bachelet et al. (1982), Troullier and Martins (1991), and Vanderbilt's ultrasoft pseudopotential (Vanderbilt 1990) were developed. Currently, the pseudopotential method is one of the most efficient methods to calculate the band structure. It can handle materials having a transition metal or a rare earth element. In 1994, a projector-augmented wave (PAW) method (Blöchl 1994) was developed that is an all-electron method but retains the efficiency of the pseudopotential method. Now, this method is one of the most commonly used methods.

Once the band structure is obtained, the next step is to calculate the total energy of a material for a given structure. The total energy of a material is an important quantity because by changing the structure or lattice constants, the total energy will change. Thus, the total energy of a material can be optimized with respect to its structure and one can then predict the structure and lattice constants from a band structure calculation. The structures predicted by using the LDA come out to be correct for a large number of materials but the lattice constants are underestimated. Better exchange-correlation functionals improve the agreement with the experimental values. It is also possible to deal with nonperiodic systems such as surfaces, nanowires, defects, and amorphous systems using supercell geometry. A supercell is a large unit cell containing a large number of atoms and is periodically repeated in space. The idea is to approximate a nonperiodic system by a periodic system but with a large unit cell so that the usual band structure methods could be applied.

Another breakthrough in the field of electronic structure occurred in 1985 when first-principles molecular dynamics was developed and realized in practice by Car and Parrinello (1985). In this method, electrons are treated quantum mechanically using DFT to obtain total energy but the ions are treated classically. The force on an ion is obtained using the total energy obtained from the DFT calculation. In the Car–Parrinello method, electrons and ions can be simultaneously moved during the dynamics and at each time step, it is not necessary to solve the Kohn–Sham equations. This makes it a very efficient method for finding the ground-state structure, for studying systems at finite temperature, dynamics, chemical reactions, and so on. Thus, the development of first-principles molecular dynamics combined the first-principles electronic structure calculations with molecular dynamics and paved the way for studying statistical mechanics from first principles.

All the electronic structure methods we have mentioned so far, assume translational symmetry. Disordered systems do not have translational symmetry and as a result, Bloch's theorem is not applicable. Therefore, all the above-mentioned methods cannot be used to obtain the electronic structure of disordered alloys. Substitutional disordered alloys are the simplest disordered systems that have only compositional disorder and there is an underlying periodic lattice. The simplest way to handle these systems is to use some approximation similar to the mean-field approximation. Essentially, one replaces a disordered alloy by an ordered alloy of effective atoms. Depending on how one chooses the effective atom, various approximations such as virtual crystal approximation (VCA), average t-matrix approximation (ATA), and coherent potential approximation (CPA) can be obtained. The development of CPA by Soven (1967) was an important breakthrough. Later, these ideas were developed in the KKR framework into what is known as the KKR–CPA method (Bansil 1978, Stocks et al. 1978). These methods are single-site approximations and, therefore, cannot deal with short-range ordering or clustering in alloys. To deal with such problems, approximations such as cluster CPA (CCPA) and KKR–CCPA were developed (Mookerjee and Prasad 1993, Razee and Prasad 1993a,b, Prasad 1994). The recursion method in the augmented space was also used (Dasgupta et al. 1995). Recently, another method called nonlocal CPA (NLCPA) has been developed that is supposed to be computationally more efficient (Rowlands 2009). A locally self-consistent Green's function (LSGF) method has also been developed that is analytic and computationally efficient (Ruban and Abrikosov 2008).

Amorphous materials form another class of disordered materials in which there is no underlying periodic lattice. Examples of such materials are amorphous semiconductors and metallic glasses. Amorphous semiconductors such as amorphous silicon and hydrogenated amorphous silicon have been the subject of intense research for a long time due to both scientific and technological interests. The disorder in these materials can cause a phenomenon known as Anderson localization (Anderson 1958). It was shown that the disorder can localize an electron. In one and two dimensions, even a small disorder in a material can localize all electronic states making it an insulator but in three dimensions, this depends on the strength of the disorder.

Computers have played an extremely important role in the growth of this field. In earlier days, computers were not available and calculations were done by hand. Therefore, band structure calculations for only simple systems such as Na could be done (Slater 1934, Wigner and Seitz 1934). In the 1950s, band structure calculations of more complicated systems such as semiconductors were possible (Herman and Callaway 1953, Herman 1958, 1984). Until 1960, the progress in this field was slow although many methods of calculating the band structure were well known. In the 1960s and 1970s, there was rapid growth in this field due to the availability of powerful computers. For example, Moruzzi et al. (1978) calculated band structure, equilibrium volume, bulk modulus, cohesive energy for a large number of elements using the KKR method and the LDA. They were able to reproduce these values within a few percent of experimental values. This was a remarkable achievement of the band theory that demonstrated the predictive power of the first-principles electronic structure calculations. Currently, computers have become much faster and computing cost has become very cheap. Also, very fast and efficient computer codes have been developed for electronic structure calculations. Some such codes have been listed in Appendix A. Thus, it is now possible to handle materials with large unit cells with these codes. As a result of these developments, there has been a fast growth in the applications of the first-principles methods to study complex materials such as multiferroics, amorphous systems, high T_c, superconductors such as cuprates and pnictides, heterostructures, surfaces, interfaces, etc. and design new materials such as topological insulators and lithium-ion battery materials.

The formulation for the calculation of polarization from first principles in terms of Berry's phase has been another important development in the field in the 1990s (King-Smith and Vanderbilt 1993). It turns out that the polarization for an infinite solid is an ill-defined quantity. However, the change in polarization can be expressed as an integral over Berry's phase. Using this formulation, one can calculate the experimentally measurable quantities such as spontaneous polarization and Born effective charges. These quantities have been calculated for a number of ferroelectrics and are in good agreement with the corresponding observed quantities. The theory can also be applied to explore multiferroics that are materials, which exhibit ferroelectric as well as magnetic behavior. These materials are interesting because their properties can be controlled by the application of electric as well as magnetic fields.

The study of low-dimensional materials such as atomic clusters, nanowires, and surfaces has remained a very active field during the last three decades. Two-dimensional electron gas, for example, shows a phenomenon such as the quantum Hall effect that has made a great impact on the condensed matter physics. The discovery of the two-dimensional material, graphene in 2004, has revolutionized the whole field. It can have electron states that survive only on the surface but decay away from it. Graphene surfaces can show many interesting phenomena such as surface relaxation, surface reconstructions, catalysis, corrosion, and so on. Interfaces also show many interesting phenomena such as band offsets. One-dimensional materials such as nanowires show amazing phenomena such as quantization of conductance. Zero-dimensional materials such as atomic clusters have attracted a lot of attention after the discovery of fullerenes. There exist magic clusters such as fullerenes that are highly stable. Nanotubes also have attracted a lot of attention during the last decade presumably due to their potential for applications in nanotechnology.

The predictive power of the first-principles electronic structure calculations makes it an ideal tool for materials design and for discovery of new and exotic materials. The discovery of the topological insulator is a good example. All topological insulators known so far were predicted by using theoretical calculations. The first-principles calculations have played an important role in finding appropriate materials for lithium-ion batteries and for spintronic applications. There has been an intense search for appropriate spintronic materials such as half-metallic ferromagnets, dilute magnetic semiconductors, and Heusler alloys. Another very interesting application of the first-principles calculations is to study materials under extreme conditions such as high pressure and temperature that may not be attainable in the laboratory. In such applications, the computer acts like a virtual laboratory where you can change the parameters such as temperature or pressure to very high values without the fear of explosion, and so on.

The applications that we have mentioned constitute a very small subset of possible applications of the electronic structure methods, and it seems that the possibilities are unlimited. The field of electronic structures has now come of age and is ready to tackle challenging problems in materials science. Thus, we are likely to see much more vigorous activity in the near future.

Further Reading

Ashcroft N. and Mermin N. 1976. *Solid State Physics*. New York, NY: W. B. Saunders Company.

Hoddeson L., Braun E., Teichmann and Weart S. 1992. *Out of the Crystal Maze, Chapters from the History of Solid State Physics*. New York, NY: Oxford University Press.

Kaxiras E. 2003. *Atomic and Electronic Structure of Solids*. Cambridge: Cambridge University Press.

Kittel C. 1986. *Introduction to Solid State Physics* (6th edition). New York, NY: John Wiley & Sons.

Mardar M. 2010. *Condensed Matter Physics*. New York, NY: John Wiley and Sons.

Martin R. M. 2004. *Electronic Structure, Basic Theory and Practical Methods*. Cambridge: Cambridge University Press.

Seitz F. 1940. *The Modern Theory of Solids*. New York, NY: McGraw-Hill Book Company.

Ziman J. M. 1969. *Principles of the Theory of Solids*. Cambridge: Cambridge University Press.

2

Quantum Description of Materials

2.1 Introduction

We know that all materials are made of atoms that are in turn made of electrons and nuclei. In other words, a material is made of electrons and nuclei that interact with each other in accordance with Coulomb's law. Classical mechanics fails to describe the behavior of electrons; therefore, only quantum mechanics can give us a correct description of materials. Thus, to understand it, we have to solve a quantum mechanical many-body problem, that is, Schrödinger's equation for interacting electrons and nuclei. Although not necessary, but for convenience and simplicity, we shall divide the electrons into valence and core electrons. For example, in sodium, the 3*s* electron, which is the outermost electron, will be called a valence electron and the remaining will be called core electrons. When a solid is formed, the valence electrons get unbound from atoms and can move in the solid. Since core electrons do not play much of a role in solid-state properties, we can combine core electrons with their nucleus and call the combined entity an ion. Thus, a condensed matter system can be described as a many-body system of interacting electrons and ions and we are interested in finding a solution of Schrödinger's equation for this system. In Section 2.2, we shall discuss how to solve this problem approximately by separating the electronic problem from the ionic problem using the Born–Oppenheimer approximation. In the remaining sections, we focus on the electronic problem. In Section 2.3, we discuss the simplest solution using the Hartree approximation and in Section 2.4, we discuss the Hartree–Fock (H–F) approximation. In Section 2.5, we briefly discuss the configuration interaction (CI) method. In Sections 2.6 and 2.7, we discuss the application of the Hartree and H–F approximations to homogeneous electron gas (HEG).

2.2 Born–Oppenheimer Approximation

The Born–Oppenheimer approximation allows us to decouple the electronic and ionic motions. As a result, electrons and ions can be treated as nearly independent entities. Since the number of electrons and ions is very large ~10^{24}, an exact quantum mechanical solution of this problem is almost impossible. The Born–Oppenheimer approximation (Born and Oppenheimer 1927) (Figure 2.1a and b) helps us in finding an approximate solution of this problem by separating electronic and ionic wave functions. The essential idea is that the ions, being 10^3–10^5 times heavier than the electrons, move much more slowly than the electrons. In other words, the timescale of electronic motion is much smaller than that of ionic motion. For example, the electronic velocities in a sodium metal are of the order of 10^6 m/s while the ions move with velocities of the order of 10^2 m/s (see Exercise 2.1). As a result, the electrons adjust very quickly to the motions of ions. Thus, when solving the electronic problem, we can assume that the ions are fixed at their instantaneous positions (see below) and the electrons follow the ionic motion adiabatically. We further assume that the electrons at any moment will be in their ground state for that particular instantaneous ionic configuration (this will be clear later). In other words, the electrons are not allowed to make transitions from one state to another. Rather, they are assumed in their ground state and the ground state is deformed progressively by the ionic displacements.

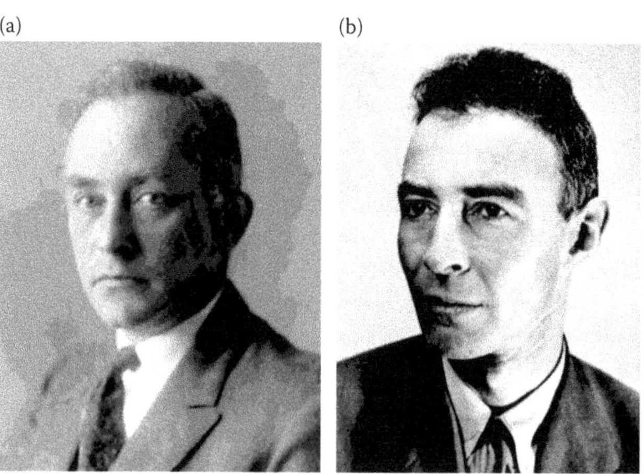

FIGURE 2.1
Born and Oppenheimer. (a) Max Born (1862–1970) played a key role in developing quantum mechanics and is known for many contributions. He was awarded the Nobel Prize in 1954. (b) J. R. Oppenheimer (1904–1967) is known for many contributions including the Born–Oppenheimer approximation. He headed the Manhattan project and is considered as the "father of the atomic bomb." (Oppenheimer photo courtesy of Los Alamos Scientific Laboratory, *Los Alamos: Beginning of an Era*, 1943–1945, LASL, Los Alamos, 1986.)

Let $\vec{r}_i$ denote the position of an electron, m_e the electronic mass, $\vec{R}_\ell$ the ionic position, and M the ionic mass (Figure 2.2). The Hamiltonian of the whole system of ions and electrons can be written as

$$H = T_E + T_I + V_{II} + V_{EE} + V_{IE} \tag{2.1}$$

where T_E is the kinetic energy operator for electrons and can be written as

$$T_E = -\sum_i \frac{\hbar^2}{2m_e} \frac{\partial^2}{\partial \vec{r}_i^{\,2}}$$

T_I is the kinetic energy operator for ions and can be written as

$$T_I = -\sum_\ell \frac{\hbar^2}{2M} \frac{\partial^2}{\partial \vec{R}_\ell^{\,2}}$$

V_{II} is the ion–ion interaction

$$V_{II} = \frac{1}{2} \sum_{\substack{\ell,\ell' \\ \ell \neq \ell'}} \frac{e^2}{4\pi\varepsilon_0} \frac{Z_\ell Z_{\ell'}}{\left|\vec{R}_\ell - \vec{R}_{\ell'}\right|}$$

where $Z_\ell e$ is the ionic charge at site ℓ. Here, the double summation is implied and the factor of 1/2 is for avoiding double counting in the sum. V_{EE} is the electron–electron interaction and can similarly be written as

$$V_{EE} = \frac{1}{2} \sum_{\substack{i,j \\ i \neq j}} \frac{1}{4\pi\varepsilon_0} \frac{e^2}{\left|\vec{r}_i - \vec{r}_j\right|}$$

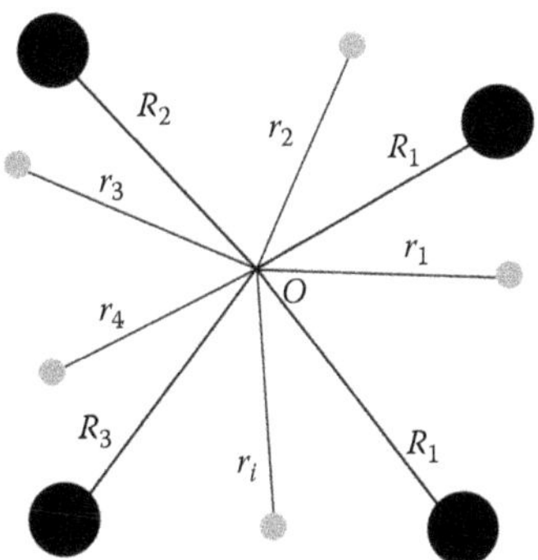

FIGURE 2.2
Filled circles and dots denote ions and electrons, respectively. O denotes the origin.

V_{IE} is the ion–electron interaction. We shall use $\vec{r}$ as the collective symbol for electronic coordinates, that is, $\vec{r} \equiv (\vec{r}_1, \vec{r}_2, \ldots)$. Similarly, $\vec{R}$ is the collective symbol for the ionic coordinates. We have assumed that all the ions are of the same kind and have the same mass but the formulation can be easily generalized to the case when the ions have different masses.

We are interested in finding the solution of the time-independent Schrödinger equation for the entire system of ions and electrons

$$H\phi = E\phi \tag{2.2}$$

where E is the energy of the entire system of interacting electrons and ions and ϕ is the eigenfunction of H. The Hamiltonian (2.1) contains all the interactions and would provide the theory of everything in materials science provided we could solve it. This is a very complicated problem and could be considerably simplified if somehow we could separate electronic and ionic motions. To do this, let us try the following kind of decompositions of ϕ

$$\phi = \chi_I(\vec{R})\ \psi_E(\vec{r}, \sigma, \vec{R}) \tag{2.3}$$

where σ is a composite symbol for spins of the electrons. In the remaining part of this section, the electronic spin does not play any role; so, we will suppress symbol σ in the discussion below. In Equation 2.3, ψ_E represents the many-body electron wave function that satisfies the Schrödinger equation with ions at frozen positions $\vec{R}$, that is

$$\left[T_E + V_{EE}(\vec{r}) + V_{IE}(\vec{r}, \vec{R})\right]\psi_E(\vec{r}, \vec{R}) = E_E(\vec{R})\psi_E(\vec{r}, \vec{R}) \tag{2.4}$$

where $E_E(\vec{R})$ is energy of the electrons when the ions are frozen at position $\vec{R}$ and $\chi_I(\vec{R})$ is some function of ionic position $\vec{R}$, which is yet to be determined. As mentioned at the beginning of this section, the physics behind this decomposition is that the electrons are light and, therefore, move much faster than the ions. As a result, the electrons adjust their motion almost instantaneously to the motion of ions. This idea can be understood better with the help of a cartoon shown in Figure 2.3 that shows an elephant walking. The elephant has some honey on its head so that the bees are flying around it. Now, if you concentrate on the motion of bees, you see that they adjust to the motion of the elephant almost instantaneously. This is because the bees are very light and move much faster than the elephant. In a similar way, the ions are like elephants and electrons are like bees. Thus, a decomposition given in Equation 2.3 seems to be a good approximation.

FIGURE 2.3
An elephant and bees.

Let us examine if this kind of decomposition is possible by substituting Equation 2.3 for ϕ in Equation 2.2 and check if it is an eigenfunction of H

$$\begin{aligned} H\phi &= \left[T_E(\vec{r}) + T_I(\vec{R}) + V_{II}(\vec{R}) + V_{EE}(\vec{r}) + V_{IE}(\vec{r},\vec{R})\right] \\ &\quad \times \psi_E(\vec{r},\vec{R})\chi_I(\vec{R}) \\ &= \left[T_E(\vec{r}) + V_{EE}(\vec{r}) + V_{IE}(\vec{r},\vec{R})\right]\psi_E(\vec{r},\vec{R})\chi_I(\vec{R}) \\ &\quad + \left[-\frac{\hbar^2}{2M}\sum_\ell \frac{\partial^2}{\partial\vec{R}_\ell^2} + V_{II}(\vec{R})\right]\psi_E(\vec{r},\vec{R})\chi_I(\vec{R}) \\ &= E\psi_E(\vec{r},\vec{R})\chi_I(\vec{R}) \end{aligned} \tag{2.5}$$

Now, the second term in Equation 2.5 is

$$\begin{aligned} &\left[-\frac{\hbar^2}{2M}\sum_\ell\left\{\frac{\partial}{\partial\vec{R}_\ell}\cdot\left(\frac{\partial\psi_E}{\partial\vec{R}_\ell}\chi_I(\vec{R}) + \psi_E\frac{\partial\chi_I}{\partial\vec{R}_\ell}\right)\right\} + V_{II}\psi_E\chi_I\right] \\ &= -\frac{\hbar^2}{2M}\sum_\ell\left\{\frac{\partial^2\psi_E}{\partial\vec{R}_\ell^2}\chi_I + 2\frac{\partial\psi_E}{\partial\vec{R}_\ell}\cdot\frac{\partial\chi_I}{\partial\vec{R}_\ell} + \psi_E\frac{\partial^2\chi_I}{\partial\vec{R}_\ell^2}\right\} + V_{II}\psi_E\chi_I \end{aligned} \tag{2.6}$$

Thus, Equation 2.5 can be written as

$$E(\vec{R})\psi_E(\vec{r},\vec{R})\chi_I(\vec{R}) = \psi_E(\vec{r},\vec{R})\left[-\frac{\hbar^2}{2M}\sum_\ell \frac{\partial^2}{\partial\vec{R}_\ell^2} + V_{II}(\vec{R}) + E_E(\vec{R})\right]\chi_I(\vec{R})$$

$$-\frac{\hbar^2}{2M}\sum_\ell\left\{2\frac{\partial\psi_E}{\partial\vec{R}_\ell}\cdot\frac{\partial\chi_I}{\partial\vec{R}_\ell} + \frac{\partial^2\psi_E}{\partial\vec{R}_\ell^2}\chi_I\right\} \quad (2.7)$$

Now, if somehow the items in the curly brackets in Equation 2.7 could be ignored, $\chi_I(\vec{R})$ could be made to satisfy the following Schrödinger-type eigenvalue equation:

$$\left[-\frac{\hbar^2}{2M}\sum_\ell \frac{\partial^2}{\partial\vec{R}_\ell^2} + V_{II}(\vec{R}) + E_E(\vec{R})\right]\chi_I(\vec{R}) = E\chi_I(\vec{R}) \quad (2.8)$$

and then

$$H\psi_E\chi_I = E\psi_E\chi_I \quad (2.9)$$

Equation 2.8 is an equation for a wave function of ions alone. It is what we should have written if we were solving the ionic problem by quantum mechanics except that now we must add to V_{II} the term $E_E(\vec{R})$, which is the total energy of the electron system as a function of ionic positions. We shall see in Chapter 3 that $E_E(\vec{R})$ can be easily computed by using the DFT if the electrons are in their ground state. It is clear from Equation 2.8 that the binding of ions is provided by $E_E(\vec{R})$ as V_{II} is a repulsive potential between the ions, which will fly away otherwise. Thus, electrons form a glue that binds these ions.

Now, we shall show that the two terms in the items in the curly brackets of Equation 2.7 can indeed be neglected. These terms do not contribute much to the expectation value of the energy. The contributions of the first term would involve a term such as

$$\frac{\partial\chi_I}{\partial\vec{R}_\ell}\cdot\int\psi_E^*\frac{\partial\psi_E}{\partial\vec{R}_\ell}d^3r \approx \frac{\partial\chi_I}{\partial\vec{R}_\ell}\cdot\int\frac{1}{2}\frac{\partial}{\partial\vec{R}_\ell}\left[\psi_E^*\psi_E\right]d^3r$$

$$= \frac{\partial\chi_I}{\partial\vec{R}_\ell}\cdot\frac{1}{2}\frac{\partial}{\partial\vec{R}_\ell}\int\psi_E^*\psi_E d^3r$$

$$= \frac{\partial\chi_I}{\partial\vec{R}_\ell}\cdot\frac{1}{2}\frac{\partial}{\partial\vec{R}_\ell}(N_E) = 0 \quad (2.10)$$

where $N_E = \int\psi_E^*\psi_E d^3r$ is the total number of electrons in the system that does not change as a result of ionic motion.

The second term in the second line of Equation 2.7 may also be shown to be quite small. Consider the extreme case, for example, when the electron–ion interaction is so strong that electrons are tightly bound to their ions. In this case

$$\psi(\{\vec{r}_i, \vec{R}_\ell\}) = \psi(\{\vec{r}_i - \vec{R}_\ell\}) \tag{2.11}$$

where $\{\vec{r}_i, \vec{R}_\ell\}$ means $(\vec{r}_1, \vec{r}_2 \ldots, \vec{R}_1, \vec{R}_2, \ldots)$. This would give:

$$\frac{\partial^2 \psi(\{\vec{r}_i - \vec{R}_\ell\})}{\partial \vec{R}_\ell^2} = \frac{\partial^2 \psi(\{\vec{r}_i - \vec{R}_{\bar{\ell}}\})}{\partial \vec{r}_i^{\,2}} \tag{2.12}$$

Using this, the contributions of this term to the expectation value of the energy involves

$$\begin{aligned} -\int \psi_E^* \frac{\hbar^2}{2M} \frac{\partial^2 \psi_E}{\partial \vec{R}_\ell^2} d^3r &= -\int \psi_E^* \frac{\hbar^2}{2M} \frac{\partial^2 \psi_E}{\partial \vec{r}_i^{\,2}} d^3r \\ &= -\frac{m_e}{M} \int \psi_E^* \frac{\hbar^2}{2m_e} \frac{\partial^2 \psi_E}{\partial \vec{r}_i^{\,2}} d^3r \end{aligned} \tag{2.13}$$

This is m_e/M times the kinetic energy of the electrons. Since m_e/M is of the order of 10^{-4} or 10^{-5}, this can also be neglected. Note that this is only the upper limit; generally, the contribution will be smaller than this.

Thus, we have shown that the decomposition of ϕ as proposed in Equation 2.3 is indeed a good approximation. Let us look at the form of this decomposition more closely. The first factor $\chi_I(\vec{R})$ describes the ionic motion and the second factor shows that the electrons move as if the ions were fixed in their instantaneous positions. The electron follows the ionic motion adiabatically. This decomposition is an important approximation because it allows us to separate electronic and ionic motions and to a good approximation we can treat electrons and ions as independent entities.

Note that in Equation 2.3, we have taken only one electronic wave function. This could be generalized to include the complete set of wave functions in the following way:

$$\phi = \sum_n \chi_I^n(\vec{R}) \psi_E^n(\vec{r}, \vec{R}) \tag{2.14}$$

where $n = 0$ corresponds to the ground state of the electrons. Using Equation 2.14, one can proceed along the same lines. The ionic motion will now lead to a term containing off-diagonal transitions between these states. In the first-order perturbation theory, this term will not contribute to the total energy but we will obtain contributions from this using the second-order perturbation theory. Generally, these contributions are going to be quite small. Thus, in the Born–Oppenheimer approximation, we neglect the off-diagonal transitions. In other words, we assume that nuclear motion cannot cause electronic transitions and the electrons always remain in their ground state.

We see that in Equation 2.8, $E_E^n(\vec{R}) + V_{II}(\vec{R})$ is the potential in which the ions move. It defines a surface in a space spanned by nuclear coordinates, which is called Born–Oppenheimer potential energy surface (BOPES). For each n, there is a BOPES. Normally, we assume $n = 0$, that is, the electrons are in the ground state. Thus, if the dynamics are adiabatic, the electrons will stay on the BOPES given by $E_E^0(\vec{R}) + V_{II}(\vec{R})$ and will not make any transitions to other surfaces corresponding to $n \neq 0$.

The Born–Oppenheimer approximation is an excellent approximation for semiconductors and insulators as there is an energy gap in the electronic spectrum. This gap is of the order of electron volt, whereas the typical phonon energies are of the order of 0.01 eV. Thus, in these materials, the nuclear (ionic) motion cannot excite electrons to higher energies and consequently the off-diagonal terms are zero. On the other hand, in metallic systems, the energy gap between the highest occupied and lowest unoccupied states is zero and ionic motion can cause transitions of vanishing energy. In that case, the Born–Oppenheimer approximation can break down. Thus, in first-principles simulations of the metallic system, one has to be very watchful.

2.3 Hartree Method

In the previous section, we saw that the electronic motion can be separated from the ionic motion using the Born–Oppenheimer approximation. Now, we shall focus on solving the electronic problem and return to the ionic problem later. To do this, we shall assume that the ions are fixed at their positions and try to solve Equation 2.4, which can be written as

$$\left(-\frac{\hbar^2}{2m_e}\sum_i \nabla_i^2 + V_{IE} + V_{EE}\right)\psi(\vec{r}_1,\sigma_1,\vec{r}_2,\sigma_2,\ldots,\vec{r}_N,\sigma_N)$$
$$= E\psi(\vec{r}_1,\sigma_1,\vec{r}_2,\sigma_2,\ldots,\vec{r}_N,\sigma_N) \quad (2.15)$$

For simplicity, we have dropped the subscript E of energy E of Equation 2.4. Equation 2.15 is a complicated equation for many-electron wave function ψ, which is a function of all electronic coordinates (position as well as spin). The simplest approximation is due to Hartree (1928) who wrote it as the product of single-particle functions as

$$\psi_H = \psi_1(\vec{r}_1,\sigma_1)\psi_2(\vec{r}_2,\sigma_2)\cdots\psi_N(\vec{r}_N,\sigma_N) \tag{2.16}$$

where $\psi_i(\vec{r},\sigma)$ denotes the wave function of a single electron i with spin σ and is also called orbital. It has the usual interpretation that $|\psi_i|^2 d^3r$ gives the probability of finding the electron with spin σ in volume element d^3r.

In this section, we shall be neglecting the spin of the electrons as it plays no role in the Hartree method. Now, we shall use the variational principle for ground-state energy to find E and ψ_i. The variational principle states that the expectation value of the energy

$$E = \langle \psi_H | H | \psi_H \rangle \geq E_0 \tag{2.17}$$

where E_0 is the ground-state energy of the system. We shall assume that ψ_i are orthonormal. Now, substituting Equation 2.16 and

$$H = -\sum_i \frac{\hbar^2}{2m_e}\nabla_i^2 + V_{IE} + V_{EE}$$

in Equation 2.17, where

$$V_{IE} = -\frac{1}{4\pi\varepsilon_0}\sum_i\sum_\ell \frac{Ze^2}{|\vec{r}_i - \vec{R}_\ell|}$$

and

$$V_{EE} = \frac{1}{8\pi\varepsilon_0}\sum_{\substack{ij\\ i\neq j}} \frac{e^2}{|\vec{r}_i - \vec{r}_j|}$$

we get

$$E = -\sum_i \int \psi_i^* \frac{\hbar^2}{2m_e}\nabla_i^2\psi_i d^3r_i - \frac{1}{4\pi\varepsilon_0}\sum_i\sum_\ell \int \psi_i^* \frac{e^2 Z}{|\vec{r}_i - \vec{R}_\ell|}\psi_i d^3r_i$$
$$+ \frac{1}{8\pi\varepsilon_0}\sum_i\sum_{\substack{j\\ i\neq j}} \int \psi_i^*\psi_j^* \frac{e^2}{|\vec{r}_i - \vec{r}_j|}\psi_i\psi_j d^3r_i d^3r_j \tag{2.18}$$

where we have assumed that ψ_i are orthonormal. We note that E is a functional of ψ (see Appendix D).

Now, we vary the wave functions ψ_i so as to minimize E subject to constraint

$$\int \psi_i^* \psi_i d^3 r_i = 1 \tag{2.19}$$

A more general treatment would be to demand orthonormality during the variational procedure, but this would add more constraint equations and complicate the algebra. However, the above treatment using Equation 2.19 also gives the same result. This is done, most easily, by using the Lagrange multiplier method. Thus, we minimize

$$E - \sum_i \varepsilon_i \left[\int \psi_i^* \psi_i d^3 r_i - 1 \right] \tag{2.20}$$

without any constraint. Here, ε_i are the Lagrange multipliers that are yet to be determined. Taking the functional derivative of Equation 2.20 (see Appendix D), we get

$$\frac{\delta E}{\delta \psi_i^*} - \varepsilon_i \psi_i = 0$$

where $\delta E/\delta \psi_i^*$ is the functional derivative of E.

Thus, we get

$$-\frac{\hbar^2}{2m_e}\nabla_i^2 \psi_i - \frac{1}{4\pi\varepsilon_0}\sum_\ell \frac{Ze^2}{\left|\vec{r}_i - \vec{R}_\ell\right|}\psi_i + \frac{1}{4\pi\varepsilon_0}\sum_{j \neq i}\int \frac{e^2 \left|\psi_j\right|^2}{\left|\vec{r}_i - \vec{r}_j\right|} d^3 r_j \psi_i = \varepsilon_i \psi_i \tag{2.21}$$

Equation 2.21 is known as Hartree equation and by solving this equation, we obtain all ε_i and ψ_i. Now, we populate the lowest N levels in accordance with the Pauli exclusion principle at absolute zero temperature. The treatment can be generalized to finite temperature by using the Fermi–Dirac statistics.

Now, the number density $n(\vec{r})$ can be written as

$$n(\vec{r}) = \sum_{\substack{i \\ \text{occupied}}} \psi_i^* \psi_i = \sum_{\substack{i \\ \text{occupied}}} \left|\psi_i\right|^2 \tag{2.22}$$

Also, we define Hartree potential as

$$v_H(\vec{r}_i) = \frac{1}{4\pi\varepsilon_0} \sum_{j \neq i} \int \frac{e^2 |\psi_j|^2}{|\vec{r}_i - \vec{r}_j|} d^3 r_j$$

and

$$v_I(\vec{r}_i) = -\frac{1}{4\pi\varepsilon_0} \sum_{\ell} \frac{Ze^2}{|\vec{r}_i - \vec{R}_\ell|} \tag{2.23}$$

The Hartree potential can be rewritten as

$$\begin{aligned} v_H(\vec{r}_i) &= \frac{1}{4\pi\varepsilon_0} \iint \sum_{j \neq i} \frac{e^2 |\psi_j|^2}{|\vec{r}_i - \vec{r}|} \delta(\vec{r} - \vec{r}_j) d^3 r \, d^3 r_j \\ &= \frac{1}{4\pi\varepsilon_0} \int \frac{d^3 r}{|\vec{r}_i - \vec{r}|} e^2 \sum_{j \neq i} |\psi_j(\vec{r})|^2 \end{aligned} \tag{2.24}$$

We see that the Hartree potential is the average electrostatic potential created by the rest of the electrons. For a system containing a large number of electrons such as a solid, the $j = i$ term in Equation 2.24 will have negligible contribution in comparison to the total contribution. For such systems, $v_H(\vec{r})$ can be approximated by

$$v_H(\vec{r}) \simeq \frac{1}{4\pi\varepsilon_0} \int \frac{e^2 n(\vec{r}')}{|\vec{r}' - \vec{r}|} d^3 r' \tag{2.25}$$

Thus, the Hartree equation can be written as

$$\left[-\frac{\hbar^2}{2m_e} \nabla^2 + v_I(\vec{r}) + v_H(\vec{r}) \right] \psi_i(\vec{r}) = \varepsilon_i \psi_i(\vec{r}) \tag{2.26}$$

Equation 2.26 has a very familiar form and is similar to Schrödinger's equation for a single electron moving in a potential $v_I(\vec{r}) + v_H(\vec{r})$. However, there is a big difference: in the Hartree equation, the potential depends on the solution itself (see Equation 2.24). The way this equation is solved is as follows. First, a guess is made for $v_H(\vec{r})$, and Equation 2.26 is solved. Then, using the solutions, $v_H(\vec{r})$ is computed and the equation is solved again and the potential is computed. This procedure is repeated until the input and the output potentials are very close, within some tolerance. This procedure is called a self-consistent procedure.

Under assumption (2.16), the total wave function can be written as a simple product of single-particle wave functions and leads to Equation 2.26, which is an equation for a single electron moving in an effective potential $v_I(\vec{r}) + v_H(\vec{r})$. Thus, we see that the Hartree approximation reduces the many-electron problem to a one-electron problem. This kind of approximation is also known as one-electron approximation or independent particle approximation. We shall see that this is the common philosophy followed in the Hartree, H–F, and the DFT. In all such theories, the problem is reduced to the equation of the form given in Equation 2.26. One-electron approximation has been extremely successful in understanding and predicting a materials' properties.

Although one-electron approximation is very successful, it neglects correlations between electrons that arise due to the Coulomb repulsion between electrons. Because of the Coulomb repulsion, electrons move in such a way as to avoid each other as far as possible. For example, if one electron moves, it affects the motion of all other electrons. Thus, owing to the Coulomb repulsion, the motion of electrons is coupled or correlated. Such correlations between electronic motions are called Coulomb or electron correlations. But in one-electron theories, an electron moves like an independent particle and thus, such theories miss out the effect of correlated motion or electron correlations. Thus, electron correlation is a many-body effect, and to understand it, one has to go beyond one-electron theories.

From Equation 2.18, total energy can be written as

$$E = \sum_i \varepsilon_i - \frac{e^2}{2} \sum_{\substack{ij \\ i \neq j}} \iint \frac{1}{4\pi\varepsilon_0} \frac{|\psi_i|^2 |\psi_j|^2}{|\vec{r}_i - \vec{r}_j|} d^3r_i\, d^3r_j \tag{2.27}$$

Here, the sum is only over occupied levels ($T = 0$). Using Equation 2.23, the second term in Equation 2.18 can be written as

$$\int \sum_i v_I(\vec{r}_i) |\psi_i(\vec{r}_i)|^2 d^3r_i$$

$$= \iint \sum_i v_I(\vec{r}_i) \delta(\vec{r} - \vec{r}_i) |\psi_i(\vec{r}_i)|^2 d^3r_i d^3r$$

$$= \int \sum_i v_I(\vec{r}) |\psi_i(\vec{r})|^2 d^3r \tag{2.28}$$

$$= \int v_I(\vec{r}) n(\vec{r}) d^3r \tag{2.29}$$

Although the Hartree approximation introduces great simplification and reduces the many-body problem to a one-body problem, it has two major shortcomings. First, it includes the electron–electron interaction only in an average manner. Second, the many-body Hartree wave function violates the antisymmetry requirement.

2.3.1 Interpretation of ε_i

We had introduced ε_i in Equation 2.20 as a Lagrange multiplier. To interpret it physically, let us remove the kth electron from the system and calculate the total energy. We assume that by removing one electron from the system, the orbitals ψ_i remain unchanged. The assumption is valid if the system is very large such as a solid.

Let us call the energy of $(N-1)$ electrons as E'. Then, from Equation 2.18, E' can be written as

$$E' = -\sum_{i\neq k}\int \psi_i^* \frac{\hbar^2}{2m_e}\nabla_i^2\psi_i\, d^3r_i - \sum_{i\neq k}\sum_{\ell}\int \psi_i^* \frac{1}{4\pi\varepsilon_0}\frac{Ze^2}{\left|\vec{r}_i - \vec{R}_\ell\right|}\psi_i\, d^3r_i$$

$$+\frac{1}{2}\sum_{\substack{i\neq k\\ i\neq j}}\sum_{j\neq k}\iint \psi_i^*\psi_j^* \frac{1}{4\pi\varepsilon_0}\frac{e^2}{\left|\vec{r}_i - \vec{r}_j\right|}\psi_i\psi_j d^3r_i d^3r_j \tag{2.30}$$

$$E - E' = -\frac{\hbar^2}{2m_e}\int \psi_k^*\nabla_k^2\psi_k\, d^3r_k - \sum_{\ell}\int \psi_k^* \frac{1}{4\pi\varepsilon_0}\frac{Ze^2}{\left|\vec{r}_k - \vec{R}_\ell\right|}\psi_k d^3r_k$$

$$+\frac{1}{2}\sum_{j\neq k}\iint \psi_k^*\psi_j^* \frac{1}{4\pi\varepsilon_0}\frac{e^2}{\left|\vec{r}_k - \vec{r}_j\right|}\psi_k\psi_j d^3r_k d^3r_j$$

$$+\frac{1}{2}\sum_{i\neq k}\iint \psi_i^*\psi_k^* \frac{1}{4\pi\varepsilon_0}\frac{e^2}{\left|\vec{r}_i - \vec{r}_k\right|}\psi_k\psi_i d^3r_k d^3r_j$$

$$= -\frac{\hbar^2}{2m_e}\int \psi_k^*\nabla_k^2\psi_k\, d^3r_k - \sum_{\ell}\int \left|\psi_k\right|^2 \frac{1}{4\pi\varepsilon_0}\frac{Ze^2}{\left|\vec{r}_k - \vec{R}_\ell\right|}d^3r_k$$

$$+\sum_{j\neq k}\iint \left|\psi_k\right|^2\left|\psi_j\right|^2 \frac{1}{4\pi\varepsilon_0}\frac{e^2}{\left|\vec{r}_k - \vec{r}_j\right|}d^3r_k\, d^3r_j$$

$$= \varepsilon_k \text{ from Equation 2.21} \tag{2.31}$$

or

$$E' - E = -\varepsilon_k \tag{2.32}$$

that is, $-\varepsilon_k$ is the energy required to remove the kth electron from the system. Thus, ε_k can be interpreted as single-particle energy.

2.4 Hartree–Fock (H–F) Method

There are two main problems with the Hartree method:

1. Since electrons are fermions, the many-body wave function should be antisymmetric with respect to exchange of two electrons. Hartree's assumption (Equation 2.16) does not satisfy this requirement.
2. The electrostatic interaction between electrons is treated in an average way.

In the H–F method (Fock 1930), the first problem is addressed. We want the many-body wave function to satisfy the requirement

$$\psi(\vec{r}_1\sigma_1,\ldots\vec{r}_i\sigma_i,\ldots,\vec{r}_j\sigma_j,\ldots) = -\psi(\vec{r}_1\sigma_1,\ldots\vec{r}_j\sigma_j,\ldots,\vec{r}_i\sigma_i,\ldots) \tag{2.33}$$

If we have our trial wave function as the determinant, we can easily satisfy this requirement

$$\psi(\vec{r}_1\sigma_1,\ldots,\vec{r}_N\sigma_N) = \frac{1}{\sqrt{N!}}\begin{vmatrix} \psi_1(\vec{r}_1\sigma_1) & \psi_1(\vec{r}_2\sigma_2) & \ldots & \psi_1(\vec{r}_N\sigma_N) \\ \psi_2(\vec{r}_1\sigma_1) & \psi_2(\vec{r}_2\sigma_2) & \ldots & \\ & & & \\ \psi_N(\vec{r}_1\sigma_1) & \psi_N(\vec{r}_2\sigma_2) & \ldots & \psi_N(\vec{r}_N\sigma_N) \end{vmatrix} \tag{2.34}$$

where N is the number of electrons. The determinantal form of the many-body wave function was first proposed by Slater (1929) and is known as the Slater determinant. The determinantal wave function ψ can be written as

$$\psi = \frac{1}{\sqrt{N!}} \sum_P (-1)^p P\psi_1(x_1)\psi_2(x_2),\ldots\ldots,\psi_N(x_N) \tag{2.35}$$

where $x \equiv (\vec{r},\sigma)$. Here, P is the permutation operator and p is the number of interchanges making up this permutation.

Now, the expectation value of the energy can be written as

$$E = \langle \psi | H | \psi \rangle \tag{2.36}$$

where

$$H = -\sum_i \frac{\hbar^2}{2m_e}\nabla_i^2 - \sum_i \sum_\ell \frac{1}{4\pi\varepsilon_0} \frac{Ze^2}{|\vec{r}_i - \vec{R}_\ell|} + \frac{1}{2}\sum_{\substack{ij \\ i \neq j}} \frac{1}{4\pi\varepsilon_0} \frac{e^2}{|\vec{r}_i - \vec{r}_j|} \tag{2.37}$$

Substituting Equation 2.35 into Equation 2.36, we get

$$E = \langle \psi | H | \psi \rangle = \int \psi^* H \psi \, d\tau \tag{2.38}$$

where

$$d\tau = dx_1 dx_2 \cdots dx_N$$

$$E = \frac{1}{\sqrt{N!}} \int \psi^* H \sum_P (-1)^p P\psi_1(x_1)\cdots\psi_N(x_N) d\tau \tag{2.39}$$

$$= \frac{1}{\sqrt{N!}} \sum_P (-1)^p P \int (P^{-1}\psi^*) H \psi_1(x_1)\cdots\psi_N(x_N) d\tau \quad [\text{as } PH = H]$$

$$= \frac{1}{\sqrt{N!}} \sum_P P \int \psi^* H \psi_1(x_1)\cdots\psi_N(x_N) d\tau \tag{2.40}$$

Note that all x_i have been integrated out in Equation 2.40 and the integral in Equation 2.40 is independent of any x_i. Thus, the operation of P over the integral gives the same number and, therefore, the sum over all the permutations gives just $N!$ times the integral. Thus, we get

$$E = \sqrt{N!}\int \psi^* H\psi_1(x_1)\cdots\psi_N(x_N)d\tau$$

$$= \sum_P \int (-1)^p P\psi_1^*(x_1)\cdots\psi_N^*(x_N)\left[-\sum_i \frac{\hbar^2}{2m_e}\nabla_i^2 - \sum_{i\ell}\frac{1}{4\pi\varepsilon_0}\frac{Ze^2}{|\vec{r}_i - \vec{R}_\ell|}\right.$$

$$\left. + \frac{1}{2}\sum_{\substack{ij \\ i\neq j}} \frac{e^2}{|\vec{r}_i - \vec{r}_j|}\right]\psi_1(x_1)\cdots\psi_N(x_N)d\tau$$

$$= \sum_i \int \psi_i^*(x_i)\left[-\frac{\hbar^2}{2m_e}\nabla_i^2 - \sum_{i\ell}\frac{1}{4\pi\varepsilon_0}\frac{Ze^2}{|\vec{r}_i - \vec{R}_\ell|}\right]\psi_i(x_i)dx_i$$

$$+ \frac{1}{2}\sum\sum_{i\neq j}\iint \left[\psi_i^*(x_i)\psi_j^*(x_j) - \psi_j^*(x_i)\psi_i^*(x_j)\right]$$

$$\times \frac{1}{4\pi\varepsilon_0}\frac{e^2}{|\vec{r}_i - \vec{r}_j|}\psi_i(x_i)\psi_j(x_j)dx_i dx_j \tag{2.41}$$

$$E = \sum_i \int \psi_i^*(x_i)\left[-\frac{\hbar^2}{2m_e}\nabla_i^2 - \sum_{i\ell}\frac{Ze^2}{|\vec{r}_i - \vec{R}_\ell|}\right]\psi_i(x_i)dx_i$$

$$+ \frac{1}{2}\sum\sum_{i\neq j}\iint \frac{1}{4\pi\varepsilon_0}\frac{e^2|\psi_i(x_i)|^2|\psi_j(x_j)|^2}{|\vec{r}_i - \vec{r}_j|}dx_i dx_j$$

$$- \frac{1}{2}\sum_{i,j}\sum_{j\neq i}\iint \frac{1}{4\pi\varepsilon_0}\frac{e^2\psi_j^*(x_i)\psi_i^*(x_j)\psi_i(x_i)\psi_j(x_j)}{|\vec{r}_i - \vec{r}_j|}dx_i dx_j \tag{2.42}$$

This differs from the Hartree expression for E due to the presence of the last term known as exchange energy because it arises due to the exchange of electronic coordinates. We shall now show that this is finite only for parallel spins.

Let us introduce two orthonormal functions, $\alpha(\sigma)$ and $\beta(\sigma)$, corresponding to spin up and spin down. Thus, $\psi_i(\vec{r}_i\sigma_i)$ can be written as the product of spatial part and spin part as

$$\psi_i(\vec{r}_i\sigma_i) = \psi_i(\vec{r}_i)\alpha(\sigma_i) \quad \text{if } \sigma_i = +1, \text{ spin up}$$

$$= \psi_i(\vec{r}_i)\beta(\sigma_i) \quad \text{if } \sigma_i = -1\text{, spin down}$$

$$\alpha(1) = 1, \quad \alpha(-1) = 0$$

$$\beta(1) = 0, \quad \beta(-1) = 1$$

$$\int |\alpha(\sigma)|^2 \, d\sigma = \sum_{\sigma=\pm1} |\alpha(\sigma)|^2 = 1$$

$$\int \alpha(\sigma)\beta(\sigma) d\sigma = \sum_{\sigma=\pm1} \alpha(\sigma)\beta(\sigma) = 0$$

In the first two terms of Equation 2.42, the integration over spin variables is easily performed, and the result is that x_i is replaced by $\vec{r}_i$. Let us consider the last term of Equation 2.42. Assuming ψ_i and ψ_j have opposite spins, we have

$$\sum_{i,j}\sum_{j\neq i}\iint \frac{e^2}{|\vec{r}_i - \vec{r}_j|} \psi_i^*(\vec{r}_j)\alpha(\sigma_j)\psi_j^*(\vec{r}_i)\beta(\sigma_i)\psi_i(\vec{r}_i)\alpha(\sigma_i)\psi_j(\vec{r}_j)\beta(\sigma_j)$$

$$d^3r_i \, d^3r_j \, d\sigma_i \, d\sigma_j = 0$$

Therefore, the last term gives finite contribution only if ψ_i and ψ_j have the same spins. Thus, the exchange energy can be written as

$$-\frac{1}{2}\sum_{\substack{i\neq j \\ (\| \text{ spins})}}\sum \iint \frac{1}{4\pi\varepsilon_0} \frac{e^2\psi_i^*(\vec{r}_j)\psi_j^*(\vec{r}_i)\psi_i(\vec{r}_i)\psi_j(\vec{r}_j)}{|\vec{r}_i - \vec{r}_j|} d^3r_i d^3r_j$$

$$= -\frac{1}{2}\sum_{\substack{i\neq j \\ (\| \text{ spins})}}\sum \iint \frac{1}{4\pi\varepsilon_0} \frac{e^2\psi_i^*(\vec{r})\psi_j^*(\vec{r}\,')\psi_i(\vec{r}\,')\psi_j(\vec{r})}{|\vec{r} - \vec{r}\,'|} d^3r \, d^3r' \tag{2.43}$$

Let us define $v_I(\vec{r})$ as

$$v_I(\vec{r}) = -\frac{1}{4\pi\varepsilon_0}\sum_{\ell} \frac{Ze^2}{|\vec{r} - \vec{R}_\ell|} \tag{2.44}$$

E can be written from Equation 2.42 as

$$E = \sum_i \int \psi_i^*(\vec{r})\left[-\frac{\hbar^2}{2m_e}\nabla^2 + v_I(\vec{r})\right]\psi_i(\vec{r})d^3r$$

$$+\frac{1}{2}\sum\sum_{i\neq j} \iint \frac{1}{4\pi\varepsilon_0}\frac{e^2}{|\vec{r}-\vec{r}'|}|\psi_i(\vec{r})|^2|\psi_j(\vec{r}')|^2 d^3r\,d^3r'$$

$$-\frac{1}{2}\sum\sum_{\substack{i\neq j\\(\parallel \text{ spins})}} \iint \frac{1}{4\pi\varepsilon_0}\frac{e^2}{|\vec{r}-\vec{r}'|}\psi_i^*(\vec{r})\psi_j^*(\vec{r}')\psi_i(\vec{r}')\psi_j(\vec{r})d^3r\,d^3r' \quad (2.45)$$

Note that $i = j$ contribution cancels out in the last two terms; therefore, we shall drop this restriction.

Now, as in the Hartree's case, minimize E, subject to constraint that

$$\int \psi_i^*(\vec{r})\psi_i(\vec{r})d^3r = 1$$

that is, minimize

$$E - \sum_i \varepsilon_i\left[\int \psi_i^*(\vec{r})\psi_i(\vec{r})d^3r - 1\right] \quad (2.46)$$

Similar to Hartree's case, this gives

$$\frac{\delta E}{\delta\psi_i^*(\vec{r})} - \varepsilon_i\psi_i(\vec{r}) = 0$$

or

$$\left[-\frac{\hbar^2}{2m_e}\nabla^2 + v_I(\vec{r})\right]\psi_i(\vec{r}) + \sum_j e^2\frac{1}{4\pi\varepsilon_0}\int\frac{|\psi_j(\vec{r}')|^2}{|\vec{r}-\vec{r}'|}d^3r'\psi_i(\vec{r})$$

$$-\sum_{\substack{j\\(\text{spin } j=\text{spin } i)}} \int \frac{1}{4\pi\varepsilon_0}\frac{e^2}{|\vec{r}-\vec{r}'|}\psi_j^*(\vec{r}')\psi_i(\vec{r}')\psi_j(\vec{r})d^3r' = \varepsilon_i\psi_i(\vec{r}) \quad (2.47)$$

This is known as the H–F equation. This can also be written as

$$\left[-\frac{\hbar^2}{2m_e}\nabla^2+v_I(\vec{r})+v_H(\vec{r})\right]\psi_i(\vec{r})$$

$$-\int \sum_{\substack{j\neq i \\ (\text{spin } j=\text{spin } i)}} \frac{1}{4\pi\varepsilon_0}\frac{e^2}{|\vec{r}-\vec{r}'|}\psi_j^*(\vec{r}')\psi_i(\vec{r}')d^3r'\psi_j(\vec{r}) = \varepsilon_i\psi_i(\vec{r}) \qquad (2.48)$$

where $v_H(\vec{r})$ is the Hartree potential given by Equation 2.24.

The H–F method introduces an improvement over Hartree energy, and it is due to the appearance of exchange energy, which is caused by taking the antisymmetry of the wave function into account. This exchange interaction is a real quantum mechanical effect and must appear in any one-electron model.

The H–F method is a mean-field theory, that is, the many-body problem is reduced to one-electron problem in which the electron moves in an average field generated by other electrons. It ignores Coulomb correlations that arise because electrons repel each other and move in such a way as to avoid each other. However, the H–F method does include some correlations between the parallel spin electrons, which give rise to the exchange interaction. The Coulomb correlations actually reduce the exchange interaction between electrons with parallel spins, so that the H–F equation greatly overestimates the importance of the exchange interaction. This equation describes compact systems such as atoms well but it predicts totally incorrect behavior for the conduction electrons in solids.

As in the Hartree case, E can also be expressed in terms of ε_i. Multiply Equation 2.47 by $\psi_i^*(\vec{r})$ from the left and integrate. This gives

$$\varepsilon_i = \int \psi_i^*(\vec{r})\left[-\frac{\hbar^2}{2m_e}\nabla^2+v_I(\vec{r})\right]\psi_i(\vec{r})d^3r$$

$$+\sum_j e^2\frac{1}{4\pi\varepsilon_0}\int\frac{|\psi_j(\vec{r}')|^2|\psi_i(\vec{r})|^2}{|\vec{r}-\vec{r}'|}d^3r\,d^3r'$$

$$-\sum_{\substack{j \\ (\text{spin } j=\text{spin } i)}}\int\frac{1}{4\pi\varepsilon_0}\frac{e^2}{|\vec{r}-\vec{r}'|}\psi_i^*(\vec{r})\psi_j^*(\vec{r}')\psi_j(\vec{r})\psi_i(\vec{r}')d^3r\,d^3r' \qquad (2.49)$$

Comparing with Equation 2.45, this gives

$$E = \sum_i \varepsilon_i - \frac{1}{2}\sum_i \sum_j \iint \frac{1}{4\pi\varepsilon_0} \frac{e^2}{|\vec{r} - \vec{r}'|} |\psi_i(\vec{r})|^2 |\psi_j(\vec{r}')|^2 \, d^3r \, d^3r'$$

$$+ \frac{1}{2}\sum_i \sum_{\substack{j \\ (\parallel \text{spins})}} \iint \frac{1}{4\pi\varepsilon_0} \frac{e^2}{|\vec{r} - \vec{r}'|} \psi_i^*(\vec{r})\psi_j^*(\vec{r}')\psi_i(\vec{r}')\psi_j(\vec{r}) d^3r \, d^3r' \quad (2.50)$$

2.4.1 Interpretation of ε_i: Koopmans' Theorem

In Equation 2.46, ε_i were introduced as Lagrange multipliers. As in the Hartree case, we can show that $-\varepsilon_i$ is the energy to remove the ith electron (of energy ε_i) from the system (Exercise 2.6). This result is known as Koopmans' theorem.

2.5 Configuration Interaction (CI) Method

The most straightforward way to introduce Coulomb correlations is to write the many-body wave function as

$$\psi = c_1\psi_{HF}^{(1)} + c_2\psi_{HF}^{(2)} + c_3\psi_{HF}^{(3)} + \cdots \quad (2.51)$$

where $\psi_{HF}^{(i)}$ are H–F wave functions for different configuration of electrons. Then, a variational calculation is done to evaluate constants c_i's and the method is called the CI method. For atoms and molecules, this method has been applied and gives very good results. However, for solids or large molecules, this method is very difficult to apply. Also, conceptual advantages of the one-electron model have been lost in the CI wave function. For details, see for example, the books by Szabo and Ostlund (1996) and Springborg (2000).

2.6 Application of Hartree Method to Homogeneous Electron Gas (HEG)

To get a feel for Hartree and H–F methods, we shall apply these methods to the HEG, also known as uniform electron gas or jellium. A great advantage of this system is that we can obtain analytical results; otherwise, in general, we have to solve Hartree or H–F equations numerically.

We shall solve Hartree Equation 2.26 for the jellium model. We shall assume that electrons interact via Coulomb interaction (the term free electron gas is still used in the literature!). We also assume a uniform positive charge background to neutralize the total charge of the electrons. Thus, the total potential in Hartree Equation 2.26 is

$$v_I(\vec{r}_i) + v_H(\vec{r}_i) = -\int \frac{1}{4\pi\varepsilon_0} \frac{Ze^2\rho(\vec{R})}{|\vec{r}_i - \vec{R}|} d^3R + \int \frac{1}{4\pi\varepsilon_0} \frac{e^2 n(\vec{r})}{|\vec{r} - \vec{r}_i|} d^3r \tag{2.52}$$

where $\rho(\vec{R})$ is the charge density at $\vec{R}$ due to ions.

Since at every point $Ze^2\rho = e^2 n, v_I(\vec{r}_i) + v_H(\vec{r}_i) = 0$. Thus, Hartree Equation 2.26 is simply

$$-\frac{\hbar^2}{2m_e}\nabla^2\psi_i = \varepsilon_i\psi_i \tag{2.53}$$

The solution for this equation is well known

$$\psi_{\vec{k}}(\vec{r}) = \frac{1}{\sqrt{\Omega}} e^{i\vec{k}\cdot\vec{r}}$$

where the periodic boundary condition has been used and Ω is the volume of the system. The one-electron energy eigenvalues are given as

$$\varepsilon_{\vec{k}} = \frac{\hbar^2 k^2}{2m_e} \tag{2.54}$$

A plot of $\varepsilon_{\vec{k}}$ versus k is given in Figure 2.4, which has a parabolic shape. At absolute temperature $T = 0$, the topmost filled energy level is ε_F, called the Fermi energy. From Equation 2.54, the constant energy surfaces in $\vec{k}$ space corresponding to $\varepsilon_{\vec{k}}$ are spheres. The constant energy surface in $\vec{k}$ space corresponding to ε_F is called the Fermi surface, which is the hallmark of a

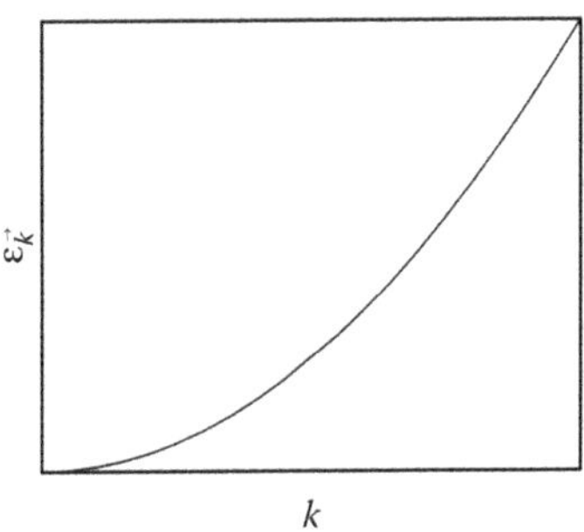

FIGURE 2.4
Free electron energy dispersion.

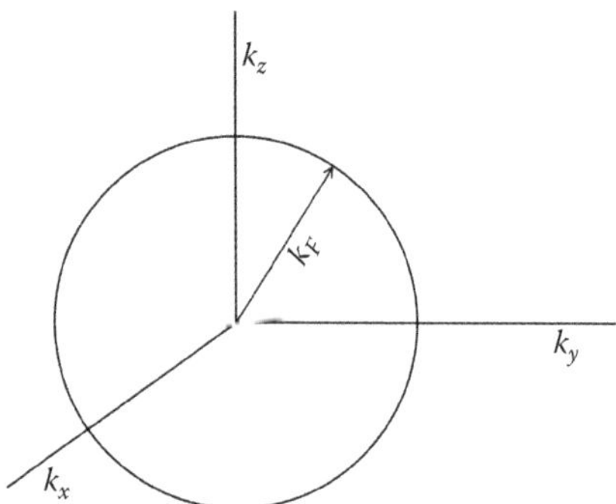

FIGURE 2.5
Free electron Fermi surface.

metallic state. For the free electron model, the Fermi surface is a sphere and is shown in Figure 2.5.

The volume of the Fermi sphere is given as $(4/3)\pi k_F^3$, where k_F is the Fermi radius. Since all electrons must be accommodated within this sphere at $T = 0$, we can calculate the total number of electrons N as

$$N = 2 \cdot \frac{4}{3}\pi k_F^3/(8\pi^3/\Omega)$$
$$= \frac{\Omega}{3\pi^2} k_F^3$$

where we have assumed the nonmagnetic ground state of the system so that each energy level can be occupied by two electrons of opposite spins. We can define electron density n_0 as

$$n_0 = \frac{N}{\Omega} = k_F^3/3\pi^2 \tag{2.55}$$

Let us associate a spherical volume of radius r_s to a volume associated with one conduction electron, which is given as $(\Omega/N) = (1/n_0) = (4/3)\pi r_s^3$ or

$$\frac{4}{3}\pi r_s^3 = \frac{3\pi^2}{k_F^3}$$

$$k_F = \left(\frac{9\pi}{4}\right)^{1/3} \frac{1}{r_s} = \frac{1.92}{r_s} \tag{2.56}$$

We shall see in the next chapter that r_s is a very useful quantity for characterizing the density of the electron gas.

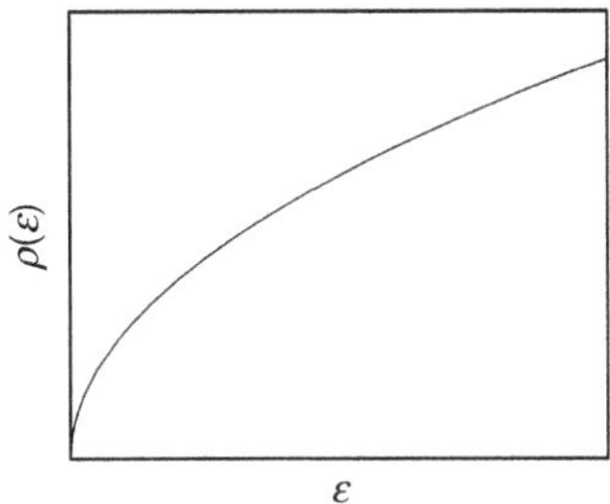

FIGURE 2.6
Free electron DOS.

The total number of states up to energy ε is

$$N(\varepsilon) = \frac{\Omega}{3\pi^2}\left(\frac{2m_e}{\hbar^2}\varepsilon\right)^{3/2} \tag{2.57}$$

$N(\varepsilon)$ is called the integrated density of states (DOS). DOS, $\rho(\varepsilon)$, is defined such that $\rho(\varepsilon)d\varepsilon$ gives the number of states between energy ε and $\varepsilon + d\varepsilon$, that is

$$\begin{aligned}\rho(\varepsilon) &= \frac{dN}{d\varepsilon} = \frac{3}{2}\frac{\Omega}{3\pi^2}\left(\frac{2m_e}{\hbar^2}\right)^{3/2}\varepsilon^{1/2} \\ &= \frac{\Omega}{2\pi^2}\left(\frac{2m_e}{\hbar^2}\right)^{3/2}\varepsilon^{1/2}\end{aligned} \tag{2.58}$$

The DOS for a three-dimensional free electron model is shown schematically in Figure 2.6. The DOS strongly depends on dimensionality although the free electron dispersion relation in one, two, and three dimensions has the same form as Equation 2.54. For free electron model in two dimensions, the DOS is a constant and in one dimension, it goes as $\varepsilon^{-1/2}$ (Exercise 2.8).

2.7 Application of H–F Method to HEG

Recall that the H–F equation can be written as

$$\begin{aligned}&\left[-\frac{\hbar^2}{2m_e}\nabla^2 + v_I(\vec{r}) + v_H(\vec{r})\right]\psi_i(\vec{r}) \\ &- \sum_{\substack{j\neq i \\ (\text{spin } j=\text{spin } i)}} \int \frac{1}{4\pi\varepsilon_0}\frac{e^2}{|\vec{r}-\vec{r}'|}\psi_j^*(\vec{r}')\psi_i(\vec{r}')d^3r'\psi_j(\vec{r}) = \varepsilon_i\psi_i(\vec{r})\end{aligned} \tag{2.59}$$

For a solid, we shall remove the $j \neq i$ restriction. In fact, the $j = i$ term of v_H and the exchange term cancel out. Also, we shall suppress spin j = spin i label. Now, it will be shown that

$$\psi_{\vec{k}}(\vec{r}) = \frac{1}{\sqrt{\Omega}} e^{i\vec{k}\cdot\vec{r}} \tag{2.60}$$

satisfies H–F Equation 2.59.

First, we note that $v_I + v_H = 0$, as was shown in the Hartree case for the free electron model. Let us rewrite Equation 2.59 by putting $i = \vec{k}$ and $j = \vec{k}'$ as

$$-\frac{\hbar^2}{2m_e}\nabla^2\psi_{\vec{k}}(\vec{r}) - \sum_{\vec{k}'}\int \frac{1}{4\pi\varepsilon_0}\frac{e^2}{|\vec{r}-\vec{r}'|}\frac{\psi^*_{\vec{k}'}(\vec{r}')\psi_{\vec{k}}(\vec{r}')d^3r'\psi_{\vec{k}'}(\vec{r})\psi_{\vec{k}}(\vec{r})}{\psi_{\vec{k}}(\vec{r})} = \varepsilon_{\vec{k}}\psi_{\vec{k}}(\vec{r}) \tag{2.61}$$

If we substitute Equation 2.60 in Equation 2.61, the exchange term becomes

$$-\frac{e^2}{\Omega}\sum_{\vec{k}'}\int \frac{1}{4\pi\varepsilon_0}\frac{e^{-i\vec{k}'\cdot\vec{r}'}e^{i\vec{k}\cdot\vec{r}'}e^{i\vec{k}'\cdot\vec{r}}}{e^{+i\vec{k}\cdot\vec{r}}|\vec{r}-\vec{r}'|}d^3r'\psi_{\vec{k}}(\vec{r})$$

$$= -\frac{e^2}{\Omega}\sum_{\vec{k}'}\int \frac{1}{4\pi\varepsilon_0}\frac{e^{-i(\vec{k}-\vec{k}')\cdot(\vec{r}-\vec{r}')}}{|\vec{r}-\vec{r}'|}d^3r'\psi_{\vec{k}}(\vec{r}) \tag{2.62}$$

In evaluating the integral in Equation 2.62, we use the following theorem:

$$\frac{1}{r} = \frac{4\pi}{\Omega}\sum_{\vec{q}}\frac{1}{q^2}e^{i\vec{q}\cdot\vec{r}} \tag{2.63}$$

or

$$\int \frac{e^{-i\vec{q}\cdot\vec{r}}}{r}d^3r = \frac{4\pi}{q^2} \tag{2.64}$$

Thus, using Equation 2.64, Equation 2.62 becomes

$$-\frac{e^2}{\Omega}\sum_{\vec{k}'}\frac{1}{4\pi\varepsilon_0}\frac{4\pi}{|\vec{k}-\vec{k}'|^2}\psi_{\vec{k}}(\vec{r}) = \varepsilon_x(\vec{k})\psi_{\vec{k}}(\vec{r}) \tag{2.65}$$

where

$$\varepsilon_x(\vec{k}) = -\frac{e^2}{\Omega}\sum_{\vec{k}'}\frac{1}{\varepsilon_0}\frac{1}{\left|\vec{k}-\vec{k}'\right|^2} \tag{2.66}$$

Thus, we see that $\psi_{\vec{k}}(\vec{r})$ given by Equation 2.60 are indeed eigenfunctions of Equation 2.59 with an eigenvalue given by

$$\varepsilon_{\vec{k}} = \frac{\hbar^2 k^2}{2m_e} + \varepsilon_x(\vec{k}) \tag{2.67}$$

$\varepsilon_x(\vec{k})$ can be written as

$$\varepsilon_x(\vec{k}) = \frac{-e^2}{\Omega}\frac{\Omega}{8\pi^3}\int\frac{1}{\varepsilon_0}\frac{1}{\left|\vec{k}-\vec{k}'\right|^2}\,d^3k'$$

The integration is over the Fermi sphere. This can be evaluated by fixing k_z' axis along $\vec{k}$ and using the spherical polar coordinates for $\vec{k}'$. The result is

$$\varepsilon_x(\vec{k}) = \frac{-e^2 k_F}{8\pi^2}\frac{1}{\varepsilon_0}\left[2 + \frac{k_F^2 - k^2}{kk_F}\ell n\left|\frac{k_F + k}{k_F - k}\right|\right] \tag{2.68}$$

where k_F is the Fermi radius. Thus

$$\varepsilon_{\vec{k}} = \frac{\hbar^2 k^2}{2m_e} - \frac{e^2 k_F}{8\pi^2}\frac{1}{\varepsilon_0}\left[2 + \frac{k_F^2 - k^2}{k\,k_F}\ell n\left|\frac{k_F + k}{k_F - k}\right|\right] \tag{2.69}$$

In Figure 2.7, we show Hartree and H–F bands for parameters corresponding to sodium together with the experimental result obtained from the ARPES (Lyo and Plummer 1988). Surprisingly, we see that the experimental data is very close to the Hartree band and shows a large deviation from the H–F result. The H–F band is pulled down in energy and the exchange interaction term has led to much increased bandwidth.

Now, let us look at the DOS. The Fermi surface is still a sphere and so are the constant energy surfaces. Therefore

$$N(\varepsilon) = 2\cdot\frac{4}{3}\pi k^3\cdot\frac{\Omega}{8\pi^3} = \frac{\Omega k^3}{3\pi^2}$$

$$\rho(\varepsilon) = \frac{dN(\varepsilon)}{d\varepsilon} = \frac{\Omega k^2}{\pi^2}\frac{dk}{d\varepsilon} = \frac{\Omega k^2}{\pi^2}\bigg/\frac{d\varepsilon}{dk} \tag{2.70}$$

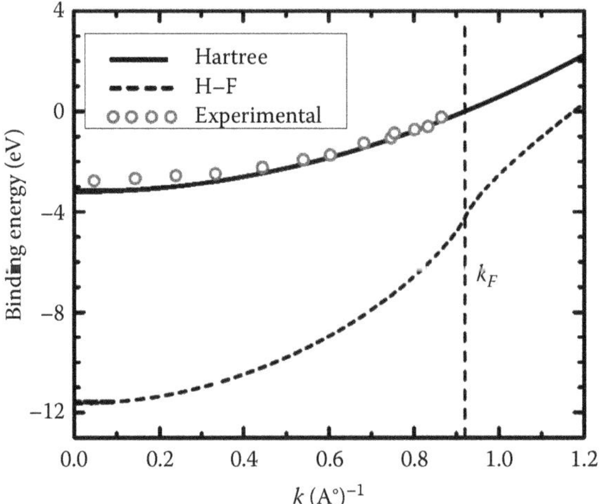

FIGURE 2.7
Hartree (solid line) and H–F (dashed line) bands corresponding to parameters for Na (lattice constant $a = 4.225$ Å). Also shown are experimental bands using ARPES. (Adapted from Lyo I. W. and Plummer E. W. 1988. *Phys. Rev. Lett.* 60: 158, 1561.)

$$\frac{d\varepsilon}{dk} = \frac{\hbar^2 k}{m_e} - \frac{e^2}{8\pi^2}\frac{1}{\varepsilon_0}\left\{\frac{2k_F}{k} - \left(1 + \frac{k_F^2}{k^2}\right)\ell n\left|\frac{k_F + k}{k_F - k}\right|\right\} \tag{2.71}$$

We see that $d\varepsilon/dk$ is infinite at $k = k_F$ and therefore, the DOS goes to zero at ε_F.

In Figure 2.8, we show the DOS for sodium using Hartree and H–F theory as discussed above. Here too, the Hartree theory does much better than the H–F theory. We see that the H–F DOS is singular and approaches zero at the Fermi energy. Indeed, $\rho(\varepsilon_F)$ going to zero for a metal is an alarming feature, which indicates that something is seriously wrong with the H–F theory. Recall that the Hartree theory neglects Coulomb correlations as well as correlations between parallel spins (exchange correlations). The H–F theory, on the other hand, includes exchange correlations but neglects Coulomb correlations. It seems that for the free electron model, the contributions from Coulomb and exchange correlations largely cancel out and thus, the Hartree theory gives better results. However, the H–F theory does give lower total energy than the Hartree theory. Thus, the message from this exercise is that the H–F theory is an improvement over Hartree theory as it includes exchange interaction, but it is not enough. One must find a way to include Coulomb correlations.

Let us go a little deeper regarding the origin of the pathological behavior of the H–F DOS. The electron correlations give rise to the screening of Coulomb interaction between the electrons. One can see from Equation 2.64 that the bare Coulomb potential has $q = 0$ singularity in its Fourier transform. If, instead of

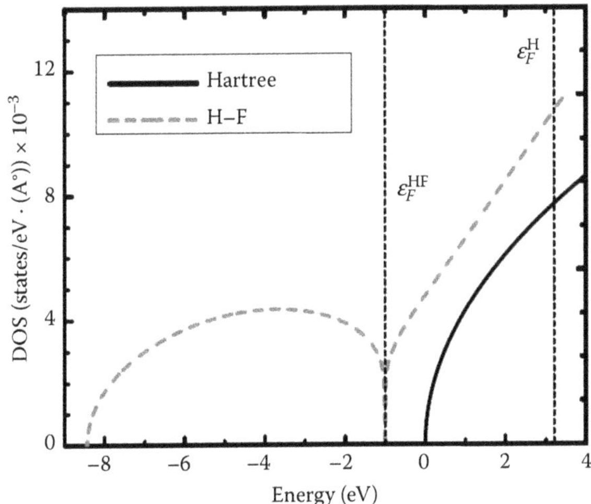

FIGURE 2.8
DOS using Hartree (solid line) and H–F (dashed line) approximations using parameters corresponding to Na (lattice constant $a = 4.225$ Å). Note that the Hartree DOS has been scaled down by a factor of 3.

bare Coulomb potential, a screened Coulomb potential is used; this singularity disappears and so does this unphysical feature in the DOS. It is clear from this discussion that to improve H–F results, one has to take into account electron correlations, which will in turn introduce screening. We shall see that this is the case in the DFT or GW theory, which we shall discuss later. Thus, because of the neglect of the screening, the energy gaps in the H–F theory come out to be much larger than that obtained in the DFT or GW theory.

2.8 Beyond the H–F Theory for HEG

We saw in the previous section that the inclusion of the exchange in the H–F theory did not improve the Hartree result for the HEG. Thus, it is important to treat the correlation between electrons, which tries to keep them away from coming close to each other and is a many-body effect. Correlations are more pronounced between the opposite spin electrons since the exchange already keeps the parallel spin electron coming very close to each other.

2.8.1 Correlation Energy

Correlation energy is the total energy calculated with proper allowance for Coulomb correlations minus the H–F energy. The correlation energy is always negative because the H–F energy is an upper bound to the exact ground-state energy (see Exercise 2.10). Several attempts have been made to improve the H–F theory for HEG by including correlations such as by Wigner (1934), Gell-Mann and Brueckner (1957), Hubbard (1958), Hedin and Lundqvist (1969), Singwi et al. (1968), Vashishta and Singwi (1972), and so on who calculated the correlation energy. The interacting HEG has also been studied by using quantum Monte Carlo simulations (e.g., Ceperley and Alder 1980), and their results are considered to be most accurate at present. We shall not discuss these works here and refer to the original papers and the books by Mahan (2000) and Martin (2004).

EXERCISES

2.1 a. Find the Fermi velocity in Na metal.

b. Find an approximate ionic velocity of Na ion of mass M.

c. Show that the ratio of the ionic velocity and Fermi velocity is of the order m/M.

2.2 Using the variational principle, prove that the ground-state energy of a system, using the Born–Oppenheimer approximation, is a lower bound to the true ground-state energy.

2.3 Apply the Hartree method to He atom and write the Hartree equation when both electrons are in their lowest energy state. Repeat it with the H–F method.

2.4 Apply the Hartree method to H_2 molecule and write the Hartree equation when both electrons are in their lowest energy state. Repeat it with the H–F method.

2.5 Find the normalization constant for the trial wave function in the H–F method.

2.6 Prove Koopmans' theorem.

2.7 Express the total energy E in the H–F method in terms of single-particle energies.

2.8 Show that for the free electron model in two dimensions, the DOS is a constant and in one dimension, it goes as $\varepsilon^{-1/2}$.

2.9 Evaluate $\varepsilon_x(\vec{k}) = -\dfrac{e^2}{\Omega}\dfrac{1}{\varepsilon_0}\sum_{\vec{k}'}\dfrac{1}{|\vec{k}-\vec{k}'|^2}$.

2.10 Using the variational principle, prove that the H–F energy is an upper bound to the exact energy of an electronic system.

2.11 Plot $\varepsilon(k)$ versus k when the H–F and Hartree methods are applied to the jellium model. Use parameters corresponding to Na.

2.12 Evaluate DOS when the H–F method is applied to the jellium model and plot it. Use parameters corresponding to Na.

2.13 Show that the total kinetic energy of the free electron gas is

$$T = (3/5)\,\varepsilon_F N.$$

Further Reading

Szabo A. and Ostlund N. S. 1996. *Modern Quantum Chemistry, Introduction to Advanced Electronic Structure Theory*. New York, NY: Dover.

3

Density Functional Theory

3.1 Introduction

Until now we have been trying to solve the many-body problem by approximating the many-body wave function such as in Hartree or Hartree–Fock (H–F) methods. Since the many-body wave function involves the coordinates of all the electrons, it is a function of 3N variables, where N is the number of electrons. Obviously, such an approach is complex and numerically very demanding. It would be much simpler if the electronic system could be described in terms of electron density $n(\vec{r})$, which is a function of only three variables (e.g., x, y, z). A firm and exact theoretical foundation for such description was provided by Hohenberg and Kohn in 1964 and is known as the density functional theory (DFT). We shall see that in the local density approximation (LDA) it is as simple as the Hartree theory but gives much better results than that given by the H–F theory. The reason is that even the simplest version of the theory includes both exchange and correlations, fairly accurately, while the H–F equation includes exchange exactly but neglects correlations completely. Owing to its simplicity, most of the electronic structure calculations on materials are done using the DFT, and it has made a big impact in understanding the atoms, molecules, atomic clusters, and solids. Recognizing this achievement Walter Kohn (Figure 3.1), who played the leading role in the development of DFT, was awarded the Nobel Prize in 1998.

We shall start with the Thomas–Fermi theory in Section 3.2, which first used electronic density as the basic variable. We shall see that although the Thomas–Fermi theory is a very approximate theory, it contained the seeds for the development of the DFT (for definition of functional, see Appendix D). In Section 3.3, we shall discuss an application of the Thomas–Fermi theory to the problem of screening. In Section 3.4, we shall prove Hohenberg–Kohn theorems, which laid down the foundation of the DFT. Kohn–Sham (KS) equation, which is the main equation of the DFT, will be derived Section 3.5, We shall then discuss the LDA in Section 3.6, which has proved most useful in calculating the electronic structure of solids. In Section 3.7, we briefly discuss the comparison of the DFT with the Hartree and H–F theories and

FIGURE 3.1
Walter Kohn (1923–) played the leading role in the development of the density functional theory, for which he was awarded the Nobel Prize in 1998. (Courtesy of Jtk33, Wikimedia Commons.)

in Section 3.8, we make some comments on KS eigenvalues and KS orbitals. In Section 3.9, we briefly discuss the extension of DFT to magnetic systems known as spin DFT. Following this, some successes and failures of the LDA are mentioned in Section 3.10, In Section 3.11, other approximate schemes such as GGA, SIC, LDA + U, and GW method are briefly discussed. Finally, in Section 3.12, we briefly discuss the idea of the time-dependent density functional theory (TDDFT).

Before we proceed further, we would like to introduce a few definitions that will be used in this chapter. Recall that the Hamiltonian for the interacting inhomogeneous electron gas is

$$H = T + V + V_{EE} \tag{3.1}$$

where

$$T = \sum_i -\frac{\hbar^2}{2m_e}\nabla_i^2$$

$$V = \sum_i v(\vec{r}_i)$$

$$V_{EE} = \frac{1}{2}\sum_{\substack{i,j \\ i\neq j}} \frac{1}{4\pi\varepsilon_0}\frac{e^2}{|\vec{r}_i - \vec{r}_j|}$$

The electron density operator is defined as

$$\hat{n}(\vec{r}) = \sum_{i=1}^{N} \delta(\vec{r} - \vec{r}_i) \tag{3.2}$$

The electron density is then

$$\begin{aligned} n(\vec{r}) &= \langle \psi | \hat{n}(\vec{r}) | \psi \rangle \\ &= \int \psi^*(\vec{r}_1 \cdots \vec{r}_N)\hat{n}(\vec{r})\psi(\vec{r}_1 \cdots \vec{r}_N) d^3r_1 \cdots d^3r_N \end{aligned} \tag{3.3}$$

where ψ is the many-electron state and ∫ stands for multiple integrations.

Similarly, $\langle\psi|V|\psi\rangle$ can be written as

$$\begin{aligned} \langle \psi | V | \psi \rangle &= \int \psi^* \sum_i v(\vec{r}_i)\psi d^3r_1 \cdots d^3r_N \\ &= \int \psi^* \sum_i v(\vec{r})\delta(\vec{r} - \vec{r}_i)\psi d^3r_1 \cdots d^3r_N d^3r \\ &= \int \psi^* v(\vec{r})\hat{n}(\vec{r})\psi d^3r_1 \cdots d^3r_N d^3r \\ &= \int v(\vec{r})n(\vec{r}) d^3r \end{aligned} \tag{3.4}$$

3.2 Thomas–Fermi Theory

The method was proposed by Thomas (1927) and Fermi (1927). The basic idea is to describe electronic systems such as atoms, molecules, or solids in terms of electron density. Hartree and H–F theories are formulated in terms of single-particle wave functions or orbitals $\psi_i(\vec{r})$. We shall see that the use of electron density results in great simplification.

First, consider homogeneous electron gas of volume Ω, containing N electrons.

$$N = 2 \cdot \frac{4}{3}\pi k_F^3/(8\pi^3/\Omega)$$

or

$$\frac{N}{\Omega} = \frac{1}{3}\frac{k_F^3}{\pi^2} = n_0 = \text{electron density} \tag{3.5}$$

The kinetic energy is given by $T = (3/5)\varepsilon_F N$. Thus, we can define kinetic energy density t_0 as

$$\begin{aligned} t_0 &= \frac{T}{\Omega} = \frac{3}{5}\varepsilon_F \frac{N}{\Omega} \\ &= \frac{3}{5}\frac{\hbar^2 k_F^2}{2m_e} n_0 \\ &= \frac{3}{5}\frac{\hbar^2}{2m_e}(3\pi^2 n_0)^{2/3} n_0 \\ &= \frac{3}{10}\frac{\hbar^2}{m_e}(3\pi^2)^{2/3} n_0^{5/3} = C_k n_0^{5/3} \end{aligned} \tag{3.6}$$

where we have used the relation $k_F = (3\pi^2 n_0)^{1/3}$. In solids, the electron density is not homogeneous but inhomogeneous. The idea of the Thomas–Fermi theory is to use results for the homogeneous electron gas for the inhomogeneous one. So, if the electron density $n(\vec{r})$ is varying very slowly in space, we can assume that locally the gas is uniform and can use results (3.5) and (3.6). Thus

$$n(\vec{r}) = \frac{k_F^3(\vec{r})}{3\pi^2}$$

and

$$t(\vec{r}) = C_k n(\vec{r})^{5/3}$$

Thus, the total kinetic energy T is

$$T = \int t(\vec{r}) d^3r = C_k \int n(\vec{r})^{5/3} d^3r$$

The total energy E is given by

$$\begin{aligned} E &= T + V_{IE} + V_{EE} \\ &= C_k \int n(\vec{r})^{5/3} d^3r + \int v_{IE}(\vec{r}) n(\vec{r}) d^3r \\ &\quad + \frac{1}{2}\iint \frac{e^2}{4\pi\varepsilon_0}\frac{n(\vec{r})\, n(\vec{r}')}{|\vec{r} - \vec{r}'|} d^3r\, d^3r' \end{aligned}$$

where we have included the classical electron–electron interaction and neglected the exchange–correlation energy. We see that in the Thomas–Fermi theory, the total energy E can be explicitly written as a functional of density only and thus it contained the basic feature of DFT (for definition of functional, see Appendix D).

Now, minimize E subject to constraint that

$$\int n(\vec{r})d^3r = N = \text{constant}$$

or minimize $E - \mu N$ without any constraint, that is

$$\frac{\delta(E - \mu N)}{\delta n(\vec{r})} = 0 \tag{3.7}$$

or

$$\frac{\delta E}{\delta n(\vec{r})} - \mu = 0$$

or

$$\frac{5}{3}C_k n(\vec{r})^{2/3} + v_{IE}(\vec{r}) + \int \frac{e^2}{4\pi\varepsilon_0} \frac{n(\vec{r}')}{|\vec{r} - \vec{r}'|} d^3r' = \mu$$

or

$$\mu = \frac{5}{3}C_k n(\vec{r})^{2/3} + v(\vec{r}) \tag{3.8}$$

where

$$v(\vec{r}) = v_{IE}(\vec{r}) + \int \frac{e^2}{4\pi\varepsilon_0} \frac{n(\vec{r}')}{|\vec{r} - \vec{r}'|} d^3r'$$

We can see that the Lagrange multiplier μ from Equation 3.7 is nothing but the chemical potential

$$\mu = \frac{\delta E}{\delta n(\vec{r})} = \frac{\delta E}{\delta N}$$

Equation 3.8 is the basic equation of the Thomas–Fermi theory. We see that this equation is extremely simple to solve as we have an equation in terms of electron density, which only depends on $\vec{r}$. The method has been extended by Dirac (1930) to include exchange interaction and by Weizsäcker (1935) to include correction to kinetic energy functional. For more details, we refer to the book by Parr and Yang (1989). However, this approach is very approximate and fails to explain many important phenomena such as shell structure of atoms.

3.3 Screening: An Application of Thomas–Fermi Theory

As we have mentioned in our earlier section, screening is an important phenomenon arising due to the electron–electron interaction. As an application of the Thomas–Fermi method, we shall discuss the screening of a positive point charge by the free electron gas. Suppose a positive charge Q is placed in the free electron gas at a fixed position; it will attract excess electrons in its vicinity in such a way that its field gets screened. We shall show by using the Thomas–Fermi theory that the potential of such a charge, at a distance r, from the charge in the electron gas is given by

$$\phi(r) = \frac{1}{4\pi\varepsilon_0}\frac{Q}{r}e^{-k_s r}$$

where k_s is a positive constant.

Let n_0 be the uniform electron density of the free electron gas before introducing the point charge Q. Let us now introduce the charge at the origin. The electrons will interact with this charge and as a result there will be redistribution of the electron density. Let $n(\vec{r})$ be the electron density at a point $\vec{r}$. Then the net charge density at point $\vec{r}$ will be $-e(n(\vec{r}) - n_0)$ because there is a uniform positive charge density background of en_0. Then, the potential $\phi(\vec{r})$ at $\vec{r}$ is given by Poisson's equation

$$\nabla^2\phi(\vec{r}) = \frac{e}{\varepsilon_0}(n(\vec{r}) - n_0) \tag{3.9}$$

For a large system like a solid, we can take $\phi(\vec{r})$ and $n(\vec{r})$ to be spherically symmetric around Q, provided we neglect the surface effects. Hence, $\phi(\vec{r}) = \phi(r)$ and $n(\vec{r}) = n(r)$.

We obtain another relation between $\phi(r)$ and $n(r)$ by using the Thomas–Fermi equation

$$\mu = \frac{5}{3}C_k n(r)^{2/3} - e\phi(r) \tag{3.10}$$

where μ is the chemical potential.

As mentioned above, introduction of the charge +Q in the free electron gas causes the redistribution of electrons. In the vicinity of the charge +Q, there will be excess electronic charge. This excess electronic charge will screen the charge Q in such a way that its field at large distances will be zero, but there will be very little effect near Q. Thus, the boundary conditions on $\phi(r)$ are

$$\phi(r) \to 0 \quad \text{as } r \to \infty \tag{3.11}$$

and

$$\phi(r) \to \frac{1}{4\pi\varepsilon_0}\frac{Q}{r} \quad \text{as } r \to 0 \tag{3.12}$$

According to condition (3.11), at large distances from the charge, one cannot detect any change due to Q and hence $n(r) = n_0$. Thus, as $r \to \infty$, Equation 3.10 implies

$$\begin{aligned}\mu &= \frac{5}{3}C_k\, n_0^{2/3} = \frac{5}{3}\cdot\frac{3}{10}\frac{\hbar^2}{m_e}(3\pi^2)^{2/3}\, n_0^{2/3} \\ &= \frac{\hbar^2}{2m_e}(3\pi^2 n_0)^{2/3} = \varepsilon_F \end{aligned}\tag{3.13}$$

We can also rewrite Equation 3.10 as

$$\mu = \frac{\hbar^2}{2m_e}(3\pi^2\, n(r))^{2/3} - e\,\phi(r) \tag{3.14}$$

Equations 3.9 and 3.14 are coupled equations and their exact solution is difficult to obtain. But we can find an approximate solution quite easily. If we assume that $n(r)$ is slowly varying, which is true if one is away from the point charge, we can Taylor expand the first term in Equation 3.14 around n_0 and retain only the first-order term, that is

$$\begin{aligned}\frac{\hbar^2}{2m_e}(3\pi^2 n(r))^{2/3} &= \frac{\hbar^2}{2m_e}(3\pi^2 n_0)^{2/3} + \frac{\hbar^2}{2m_e}(n(r)-n_0)\frac{d}{dn}(3\pi^2 n)^{2/3}\Big|_{n_0} \\ &= \frac{\hbar^2}{2m_e}(3\pi^2 n_0)^{2/3} + \frac{\hbar^2}{2m_e}(n(r)-n_0)(3\pi^2)^{2/3}\frac{2}{3}n_0^{-1/3} \\ &= \varepsilon_F + \frac{\hbar^2}{2m_e}(n(r)-n_0)(3\pi^2 n_0)^{2/3}\frac{2}{3n_0} \\ &= \varepsilon_F + \frac{2}{3}(n(r)-n_0)\frac{\varepsilon_F}{n_0} \end{aligned}\tag{3.15}$$

Substituting Equation 3.15 into Equation 3.14 and using Equation 3.13, we get

$$\varepsilon_F = \varepsilon_F + \frac{2}{3}(n(r) - n_0)\frac{\varepsilon_F}{n_0} - e\phi(r)$$

or

$$n(r) - n_0 = \frac{3}{2}\frac{n_0}{\varepsilon_F} e\phi(r) \tag{3.16}$$

Substituting Equation 3.16 into Equation 3.9, we get

$$\begin{aligned} \nabla^2\phi &= e^2 \frac{3}{2\varepsilon_0} \cdot \frac{n_0}{\varepsilon_F} \phi(r) \\ &= k_s^2 \phi(r) \end{aligned} \tag{3.17}$$

where

$$\begin{aligned} k_s &= \left[\frac{3n_0 e^2}{2\varepsilon_0\varepsilon_F}\right]^{1/2} \\ &= \left[\frac{3n_0 e^2}{2\varepsilon_0\hbar^2/2m_e\,(3\pi^2 n_0)^{2/3}}\right]^{1/2} \\ &= \left[\frac{3m_e e^2}{\varepsilon_0\hbar^2(3\pi^2)^{2/3}} n_0^{1/3}\right]^{1/2} \end{aligned} \tag{3.18}$$

Since $\phi(r)$ is spherically symmetric, we get

$$\frac{1}{r}\frac{\partial^2}{\partial r^2}(r\phi) = k_s^2 \phi$$

or

$$\frac{\partial^2}{\partial r^2}(r\phi) = k_s^2 (r\phi) \tag{3.19}$$

This gives

$$r\phi = A e^{-k_s r}$$

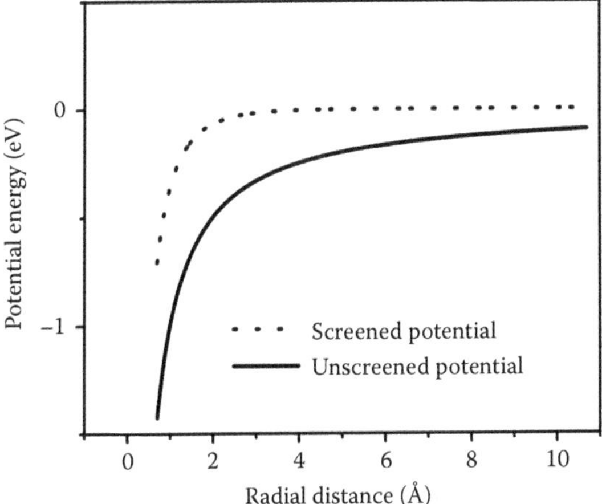

FIGURE 3.2
Schematic plot of the potential energy of an electron in the screened and unscreened potentials of a positive charge as a function of radial distance.

where A is a constant, which from Equation 3.12 is equal to $(1/4\pi\varepsilon_0)Q$. Thus, we get

$$\phi(r) = \frac{1}{4\pi\varepsilon_0}\frac{Q}{r}e^{-k_s r} \tag{3.20}$$

Thus, the potential at r is Coulomb potential times an exponential factor and is of the form of Yukawa potential. This is called screened Coulomb potential. The potential energy of an electron in the potential of a positive charge is shown schematically in Figure 3.2. The potential energy for the screened potential approaches zero much faster than for the unscreened potential. The length $1/k_s$ is a measure of the effective range of the potential and is called the screening distance. In metals, the screening distance is of the order of interatomic spacing but in semiconductors, it is much larger.

3.4 Hohenberg–Kohn Theorems

We saw in the Thomas–Fermi theory that electronic systems could be described in terms of electron density $n(\vec{r})$. A firm and exact theoretical foundation for such description was provided by Hohenberg and Kohn (1964) who gave two theorems that laid the foundation of the DFT.

1. The external potential $v(\vec{r})$ is a unique functional of the electron density $n(\vec{r})$ (for definition of functional, see Appendix D). As a result, the total ground-state energy E of any many-electron system is also a unique functional of $n(\vec{r})$, that is, $E = E[n]$.
2. The functional $E[n]$ for the total energy has a minimum equal to the ground-state energy at the ground-state density.

To prove the first theorem, we assume that there are two different potentials $v(\vec{r})$ and $v'(\vec{r})$ with ground states ψ and ψ' producing the same density $n(\vec{r})$. We show that this assumption leads to an absurd result. We assume the ground state to be nondegenerate. The proof also exists for the degenerate case but is more complicated and, therefore, will not be discussed here.

Since $v'(\vec{r}) \neq v(\vec{r})$, $\psi' \neq \psi$ because these states are the eigenstates of different Hamiltonians. Let us denote the Hamiltonians and ground-state energies associated with ψ and ψ' by H, H' and E, E', respectively. Then, from Raleigh–Ritz variational principle, we have

$$E = \langle \psi | H | \psi \rangle < \langle \psi' | H | \psi' \rangle \tag{3.21}$$

Now

$$\begin{aligned} \langle \psi' | H | \psi' \rangle &= \langle \psi' | H' + (H - H') | \psi' \rangle \\ &= \langle \psi' | H' | \psi' \rangle + \langle \psi' | V - V' | \psi' \rangle \\ &= E' + \int (v(\vec{r}) - v'(\vec{r})) n(\vec{r}) d^3 r \end{aligned} \tag{3.22}$$

where we have used Equation 3.4. Thus, from Equation 3.21, we get

$$E < E' + \int (v(\vec{r}) - v'(\vec{r})) n(\vec{r}) d^3 r \tag{3.23}$$

Similarly, by interchanging primed and unprimed quantities, we get

$$E' < E + \int (v'(\vec{r}) - v(\vec{r})) n(\vec{r}) d^3 r \tag{3.24}$$

by adding Equations 3.23 and 3.24, we get

$$E + E' < E + E'$$

which is absurd.

Thus, we conclude that electron density $n(\vec{r})$ associated with the ground state corresponding to an external potential $v(\vec{r})$ cannot be reproduced by the ground-state density corresponding to a different potential $v'(\vec{r})$ (except when $v'(\vec{r}) - v(\vec{r})$ = constant). Thus, it follows that the ground-state electron density $n(\vec{r})$ determines $v(\vec{r})$ uniquely. This further implies that $n(\vec{r})$ is produced by unique H. Since H uniquely determines the nondegenerate ground state of the system, $n(\vec{r})$ also determines the ground-state wave function uniquely. Hence, $E = \langle\psi|H|\psi\rangle = E[n(\vec{r})]$ is uniquely determined by $n(\vec{r})$ only. Note that since $n(\vec{r})$ determines H uniquely, therefore, it determines not only the ground state but all the excited states as well. However, here we shall focus on the ground state of the system.

To prove the second theorem, let us write the ground-state energy $E[n(\vec{r})]$ as

$$E[n(\vec{r})] = \langle\psi|V|\psi\rangle + \langle\psi|T + V_{EE}|\psi\rangle$$
$$= \int v(\vec{r})n(\vec{r})d^3r + F[n(\vec{r})] \tag{3.25}$$

Since ψ is functional of $n(\vec{r})$, $F = \langle\psi|T + V_{EE}|\psi\rangle$ is also functional of $n(\vec{r})$.

Now, consider a many-electron system with N electrons in a given external potential $v(\vec{r})$. This means that we now keep $v(\vec{r})$ *fixed* and change $n(\vec{r})$ ($n(\vec{r})$ = ground-state electron density). Let us define the following energy functional for the density $n'(\vec{r})$:

$$E_v[n'(\vec{r})] = \int v(\vec{r})n'(\vec{r})d^3r + F[n'(\vec{r})] \tag{3.26}$$

Now, we know that

$$\tilde{E}_v[\psi'] = \langle\psi'|V|\psi'\rangle + \langle\psi'|T + V_{ee}|\psi'\rangle$$

has lower value for the correct nondegenerate ground-state ψ than for any other ψ'. Let ψ' be the state associated with density $n'(\vec{r})$. Then

$$\tilde{E}_v[\psi'] = \int v(\vec{r})n'(\vec{r})d^3r + F[n'(\vec{r})]$$
$$> \tilde{E}_v[\psi] = \int v(\vec{r})n(\vec{r})d^3r + F[n(\vec{r})] \tag{3.27}$$

But

$$\tilde{E}_v[\psi'] = E_v[n']$$

and

$$\tilde{E}_v[\psi] = E_v[n]$$

Therefore

$$E_v[n] < E_v[n']$$

This proves the second theorem.

The functional $F[n]$ introduced in Equation 3.25 is a functional of density $n(\vec{r})$ and is independent of the external potential $v(\vec{r})$. It contains kinetic and interaction energies and no exact expression for it is known. However, several good approximations for $F[n]$ have been developed. Let us first split $F[n]$ into three parts:

$$F[n] = T_s[n] + \frac{e^2}{2}\iint \frac{1}{4\pi\varepsilon_0}\frac{n(\vec{r})n(\vec{r}')}{|\vec{r}-\vec{r}'|}d^3r d^3r' + E_{xc}[n] \tag{3.28}$$

where $T_s[n]$ is the kinetic energy of *noninteracting* electron gas of density $n(\vec{r})$ in its ground state, the second term is the classical Coulomb energy, and the remainder $E_{xc}[n]$ is called exchange–correlation energy. Note that $E_{xc}[n]$ contains part of the total kinetic energy. The reason why $T_s[n]$ and not $\langle T\rangle$ has been separated in Equation 3.28 will become clear in the following section.

3.5 Derivation of Kohn–Sham (KS) Equations

Using Equations 3.25 and 3.28, we can write the total energy functional as

$$E[n] = \int v(\vec{r})n(\vec{r})d^3r + T_s[n] + \frac{e^2}{2}\iint \frac{1}{4\pi\varepsilon_0}\frac{n(\vec{r})n(\vec{r}')}{|\vec{r}-\vec{r}'|}d^3r d^3r' + E_{xc}[n] \tag{3.29}$$

We have to minimize Equation 3.29 subject to the condition

$$\int n(\vec{r})d^3r = N$$

This gives

$$\frac{\delta E[n]}{\delta n(\vec{r})} - \mu = 0 \tag{3.30}$$

or

$$\frac{\delta T_s[n]}{\delta n(\vec{r})} + v(\vec{r}) + \int \frac{e^2}{4\pi\varepsilon_0} \frac{n(\vec{r}')}{|\vec{r} - \vec{r}'|} d^3r' + \frac{\delta E_{xc}[n]}{\delta n(\vec{r})} - \mu = 0$$

Let us write

$$v_H(\vec{r}) = \int \frac{e^2}{4\pi\varepsilon_0} \frac{n(\vec{r}')}{|\vec{r} - \vec{r}'|} d^3r' \tag{3.31}$$

and

$$v_{xc}(\vec{r}) = \frac{\delta E_{xc}[n]}{\delta n(\vec{r})} \tag{3.32}$$

where $v_{xc}(\vec{r})$ incorporates the effects of exchange and correlations and is known as the exchange–correlation potential. Thus, we get

$$\frac{\delta T_s[n]}{\delta n(\vec{r})} + v(\vec{r}) + v_H(\vec{r}) + v_{xc}(\vec{r}) - \mu = 0 \tag{3.33}$$

where μ is the Lagrange multiplier. Equation 3.33 is the required equation only if it could be solved. Then, it would give us the electron density $n(\vec{r})$, which in turn would give us the total energy E from Equation 3.29. However, there are two problems: (i) functional $T_s[n]$ is not known and (ii) $v_{xc}(\vec{r})$ is also unknown. Thus, Equation 3.33 can be solved only approximately. We first note that Equation 3.33 reduces to the Thomas–Fermi equation if one neglects $v_{xc}(\vec{r})$ and uses

$$T_s[n] = C_k \int n(\vec{r})^{5/3} d^3r$$

Better functionals for $T_s[n]$ have been proposed such as Thomas–Fermi–Weizsäcker functional (Weizsäcker 1935). But such theories had only limited success so far and could not predict the shell structure for atoms.

Another approach to solve Equation 3.33 is due to Kohn and Sham (1965) who had to introduce orbitals in the DFT. They noted that if the last term in Equation 3.29, that is, $E_{xc}[n]$, was absent, then the expression would be identical to Hartree's expression for energy. Then, $v_{xc}(\vec{r})$ would be absent in Equation 3.33 too and this equation could be solved by using

$$n(\vec{r}) = \sum_{i=1}^{N} |\psi_i(\vec{r})|^2 \tag{3.34}$$

and by solving

$$\left(-\frac{\hbar^2}{2m_e}\nabla^2 + v(\vec{r}) + v_H(\vec{r})\right)\psi_i(\vec{r}) = \varepsilon_i\psi_i(\vec{r}) \quad (3.35)$$

Thus, if one could write $n(\vec{r})$ in terms of single-particle orbitals as in Equation 3.34 even when $v_{xc}(\vec{r})$ is present, Equation 3.33 could be solved in a similar way. Thus, we define

$$v_{eff}(\vec{r}) = v(\vec{r}) + v_H(\vec{r}) + v_{xc}(\vec{r}) \quad (3.36)$$

and use

$$n(\vec{r}) = \sum_{i=1}^{N} |\psi_i(\vec{r})|^2 \quad (3.37)$$

Then, in analogy with Equation 3.35, $\psi_i(\vec{r})$ can be obtained by solving

$$\left[-\frac{\hbar^2}{2m_e}\nabla^2 + v_{eff}(\vec{r})\right]\psi_i(\vec{r}) = \varepsilon_i\psi_i(\vec{r}) \quad (3.38)$$

Equation 3.38 is known as KS equations and $\psi_i(\vec{r})$, ε_i, and $v_{eff}(\vec{r})$ as KS orbitals, KS energies, and KS potential, respectively. The Hamiltonian

$$H = -\frac{\hbar^2}{2m_e}\nabla^2 + v_{eff}(\vec{r})$$

is called as the KS Hamiltonian.

The KS equations have to be solved self-consistently. Note that the orbitals in the DFT are introduced via Equation 3.37, that is, we are assuming that $n(\vec{r})$ can always be expressed in terms of orbitals by Equation 3.37. In other words, we are mapping our interacting electron system onto an auxiliary system of independent particles, which has the same density as our original system. These independent particles move in a potential given by Equation 3.36. This (Equation 3.37) is an ansatz that enabled Kohn and Sham to arrive at Equation 3.38 and hence it provided a way to solve the many-electron problem. Note the simplicity of Equation 3.38, which reduces the many-electron problem to a one-electron problem. This is similar to Hartree or H–F methods but unlike these methods, this includes the effects of exchange and correlations. Note that the exchange–correlation potential $v_{xc}(\vec{r})$ is an unknown function for which some approximation has to be made if we want to solve the problem. This will be done in the next section.

Let us now express the total energy in terms of ε_i. To do this, multiply Equation 3.38 by $\psi_i^*(\vec{r})$ from the left and integrate over $\vec{r}$. This gives

$$\varepsilon_i = \int \psi_i^*(\vec{r}) \left[-\frac{\hbar^2}{2m_e} \nabla^2 + v_{eff}(\vec{r}) \right] \psi_i(\vec{r}) d^3r$$

or

$$\begin{aligned}
\sum_i \varepsilon_i &= \sum_i \int \psi_i^*(\vec{r}) \left[-\frac{\hbar^2}{2m_e} \nabla^2 \right] \psi_i(\vec{r}) d^3r \\
&\quad + \sum_i \int \psi_i^*(\vec{r}) \psi(\vec{r}) v(\vec{r}) d^3r + \sum_i e^2 \int \frac{1}{4\pi\varepsilon_0} \frac{\psi_i^* \psi_i n(\vec{r}')}{|\vec{r} - \vec{r}'|} d^3r d^3r' \\
&\quad + \sum_i \int v_{xc}(\vec{r}) \psi_i^*(\vec{r}) \psi_i(\vec{r}) d^3r \\
&= T_s[n] + \int v(\vec{r}) n(\vec{r}) d^3r + e^2 \iint \frac{1}{4\pi\varepsilon_0} \frac{n(\vec{r}) n(\vec{r}')}{|\vec{r} - \vec{r}'|} d^3r d^3r' \\
&\quad + \int v_{xc}(\vec{r}) n(\vec{r}) d^3r
\end{aligned} \tag{3.39}$$

By comparing Equations 3.29 and 3.39, we can write

$$\begin{aligned}
E &= \sum_i \varepsilon_i - \frac{e^2}{2} \iint \frac{1}{4\pi\varepsilon_0} \frac{n(\vec{r}) n(\vec{r}')}{|\vec{r} - \vec{r}'|} d^3r d^3r' + E_{xc}[n] \\
&\quad - \int v_{xc}(\vec{r}) n(\vec{r}) d^3r
\end{aligned} \tag{3.40}$$

Note that in the KS theory, $T_s[n]$ is treated exactly.

3.6 Local Density Approximation (LDA)

The exact form of $F[n]$ or $E_{xc}[n]$ (or $v_{xc}(\vec{r})$) is not known, but approximations have been developed to calculate these quantities. The first such approximation was developed by Kohn and Sham (1965) and is known as the LDA, which has been highly successful in calculating the electronic structure of atoms, molecules, and solids. It is assumed that the electron density $n(\vec{r})$ varies very slowly in space, so that the electron gas in a small-volume element

d^3r could be considered locally uniform. Therefore, one could use the exchange–correlation energy of the homogeneous electron gas to evaluate $E_{xc}[n]$ as follows. If $\varepsilon_{xc}(n)$ is the exchange–correlation energy per particle of a uniform electron gas of density n, the exchange–correlation energy of the uniform gas of density $n(\vec{r})$ in volume d^3r is $\varepsilon_{xc}(n(\vec{r}))n(\vec{r})d^3r$. Therefore, the exchange–correlation energy of the system with slowly varying density can be written as

$$E_{xc}[n] = \int \varepsilon_{xc}\left(n(\vec{r})\right)n(\vec{r})d^3r \tag{3.41}$$

This is known as the LDA.

In the LDA, $v_{xc}(\vec{r})$ can be found by using Equations 3.32 and 3.41 as

$$\begin{aligned} v_{xc}(\vec{r}) &= \frac{\delta E_{xc}[n]}{\delta n(\vec{r})} = \frac{d}{dn}(\varepsilon_{xc}(n(\vec{r}))n(\vec{r})) \\ &= \mu_{xc}(n(\vec{r})) \end{aligned} \tag{3.42}$$

where $\mu_{xc}(n)$ is the exchange and correlation contributions to the chemical potential of the uniform electron gas.

Using Equations 3.40 through 3.42, the total energy can be written in the LDA as

$$\begin{aligned} E &= \sum_i \varepsilon_i - \frac{e^2}{2}\iint \frac{1}{4\pi\varepsilon_0}\frac{n(\vec{r})n(\vec{r}')}{|\vec{r}-\vec{r}'|}d^3rd^3r' \\ &\quad + \int \left(\varepsilon_{xc}(n(\vec{r})) - \mu_{xc}(n(\vec{r}))\right)n(\vec{r})d^3r \end{aligned} \tag{3.43}$$

As mentioned in Chapter 2, the homogeneous electron gas has been studied by various workers. Several analytic fits to theoretically calculated $\varepsilon_{xc}(n)$ have been proposed. The one proposed by Gunnarsson and Lundqvist (1976) in atomic units (see Appendix C) is

$$\varepsilon_{xc}(n) = -\frac{0.458}{r_s} - 0.0666\,G\left(\frac{r_s}{11.4}\right) \tag{3.44}$$

where $G(x)$ is defined as

$$G(x) = \frac{1}{2}[(1+x^3)ln(1+x^{-1}) - x^2 + x/2 - 1/3]$$

The correlation energy of the homogeneous electron gas has also been calculated by using the quantum Monte Carlo method by Ceperley and Alder

(1980), which is considered to be a very accurate calculation. This has been fitted to analytical forms by Perdew and Zunger (1981) and Vosko et al. (1980). These fitted forms are used in most of the current electronic structure calculations.

As can be seen from the above discussion, the LDA should be a good approximation if the electron density $n(\vec{r})$ is slowly varying. However, it has yielded surprisingly good results even in cases where the density is not slowly varying. For metals, the theory does extremely well. In semiconductors, the band gap is underestimated by a 1/2–2/3 factor. Insulators, particularly Mott insulators, are not described well. Intense efforts are going on to remedy these defects of the theory. We shall briefly discuss these effects later.

3.7 Comparison of the DFT with the Hartree and H–F Theories

To make the comparison, let us split the exchange–correlation energy $E_{xc}[n]$ in Equation 3.41 into exchange and correlations parts, that is

$$E = \int v(\vec{r})n(\vec{r})d^3r + T_s[n] + \frac{e^2}{2}\iint \frac{1}{4\pi\varepsilon_0}\frac{n(\vec{r})n(\vec{r}')}{|\vec{r}-\vec{r}'|}d^3rd^3r' + E_x[n] + E_c[n] \tag{3.45}$$

The exact equations of Hartree theory are obtained from Equation 3.45 by dropping $E_x + E_c$. The H–F equations are obtained by dropping E_c. Obviously, the inclusion of any reasonable approximation will improve these equations, which is done in the LDA. The advantage of using the KS equations in the LDA is that these equations are as simple as the Hartree equations but in most cases they give very good account of exchange and correlation effects. In practice, it is found that for solids the KS results are better than H–F results, although the latter are much more computationally demanding. The reason is that the KS equation includes fairly accurately, both exchange and correlations, while the H–F equation includes exchange only but neglects correlations completely.

3.8 Comments on the KS Eigenvalues and KS Orbitals

Although the DFT provides a scheme to reduce the entire many-body problem to a Schrödinger-like single-particle equation, the physical meaning of ε_i is not clear. Often, these eigenvalues have been used to interpret excitation spectra and in most cases good agreement with experimental results was found. But there are problematic cases too.

There is a special case when ε_i does have a physical meaning. For infinite systems, the highest occupied level ε_N equals the chemical potential, that is, $\varepsilon_N = \mu$, provided highest occupied orbitals are extended. This is a density functional analog of Koopmans' theorem, which we shall now prove.

Let us consider an infinite system in which eigenfunction ψ_i at the maximum occupied single-particle energy ε_F is all extended. We shall calculate the difference in ground-state energies of N and $(N+m)$ particles for $m \ll N$.

Change in $n(\vec{r})$ as a result of adding m particle is

$$\delta n(\vec{r}) = \sum_{i=N+1}^{N+m} |\psi_i(\vec{r})|^2 \tag{3.46}$$

Also, the change in ground-state energy

$$\begin{aligned}
\delta E[n] &= T_s\left[n(\vec{r}) + \delta n(\vec{r})\right] - T_s[n(\vec{r})] \\
&\quad + \int \delta n(\vec{r}) \left[v(\vec{r}) + \int \frac{e^2}{4\pi\varepsilon_0} \frac{n(\vec{r}\,')}{|\vec{r} - \vec{r}\,'|} d^3r' + \frac{\delta E_{xc}}{\delta n(\vec{r})} \right] d^3r \\
&= \sum_{i=N+1}^{N+m} \int -\frac{\hbar^2}{2m_e} \psi_i^*(\vec{r}) \nabla^2 \psi_i(\vec{r}\,') d^3r \\
&\quad + \sum_{i=N+1}^{N+m} |\psi_i(\vec{r}\,')|^2 \left[v(\vec{r}) + \int \frac{e^2}{4\pi\varepsilon_0} \frac{n(\vec{r}\,')}{|\vec{r} - \vec{r}\,'|} d^3r' + \frac{\delta E_{xc}}{\delta n(\vec{r}\,')} \right] d^3r \\
&= \sum_{i=N+1}^{N+m} \varepsilon_i = m\varepsilon_F
\end{aligned} \tag{3.47}$$

Now

$$\int \delta n(\vec{r})\, d^3r = m$$

$$\therefore\ \delta E\left[n(\vec{r})\right] = \varepsilon_F \int \delta n(\vec{r}) d^3r$$

or

$$\frac{\delta E[n]}{\delta n(\vec{r})} = \varepsilon_F$$

But from Equation 3.30

$$\frac{\delta E[n]}{\delta n(\vec{r})} = \mu$$

$$\therefore \ \varepsilon_F = \mu \tag{3.48}$$

This proves the theorem.

Similarly, one should remember that KS orbital $\psi_i(\vec{r})$ enters in the DFT via Equation 3.37, through its relation with electron charge density. Thus, there is no reason to regard the Slater determinant constructed from ψ_i's as the true many-electron wave function, that is, $\psi_i(\vec{r})$ is not the one-electron wave function in the sense of the H–F approximation.

3.9 Extensions to Magnetic Systems

So far, we have only discussed nonmagnetic systems at absolute zero temperature. The DFT has been generalized to the ground state of magnetic systems by von Barth and Hedin (1972), Pant and Rajagopal (1972), and Rajagopal and Callaway (1973) and to finite temperatures by Mermin (1965). Here, we shall briefly discuss the extension to magnetic systems. This extension of the theory to magnetic systems is known as spin DFT. For simplicity, we assume that the system is in a magnetic field $B(\vec{r})$, which is pointing in only one direction that we assume our z direction. Also, we assume that the electronic spin is either up ($\sigma = +1$ along the magnetic field) or down ($\sigma = -1$). Thus, the total density $n(\vec{r})$ is equal to the sum of two densities $n_\uparrow(\vec{r})$ and $n_\downarrow(\vec{r})$

$$n(\vec{r}) = n_\uparrow(\vec{r}) + n_\downarrow(\vec{r}) = \sum_\sigma n_\sigma(\vec{r})$$

The exchange–correlation function is now functional of $n_\uparrow(\vec{r})$ and $n_\downarrow(\vec{r})$

$$E_{xc} = E_{xc}[n_\uparrow, n_\downarrow]$$

Following similar arguments as discussed earlier, we obtain the KS equation as

$$\left[-\frac{\hbar^2}{2m_e}\nabla^2 + v_{eff}^\sigma(\vec{r})\right]\psi_{i\sigma}(\vec{r}) = \epsilon_{i\sigma}\psi_{i\sigma}(\vec{r}) \quad [\sigma = \pm 1] \tag{3.49}$$

where

$$v_{eff}^{\sigma}(\vec{r}) = v(\vec{r}) - \sigma\mu_B B(\vec{r}) + \frac{e^2}{4\pi\varepsilon_0}\int \frac{n(\vec{r}')}{|\vec{r} - \vec{r}'|} d^3r'$$
$$+ \frac{\delta E_{xc}[n_\uparrow, n_\downarrow]}{\delta n_\sigma(\vec{r})} \quad [\sigma = \pm 1] \tag{3.50}$$

and

$$n_\sigma(\vec{r}) = \sum_{\text{occupied}} |\psi_{i\sigma}|^2 \tag{3.51}$$

$\mu_B = e\hbar/2m_e$ is the Bohr magneton and $\epsilon_{i\sigma}$ are the Lagrange multipliers. As was the case for nonmagnetic systems, the exact form of the exchange–correlation functional $E_{xc}[n_\uparrow, n_\downarrow]$ is not known and approximations have been developed for it. Similar to the LDA, one can make local approximation, known as local spin density approximation (LSDA). For more details, we refer to the book by Parr and Yang (1989).

3.10 Performance of the LDA/LSDA

The LDA and LSDA have proved to be remarkably good approximations. Considering that these approximations were derived by assuming slowly varying densities, they give very good results for atoms, molecules, and solids. Some of the successes of LDA/LSDA are

1. Ground-state geometries are given accurately.
2. Charge densities are better than 2%.
3. Most of the time, LDA results are better than H–F results.
4. Physical trends are given correctly (e.g., see Moruzzi et al. 1978).
5. Equilibrium distances are given reasonably well typically with an accuracy of 1% although somewhat underestimated.

However, LDA/LSDA fails in many cases. Some of their failures are

1. LDA underestimates band gaps in semiconductors. For example, LDA gives a band gap of Si about 0.5 eV compared to the experimental result of about 1.1 eV.
2. FeO and CoO, which are insulators, are predicted to be metallic.

3. Fe is predicted to be an fcc paramagnet rather than bcc ferromagnet at low temperatures.
4. LDA predicts incorrect ground state for many atoms.
5. LDA predicts unstable negative atomic ions.
6. It may fail in weak molecular bonds, for example, in hydrogen bonds as in the bonding region the charge density is very small. It may fail in van der Waal's systems too.
7. LDA overbinds molecules and solids.

3.11 Beyond LDA

We saw that in most cases, the LDA/LSDA is adequate. However, in cases where higher accuracy is desired, one must go beyond the LDA. For example, an error of 1 eV in binding energy is not acceptable in the study of chemical reactions. Similarly, higher accuracy is needed in the study of phase transformations or if one wants to use the theory for materials design. What one would like is the accuracy of 0.1 eV or less in the calculation of binding energy. Several attempts have been made in this direction although success has been rather limited. We shall summarize few such attempts in this direction.

3.11.1 Generalized Gradient Approximations (GGAs)

We had seen that the LDA/LSDA was derived by assuming that the electron density varies very slowly in space. Thus, it was argued that LDA/LSDA could be improved if we expanded the exchange–correlation functional in terms of gradient of the density in Taylor series and truncated it at some order. Such an approximation is known as gradient expansion approximation (GEA) and was implemented by Herman et al. (1969). It turned out that such an approximation often gave worse results and did not provide any improvement over LDA. The reason for this seems to be that gradients of the density in real system become very large; as a result, such an expansion breaks down. Also, it was found that the GEA does not satisfy certain sum rules. Later on, it was realized that there was no need of such an expansion and it was possible to construct exchange–correlation functional, which was a functional of density as well as its gradient and satisfied the sum rules. This could be written as

$$E_{xc}^{GGA}[n_\uparrow, n_\downarrow] = \int d^3r\, f(n_\uparrow(\vec{r}), n_\downarrow(\vec{r}), \nabla n_\uparrow, \nabla n_\downarrow) \tag{3.52}$$

In contrast to LSDA, there are many functions f that are in use. One that has been widely used is by Perdew and Wang, known as PW91 (Perdew

1991). Later on, Perdew et al. (1996) proposed another functional (PBE), which retains important features of PW91.

The GGA often provides results for structural and magnetic properties of materials that are better than LDA results and are closer to experimental values. For example, the GGA gives lattice constants that are closer to experimental values and also gives the correct ground state for ferromagnetic Fe. However, it also fails sometimes. At times it overcorrects LDA bond lengths and gives values that are up to 2% larger than experimental values. Also, sometimes cohesive energy is 10–20% too small. The GGA gives poor results for ferroelectrics, and also like LDA it does not give correct band gaps for semiconductors. It does not give good results for strongly correlated systems and present research is still continuing in search of a better functional.

3.11.2 LDA + U Method

We had seen that LSDA and GGA fail to give the correct ground state of strongly correlated systems such as transition metal oxides. LDA + U or GGA + U methods have been designed to correct this situation by constructing an orbital-dependent functional. The idea has been taken from the Hubbard model (Hubbard 1963). Suppose we have an orbital that already has one electron. If we put another electron in it, it will cost an energy U because of the electron–electron repulsion in that orbital. This idea when incorporated in LSDA or GGA produces LDA + U method.

To implement the above idea, the electrons are subdivided into two subsystems; *d* or *f* electrons, which interact strongly with Coulomb repulsion U and s, and *p* electrons, which are well described by the LDA or GGA. For *d* (or *f*) electrons, an extra term is added to the LDA functional equal to (1/2) $U\Sigma_{i,j\ i\neq j}\, n_i\, n_j$, where n_i, n_j are *d* occupancies. Thus, the LDA + U functional can be written as

$$E_{\text{LDA+U}} = E_{\text{LDA}} - \frac{U}{2} N_d(N_d - 1) + \frac{1}{2} U \sum_{i,j i\neq j} n_i n_j \tag{3.53}$$

where $N_d = \Sigma_i\, n_i$ is total number of *d* electrons.

This orbital-dependent functional gives upper and lower Hubbard band separation equal to U and thus correctly describes the physics of Mott insulators. For details of the method, we refer to the review article by Animosov et al. (1997). This approach has been further extended to include dynamical correlations within dynamical mean-field theory (DMFT) and known as LDA + DMFT (see, e.g., Savrasov et al. 2000).

The LDA + U method has proved quite successful in reproducing the correct ground state of strongly correlated systems. For example, it correctly gives the antiferromagnetic insulating ground states for many transition metal oxides and cuprate superconductors, which turn out to be metallic using LSDA or GGA. The main problem of the LDA + U method is the choice of the value of U. Although U can be found *ab initio* by using the constrained DFT approach, it is overestimated by this method. Thus, the common way of finding U is to treat it as an adjustable parameter so that one finds good agreement between some calculated and experimental results.

3.11.3 Self-Interaction Correction (SIC) Method

In any theory, there should be no self-interaction as an electron cannot interact with itself. Recall that in H–F approximations, the Coulomb interaction energy of an electron in a given state with itself is exactly canceled by its exchange interaction. Thus, there was no problem in H–F theory as far as the self-interaction was concerned. However, in the LDA, the self-interaction is not canceled properly and, therefore, a self-interaction correction is required.

Perdew and Zunger (1981) proposed a self-interaction corrected exchange–correlation energy functional, which is also an orbital-dependent functional. Results for atoms show that SIC corrects most of the errors of the LDA. Also, using SIC, stable solutions for negative ions can be obtained. Thus, SIC leads to highly accurate results for atomic systems. However, the method is easily applicable to systems where electronic orbitals are localized, but for extended orbitals, it is very difficult to apply. Calculations for band gaps of insulators show a marked improvement over the LDA results.

3.11.4 GW Method

We saw that neither the LDA nor the GGA gave correct band gaps for semiconductors. This is because the calculation of band gap involves an excited state and the LDA and GGA do not give it correctly because both of them are ground-state theories. Thus, to get the band gap correctly, we have to develop a theory that can give excited states correctly and this is what the GW method does. It has been shown that single-particle energies can be obtained by solving the equation

$$\left(-\frac{\hbar^2}{2m_e}\nabla^2 + v(\vec{r}) + v_H(\vec{r})\right)\psi_i(\vec{r}) + \int d^3r'\,\Sigma(\vec{r},\vec{r}',\varepsilon)\psi_i(\vec{r}') = \varepsilon_i\psi_i(\vec{r}) \tag{3.54}$$

where $v_H(\vec{r})$ is the Hartree potential and Σ is called self-energy, which includes the effects of exchange and correlations. A comparison with H–F equation shows that in the H–F theory, this corresponds to the exchange interaction. Thus, in any theory better than H–F, this term should contain exchange as well as correlation effects. The problem is that Σ is unknown and has to be approximated. In the GW method, Σ is approximated by a product involving the Green's function G and screened Coulomb interaction W, which can be written symbolically as

$$\Sigma = iGW \tag{3.55}$$

This factorization gives the "GW" name to the method. Note that if instead of W the bare Coulomb interaction is taken, one obtains the H–F method that gives pathological results for the density of states for the free electron gas. We see that this behavior was linked to the long-range tail of the bare Coulomb interaction. Since in the GW method, screened Coulomb interaction is used, such a situation does not arise. Formulation of the GW method is quite complicated and for details we refer to the review article by Aryasetiawan and Gunnarsson (1998).

One of the great successes of the GW method is that it gives the band gaps of semiconductors in good agreement with experiments. However, the method is difficult to implement and is computationally very expensive. Another method that has been developed by Harbola and collaborators based on splitting the k-space also gives correct band gaps for a large number of semiconductors but is much simpler than the GW method (Samal and Harbola 2006, Rahaman et al. 2009). A recently developed functional TB-mBJ also gives a band gap of semiconductors and insulators in agreement with experiments (Tran and Blaha 2009).

3.12 Time-Dependent Density Functional Theory (TDDFT)

The DFT has been generalized to include time-dependent potentials and is known as TDDFT. This theory can be thought of as a reformulation of the time-dependent quantum mechanics in terms of time-dependent electron density $n(\vec{r}, t)$. This approach has a great advantage over the wave function approach, as the many-body wave function is a function of N positions, where N is the number of electrons while the electron density is a function of one position only.

The TDDFT was developed due to the pioneering work of Runge and Gross (1984) who gave a theorem known as Runge–Gross theorem, which is the analog of the Hohenberg–Kohn theorem for the time-dependent electron density $n(\vec{r}, t)$. The theorem establishes a one-to-one correspondence between $n(\vec{r}, t)$ and time-dependent external potential $v_{ext}(\vec{r}, t)$ for a given initial state. Unlike in DFT, there is no energy minimum principle in the TDDFT. Instead, the formulation of TDDFT is based on the stationary action principle. Similar to DFT, one can define a fictitious system of noninteracting electrons that has the same density $n(\vec{r}, t)$ of the real system and which moves in a time-dependent effective potential $v_{eff}(\vec{r},t)$. This leads to an equation similar to the time-dependent Schrödinger equation, known as the time-dependent KS equation:

$$i\hbar\frac{\partial}{\partial t}\psi_i(\vec{r},t) = \left[-\frac{\hbar^2}{2m_e}\nabla^2 + v_{eff}(\vec{r},t)\right]\psi_i(\vec{r},t) \tag{3.56}$$

where

$$v_{eff}(\vec{r},t) = v_{ext}(\vec{r},t) + v_{xc}(\vec{r},t) + e^2\int\frac{1}{4\pi\varepsilon_0}\frac{n(\vec{r}',t)}{|\vec{r}-\vec{r}'|}d^3r' \tag{3.57}$$

Here, $v_{xc}(\vec{r},t)$ is the time-dependent exchange–correlation potential and $n(\vec{r},t)$ is given by

$$n(\vec{r},t) = \sum_i |\psi_i(\vec{r},t)|^2 \tag{3.58}$$

The exchange–correlation potential $v_{xc}(\vec{r},t)$ is an unknown functional of $n(\vec{r},t)$, which has to be approximated. The simplest approximation is the adiabatic local density approximation (ALDA) in which exchange–correlation potential of the uniform gas with the instantaneous density is used to evaluate $v_{xc}(\vec{r},t)$, that is

$$v_{xc}^{\mathrm{ALDA}}(\vec{r},t) = v_{xc}^{\mathrm{unif}}(n(\vec{r},t)) \tag{3.59}$$

This formalism has been applied to calculate excited-state energies and photo-absorption cross sections of molecules and clusters. The excited-state energies of the system can be found by calculating the linear response function, which has poles at the excitation energies. This field is still developing and has several challenges ahead, for example, developing theory beyond ALDA, and so on. For further details, we refer to the book by Gross and Dreizler (1995).

EXERCISES

3.1 a. Dirac extended the Thomas–Fermi method by including the approximate form of the exchange energy given by

$$-\frac{3}{4}\left(\frac{3}{\pi}\right)^{1/3}\int n(\vec{r})^{4/3}d^3r$$

Using the variational principle, derive the equation satisfied by $n(\vec{r})$.

b. The above model was further extended by Weizsäcker who proposed the correction to the kinetic energy functional given by

$$\frac{1}{8}\int \frac{|\nabla n(\vec{r})|^2}{n(\vec{r})} d^3r$$

Using the variational principle, derive the equation satisfied by $n(\vec{r})$ (see Parr and Yang (1989) for taking the functional derivative of the above term).

3.2 Suppose we have two types of particles with densities $n_1(\vec{r})$ and $n_2(\vec{r})$ with internal energy $U[n_1,n_2]$ such that the total energy E can be written as

$$E[n_1,n_2] = U[n_1,n_2] + \int v_{ext}(\vec{r})n(\vec{r})d^3r$$

where $n = n_1(\vec{r}) + n_2(\vec{r})$. Show that the total energy is a functional of only n.

3.3 a. Write the Kohn–Sham equation for the electron in a hydrogen atom.

b. Estimate the LDA exchange energy for the hydrogen atom and compare it with the exact result.

c. Estimate the LDA total energy of the hydrogen atom and compare it with the exact energy.

3.4 Write the Kohn–Sham equation for an electron in the jellium model. What are the Kohn–Sham orbitals and energies? Plot the energy eigenvalues taking parameters corresponding to Na and compare your result with the Hartree and H–F results.

3.5 Show that the self-interaction term is exactly canceled in the H–F approximation.

Further Reading

Dreizler R. M. and Gross E. K. U. 1990. *Density Functional Theory: An Approach to the Quantum Many-Body Problem*. Berlin: Springer.

Kohn W. 1999. Nobel lecture: Electronic structure of matter wave functions and density functionals. *Rev. Mod. Phys.* 71:1253–1266.

Lundqvist S. and March N. H. Eds. 1983. *Theory of the Inhomogeneous Electron Gas.* New York: Plenum Press.

Martin R. M. 2004. *Electronic Structure, Basic Theory and Practical Methods.* Cambridge: Cambridge University Press.

Parr R. G. and Yang W. 1989. *Density Functional Theory of Atoms and Molecules.* New York: Oxford University Press.

4

Energy Band Theory

4.1 Introduction

In the last chapter, we had reduced the many-body problem of interacting electrons and ions to a one-body problem using the DFT. This led to the KS equation

$$\left[\frac{-\hbar^2}{2m_e}\nabla^2 + v_{eff}(\vec{r})\right]\psi_i(\vec{r}) = \varepsilon_i\psi_i(\vec{r}) \tag{4.1}$$

The knowledge of eigenvalues ε_i and eigenfunctions ψ_i is necessary to understand various properties of solids such as transport, optical, magnetic, and superconducting properties. In the next three chapters, we shall discuss how to solve Equation 4.1 for the periodic or crystalline solids. In this chapter, we shall discuss general consequences of a periodic potential on eigenfunctions and eigenvaules of Equation 4.1. We shall also discuss the effect of various symmetries such as inversion symmetry, time-reversal symmetry, and so on. In this chapter, we shall introduce the basic terminology used in discussing the electronic structure of solids. In Section 4.2, we start with the crystal potential and then in Section 4.3, we discuss the Bloch's theorem. In Section 4.4, we introduce the concept of Brillouin zone. In Section 4.5, we discuss spin–orbit interaction and in Section 4.6, symmetry in crystals. In Section 4.7, inversion symmetry, time-reversal symmetry, and Kramers' theorem are discussed. In Section 4.8, we introduce the concepts of band structure and Fermi surface. In Section 4.9, we define and develop the concepts of density of states, local density of states, and projected density of states, and in Section 4.10, we discuss the charge density. Finally, in Section 4.11, we briefly discuss the Brillouin zone integration. This chapter only covers periodic solids at absolute zero temperature. Calculations for nonperiodic solids are much more complicated and will be discussed in Chapters 8 and 11.

4.2 Crystal Potential

For a periodic solid, the potential $v_{eff}(\vec{r})$ is periodic, that is

$$v_{eff}(\vec{r} + \vec{R}_\ell) = v_{eff}(\vec{r}) \tag{4.2}$$

where $\vec{R}_\ell$ is a lattice vector. It depends on the crystal structure and on the atoms constituting the solid and is called the crystal potential. Schematically, it is shown in Figure 4.1 in one dimension. The origin for $\vec{r}$ is generally placed at a lattice point so that one can take full advantage of the point group symmetry of the lattice (see Section 4.6).

It is clear that before we attempt solving the KS equation, we must know $v_{eff}(\vec{r})$. Recall that $v_{eff}(\vec{r})$ not only includes the electron–ion potential, but also has Hartree and exchange correlation term, that is

$$\begin{aligned} v_{eff}(\vec{r}) &= v(\vec{r}) + v_H(\vec{r}) + v_{xc}(\vec{r}) \\ &= -\sum_\ell \frac{1}{4\pi\varepsilon_0}\frac{Ze^2}{|\vec{r} - \vec{R}_\ell|} + \int \frac{1}{4\pi\varepsilon_0}\frac{e^2 n(\vec{r}\,')}{|\vec{r} - \vec{r}\,'|} d^3r' + v_{xc}(\vec{r}) \end{aligned} \tag{4.3}$$

where

$$n(\vec{r}) = \sum_{\substack{i \\ \text{occupied}}} |\psi_i(\vec{r})|^2 \tag{4.4}$$

For constructing $v_{eff}(\vec{r})$, we need $n(\vec{r})$, which in turn needs the solutions of the KS Equation 4.1, which we have not solved yet. This problem is generally solved using the following steps:

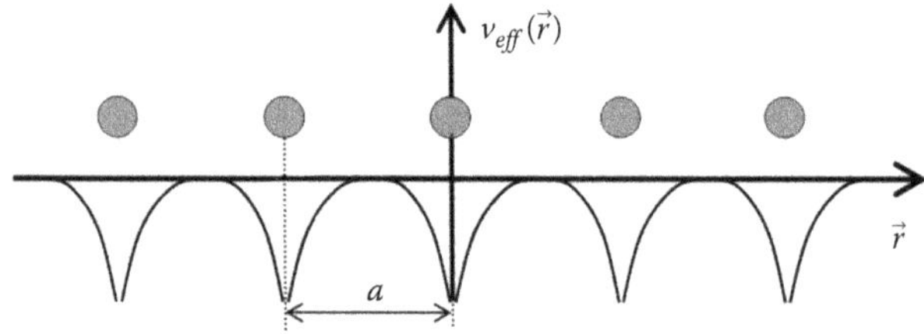

FIGURE 4.1
Crystal potential in one dimension.

i. Start with some guess for $v_{eff}(\vec{r})$ and then solve the KS Equation 4.1.
ii. Once $\psi_i(\vec{r})$ are known, then obtain $n(\vec{r})$ using Equation 4.4 and $v_{eff}(\vec{r})$ using Equation 4.3.
iii. Using the new potential, solve Equation 4.1 again. One repeats the procedure until convergence is achieved, that is, output potential is the same as the input potential within some tolerance.

In our subsequent discussion, we shall assume that $v_{eff}(\vec{r})$ is a given function and focus on how to solve the KS equation with this given potential.

4.3 Bloch's Theorem

The solution of the KS equation for periodic potential is greatly facilitated by Bloch's theorem, which was given by Felix Bloch (Figure 4.2) in 1928 (Bloch 1928). The theorem imposes a condition on the wave function of an electron moving in a periodic potential and is a consequence of the translational symmetry. The theorem can be stated as follows:

For any wave function that satisfies the Schrödinger equation (KS equation) for periodic potential $v_{eff}(\vec{r})$ such that $v_{eff}(\vec{r}+\vec{R}_\ell) = v_{eff}(\vec{r})$, there exists a vector $\vec{k}$ such that translation by a lattice vector $\vec{R}_\ell$ is equivalent to multiplying by the phase factor $\exp(i\vec{k}\cdot\vec{R}_\ell)$, that is

$$\psi_{\vec{k}}(\vec{r}+\vec{R}_\ell) = \exp(i\vec{k}\cdot\vec{R}_\ell)\psi_{\vec{k}}(\vec{r}) \tag{4.5}$$

FIGURE 4.2
Felix Bloch (1905–1983) is known for Bloch's theorem, Bloch wall, and NMR. He was awarded the Nobel Prize in 1952. (Courtesy of Stanford News Service.)

Alternatively, the theorem can also be stated as follows: The wave function should have a form

$$\psi_{\vec{k}}(\vec{r}) = e^{i\vec{k}\cdot\vec{r}}\, u_{\vec{k}}(\vec{r}) \tag{4.6}$$

where $u_{\vec{k}}$ is a periodic function, that is

$$u_{\vec{k}}(\vec{r} + \vec{R}_\ell) = u_{\vec{k}}(\vec{r}) \tag{4.7}$$

To prove the theorem, let us define a translational operator, T_ℓ, which when operated on a function $f(\vec{r})$ translates $\vec{r}$ by $\vec{R}_\ell$

$$T_\ell f(\vec{r}) = f(\vec{r} + \vec{R}_\ell) \tag{4.8}$$

It can be easily seen that T_ℓ commutes with Hamiltonian $H(\vec{r})$.

$$H(\vec{r}) = \left[\frac{-\hbar^2}{2m_e}\nabla^2 + v_{eff}(\vec{r})\right]$$

that is

$$\begin{aligned} T_\ell H(\vec{r})\psi(\vec{r}) &= H(\vec{r} + \vec{R}_\ell)\, T_\ell \psi(\vec{r}) \\ &= H T_\ell \psi(\vec{r}) \end{aligned}$$

because $H(\vec{r})$ has translational symmetry. Thus

$$T_\ell H = H T_\ell \tag{4.9}$$

This implies that eigenfunctions of H can be chosen such that those are also eigenfunctions of T_ℓ. Therefore, if ψ is a common eigenfunction

$$H\psi = \varepsilon\psi \tag{4.10}$$

and

$$T_\ell \psi = C(\vec{R}_\ell)\psi \tag{4.11}$$

where ε and $C(\vec{R}_\ell)$ are eigenvalues of H and T_ℓ, respectively. Also

$$\begin{aligned} T_\ell T_{\ell'} \psi(\vec{r}) &= \psi(\vec{r} + \vec{R}_\ell + \vec{R}_{\ell'}) \\ &= T_{\ell+\ell'}\psi(\vec{r}) = T_{\ell'} T_\ell \psi(\vec{r}) \end{aligned} \tag{4.12}$$

Using Equations 4.11 and 4.12, we can write

$$C(\vec{R}_\ell)C(\vec{R}_{\ell'}) = C(\vec{R}_\ell + \vec{R}_{\ell'}) \tag{4.13}$$

Let $\vec{a}_1, \vec{a}_2, \vec{a}_3$ be the primitive vectors of the direct lattice. Let us choose

$$\vec{R}_\ell = \vec{R}_{\ell'} = \vec{a}_1$$

Using Equation 4.13, we can write

$$C(\vec{a}_1)C(\vec{a}_1) = C(2\vec{a}_1) = C^2(\vec{a}_1)$$

or

$$C^n(\vec{a}_1) = C(n\vec{a}_1) \tag{4.14}$$

where n is a positive integer. The form of Equation 4.14 suggests that

$$C(\vec{a}_1) = e^{2\pi i x_1} \tag{4.15}$$

where x_1 is a real number. This is because the modulus of $C(\vec{a}_1)$ must be unity, otherwise Equation 4.14 will imply an unbounded function. Let us now use the periodic boundary condition

$$\psi(\vec{r} + N_1\vec{a}_1) = \psi(\vec{r})$$

where N_1, N_2, N_3 are the number of atoms in the solid along the three basis vectors of the direct lattice such that the total number of atoms in the solid $N = N_1 N_2 N_3$. The above equation can be written as

$$T_{N_1\vec{a}_1}\psi(\vec{r}) = \psi(\vec{r}) \tag{4.16}$$

or

$$C(N_1\vec{a}_1) = 1$$

This implies from Equation 4.15

$$e^{2\pi N_1 i x_1} = 1 = e^{2\pi i m_1}$$

or

$$x_1 = \frac{m_1}{N_1} \tag{4.17}$$

where m_1 is an integer.

In general

$$x_i = \frac{m_i}{N_i} \quad (i = 1,2,3) \tag{4.18}$$

Now, $\vec{R}_\ell$ can be written as

$$\vec{R}_\ell = \ell_1\vec{a}_1 + \ell_2\vec{a}_2 + \ell_3\vec{a}_3 \tag{4.19}$$

where ℓ_1, ℓ_2, ℓ_3 are integers. Using Equation 4.15, we can write

$$C(\vec{R}_\ell) = \exp\left[2\pi i(\ell_1 x_1 + \ell_2 x_2 + \ell_3 x_3)\right] \tag{4.20}$$

Let us introduce reciprocal lattice vectors $\vec{b}_1, \vec{b}_2, \vec{b}_3$

$$\vec{b}_1 = 2\pi \frac{\vec{a}_2 \times \vec{a}_3}{\vec{a}_1 \cdot (\vec{a}_2 \times \vec{a}_3)} \quad \text{etc.} \tag{4.21}$$

which have the property

$$\vec{b}_i \cdot \vec{a}_j = 2\pi\delta_{ij} \tag{4.22}$$

Let us now define a vector in reciprocal space

$$\vec{k} = x_1\vec{b}_1 + x_2\vec{b}_2 + x_3\vec{b}_3 \tag{4.23}$$

We see that

$$\vec{k} \cdot \vec{R}_\ell = 2\pi(\ell_1 x_1 + \ell_2 x_2 + \ell_3 x_3) \tag{4.24}$$

Therefore, Equation 4.20 can be expressed as

$$C(\vec{R}_\ell) = e^{i\vec{k}\cdot\vec{R}_\ell}$$

or

$$T_\ell\psi(\vec{r}) = \psi(\vec{r} + \vec{R}_\ell) = e^{i\vec{k}\cdot\vec{R}_\ell}\psi(\vec{r}) \tag{4.25}$$

This proves Bloch's theorem.

Here, a few remarks are in order.

1. The vector $\vec{k}$ defined in Equation 4.23, also called a wave vector, is a good quantum number for a periodic system. Thus, the wave functions and energy eigenvalues of an electron in a periodic solid can be labeled by $\vec{k}$ and written as $\psi_{\vec{k}}$ and $\varepsilon_{\vec{k}}$.
2. As is clear from Equation 4.5 or 4.6, if $\vec{k}$ satisfies Bloch's theorem, $\vec{k}+\vec{G}$ also satisfies Bloch's theorem, where $\vec{G}$ is a reciprocal lattice vector. Therefore, all possible values of $\vec{k}$ can be confined to the unit cell of the reciprocal lattice (the first BZ, see Section 4.4 below).
3. The vector $\vec{k}$ defined by Bloch's theorem should not be confused with the vector $\vec{k}$ occurring in the free electron theory although they play a similar role. The free electron wave vector is given by $\vec{k} = \vec{p}/\hbar$, where $\vec{p}$ is the momentum of the electron, while the Bloch wave vector $\vec{k}$ has no such relationship with the momentum. However, in many processes such as transport and optical absorption, $\hbar\vec{k}$ behaves as momentum and is called the crystal momentum.
4. For each $\vec{k}$, there are several energy eigenvalues that are labeled as $\varepsilon_{n\vec{k}}$, where n is called the band index. A detailed plot of $\varepsilon_{n\vec{k}}$ versus $\vec{k}$ is referred to as the band structure of the solid.

4.4 Brillouin Zone (BZ)

We saw in Section 4.3 that all values of $\vec{k}$ can be chosen to lie in the primitive unit cell of the reciprocal lattice. Since the primitive unit cell is not unique, it is chosen to be the Wigner–Seitz cell of the reciprocal lattice, which is called the first BZ and is unique. As the name suggests, higher zones can also be constructed, but we shall limit ourselves to the first Brillouin zone or simply Brillouin zone. It is constructed in the following way: (i) Choose a reciprocal lattice point as origin. (ii) Connect all the remaining points to the origin, drawing all the reciprocal lattice vectors and label them as $\vec{G}_1, \vec{G}_2, \vec{G}_3$, and so on. For the first BZ, there is no need to go beyond the third next neighbor. (iii) Draw perpendicular bisector planes to $\vec{G}_1, \vec{G}_2, \vec{G}_3$, and so on. (iv) The smallest polyhedron containing the origin is the first BZ. We first demonstrate it by an example of one-dimensional lattice constant a. Its reciprocal lattice is also a one-dimensional lattice with lattice constant of $2\pi/a$. We see that the first nearest neighbors of 0 are at $\pm 2\pi/a$ as shown in Figure 4.3. By drawing the perpendicular bisectors, we find that the first BZ lies between $\pm\pi/a$.

Another example we take is of a square lattice of lattice constant a. It can be shown that the reciprocal lattice of a square lattice is also a square lattice with lattice constant $2\pi/a$ as shown in Figure 4.4. Now, choose a point

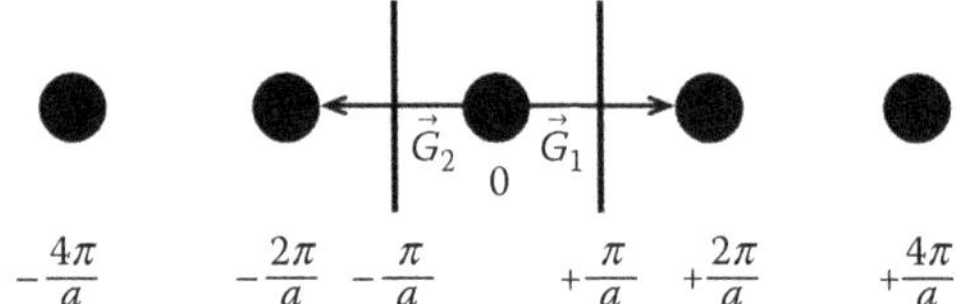

FIGURE 4.3
BZ for a one-dimensional lattice is confined between the two vertical lines as shown.

as origin and draw reciprocal lattice vectors $\vec{G}_1, \vec{G}_2, \vec{G}_3$, and so on. We see that the smallest area contained by the perpendicular bisectors is the shaded square with sides of length $2\pi/a$. Thus, all $\vec{k}$ vectors are contained within $(\pm\pi/a, \pm\pi/a)$.

In a similar way, one can construct BZ in three dimensions. The BZs for simple cubic, fcc, and bcc lattices are shown in Figures 4.5, 4.6, and 4.7 respectively. Various symmetry points have been labeled by standard symbols. For example, Γ denotes (0,0,0), M is (110), X is (100), and R is (111) for the BZ of a cubic lattice. The coordinates are with respect to a Cartesian frame fixed at Γ and are in units of $2\pi/a$. The energy bands are generally plotted along the lines joining these symmetry points.

The planes at the surface of the BZ have some special significance. A wave vector $\vec{k}$ whose tip is lying on these planes satisfies the equation

$$\vec{k} \cdot \vec{G} = \frac{1}{2} G$$

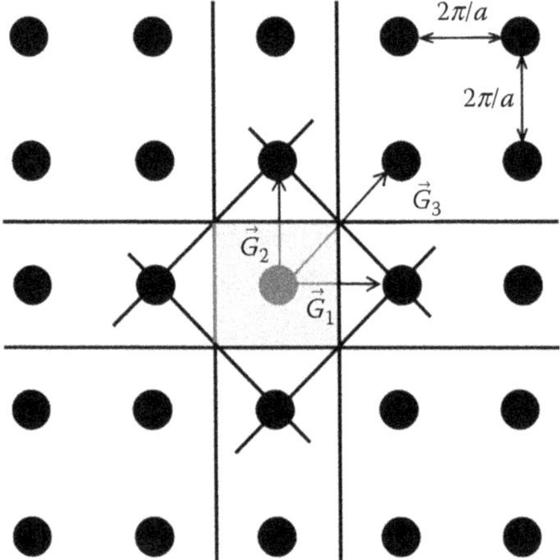

FIGURE 4.4
BZ of the square lattice is shown by the shaded region.

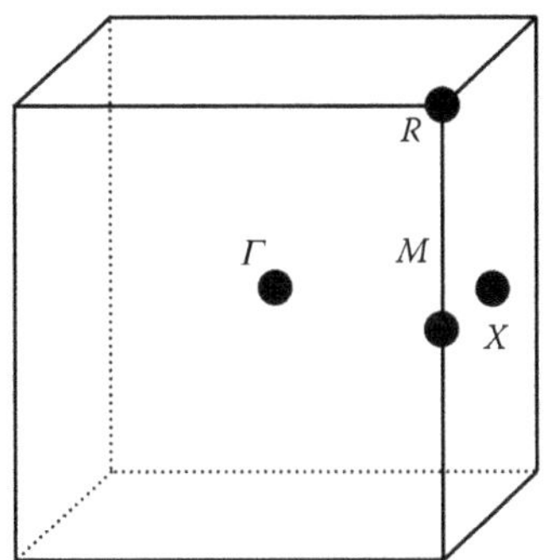

FIGURE 4.5
Brillouin zone of a simple cubic lattice.

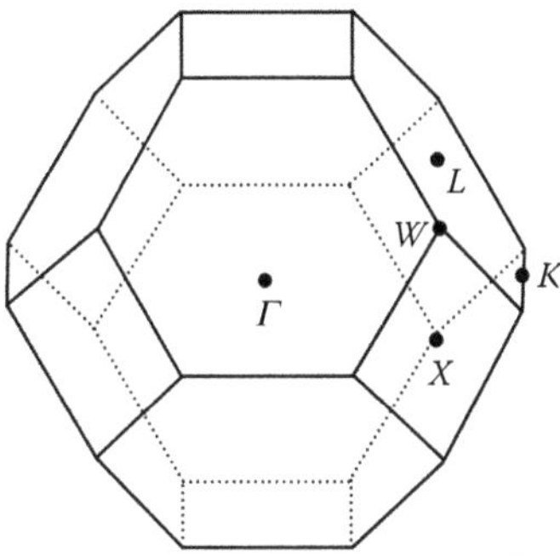

FIGURE 4.6
Brillouin zone of an fcc lattice.

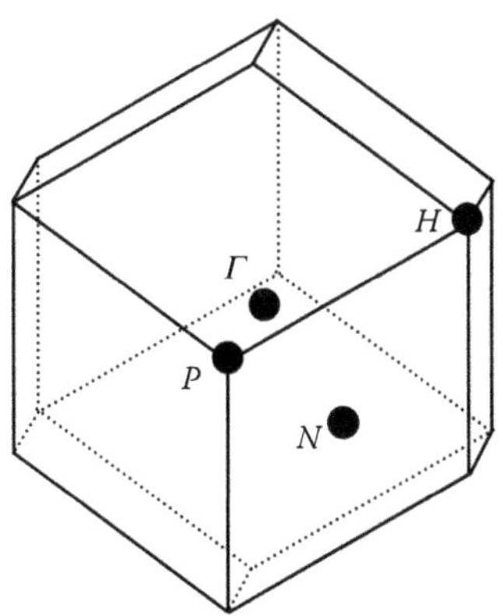

FIGURE 4.7
Brillouin zone of a bcc lattice.

where $\vec{G}$ is a reciprocal lattice vector (see Kittel (1986) or Ashcroft and Mermin (1976)). This is nothing but Bragg or Laue's condition for maxima in x-ray diffraction. Thus, the bounding planes of the BZ are the Bragg planes. We shall see in the next chapter that an electron wave with wave vector corresponding to a Bragg plane will get strongly reflected and will strongly interfere with the incoming wave. This gives rise to an energy band gap at the wave vectors corresponding to these planes.

4.5 Spin–Orbit Interaction

The spin and orbital angular momentum of an electron can couple together and give rise to an additional term in the Hamiltonian, known as the spin–orbit interaction. Although this is a relatively weak interaction, it can have pronounced effects in materials where one of the constituent atoms is a heavy element. This is basically a relativistic effect and can be derived from the Dirac equation. However, here, we shall derive it by using a simple semi-classical argument that makes it clear why there should be coupling between spin of the electron and its orbital angular momentum. Let us consider a hydrogen-like atom with an electron (charge −e) moving in a circular orbit of radius r around its nucleus with charge $+Ze$, as shown in Figure 4.8. Let us assume that its spin and orbital angular momenta are $\vec{S}$ and $\vec{L}$.

Because of spin angular momentum $\vec{S}$, the electron has a magnetic moment $\vec{\mu}$

$$\vec{\mu} = -\frac{e\vec{S}}{m_e} \tag{4.26}$$

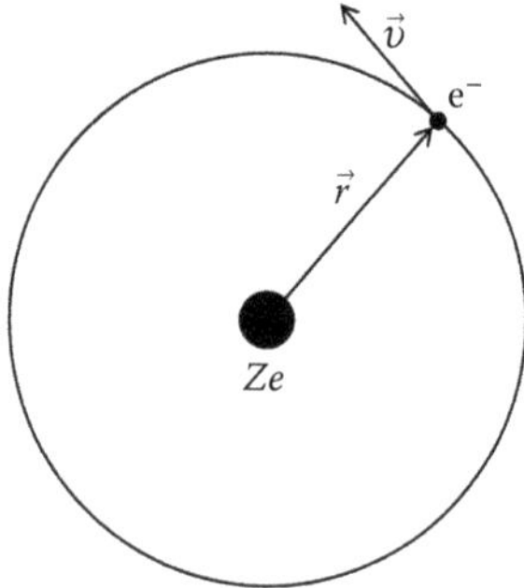

FIGURE 4.8
A hydrogen-like atom in the rest frame of the nucleus.

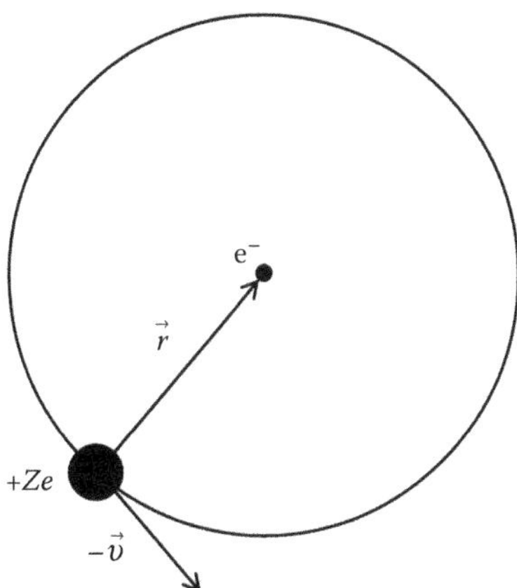

FIGURE 4.9
Motion of the nucleus in the rest frame of the electron.

where m_e is its mass (Eisberg and Resnick 1974). To see the coupling between $\vec{S}$ and $\vec{L}$, let us go to the rest frame of the electron. In this frame, the electron is at rest and the nucleus is moving with velocity $-\vec{v}$ as shown in Figure 4.9. Since the nucleus with charge +*Ze* is moving, it gives rise to a current in this frame. A current element is given by

$$\begin{aligned} i d\vec{\ell} &= \frac{dq}{dt} d\vec{\ell} = dq \frac{d\vec{\ell}}{dt} \\ &= -Ze\vec{v} \end{aligned} \tag{4.27}$$

This current element gives rise to a magnetic field $\vec{B}$ at the location of the electron, given by the Biot–Savart law as

$$\vec{B} = \frac{\mu_0}{4\pi} \frac{i d\vec{\ell} \times \vec{r}}{r^3} = -\frac{\mu_0}{4\pi} \frac{Ze\vec{v} \times \vec{r}}{r^3} \tag{4.28}$$

Let us write this in terms of the electric field created by the nucleus at $\vec{r}$

$$\vec{E} = \frac{Ze\vec{r}}{4\pi\epsilon_0 r^3} = \frac{1}{er} \frac{dV(r)}{dr} \vec{r} \tag{4.29}$$

where $V(r)$ is the potential energy of the electron at $\vec{r}$

$$V(r) = -\frac{Ze^2}{4\pi\epsilon_0 r}$$

This gives

$$\vec{B} = -\varepsilon_0 \mu_0 \vec{v} \times \vec{E}$$

$$\vec{B} = -\frac{1}{c^2} \vec{v} \times \vec{E} \tag{4.30}$$

where $c = 1/\sqrt{(\varepsilon_0 \mu_0)}$ is the speed of light in vacuum. The magnetic field Equation 4.30 interacts with the magnetic moment of the electron with interaction energy

$$\begin{aligned}
\Delta E &= -\vec{\mu} \cdot \vec{B} \\
&= -\frac{e}{m_e c^2} \vec{S} \cdot (\vec{v} \times \vec{E}) \\
&= -\frac{1}{m_e r c^2} \left(\vec{v} \times \frac{dV}{dr} \vec{r} \right) \cdot \vec{S} \\
&= \frac{1}{m_e^2 c^2} \frac{1}{r} \frac{dV(r)}{dr} (\vec{r} \times m_e \vec{v}) \cdot \vec{S}
\end{aligned} \tag{4.31}$$

where we have used Equation 4.29. Now substituting

$$\vec{L} = \vec{r} \times m_e \vec{v}$$

we get

$$\Delta E = \frac{1}{m_e^2 c^2 r} \frac{dV}{dr} \vec{L} \cdot \vec{S} \tag{4.32}$$

This is the interaction energy in the rest frame of the electron, while we are interested in the frame in which the nucleus is at rest. Because of the relativistic transformation of velocities, the transformation back to the normal frame reduces the interaction energy by a factor of 2 (Eisberg and Resnick 1974). Thus, the spin–orbit interaction can be written as

$$H_{SO} = \frac{1}{2 m_e^2 c^2} \frac{1}{r} \frac{dV(r)}{dr} \vec{S} \cdot \vec{L} \tag{4.33}$$

Although this expression has been derived semi-classically, this is in complete agreement with the expression obtained from the Dirac equation. This interaction given by Equation 4.33 is small compared to $V(r)$ and can be treated as a small perturbation. Using the hydrogen-like wave functions, it can be shown that the first-order change in energy of an atom goes as Z^4, where Z is the atomic number (Landau and Lifshitz 1977). This shows that spin–orbit interaction could be quite important for heavy elements.

It turns out that the expression (4.33), which was derived for a hydrogen-like atom, is also applicable to solids and molecules. Equation 4.33 can be rewritten as

$$\begin{aligned} H_{SO} &= \frac{e}{2m_e^2c^2}\vec{S}\cdot(\vec{E}\times\vec{p}) \\ &= -\frac{e}{2m_e^2c^2}\vec{E}\cdot(\vec{S}\times\vec{p}) \end{aligned}$$

where we have used Equation 4.29. In solids, the spin–orbit interaction can be large if the electric field, $\vec{E}$, experienced by the electron is large. This can happen if the electron is near a nucleus, a surface, or an interface. In solids, the spin–orbit interaction often splits the degenerate levels at the symmetry points of the BZ. We shall see in Chapter 15 that the spin–orbit interaction plays an important role in topological insulators.

The full one-electron Hamiltonian can be written as

$$H = H_O + \frac{e\hbar}{4m_e^2c^2}\vec{\sigma}\cdot(\vec{E}\times\vec{p}) \tag{4.34}$$

where we have written that $\vec{S} = (\hbar/2)\vec{\sigma}$ and H_O contains the kinetic energy and other potential excluding spin–orbit term. $\vec{\sigma}$ are Pauli spin matrices given by

$$\sigma_x = \begin{pmatrix} 0 & 1 \\ 1 & 0 \end{pmatrix}, \quad \sigma_y = \begin{pmatrix} 0 & -i \\ i & 0 \end{pmatrix}, \quad \sigma_z = \begin{pmatrix} 1 & 0 \\ 0 & -1 \end{pmatrix} \tag{4.35}$$

The pure spin states α and β are

$$\alpha = \begin{pmatrix} 1 \\ 0 \end{pmatrix} \quad \text{and} \quad \beta = \begin{pmatrix} 0 \\ 1 \end{pmatrix} \tag{4.36}$$

The spin–orbit interaction mixes spin states α and β.

Since Hamiltonian Equation 4.34 is translationally invariant, eigenfunctions of H can be labeled by $\vec{k}$, which can be written as

$$\psi_{\vec{k}\uparrow}(\vec{r}) = \chi_{\vec{k}\uparrow}(\vec{r})\alpha + \gamma_{\vec{k}\uparrow}(\vec{r})\beta$$

and

$$\psi_{\vec{k}\downarrow}(\vec{r}) = \chi_{\vec{k}\downarrow}(\vec{r})\alpha + \gamma_{\vec{k}\downarrow}(\vec{r})\beta \tag{4.37}$$

where ↑ denotes a state with the spin generally up, that is, its projection onto α is positive. $\chi_{\vec{k}\uparrow}$ and $\gamma_{\vec{k}\uparrow}$ are the space part of the wave function.

4.6 Symmetry

Symmetry plays an important role in the energy band theory. It allows one to classify and label the eigenstates and eigenvalues of a Hamiltonian as we have seen in Section 4.3 for translational symmetry. Symmetry can predict degeneracy of levels at symmetry points of the BZ. It can also predict certain transition probabilities and matrix elements between two states to be zero and can produce selection rules. In the electronic structure calculations, the use of symmetry can reduce the numerical effort tremendously.

In addition to the translational symmetry, a crystal can have rotational symmetries about various axes, mirror symmetries about certain planes, and inversion symmetry. These symmetries are called point group symmetries because in all point group operations, a point is kept fixed. All symmetry elements of a crystal form a group called a point group. Only 32 different point groups are possible for crystalline solids. Since the Hamiltonian is invariant under point group operations, the energy level $\varepsilon_{n\vec{k}}$ has the full point group symmetry in the reciprocal space as we shall see below.

Let S denote a point group symmetry operation that transforms $\vec{r}$ to $\vec{r'}$, that is

$$\vec{r'} = S\vec{r} \tag{4.38}$$

where S is a unitary matrix. This is because the operation does not change the magnitude of $\vec{r}$. The symmetry operation changes function $f(\vec{r})$ as

$$Sf(\vec{r}) = f(S\vec{r}) \tag{4.39}$$

Let σ denote up or down spin and $\psi_{\vec{k}\sigma}(\vec{r})$ and $\varepsilon_{\vec{k}\sigma}$ an eigenfunction and eigenvalue of the Hamiltonian H, respectively. For simplicity, we have suppressed the band index n. Since S is a symmetry operation, H remains invariant under this operation. This implies that H and S commute, that is

$$HS = SH \tag{4.40}$$

Thus

$$HS\psi_{\vec{k}\sigma}(\vec{r}) = SH\psi_{\vec{k}\sigma}(\vec{r})$$

or

$$H[S\psi_{\vec{k}\sigma}(\vec{r})] = \varepsilon_{\vec{k}\sigma}[S\psi_{\vec{k}\sigma}(\vec{r})] \tag{4.41}$$

Equation 4.41 implies that $S\psi_{\vec{k}\sigma}(\vec{r})$ and $\psi_{\vec{k}\sigma}(\vec{r})$ have the same energy $\varepsilon_{\vec{k}\sigma}$. Bloch's theorem implies that

$$\psi_{\vec{k}\sigma}(\vec{r} + \vec{R}) = e^{i\vec{k}\cdot\vec{R}}\psi_{\vec{k}\sigma}(\vec{r})$$

Thus

$$\begin{aligned} S\psi_{\vec{k}\sigma}(\vec{r} + \vec{R}) &= \psi_{\vec{k}\sigma}(S(\vec{r} + \vec{R})) = \psi_{\vec{k}\sigma}(S\vec{r} + S\vec{R}) \\ &= e^{i\vec{k}\cdot S\vec{R}}\,\psi_{\vec{k}\sigma}(S\vec{r}) \end{aligned} \tag{4.42}$$

This is because $S\vec{R}$ is a lattice vector. Also, we note that if we apply the same symmetry operation to both vectors, $\vec{k}$ and $S\vec{R}$, their scalar product remains unchanged (e.g., if we rotate both vectors by the same angle). Thus

$$\vec{k}\cdot S\vec{R} = S^{-1}\vec{k}\cdot S^{-1}S\vec{R} = S^{-1}\vec{k}\cdot\vec{R} \tag{4.43}$$

Using Equation 4.43, Equation 4.42 can be written as

$$\psi_{\vec{k}\sigma}(S(\vec{r} + \vec{R})) = e^{iS^{-1}\vec{k}\cdot\vec{R}}\psi_{\vec{k}\sigma}(S\vec{r}) \tag{4.44}$$

This implies that $\psi_{\vec{k}\sigma}(S\vec{r})$ is an eigenfunction of the translational operator with wave vector $S^{-1}\vec{k}$ and corresponds to the same energy $\varepsilon_{\vec{k}\sigma}$. Thus, we have

$$\begin{aligned} \psi_{\vec{k}\sigma}(S\vec{r}) &= \psi_{S^{-1}\vec{k}\sigma}(\vec{r}) \\ \varepsilon_{S^{-1}\vec{k}\sigma} &= \varepsilon_{\vec{k}\sigma} \end{aligned} \tag{4.45}$$

The above discussion implies that many $\vec{k}$ points in the BZ, which are connected by the point group symmetry operations, have the same energy. Thus, it is useful to define the irreducible Brillouin zone (IBZ), which is much smaller than the BZ, such that all points outside the IBZ are connected with

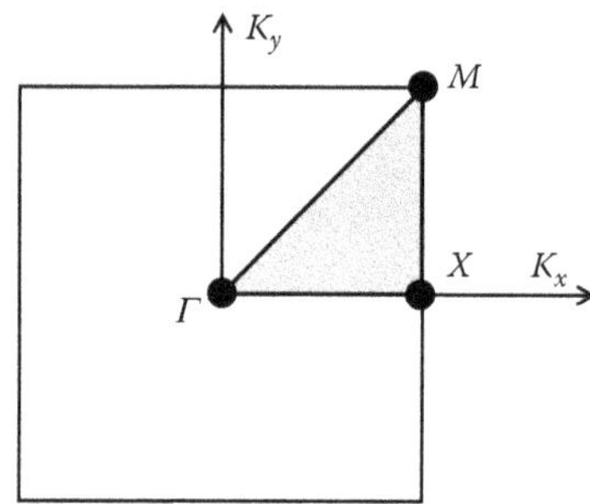

FIGURE 4.10
IBZ of the square lattice.

$\vec{k}$ points in the IBZ by some point group symmetry. For example, the IBZ of a square lattice is shown in Figure 4.10, which is 1/8th of the full zone. For a cubic system, the IBZ is 1/48th of the full zone. Thus, if one wants to integrate a function that has the cubic point group symmetry, one needs to integrate only over 1/48 of the BZ. Thus, the use of symmetry can reduce numerical effort by an enormous amount.

If we add translations to the point group operations, the resulting symmetry elements form a group called a space group. There are only 230 different space groups possible. We shall not go into the details of the group theory but refer to some excellent books such as by Heine (1960), Tinkham (1964), and Falicov (1966). These space groups have been assigned unique symbols and a number identifying them (see, e.g., Hahn 2002). All electronic structure studies use space group symmetries to simplify calculations; therefore, information about the space group is an input to these calculations.

4.7 Inversion Symmetry, Time Reversal, and Kramers' Theorem

Inversion symmetry means that if we change $\vec{r}$ to $-\vec{r}$, the system remains invariant. Let us define an operator I that performs the inversion operation. If the Hamiltonian is invariant under inversion, I and H commute, that is

$$HI = IH \tag{4.46}$$

Thus, if $\psi_{\vec{k}\sigma}$ is an eigenfunction of H with energy $\varepsilon_{\vec{k}\sigma}$ and spin σ, then

$$\begin{aligned} HI\psi_{\vec{k}\sigma} &= IH\psi_{\vec{k}\sigma} \\ &= I\varepsilon_{\vec{k}\sigma}\psi_{\vec{k}\sigma} \end{aligned}$$

or

$$H(I\psi_{\vec{k}\sigma}) = \varepsilon_{\vec{k}\sigma}(I\psi_{\vec{k}\sigma}) \tag{4.47}$$

Thus, $I\psi_{\vec{k}\sigma}$ is also an eigenfunction of H with energy $\varepsilon_{\vec{k}\sigma}$. To identify the wave vector of $I\psi_{\vec{k}\sigma}$, we use Bloch's theorem, which gives

$$I\psi_{\vec{k}\sigma}(\vec{r} + \vec{R}) = e^{-i\vec{k}\cdot\vec{R}}\ \psi_{\vec{k}\sigma}(-\vec{r}) = e^{-i\vec{k}\cdot\vec{R}}\ I\psi_{\vec{k}\sigma}(\vec{r})$$

This implies that $I\psi_{\vec{k}\sigma}$ corresponds to wave vector $-\vec{k}$ and energy $\varepsilon_{-\vec{k}\sigma}$. But according to Equation 4.47, this is equal to $\varepsilon_{\vec{k}\sigma}$. Thus, the inversion symmetry implies that

$$\varepsilon_{-\vec{k}\sigma} = \varepsilon_{\vec{k}\sigma} \tag{4.48}$$

This is an important result of inversion symmetry, which means that we do not have to calculate bands for negative values of wave vectors, if the system has inversion symmetry.

Now, we discuss the time-reversal symmetry, which means that the Hamiltonian remains invariant as time t is replaced by $-t$. We see that the Hamiltonian H (Equation 4.34) has the time-reversal symmetry. This is because on time reversal the direction of spin $\vec{S}$ as well as the momentum $\vec{p}$ reverses. Let T denote the time-reversal operator. Kramers' theorem states that if $\psi_{\vec{k}\sigma}$ is a one-electron eigenstate of H, then $T\psi_{\vec{k}\sigma}$ is also an eigenstate with the same energy in the absence of the external magnetic field. Also, $T\psi_{\vec{k}\sigma}$ is orthogonal to $\psi_{\vec{k}\sigma}$.

Since H is invariant under time-reversal, T commutes with H, that is

$$HT = TH \tag{4.49}$$

This implies that if $\psi_{\vec{k}\sigma}$ is an eigenfunction of H with energy $\varepsilon_{\vec{k}\sigma}$, then

$$\begin{aligned} HT\psi_{\vec{k}\sigma} &= TH\psi_{\vec{k}\sigma} \\ &= T\varepsilon_{\vec{k}\sigma}\psi_{\vec{k}\sigma} \end{aligned}$$

or

$$H(T\psi_{\vec{k}\sigma}) = \varepsilon_{\vec{k}\sigma}(T\psi_{\vec{k}\sigma})$$

that is, $T\psi_{\vec{k}\sigma}$ is also an eigenfunction of H with energy $\varepsilon_{\vec{k}\sigma}$.

T transforms $\psi_{\vec{k}\sigma}$ into $\psi_{-\vec{k}-\sigma}$ because by reversing t the momentum and spin are reversed. Since these two eigenvectors belong to different eigenvalues of the translational operator, they are orthogonal to each other. This proves the theorem.

Thus

$$T\psi_{\vec{k}\uparrow} = \psi_{-\vec{k}\downarrow}$$

and

$$T\psi_{\vec{k}\downarrow} = \psi_{-\vec{k}\uparrow} \tag{4.50}$$

This further implies from Kramers' theorem that

$$\varepsilon_{\vec{k}\uparrow} = \varepsilon_{-\vec{k}\downarrow}$$

and

$$\varepsilon_{\vec{k}\downarrow} = \varepsilon_{-\vec{k}\uparrow} \tag{4.51}$$

that is, there is a twofold degeneracy in the bands. If there is no spin–orbit coupling, then in the absence of a magnetic field, $\varepsilon_{\vec{k}\uparrow} = \varepsilon_{\vec{k}\downarrow}$ and all the energies in Equation 4.51 are the same. This implies that $\varepsilon_{\vec{k}\sigma} = \varepsilon_{-\vec{k}\sigma}$, which is an important consequence of the time-reversal symmetry and holds even if there is no inversion symmetry.

In the presence of the external magnetic field $\vec{B}$, the Hamiltonian will have a term like $\vec{S}\cdot\vec{B}$. Since T will reverse $\vec{S}$, it will no longer commute with H, or in other words the magnetic field breaks the time-reversal symmetry. Therefore, Kramers' theorem does not hold in the presence of the magnetic field.

4.8 Band Structure and Fermi Surface

For each $\vec{k}$, there are many eigenvalues of H that are labeled as $\varepsilon_{n\vec{k}}$, where n is a band index. A plot of $\varepsilon_{n\vec{k}}$ versus $\vec{k}$ is called band structure. There are many methods for calculating band structure, which we will discuss in the next few chapters. Here, we shall discuss some main features of band structure.

One very important result of band theory is that in the band structure of a crystalline solid, we have allowed ranges of energy, called energy bands. The forbidden ranges of energy are called energy band gaps. In the next chapter, we shall discuss in detail the origin of energy band gaps. When discussing the nearly free electron model, we shall see that the band gaps arise due to the quantum interference of electron waves moving in a periodic lattice, very similar to the interference of electromagnetic waves. Another way to understand it is to look at this from the perspective of the tight-binding model. We bring the atoms from infinity to a very small distance on a periodic lattice and see how the atomic levels evolve into bands. We shall discuss it in detail in Chapter 5.

As an example, we show the band structure of copper in Figure 4.11, along various symmetry lines in the BZ. Γ, X, W, and so on are various symmetry points in the BZ shown in Figure 4.6 and the band structure is plotted along the lines joining these points. The band structure of copper was obtained by using the Vienna *ab initio* simulation package (VASP) code (Kresse and Furthmüller 1996a,b). The lattice constant was taken to be 3.6309 Å and the energy cutoff, 400 eV with $18 \times 18 \times 18$ Monkhorst–Pack k-mesh (Monkhorst and Pack 1976). In the figure, the highest filled level, called Fermi energy, is denoted by ε_F and taken as the zero of energy. The d bands lie in the narrow energy range of approximately –2 to –4 eV. We see that the d bands are quite flat. The s–p type bands have parabolic shape and lie in the energy range of –10 to –5 eV and 0 to 2 eV. In the region of the d bands, there is a mixing of s states with the d states, and this is known as s–d hybridization. Owing to the s–d hybridization, the lowest band starting from Γ, which has parabolic shape up to about –5 eV, changes to a flatter band up to X or L. The band structure can be measured by ARPES. The experimental band structure of Cu obtained by ARPES has been found in excellent agreement with theoretical calculations (Courths and Hufner 1984).

The one-electron levels $\varepsilon_{n\vec{k}}$ are filled in accordance with the Pauli exclusion principle. As we had defined earlier, the Fermi energy, ε_F, is the highest filled

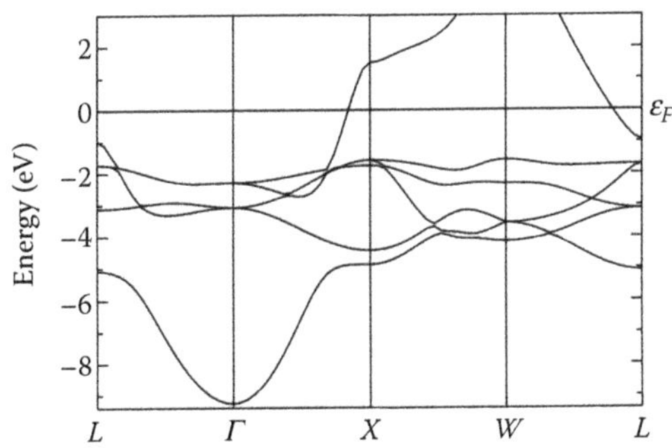

FIGURE 4.11
Band structure of copper obtained by using the VASP code (Kresse and Furthmüller 1996a,b) and GGA. The lattice constant was taken to be 3.6309 Å and the energy cutoff 400 eV with $18 \times 18 \times 18$ Monkhorst–Pack k-mesh. The zero of energy is fixed at the Fermi energy.

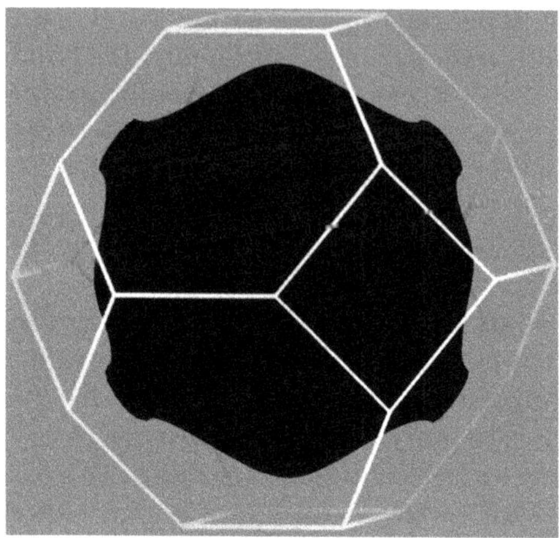

FIGURE 4.12
Fermi surface of copper obtained by using the Wien2 k code (Schwarz and Blaha 2003) and GGA. The lattice constant was taken to be 3.6309 Å.

level in the band structure at absolute zero temperature. The constant energy surface in $\vec{k}$ space corresponding to ε_F is known as the Fermi surface. The Fermi surface of copper is shown in Figure 4.12. As we can see, it is very different from a sphere. This is because there is *s*–*d* hybridization and the band intersecting the Fermi energy is not purely *s*–*p* type but also has *d* character. It has a neck-like structure around point *L* and is not a closed surface. The Fermi surface, in general, can have a very complex structure. For example, Nb, Mo, and Ni have a very complicated Fermi surface (see, e.g., Rajput et al. 1996, Prasad et al. 1977). The Fermi surface can be measured by the de Haas–van Alphen (dHvA) method. This is a very accurate method for measuring Fermi surface, but requires very pure single crystal samples and very low temperatures. Another method is the two-dimensional angular correlation of positron annihilation radiation (2D-ACAR). This method does not require pure samples and low temperatures but is not as accurate as the dHvA. The Fermi surfaces of disordered alloys have been measured by this method.

We see that in copper, bands are partially filled. Such materials with partially filled bands are metals, as a small application of electric field can produce a current. Band theory can predict whether a material is a metal or insulator at absolute zero. If the bands are partially filled, as is the case for copper, it is a metal because the application of an electric field would produce a current. It can be shown that a filled band cannot contribute to the current (see Exercise 4.8). Thus, if a material has completely filled bands, as is the case in carbon, it would be an insulator, at absolute zero. However, if the band gap between the highest filled band and lowest unoccupied band is <2 eV, the electrons can get thermally excited across the band gap at room temperature. Now, both bands are partially filled and as a result the material will conduct. Such a material is

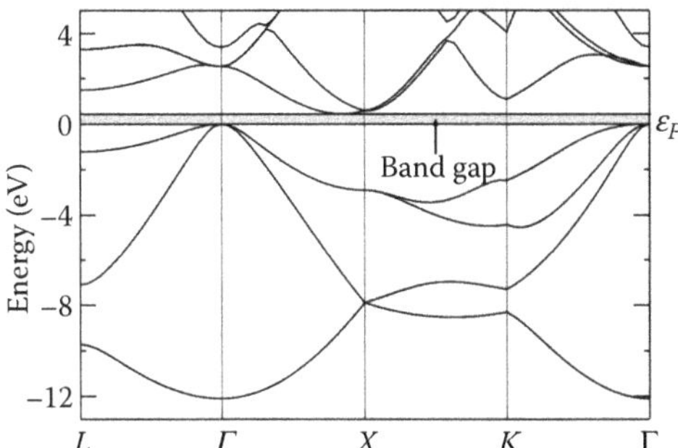

FIGURE 4.13
Band structure of silicon obtained by using the VASP code (Kresse and Furthmüller 1996a,b) and GGA. The lattice constant was taken to be 5.43 Å and the energy cutoff 250 eV with 20 × 20 × 20 Monkhorst–Pack *k*-mesh (Monkhorst and Pack 1976). The zero of energy is fixed at the highest filled level. The band gap is shown by the shaded region.

called a semiconductor. In Figure 4.13, we show the band structure of silicon, which is a typical example of a semiconductor. The band structure was calculated by using the VASP code (Kresse and Furthmüller 1996a,b). The lattice constant was taken to be 5.43 Å and the energy cut off, 250 eV with 20 × 20 × 20 Monkhorst–Pack *k*-mesh (Monkhorst and Pack 1976). We see that above the fully filled band there is a band gap shown by the shaded region. We see that the minimum band gap occurs between energy eigenvalues at two different $\vec{k}$-points. This is called indirect band gap and the corresponding material is called indirect band gap semiconductor as opposed to direct band gap semiconductor in which the minimum band gap occurs at the same $\vec{k}$-point.

The ability of band theory to predict the metallic or insulating behavior of a material has been one of its great achievements. However, it may fail if the material is a strongly correlated material such as NiO, CoO, MnO, or high-T_c cuprates.

4.9 Density of States, Local Density of States, and Projected Density of States

One very useful quantity that we will frequently encounter in electronic structure studies is the density of states (DOS). The DOS is defined as the number of states per unit energy range per unit cell

$$\rho(\varepsilon) = \frac{dN}{d\varepsilon} \tag{4.52}$$

where dN is the number of states per unit cell in the energy range ε and $\varepsilon + d\varepsilon$. An alternative definition is

$$\rho(\varepsilon) = \frac{1}{N_{cell}} \sum_{\vec{k},n} \delta(\varepsilon - \varepsilon_{n\vec{k}}) \tag{4.53}$$

where N_{cell} is the number of unit cells in the solid. Note that in case of the free electron model, the DOS was defined such that its unit was number of states per unit energy per unit volume. However, in general, it is defined using Equation 4.53 such that its unit is number of states per unit energy per unit cell. The DOS is an important quantity that appears in various calculations such as calculation of specific heat, transport properties, and so on. It shows singularities known as van Hove singularities (see Exercise 4.10).

Another important quantity of interest is local density of states (LDOS). Suppose all the atoms in the solid are not equivalent and we would like to know the contribution of one particular atom to the total DOS. For this purpose, let us assume that we know an orthonormal set of atomic wave functions $|\phi_\ell\rangle$ centered on site ℓ, so that we can expand the Bloch wave functions $|\psi_{n\vec{k}}\rangle$ as:

$$|\psi_{n\vec{k}}\rangle = \sum_{\ell} \langle\phi_\ell|\psi_{n\vec{k}}\rangle|\phi_\ell\rangle \tag{4.54}$$

where

$$\langle\phi_\ell|\,\psi_{n\vec{k}}\rangle = \int_{\ell} \phi_\ell^*(\vec{r})\psi_{n\vec{k}}(\vec{r})d^3r \tag{4.55}$$

where the integration is over the volume around atom ℓ. This implies that the probability of finding an electron with wave vector $\vec{k}$ and energy $\varepsilon_{n\vec{k}}$ in state $|\phi_\ell\rangle$ is $|\langle\phi_\ell|\psi_{n\vec{k}}\rangle|^2$. Thus, contribution of the state $|\phi_\ell\rangle$ to the DOS coming from $\psi_{n\vec{k}}$ has to be weighted by $|\langle\phi_\ell|\psi_{n\vec{k}}\rangle|^2$. Then, the LDOS, which is the contribution of site ℓ to the total DOS, is defined as

$$\rho_\ell(\varepsilon) = \frac{1}{N_{cell}} \sum_{\vec{k},n} \left|\langle\phi_\ell|\psi_{n\vec{k}}\rangle\right|^2 \delta(\varepsilon - \varepsilon_{n\vec{k}}) \tag{4.56}$$

The unit of LDOS is the number of states per unit energy per atom. Note that the total DOS is the sum of LDOS coming from all atoms

$$\rho(\varepsilon) = \sum_{\ell} \rho_\ell(\varepsilon) = \frac{1}{N_{cell}} \sum_{\vec{k},n,l} \left|\langle\phi_\ell|\psi_{n\vec{k}}\rangle\right|^2 \delta(\varepsilon - \varepsilon_{n\vec{k}})$$

$$= \frac{1}{N_{cell}} \sum_{n\vec{k}} \delta(\varepsilon - \varepsilon_{n\vec{k}}) \tag{4.57}$$

This is because

$$\sum_{\ell} \left|\langle \phi_{\ell} | \psi_{n\vec{k}} \rangle\right|^2 = 1 \tag{4.58}$$

The above idea can be generalized to define projected density of states (PDOS), which is the angular momentum decomposition of the LDOS. We expand the Bloch wave function $|\psi_{n\vec{k}}\rangle$, in terms of orthonormal set $|\phi_{\ell L}\rangle$, where L is the composite angular momentum index for (l, m). Thus

$$|\psi_{n\vec{k}}\rangle = \sum_{\ell,L} \langle \phi_{\ell L} | \psi_{n\vec{k}} \rangle |\phi_{\ell L}\rangle \tag{4.59}$$

Thus, we can define PDOS as

$$\rho_{\ell L}(\varepsilon) = \frac{1}{N_{cell}} \sum_{\vec{k},n} \left|\langle \phi_{\ell L} | \psi_{n\vec{k}} \rangle\right|^2 \delta(\varepsilon - \varepsilon_{n\vec{k}}) \tag{4.60}$$

PDOS is an important quantity to find the angular momentum character of LDOS as we shall see below. Note that the definition of PDOS and LDOS has a matrix element such as Equation 4.55, whose evaluation involves integral over volume around atom ℓ. The integral is generally evaluated over a spherical volume of radius equal to the Wigner–Seitz radius. Since one cannot unambiguously assign a volume to an atom in a solid containing different kinds of atoms, this leads to some uncertainty. Nevertheless, LDOS and PDOS are very useful quantities as we shall see later.

As examples, we show the DOS and PDOS of copper and silicon in Figures 4.14 and 4.15, respectively. For copper, we see a large density of states in the

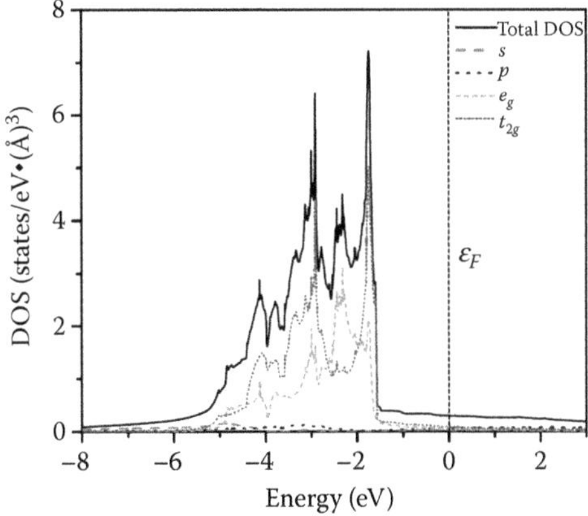

FIGURE 4.14
DOS and s, p, and d PDOS for copper obtained by using the VASP code (Kresse and Furthmüller 1996a,b) and GGA. The zero of energy is at the Fermi energy.

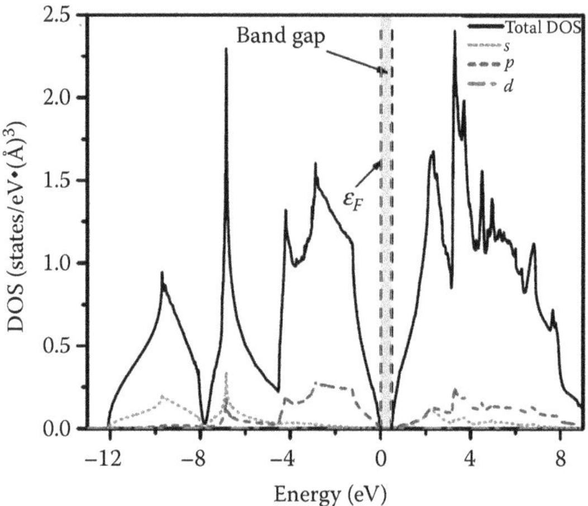

FIGURE 4.15
DOS and PDOS for silicon obtained by using the VASP code (Kresse and Furthmüller J. 1996a,b) and GGA. The zero of energy is at the highest filled level. The band gap is indicated by the shaded region.

energy range –3 to –1 eV, which is mainly contributed by *d* states (e_g and t_{2g}) as can be seen from *d* PDOS. However, at the Fermi energy, the contribution of *d* states to DOS gets reduced. In silicon, we see that in the band gap, the DOS is zero. In other words, by looking at the range of zero DOS we can find the band gap. Here, the band gap is about 0.5 eV, which is much smaller than the experimental gap of about 1.1 eV. This is due to the use of LDA, which leads to a much smaller band gap as mentioned in Chapter 3. This can be corrected by using a better functional or GW method.

4.10 Charge Density

The charge density at location $\vec{r}$ is defined as

$$\rho(\vec{r}) = -e \sum_{n,\vec{k},\text{occupied}} \left|\psi_{n\vec{k}}(\vec{r})\right|^2 = -en(\vec{r}) \tag{4.61}$$

where e is the electronic charge and $n(\vec{r})$ the electron density. Here, the sum is over all the occupied states. The charge density can be measured by using x-ray diffraction. As we have seen, the electron density $n(\vec{r})$ plays a central role in the DFT. The distribution of electronic charge in a solid can give important information about the type of bonding. For example, in metals, the

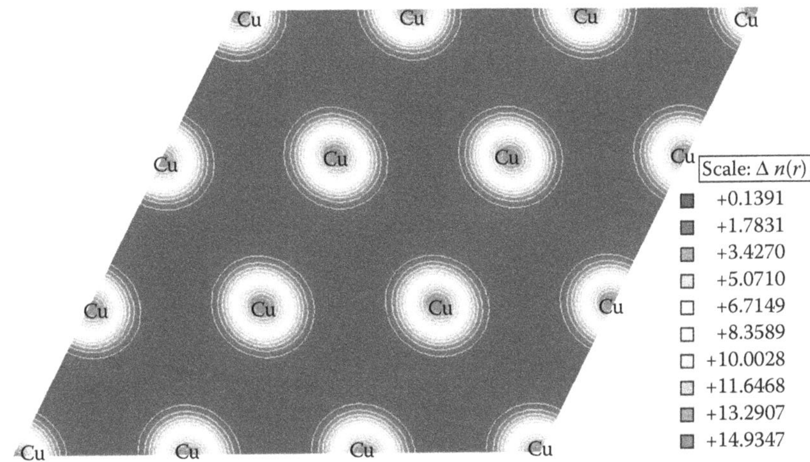

FIGURE 4.16
Charge density contours for Cu in (100) plane using the VASP code (Kresse and Furthmüller 1996a,b) and GGA. The positions of Cu atoms are indicated.

charge density is quite isotropic around an atom, while in covalent solids, it is highly directional. In Figure 4.16, we show the charge density for copper in (100) plane and we can see that the charge density in copper around an atom is quite isotropic in contrast to the charge density in silicon, which is highly directional as seen in Figure 4.17.

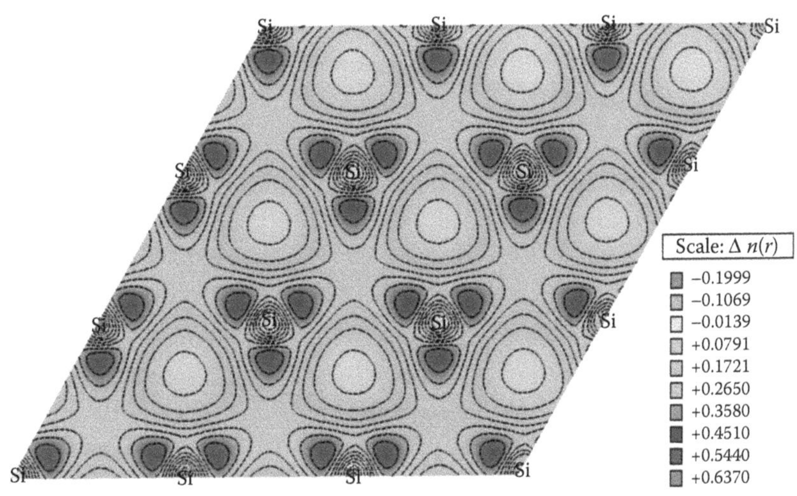

FIGURE 4.17
Charge density contours for silicon in (100) plane using the VASP code (Kresse and Furthmüller 1996a,b) and GGA. The positions of Si atoms are indicated.

4.11 Brillouin Zone Integration

For many calculations such as the calculation of the total energy, charge density, and the DOS, one has to sum over $\vec{k}$, which can be converted to a BZ integration such as

$$\sum_{\vec{k}} f(\vec{k}) = \frac{\Omega}{8\pi^3} \int_{BZ} f(\vec{k}) d^3k \tag{4.62}$$

where Ω is the volume of the solid. This integral for a real material has to be evaluated numerically. The numerical effort can be reduced considerably by exploiting the point group symmetry of the system. If the integrand $f(\vec{k})$ has point group symmetry, the integral has to be done only over the irreducible part of the Brillouin zone (IBZ), which has much smaller volume compared to the full BZ. For numerical integration, the BZ integral is evaluated by summing over discrete set of $\vec{k}$ points. For efficient calculation of integral over BZ, several methods have been developed. The choice of method depends on whether the system is an insulator or a metal. For insulators, all occupied bands are full; therefore, the function $f(\vec{k})$ is generally a smooth function of $\vec{k}$. Thus, the evaluation of such BZ integrals for insulators requires a smaller number of $\vec{k}$ points than the number required for metals. For a metal, function $f(\vec{k})$, which involves the sum over the occupied states, will have discontinuity at the Fermi surface. This requires careful handling, and in general, one will require a larger number of $\vec{k}$ points for an accurate evaluation. Various methods suggested for BZ integration are special $\vec{k}$-point methods (Baldereschi 1973, Chadi and Cohen 1973, Monkhorst and Pack 1976), the Methfessel and Paxton method (Methfessel and Paxton 1989), the tetrahedron method (Jepsen and Andersen 1971, Blochl and Andersen 1994), the special directions method (Bansil 1975, Prasad and Bansil 1980), and so on. We shall not go into the details of these methods and refer the reader to the original papers.

EXERCISES

4.1 Show that all possible values of $\vec{k}$ can be confined to the first BZ.

4.2 Show that the number of allowed wave vectors $\vec{k}$ in the BZ is equal to the number of sites in the solid.

4.3 Using the spin–orbit interaction as perturbation, show that the first-order change in energy of an atom goes as Z^4, where Z is the atomic number. You may use hydrogen-like wave functions in your calculation.

4.4 Consider a rectangular lattice with $a = 4$ Å and $b = 8$ Å. Draw the Brillouin zone and clearly show the IBZ.

4.5 Draw the BZs of simple cubic, fcc, and bcc lattices using the cube edge $a = 4$ Å. In the figure, mark the coordinates of symmetry points using standard symbols.

4.6 Show that the periodic part of the Bloch wave function $u_{\vec{k}}(\vec{r})$ satisfies the equation

$$\left(\frac{\hbar^2}{2m_e}\left(\frac{\vec{\nabla}}{i}+\vec{k}\right)^2 + v(\vec{r})\right)u_{\vec{k}}(\vec{r}) = \varepsilon_{\vec{k}}u_{\vec{k}}(\vec{r})$$

4.7 Using the result of Exercise 4.6, show that the mean velocity of an electron in level $\varepsilon_{n\vec{k}}$ is given by

$$\vec{v}_n(\vec{k}) = \frac{1}{\hbar}\vec{\nabla}_{\vec{k}}(\varepsilon_{n\vec{k}})$$

4.8 Using the result of Exercise 4.7, show that a filled band does not contribute to the current density.

4.9 Show that the density of states can be written as

$$\rho(\varepsilon) = \int_S \frac{ds}{4\pi^3}\frac{1}{|\vec{\nabla}_{\vec{k}}(\varepsilon_{\vec{k}})|}$$

where the integral is over a constant energy surface $\varepsilon = \varepsilon_{\vec{k}}$.

4.10 Exercise 4.9 implies that $\rho(\varepsilon)$ will have singularities. However, in 3D, these singularities are integrable and give finite results for $\rho(\varepsilon)$ but the slope has divergences. These singularities are known as van Hove singularities. Consider the following case when

$$\varepsilon_{\vec{k}} = \varepsilon_o + \alpha_1\left(k_x^2 + k_y^2\right) + \alpha_2 k_z^2$$

Using the result given in Exercise 4.9, show that $\rho(\varepsilon)$ is finite but $\rho'(\varepsilon)$ diverges as $\varepsilon \to \varepsilon_o$. (Hint: Exploit the cylindrical symmetry of $\varepsilon_{\vec{k}}$ and use cylindrical coordinates for the surface integration.)

Further Reading

Kittel C. 1987. *Quantum Theory of Solids (*Second Edition*)*. New York: John Wiley & Sons.

Martin R. M. 2004. *Electronic Structure, Basic Theory and Practical Methods*. Cambridge: Cambridge University Press.

Singleton J. 2001. *Band Theory and Electronic Properties of Solids*. New York: Oxford University Press.

Sutton A. P. 1996. *Electronic Structure of Materials*. Oxford: Clarendon Press.

5

Methods of Electronic Structure Calculations I

5.1 Introduction

In this chapter, we shall solve the KS equation for some simple models and try to understand the origin of the band gap. We shall then discuss the basic ideas of some well-known methods of band structure calculations such as the plane wave expansion method, the tight-binding model, the orthogonalized plane wave (OPW) method, and the pseudopotential method. We shall start with the empty lattice approximation in Section 5.2, which is the limiting case of effective potential approaching zero value. In Section 5.3, we shall then solve the KS equation for a weak effective potential, which is called the nearly free electron (NFE) model. In Sections 5.4 and 5.5, we shall discuss the plane wave expansion and the tight-binding methods. The tight-binding method provides another way to look at how the bands are formed when atoms are brought from infinite distance to form a solid. In Section 5.6, we shall briefly discuss the Hubbard model, which is an attempt to incorporate electron correlations in solids. In Section 5.7, we shall briefly discuss Wannier functions, which are orthogonal localized functions centered on atomic sites. In Section 5.8, we shall discuss the OPW method, which led to the idea of pseudopotential. In Section 5.9, we shall discuss the pseudopotential method, which has emerged as a very powerful tool for calculating band structure, total energy, and structural and other properties of solids. In this chapter, we shall only discuss the basic idea of pseudopotential and the method will be discussed in more detail in the next chapter.

5.2 Empty Lattice Approximation

If the effective potential in the KS equation is reduced to zero, eigenvalues $\varepsilon_{\vec{k}}$ of a solid approach the free electron values

$$\varepsilon_{\vec{k}} = \frac{\hbar^2 k^2}{2m_e} \tag{5.1}$$

The wave vector $\vec{k}$ in Equation 5.1 is unrestricted and can vary from $-\infty$ to $+\infty$. However, we know that if we have a periodic lattice, all values of $\vec{k}$ lie within the first BZ. Thus, if a wave vector $\vec{k}'$ lies outside the BZ boundary, it has to be folded back within the first BZ by adding (or subtracting) a suitable reciprocal vector such that

$$\vec{k}' = \vec{k} + \vec{G}$$

and

$$\varepsilon_{\vec{k}'} = \varepsilon_{\vec{k}+\vec{G}} = \frac{\hbar^2}{2m_e}(\vec{k} + \vec{G})^2 \tag{5.2}$$

Thus, $\varepsilon_{\vec{k}+\vec{G}}$ appears as a higher band at the wave vector $\vec{k}$ than $\varepsilon_{\vec{k}}$ in the first BZ. These bands are known as empty lattice bands. Here, "empty lattice" means that the effective potential is zero and in that sense the lattice is empty, that is, no atoms at the lattice points. The empty lattice approximation provides a good test for any method of calculating band structure; therefore, it is also known as the empty lattice test. The empty lattice bands provide a feel for the band structure, although the actual band structure will be quite different from it. Thus, by comparing it with the actual band structure of a solid, one can get some idea of the free electron nature of the bands.

To give an example, we first consider a one-dimensional lattice of lattice constant a. The reciprocal lattice vectors are $\pm(2\pi/a)n$. Therefore, the empty lattice bands are given by

$$\varepsilon_{n\vec{k}} = \frac{\hbar^2}{2m_e}\left(k \pm \frac{2\pi}{a}n\right)^2 \tag{5.3}$$

These bands are plotted in Figure 5.1 and compared with the free electron bands. The way in which the "empty lattice" bands are plotted on left in

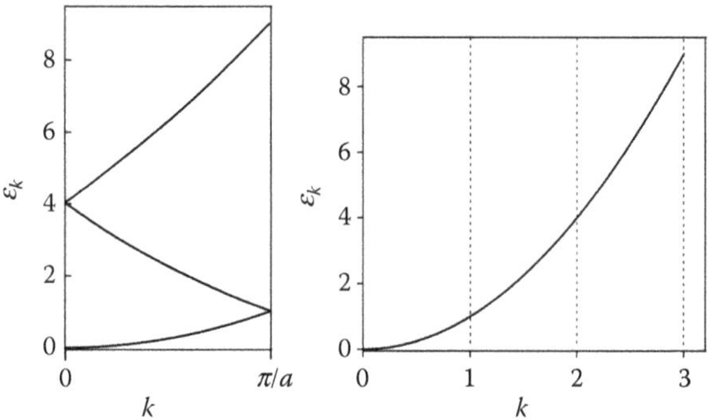

FIGURE 5.1
Left: Empty lattice bands in one dimension. Right: Free electron band is plotted for comparison. Energy is given in units of $(\hbar\pi)^2/(2m_e a^2)$ and k in units of π/a.

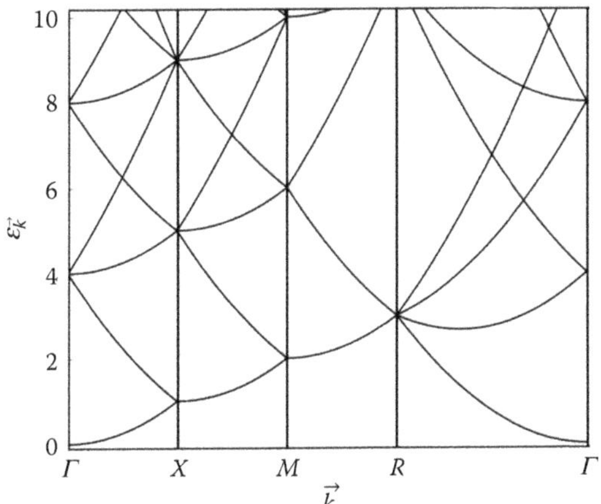

FIGURE 5.2
Empty lattice bands of a simple cubic lattice. Energy is given in units of $(\hbar\pi)^2/(2m_e a^2)$.

Figure 5.1 is known as the "reduced zone" scheme, in which the wave vector $\vec{k}$ is restricted up to the first BZ boundary. Sometimes, it may be more useful to unfold the bands and plot them like free electron bands (right side) such that $\vec{k}$ is unrestricted. This scheme is known as the "extended zone" scheme.

For simple cubic, fcc, and bcc solids, the calculation of empty lattice bands is left as an exercise but the results are shown in Figures 5.2 through 5.4. The energy is given in units of $(\hbar\pi)^2/(2m_e a^2)$, where a is the cube edge.

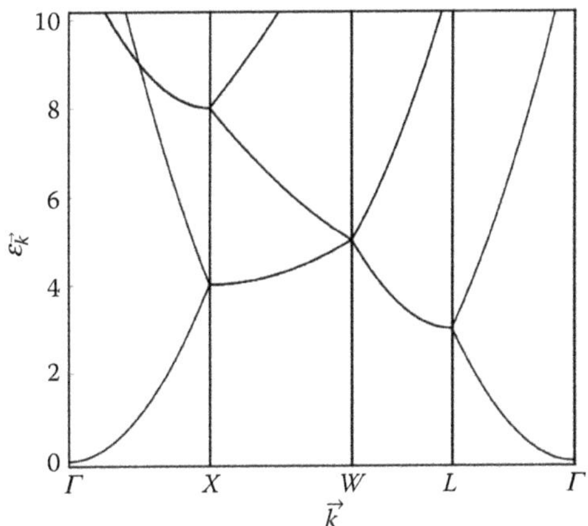

FIGURE 5.3
Empty lattice bands for an fcc lattice. Energy is given in units of $(\hbar\pi)^2/(2m_e a^2)$.

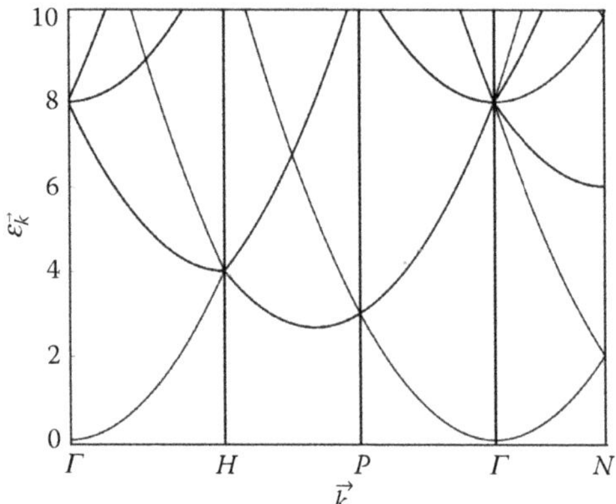

FIGURE 5.4
Empty lattice bands for a bcc lattice. Energy is given in units of $(\hbar\pi)^2/(2m_e a^2)$.

5.3 Nearly Free Electron (NFE) Model

We saw that for $v_{\text{eff}}(\vec{r}) = 0$, the solutions are plane waves. So, if the potential $v_{\text{eff}}(\vec{r})$ is very weak, we can use perturbation theory to find the solution of the KS equation.

The Hamiltonian H can be written as

$$H = -\frac{\hbar^2}{2m_e}\nabla^2 + v_{\text{eff}}(\vec{r}) = H_0 + v_{\text{eff}}(\vec{r}) \tag{5.4}$$

where H_0 is the free electron Hamiltonian. The unperturbed eigenvectors and energy eigenvalues are

$$\left|\vec{k}\right\rangle = \frac{1}{\sqrt{\Omega}}e^{i\vec{k}\cdot\vec{r}} \tag{5.5}$$

and

$$\varepsilon^0_{\vec{k}} = \frac{\hbar^2 k^2}{2m_e} \tag{5.6}$$

where Ω is the volume of the solid. As $v_{\text{eff}}(\vec{r})$ is weak, we treat it as a perturbation over H_0. Using the second-order perturbation theory, the energy and the wave function of the perturbed electron are given by

$$\varepsilon_{\vec{k}} = \varepsilon^0_{\vec{k}} + \left\langle\vec{k}\left|v_{\text{eff}}\right|\vec{k}\right\rangle + \sum_{\vec{k}'\neq\vec{k}} \frac{\left|\left\langle\vec{k}\left|v_{\text{eff}}\right|\vec{k}'\right\rangle\right|^2}{\varepsilon^0_{\vec{k}} - \varepsilon^0_{\vec{k}'}} \tag{5.7}$$

and

$$\psi_{\vec{k}}(\vec{r}) = \left|\vec{k}\right\rangle + \sum_{\vec{k}' \neq \vec{k}} \frac{\left\langle \vec{k}' \middle| v_{\text{eff}} \middle| \vec{k} \right\rangle}{\varepsilon^0_{\vec{k}} - \varepsilon^0_{\vec{k}'}} \left|\vec{k}'\right\rangle \tag{5.8}$$

where

$$\left\langle \vec{k} \middle| v_{\text{eff}} \middle| \vec{k}' \right\rangle = \frac{1}{\Omega} \int v_{\text{eff}}(\vec{r}) e^{-i(\vec{k}-\vec{k}')\cdot\vec{r}} d^3r \tag{5.9}$$

Since $v_{\text{eff}}(\vec{r})$ is periodic in $\vec{r}$, it can be expanded as

$$v_{\text{eff}}(\vec{r}) = \sum_{\vec{G}} v(\vec{G}) \exp(i\vec{G}\cdot\vec{r}) \tag{5.10}$$

where $\vec{G}$ is a reciprocal lattice vector and the sum is over all the reciprocal lattice vectors. Since $v_{\text{eff}}(\vec{r})$ is real

$$v^*(\vec{G}) = v(-\vec{G}) \tag{5.11}$$

Substituting Equation 5.10 in Equation 5.9, we get

$$\left\langle \vec{k} \middle| v_{\text{eff}} \middle| \vec{k}' \right\rangle = \frac{1}{\Omega} \sum_{\vec{G}} v(\vec{G}) \int e^{-i(\vec{k}-\vec{k}'-\vec{G})\cdot\vec{r}} d^3r \tag{5.12}$$

Using the identity

$$\int e^{i\vec{k}\cdot\vec{r}} d^3r = \Omega \delta_{\vec{k},0} \tag{5.13}$$

we get

$$\begin{aligned} \left\langle \vec{k} \middle| v_{\text{eff}} \middle| \vec{k}' \right\rangle &= v(\vec{G}) \quad \text{if } \vec{k} - \vec{k}' = \vec{G} = \text{a reciprocal lattice vector} \\ &= 0 \qquad\;\; \text{otherwise} \end{aligned} \tag{5.14}$$

Using Equations 5.7, 5.11, and 5.14, we get

$$\varepsilon_{\vec{k}} = \varepsilon^0_{\vec{k}} + v(0) + \sum_{\vec{G} \neq 0} \frac{\left|v(\vec{G})\right|^2}{\varepsilon^0_{\vec{k}} - \varepsilon^0_{\vec{k}-\vec{G}}} \tag{5.15}$$

and

$$\psi_{\vec{k}}(\vec{r}) = \left|\vec{k}\right\rangle + \sum_{\vec{G} \neq 0} \frac{v(-\vec{G})}{\varepsilon^0_{\vec{k}} - \varepsilon^0_{\vec{k}-\vec{G}}} \,|\, (\vec{k} - \vec{G})\rangle \tag{5.16}$$

Equations 5.15 and 5.16 will give meaningful results only if the expansion in the reciprocal space converges rapidly. This implies that the coefficients $v(\vec{G})$ should fall off very rapidly as $\vec{G}$ increases. Also, there should be no degeneracy between the unperturbed states, that is, $\varepsilon^0_{\vec{k}} \neq \varepsilon^0_{\vec{k}-\vec{G}}$. Thus, our perturbation expansion will break down if

$$|\vec{k}| = |\vec{k} - \vec{G}| \tag{5.17}$$

This condition can be shown geometrically as in Figure 5.5. The figure shows that condition (5.17) will hold if $\vec{k}$ lies on the perpendicular bisector plane of $\vec{G}$. But these bisector planes define the boundaries of the BZ. Thus, our perturbation expansion will break down near the zone boundary. We note that Equation 5.17 is the Laue condition of x-ray diffraction (see, e.g., Ashcroft and Mermin 1976). The Laue condition is a consequence of wave propagation in crystals. So, what we have obtained is essentially a consequence of the wave nature of electrons.

So, what is happening near the zone boundary is a strong interference of incoming wave $|\vec{k}\rangle$ and diffracted wave $|\vec{k} - \vec{G}\rangle$. In Equation 5.16, it is implicitly assumed that the coefficient of $|\vec{k} - \vec{G}\rangle$ is much smaller than the coefficient of $|\vec{k}\rangle$. But near the zone boundary, this assumption breaks down; therefore, we must treat all the coefficients on an equal footing, implying that $\psi_{\vec{k}}(\vec{r})$ should be expanded as

$$\psi_{\vec{k}}(\vec{r}) = \sum_{\vec{G}} C_{\vec{k}-\vec{G}} \left|\vec{k} - \vec{G}\right\rangle \tag{5.18}$$

$$= \frac{1}{\Omega^{1/2}} \sum_{\vec{G}} C_{\vec{k}-\vec{G}}\, e^{i(\vec{k}-\vec{G})\cdot\vec{r}} \tag{5.19}$$

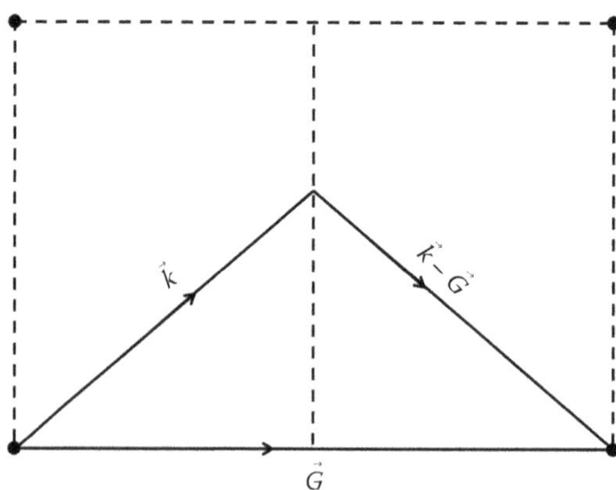

FIGURE 5.5
Geometrical representation of Equation 5.17 for a rectangular lattice. Reciprocal lattice points are shown by dots.

This is because the periodic potential $v_{\text{eff}}(\vec{r})$ can mix only those plane waves with $\vec{k}$ vectors differing by a reciprocal vector $\vec{G}$ (see Equation 5.14). For the sake of illustration, let us retain only two terms in the expansion (Equation 5.18), that is

$$\psi_{\vec{k}}(\vec{r}) = C_{\vec{k}}\left|\vec{k}\right\rangle + C_{\vec{k}-\vec{G}}\left|\vec{k} - \vec{G}\right\rangle \tag{5.20}$$

Let us substitute it in the KS equation

$$\left[\frac{-\hbar^2}{2m_e}\nabla^2 + v_{\text{eff}}(\vec{r})\right]\psi_{\vec{k}}(\vec{r}) = \varepsilon_{\vec{k}}\psi_{\vec{k}}(\vec{r}) \tag{5.21}$$

We get

$$\begin{aligned} &C_{\vec{k}}\,\varepsilon^0_{\vec{k}}\left|\vec{k}\right\rangle + C_{\vec{k}-\vec{G}}\,\varepsilon^0_{\vec{k}-\vec{G}}\left|\vec{k} - \vec{G}\right\rangle + C_{\vec{k}}\,v_{\text{eff}}(\vec{r})\left|\vec{k}\right\rangle + C_{\vec{k}-\vec{G}}\,v_{\text{eff}}(\vec{r})\left|\vec{k} - \vec{G}\right\rangle \\ &= \varepsilon_{\vec{k}}\,(C_{\vec{k}}\left|\vec{k}\right\rangle + C_{\vec{k}-\vec{G}}\left|\vec{k} - \vec{G}\right\rangle) \end{aligned} \tag{5.22}$$

By multiplying Equation 5.22 by $\langle\vec{k}'|$ from the left, we get

$$\begin{aligned} &C_{\vec{k}}\,\varepsilon^0_{\vec{k}}\,\delta_{\vec{k}',\vec{k}} + C_{\vec{k}-\vec{G}}\,\varepsilon^0_{\vec{k}-\vec{G}}\,\delta_{\vec{k}',\vec{k}-\vec{G}} + C_{\vec{k}}\,v(\vec{k}' - \vec{k}) + C_{\vec{k}-\vec{G}}\,v(\vec{k}' - \vec{k} + \vec{G}) \\ &= \varepsilon_{\vec{k}}\,C_{\vec{k}}\,\delta_{\vec{k}',\vec{k}} + \varepsilon_{\vec{k}}\,C_{\vec{k}-\vec{G}}\,\delta_{\vec{k}',\vec{k}-\vec{G}} \end{aligned} \tag{5.23}$$

Equation 5.23 gives two equations, one for $\vec{k}' = \vec{k}$ and the other for $\vec{k}' = \vec{k} - \vec{G}$. Using Equation 5.14 and setting $v(0) = 0$, we get

$$\begin{aligned} (\varepsilon^0_{\vec{k}} - \varepsilon_{\vec{k}})C_{\vec{k}} + C_{\vec{k}-\vec{G}}\,v(\vec{G}) &= 0 \\ v(-\vec{G})C_{\vec{k}} + (\varepsilon^0_{\vec{k}-\vec{G}} - \varepsilon_{\vec{k}})C_{\vec{k}-\vec{G}} &= 0 \end{aligned} \tag{5.24}$$

Equations 5.24 will have a nontrivial solution only if

$$\begin{vmatrix} \varepsilon^0_{\vec{k}} - \varepsilon_{\vec{k}} & v(\vec{G}) \\ v^*(\vec{G}) & \varepsilon^0_{\vec{k}-\vec{G}} - \varepsilon_{\vec{k}} \end{vmatrix} = 0 \tag{5.25}$$

This gives a quadratic equation in $\varepsilon_{\vec{k}}$, which has two roots given by

$$\varepsilon^{\pm}_{\vec{k}} = \frac{1}{2}\left[(\varepsilon^0_{\vec{k}} + \varepsilon^0_{\vec{k}-\vec{G}}) \pm \left\{(\varepsilon^0_{\vec{k}} - \varepsilon^0_{\vec{k}-\vec{G}})^2 + 4\left|v(\vec{G})\right|^2\right\}^{1/2}\right] \tag{5.26}$$

At the zone boundary, $|\vec{k}| = |\vec{k} - \vec{G}|$, and $\varepsilon^0_{\vec{k}} = \varepsilon^0_{\vec{k}-\vec{G}}$, so that from Equation 5.26, we get

$$\varepsilon^{\pm}_{\vec{k}} = \varepsilon^0_{\vec{k}} \pm \left| v(\vec{G}) \right| \tag{5.27}$$

So, at the zone boundary, we get a gap of $2|v(\vec{G})|$ between the energies of the states $\psi^+_{\vec{k}}(\vec{r})$ and $\psi^-_{\vec{k}}(\vec{r})$. At the zone center, that is, $\vec{k} = 0$, $\varepsilon^0_{\vec{k}} - \varepsilon^0_{\vec{k}-\vec{G}}$ is very large compared to $|v(\vec{G})|$, so that

$$\begin{aligned} \varepsilon^+_{\vec{k}} &= \varepsilon^0_{\vec{k}-\vec{G}} \\ \varepsilon^-_{\vec{k}} &= \varepsilon^0_{\vec{k}} \end{aligned} \tag{5.28}$$

Thus, the effect of perturbation is quite small at $\vec{k}$ points far away from the zone boundary; therefore, $\varepsilon_{\vec{k}}$ will remain close to a parabolic shape. For the one-dimensional case, $\varepsilon_{\vec{k}}$ is schematically shown in Figure 5.6. The same result is shown in the extended zone scheme in Figure 5.7, which shows more clearly how $\varepsilon_{\vec{k}}$ are obtained by perturbing the free electron energies $\varepsilon^0_{\vec{k}}$. From these figures, it is clear that there are some values of energy for which there are no allowed electron states and the one-electron energy spectrum has gaps. This is essentially the consequence of the wave nature of electrons, which gives rise to Bragg diffraction at the zone boundaries. We shall further discuss this from a different point of view when discussing the tight-binding method.

It turns out that the energy bands of several alkali metals such as Mg, Zn, Al, and so on resemble the NFE bands. This raises two very important questions. First, how can we view the electrons, which interact via Coulomb potential, as noninteracting particles? Second, how can we assume that the

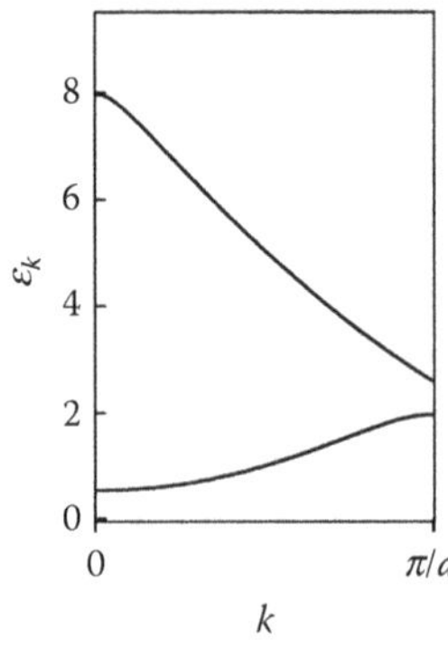

FIGURE 5.6
Energy bands in one dimension obtained from Equation 5.26 in reduced band scheme.

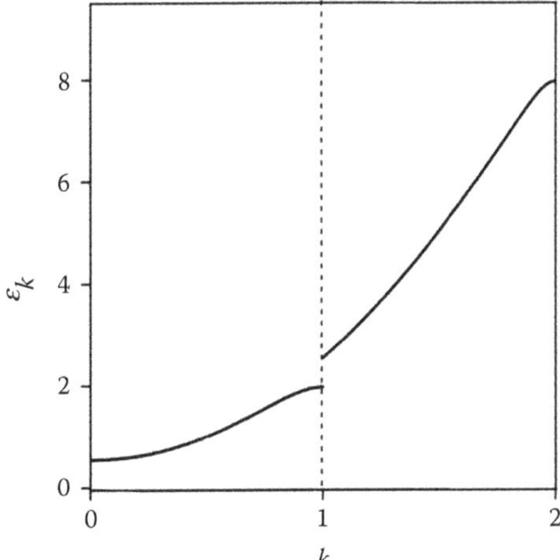

FIGURE 5.7
Energy bands obtained in one dimension from Equation 5.26 in extended zone scheme. k is given in units of π/a.

electron–ion interaction is weak? Landau answered the first question using the Fermi liquid theory (Landau 1957, 1959). The theory explains why some properties of the interacting electron gas resemble that of weakly interacting quasi-particles. We saw that the density functional theory also reduced the many-body problem by mapping it onto a system of noninteracting electrons. The second question is answered using the pseudopotential theory as we shall see later in this chapter.

5.4 Plane Wave Expansion Method

In the above treatment, we had included only two plane waves in the expansion of $\psi_k(\vec{r})$ in Equation 5.20. We can generalize the above treatment to the case where many plane waves are included in the expansion, that is

$$\psi_{\vec{k}}(\vec{r}) = \sum_{\vec{G}} C_{\vec{k}-\vec{G}} \left| \vec{k} - \vec{G} \right\rangle \tag{5.29}$$

Substituting this in Equation 5.21, we will get a set of equations, similar to Equation 5.24, that is

$$(\varepsilon^0_{\vec{k}-\vec{G}} - \varepsilon_{\vec{k}})C_{\vec{k}-\vec{G}} + \sum_{\vec{G}'} v(\vec{G}' - \vec{G})C_{\vec{k}-\vec{G}'} = 0 \tag{5.30}$$

These equations will have nontrivial solutions only if

$$\det\left|(\varepsilon^0_{\vec{k}-\vec{G}} - \varepsilon_{\vec{k}})\delta_{\vec{G},\vec{G}'} + v(\vec{G}' - \vec{G})\right| = 0 \tag{5.31}$$

We can obtain the band structure, that is, $\varepsilon_{\vec{k}}$ versus $\vec{k}$ by solving this equation. The expansion (Equation 5.29) is called the plane wave expansion and the method the plane wave expansion method. In principle, Equation 5.31 is exact but then one should take infinite number of plane waves in the expansion (Equation 5.29). Thus, the suitability of the method for practical calculations depends on the size of the determinant in Equation 5.31, which in turn depends on how quickly $v(\vec{G})$ falls as $\vec{G}$ increases. If we look at Figure 4.1, we note that $v_{\text{eff}}(\vec{r})$ has sharp variations near the ions, that is, when $\vec{r}$ is small. This would imply that its Fourier components will be large for large $\vec{G}$ and would not fall quickly as $\vec{G}$ increases, in general. Thus, the method is not suitable for practical calculation if one uses the actual potential, but if one uses the pseudopotential instead, this method is very efficient and useful, as we shall see later.

It should be noted that $\psi_{\vec{k}}(\vec{r})$ given by Equation 5.29 satisfies Bloch's theorem. This can be viewed as an alternate proof of Bloch's theorem.

5.5 Tight-Binding Method

We just saw how energy bands are formed when the electrons move in a weak periodic potential. Now, we shall look into this problem from a different point of view. We shall try to understand how energy bands are formed when atoms are brought closer, to form a crystal.

Let us suppose that we are given a large number of sodium atoms and we wish to form a solid using these atoms. We first arrange these atoms on a bcc lattice with a large lattice spacing (say 1 mm) and then slowly squeeze this system of atoms, retaining the lattice structure. When the atoms are far apart, all the electrons are localized at their respective atoms and there is a negligible overlap between the wave functions of two electrons on the neighboring sites. As we decrease the lattice spacing, the overlap between the wave functions of the outermost electrons on the neighboring atoms begins to increase. This overlap will be considerable when the lattice spacing is reduced to small values, say about the lattice constant of sodium. Now, an outermost 3*s* electron is no longer localized on a particular atom, but has a good probability of jumping on to other atoms. Therefore, a single atomic orbital $\phi_a(\vec{r})$ cannot represent this electron; rather, we should form a linear combination of atomic orbitals centered on different sites for its correct description, that is

$$\psi_{\vec{k}}(\vec{r}) = \sum_{\ell} C_{\vec{k}\ell}\, \phi_a(\vec{r} - \vec{R}_\ell) \tag{5.32}$$

For simplicity, we assume $\phi_a(\vec{r})$ to be a nondegenerate state such as s-state. Generalization to include more states will be done later.

Because $\psi_{\vec{k}}(\vec{r})$ in Equation 5.32 represents a wave function of an electron in a solid, it must satisfy Bloch's theorem, that is

$$\psi_{\vec{k}}(\vec{r} + \vec{R}) = e^{i\vec{k}\cdot\vec{R}}\, \psi_{\vec{k}}(\vec{r}) \tag{5.33}$$

This is satisfied only if $C_{\vec{k}\ell} = e^{i\vec{k}\cdot\vec{R}_\ell}$, so that

$$\psi_{\vec{k}}(\vec{r}) = \sum_{\ell} e^{i\vec{k}\cdot\vec{R}_\ell}\, \phi_a(\vec{r} - \vec{R}_\ell) \tag{5.34}$$

As a first approximation, the energy $\varepsilon_{\vec{k}}$ of the state $\psi_{\vec{k}}(\vec{r})$ is given by the expectation value of Hamiltonian H, that is

$$\varepsilon_{\vec{k}} = \frac{\int \psi^*_{\vec{k}}(\vec{r}) \left[-\frac{\hbar^2}{2m_e}\nabla^2 + v(\vec{r}) \right] \psi_{\vec{k}}(\vec{r})\, d^3r}{\int \psi^*_{\vec{k}}(\vec{r})\psi_{\vec{k}}(\vec{r})\, d^3r} \tag{5.35}$$

Using Equation 5.34, the denominator in Equation 5.35 can be written as

$$\begin{aligned} \int \psi^*_{\vec{k}}(\vec{r})\psi_{\vec{k}}(\vec{r})\, d^3r &= \sum_{\ell\ell'} e^{i\vec{k}\cdot(\vec{R}_\ell - \vec{R}_{\ell'})} \int \phi^*_a(\vec{r} - \vec{R}_{\ell'})\phi_a(\vec{r} - \vec{R}_\ell)\, d^3r \\ &= \sum_{\ell} \int \left|\phi_a(\vec{r} - \vec{R}_\ell)\right|^2 d^3r + \sum_{\substack{\ell\ell' \\ \ell' \neq \ell}} e^{i\vec{k}\cdot(\vec{R}_\ell - \vec{R}_{\ell'})} \\ &\quad \times \int \phi^*_a(\vec{r} - \vec{R}_{\ell'})\phi_a(\vec{r} - \vec{R}_\ell)\, d^3r \end{aligned} \tag{5.36}$$

Since $\phi_a(\vec{r})$ are atomic orbitals, the overlap between the atomic orbitals centered on different sites will be small as compared to $|\phi_a(\vec{r})|^2$; therefore, we neglect the second term in Equation 5.36. Also

$$\int \left|\phi_a(\vec{r} - \vec{R}_\ell)\right|^2 d^3r = \int \left|\phi_a(\vec{r})\right|^2 d^3r = 1 \tag{5.37}$$

Thus, from Equation 5.36, we get

$$\int \psi_{\vec{k}}^{*}(\vec{r})\psi_{\vec{k}}(\vec{r})d^3r = \sum_{\ell} 1 = N$$

where N is the total number of atoms in the solid. Equation 5.35 can now be written as

$$\varepsilon_{\vec{k}} = \frac{1}{N}\sum_{\ell\ell'} e^{i\vec{k}\cdot(\vec{R}_{\ell}-\vec{R}_{\ell'})}\int \phi_a^{*}(\vec{r}-\vec{R}_{\ell'})\left[-\frac{\hbar^2}{2m_e}\nabla^2 + v(\vec{r})\right]\phi_a(\vec{r}-\vec{R}_{\ell})d^3r$$

Let us write $\vec{R}_{\ell} - \vec{R}_{\ell'} = \vec{R}_j$ or $\vec{R}_{\ell'} = \vec{R}_{\ell} - \vec{R}_j$. Thus

$$\varepsilon_{\vec{k}} = \frac{1}{N}\sum_{j} e^{i\vec{k}\cdot\vec{R}_j}\sum_{\ell}\int \phi_a^{*}(\vec{r}-\vec{R}_{\ell}+\vec{R}_j)\left[-\frac{\hbar^2}{2m_e}\nabla^2 + v(\vec{r})\right]\phi_a(\vec{r}-\vec{R}_{\ell})\,d^3r \qquad (5.38)$$

Because Hamiltonian H is translationally invariant, the integral in Equation 5.38 will depend only on the relative distance between the two sites on which the orbitals are centered. Thus, for a fixed j, the integral in Equation 5.38 is the same for each ℓ. Therefore, the summation over ℓ in Equation 5.38 simply gives a multiplicative factor N. Thus, Equation 5.38 can be written as

$$\varepsilon_{\vec{k}} = \sum_{j} e^{i\vec{k}\cdot\vec{R}_j}\varepsilon_j \qquad (5.39)$$

where

$$\varepsilon_j = \int \phi_a^{*}(\vec{r}+\vec{R}_j)\left[-\frac{\hbar^2}{2m_e}\nabla^2 + v(\vec{r})\right]\phi_a(\vec{r})\,d^3r \qquad (5.40)$$

Equation 5.39 proves a very general point regarding the energy bands. It shows that the energy bands are periodic in reciprocal space, that is

$$\varepsilon_{\vec{k}+\vec{G}} = \varepsilon_{\vec{k}} \qquad (5.41)$$

where $\vec{G}$ is a reciprocal lattice vector. This means the band structure is repeated as you go from one cell of the reciprocal space to another. Sometimes, it may be useful to plot the band structure in this way, which is known as the "repeated zone scheme."

Let H_a be the Hamiltonian for an isolated single atom with atomic potential $v_a(\vec{r})$, that is

$$H_a = -\frac{\hbar^2}{2m_e}\nabla^2 + v_a(\vec{r}) \tag{5.42}$$

and $\phi_a(\vec{r})$ be the solution of the equation

$$H_a\,\phi_a(\vec{r}) = \varepsilon_a\,\phi_a(\vec{r}) \tag{5.43}$$

where ε_a is the atomic energy level. Now, let us evaluate the integral in Equation 5.40 in terms of ε_a. Thus

$$\left[-\frac{\hbar^2}{2m_e}\nabla^2 + v(\vec{r})\right]\phi_a(\vec{r}) = \left[-\frac{\hbar^2}{2m_e}\nabla^2 + v_a(\vec{r})\right]\phi_a(\vec{r}) + \left[v(\vec{r}) - v_a(\vec{r})\right]\phi_a(\vec{r})$$

$$= \varepsilon_a\,\phi_a(\vec{r}) + \left[v(\vec{r}) - v_a(\vec{r})\right]\phi_a(\vec{r}) \tag{5.44}$$

Substituting Equation 5.44 into Equation 5.39, we get

$$\varepsilon_{\vec{k}} = \sum_j e^{i\vec{k}\cdot\vec{R}_j}\int \phi_a^*(\vec{r} + \vec{R}_j)\left[\varepsilon_a\,\phi_a(\vec{r}) + \{v(\vec{r}) - v_a(\vec{r})\}\phi_a(\vec{r})\right] d^3r$$

$$= \varepsilon_0 + \sum_{j\neq 0} e^{i\vec{k}\cdot\vec{R}_j}\gamma_j \tag{5.45}$$

where

$$\varepsilon_0 = \varepsilon_a + \int \phi_a^*(\vec{r})\left[v(\vec{r}) - v_a(\vec{r})\right]\phi_a(\vec{r})\, d^3r \tag{5.46}$$

and

$$\gamma_j = \int \phi_a^*(\vec{r} + \vec{R}_j)\left[v(\vec{r}) - v_a(\vec{r})\right]\phi_a(\vec{r})\, d^3r \tag{5.47}$$

Here, γ_j is called overlap integral or hopping integral. For an s-state, $\phi_a(\vec{r})$ depends only on the magnitude of $\vec{r}$ and thus, $\gamma_j = \gamma_{-j}$. Also, for simplicity, let us neglect all the overlap integrals except the nearest-neighbor integrals. Thus, Equation 5.45 can be written as

$$\varepsilon_{\vec{k}} = \varepsilon_0 + \sum_{\substack{j\neq 0\\ nn}} \gamma_j e^{i\vec{k}\cdot\vec{R}_j} \tag{5.48}$$

where *nn* indicates that the sum is only over the nearest neighbors.

Let us evaluate Equation 5.48 for a simple cubic lattice. We have six nearest neighbors at ($\pm a$, 0, 0), (0, $\pm a$, 0), and (0, 0, $\pm a$) positions, where a is the lattice constant. Thus, $\varepsilon_{\vec{k}}$ for a simple cubic lattice are given by

$$\begin{aligned}\varepsilon_{\vec{k}} &= \varepsilon_0 + \gamma\left[e^{ik_xa} + e^{-ik_xa} + e^{ik_ya} + e^{-ik_ya} + e^{ik_za} + e^{-ik_za}\right] \\ &= \varepsilon_0 + 2\gamma(\cos k_xa + \cos k_ya + \cos k_za) \end{aligned} \tag{5.49}$$

$\varepsilon_{\vec{k}}$ versus $\vec{k}$ curve for a simple cubic lattice for $\vec{k}$ along various symmetry directions in the BZ is shown in Figure 5.8. We see that the band is flatter near the zone boundary and near the zone center as was the case with the NFE model. The bandwidth in this case is 12γ. In general, if there are z nearest neighbors, the bandwidth in tight-binding approximation is $2z\gamma$. If γ is large, we have a broadband, and if γ is small, we get a narrow band. Since γ increases by decreasing lattice spacing a, the band will get broader as we decrease a. The density of states (DOS) for a cubic single-band tight-binding model is shown in Figure 5.9. One can clearly see the van Hove singularities in the DOS around energies $\varepsilon = -1, -0.4, 0.4$, and 1.0.

Based on the above discussion, we can now answer the question with which we started, that is, how energy bands are formed when the atoms are brought together to form a solid. When the atoms are far apart, the electrons are localized at the atoms with energies ε_1, ε_2, and so on. When the atoms are brought closer, the atomic wave functions centered on different sites begin to overlap and the atomic levels ε_1, ε_2, and so on begin to broaden to form energy bands as shown in Figure 5.10. The bands are, therefore, sometimes labeled as s-band, d-band, and so on to identify the atomic levels from which they are derived. Note that each band contains N states. So, when the atoms are far apart, there will be N electrons with say energy ε_1, that is, there is

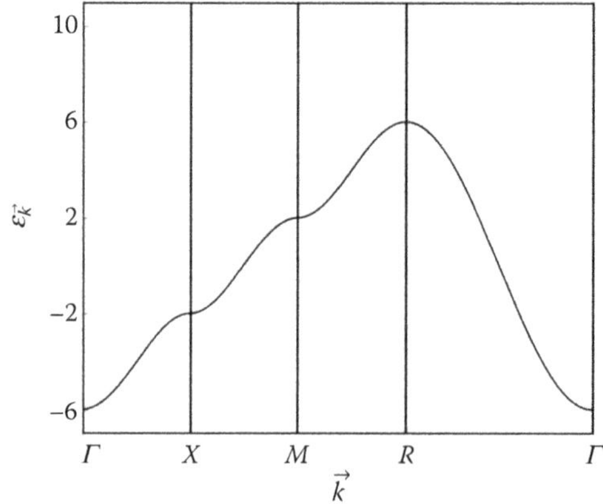

FIGURE 5.8
Energy bands for a cubic single s-band tight-binding model.

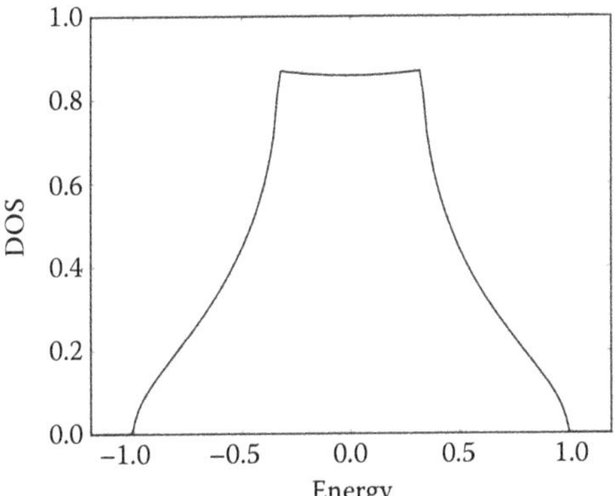

FIGURE 5.9
Schematic DOS for a cubic single-band tight-binding model.

N-fold degeneracy (although for a single atom, it is a nondegenerate state). This degeneracy is lifted when the bands are formed.

Sometimes, it may happen that by bringing the atoms closer, two different bands broaden so much as to overlap each other. Clearly, the above scheme is not applicable now because we constructed $\psi_{\vec{k}}(\vec{r})$ in Equation 5.34 using atomic orbitals corresponding to single-energy ε_a. In general, we should construct $\psi_{\vec{k}}(\vec{r})$ such that the mixing of atomic orbitals corresponding to different energies is possible. The most general Bloch function, which can be constructed using atomic orbitals, is

$$\psi_{\vec{k}}(\vec{r}) = \sum_{\ell} e^{i\vec{k}\cdot\vec{R}_\ell} \sum_{j} \alpha_j \phi_{aj}(\vec{r} - \vec{R}_\ell) \tag{5.50}$$

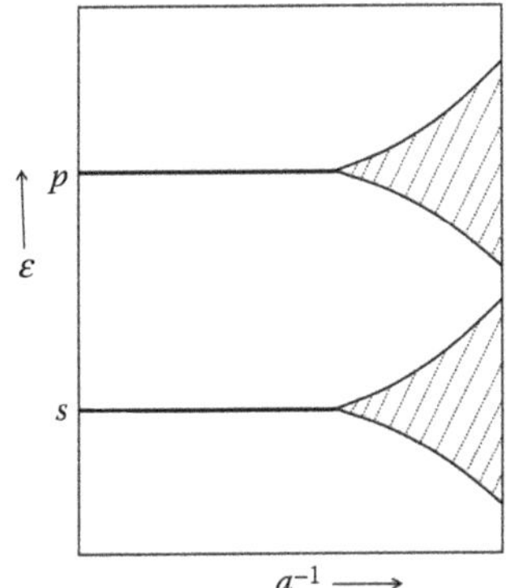

FIGURE 5.10
Schematic showing broadening of s and p levels as the lattice constant a changes.

where the sum over j runs over all the orbitals and α_j are the coefficients yet to be determined. Now, there is no assumption like nondegeneracy of the atomic orbitals as was the case in Equation 5.32. To calculate $\varepsilon_{\vec{k}}$, Equation 5.50 for $\psi_{\vec{k}}(\vec{r})$ is substituted in the KS equation

$$H\psi_{\vec{k}}(\vec{r}) = \varepsilon_{\vec{k}}\,\psi_{\vec{k}}(\vec{r}) \tag{5.51}$$

This then yields a set of linear equations in coefficients α_j in terms of integrals like Equation 5.40. By demanding that these equations have a nontrivial solution, one gets an equation for $\varepsilon_{\vec{k}}$, as was the case with the NFE method. This method of calculating energy bands is known as the linear combination of atomic orbitals (LCAO) method.

The tight-binding method would work well for electrons that are tightly bound to atoms such as core electrons or d-electrons in transition metals. This is because for such cases, one can neglect overlap integrals beyond the nearest neighbors. For extended states, this will be a poor approximation. In that case, we will have to include more and more overlap integrals making the method more complicated. Because of its simplicity, the tight-binding method is a starting point for many model calculations. This includes the study of impurity states, disorder, and electron correlations. As an example, we shall briefly discuss the Hubbard model.

5.6 Hubbard Model

The Hubbard model is the simplest model that takes into account the effect of electron–electron interaction on a lattice as studied in detail by Hubbard (1963). This model uses the tight-binding model and is expressed in second quantized notations as

$$H = \sum_{ij} t_{ij}\, c_{i\sigma}^{+}\, c_{j\sigma} + \mathrm{U}\sum_{i} n_{i\uparrow}\, n_{i\downarrow} \tag{5.52}$$

Here, t_{ij} is called the hopping parameter and is the same as the overlap integral as defined in Equation 5.47; $c_{i\sigma}^{+}$ and $c_{j\sigma}$ are, respectively, creation and annihilation operators that create or annihilate an electron of spin σ at ith or jth site. A nondegenerate s-type band is assumed. Since the orbitals are nondegenerate, two electrons of the same spin cannot stay at the same site due to the Pauli exclusion principle. U is the electron–electron repulsion between antiparallel spin electrons at the same site and $n_{i\downarrow}$ is the number operator that gives the number of electrons of $\downarrow$ spins at site i. Although the model is very simple, its exact solution does not exist except in one dimension. But many approximate solutions exist, which reveal some very interesting features of this model. Much work on magnetism and high-temperature superconductors has been done starting with this model.

To appreciate the effect of electron correlations, let us go back to the discussion of the tight-binding model, which was discussed earlier. Let us again consider a system of Na atoms separated by lattice spacing a. When a is large, the application of the tight-binding model will give a narrow *half-filled* band. This would imply that the system will be a metal and not an insulator, which should be expected for a large a. This is due to the neglect of the electron correlations in the band theory. The Hubbard model is able to explain the insulating behavior as it includes on-site electron–electron repulsion U. On the basis of the Hubbard model, this can be understood as follows. The Hubbard Hamiltonian (Equation 5.52) has two competing terms, the first corresponding to the reduction in energy due to the delocalization of one electron wave function and the second corresponding to on-site electron–electron repulsion. Both parameters t_{ij} and U change as the lattice spacing is decreased. For a large a, the second term dominates and as a result the electrons tend to remain localized at their sites and the system becomes an insulator. For a small a, the first term in Equation 5.52 dominates and the electrons gain energy by delocalizing and the system prefers to be a metal. Therefore, at some intermediate value of a, there will be a transition from insulator to metal known as Mott transition. Mott had predicted in 1949 that such a transition should occur in a solid on changing the lattice spacing (Mott 1949). For his work on the electronic structure of magnetic and disordered systems, Mott was awarded the Nobel Prize in 1977.

5.7 Wannier Functions

We saw in Section 5.5 that the Bloch wave function $\psi_{\vec{k}}(\vec{r})$ can be expanded in terms of atomic orbital ϕ_a as

$$\psi_{\vec{k}}(\vec{r}) = \frac{1}{\sqrt{N}} \sum_{\ell} e^{i\vec{k}\cdot\vec{R}_\ell} \phi_a(\vec{r} - \vec{R}_\ell) \tag{5.53}$$

The problem, which one faces while using this expression, is that $\phi_a(\vec{r} - \vec{R}_\ell)$ centered on different sites are not orthogonal functions, that is, there is a finite overlap between them. In many problems, orthogonal localized functions on different sites are needed. Gregory Wannier (1937, 1962) introduced such functions, which are known as Wannier functions. Wannier function $\phi_n^W(\vec{r} - \vec{R}_\ell)$ corresponding to band n and centered on site $\vec{R}_\ell$ is defined as

$$\phi_n^W(\vec{r} - \vec{R}_\ell) = \frac{1}{\sqrt{N}} \sum_{\vec{k}} e^{-i\vec{k}\cdot\vec{R}_\ell} \psi_{\vec{k}n}(\vec{r})$$

or

$$\psi_{\vec{k}n}(\vec{r}) = \frac{1}{\sqrt{N}} \sum_{\ell} e^{i\vec{k}\cdot\vec{R}_\ell} \phi_n^W(\vec{r} - \vec{R}_\ell) \tag{5.54}$$

where N is the number of atoms in the solid. It can be easily shown that $\phi_n^W(\vec{r} - \vec{R}_\ell)$ form an orthonormal set.

$$\begin{aligned}
\int \phi_n^{W*}(\vec{r} - \vec{R}_\ell)\phi_n^W(\vec{r} - \vec{R}_{\ell'})\, d^3r &= \frac{1}{N}\sum_{\vec{k},\vec{k}'} e^{i\vec{k}\cdot\vec{R}_\ell} e^{-i\vec{k}'\cdot\vec{R}_{\ell'}} \int \psi_{\vec{k}n}^*(\vec{r})\psi_{\vec{k}'n}(\vec{r})\, d^3r \\
&= \frac{1}{N}\sum_{\vec{k},\vec{k}'} e^{i\vec{k}\cdot\vec{R}_\ell} e^{-i\vec{k}'\cdot\vec{R}_{\ell'}} \delta_{\vec{k},\vec{k}'} \\
&= \frac{1}{N}\sum_{\vec{k}} e^{i\vec{k}\cdot(\vec{R}_\ell - \vec{R}_{\ell'})} \\
&= \delta_{\ell,\ell'}
\end{aligned} \tag{5.55}$$

However, Wannier functions are not unique. This is because the overall phase of each $\psi_{\vec{k}n}(\vec{r})$ is arbitrary, that is, all physically observable quantities such as charge density remain invariant under a gauge transformation

$$\psi_{\vec{k}n}(\vec{r}) \rightarrow e^{i\theta_{\vec{k}}}\, \psi_{\vec{k}n}(\vec{r}) \tag{5.56}$$

where $\theta_{\vec{k}}$ is arbitrary. This makes $\phi_n^W(\vec{r} - \vec{R}_\ell)$ dependent on $\theta_{\vec{k}}$ from Equation 5.54 and thus nonunique. Because of this nonuniqueness, these were not used in practical calculations for a long time. However, it is now realized that the Wannier functions can be used very effectively in electronic structure calculations in a gauge-invariant manner. Recently, these functions have been used in the calculation of polarization in solids (Resta 1994).

5.8 Orthogonalized Plane Wave (OPW) Method

We can divide all the electrons in a solid into two categories—core electrons and valence electrons. The core electrons are tightly bound to the nuclei and their wave functions are well localized about lattice sites. The valence electrons, on the other hand, are loosely bound to the nuclei and move throughout the solid. Thus, we expect the wave function of a valence electron to be plane wave-like in the interstitial region between the atoms while it must show many oscillations in the core region. Because of these oscillations, the convergence of a plane wave expansion for $\psi_{\vec{k}}(\vec{r})$ is poor as we have seen earlier. If somehow these oscillations can be taken into account in choosing the basis set, the convergence of the expansion in terms of the new basis set can be greatly improved. This is the main idea of the OPW method, due to Herring (1940).

Consider a crystal with one atom per unit cell. We first focus on core electrons and assume that we have solved the KS equation for a single atom. Let us denote the atomic wave functions by $\phi_{aj}(\vec{r})$ and atomic energies by ε_{aj}, where j stands for the usual atomic quantum numbers n, l, and m. From $\phi_{aj}(\vec{r})$, we can construct a tight-binding function $\phi_{\vec{k}j}(\vec{r})$

$$\phi_{\vec{k}j}(\vec{r}) = \frac{1}{\sqrt{N}} \sum_{\ell} e^{i\vec{k}\cdot\vec{R}_\ell} \phi_{aj}(\vec{r} - \vec{R}_\ell) \tag{5.57}$$

We saw in Section 5.5 that function (5.57) satisfies Bloch's theorem and gives a good description of the tightly bound electrons in the crystal. As the overlap between the two atomic wave functions centered on neighboring sites is negligibly small, we shall assume that function (5.57) is a solution of the KS equation of the whole crystal with energy ε_{aj}.

We know that all the solutions of the KS equation corresponding to different energies should be orthogonal to each other. This implies that the wave function $\psi_{\vec{k}}(\vec{r})$ of a valence electron should be orthogonal to all wave functions of the core electrons, that is, $\phi_{\vec{k}j}(\vec{r})$. Thus, if we can find an expansion set for $\psi_{\vec{k}}(\vec{r})$, which has been made orthogonal to all core functions $\phi_{\vec{k}j}(\vec{r})$, we shall get a rapid convergence. We know that a plane wave can be made orthogonal to $\phi_{\vec{k}j}(\vec{r})$ by the Gram–Schmidt orthogonalization process. Let us denote a function constructed in this manner by $\chi_{\vec{k}}(\vec{r})$, which we shall call an orthogonalized plane wave.

$$\chi_{\vec{k}}(\vec{r}) = \frac{1}{\sqrt{\Omega}} e^{i\vec{k}\cdot\vec{r}} - \sum_{j} \mu_{\vec{k}j} \phi_{\vec{k}j}(\vec{r}) \tag{5.58}$$

or in short

$$\left|\chi_{\vec{k}}\right\rangle = \left|\vec{k}\right\rangle - \sum_{j} \mu_{\vec{k}j} \left|\vec{k}j\right\rangle \tag{5.59}$$

where in Equation 5.59 we have used Dirac's bra and ket notations and the meaning of various symbols is quite obvious. In Equation 5.59, coefficient $\mu_{\vec{k}j}$ are determined by requiring that the OPW is orthogonal to $|\vec{k}j\rangle$, that is

$$\left\langle \vec{k}j \middle| \chi_{\vec{k}} \right\rangle = 0 \tag{5.60}$$

or

$$\left\langle \vec{k}j \middle| \vec{k} \right\rangle - \sum_{j'} \mu_{\vec{k}j'} \left\langle \vec{k}j \middle| \vec{k}j' \right\rangle = 0$$

or

$$\mu_{\vec{k}j} = \left\langle \vec{k}j \middle| \vec{k} \right\rangle \tag{5.61}$$

Here, we have assumed that $|\vec{k}j\rangle$ form an orthonormal set. Substituting Equation 5.61 into Equation 5.59, we get

$$\left| \chi_{\vec{k}} \right\rangle = \left| \vec{k} \right\rangle - \sum_{j} \left\langle \vec{k}j \middle| \vec{k} \right\rangle \left| \vec{k}j \right\rangle \tag{5.62}$$

One can check that the OPW $|\chi_{\vec{k}}\rangle$ satisfies Bloch's theorem with wave vector $\vec{k}$.

We know that $|\vec{k}j\rangle$ are highly localized functions at the lattice sites, that is, they have a large amplitude near the lattice sites, but as we move out of the core region, the amplitude becomes very small. Thus, near the lattice sites, an OPW will have many oscillations, but away from the atomic core, it will be a plane wave-like function. Thus, this is a correct type of function we were looking for, but it is not the solution of the KS equation for the solid. The solution $\psi_{\vec{k}}(\vec{r})$ can be constructed using a combination of such OPWs, which, in order to satisfy Bloch's theorem, must be of the form

$$\psi_{\vec{k}}(\vec{r}) = \sum_{\vec{G}} a_{\vec{k}-\vec{G}}\, \chi_{\vec{k}-\vec{G}}(\vec{r}) \tag{5.63}$$

We can calculate $\varepsilon_{\vec{k}}$ from Equation 5.63 in the same way as we did in the case of the plane wave expansion method. We substitute $\psi_{\vec{k}}(\vec{r})$ from Equation 5.63 in the KS equation, which gives linear algebraic equations in coefficients $a_{\vec{k}-\vec{G}}$. The requirement of a nontrivial solution then gives a secular equation, the solution of which gives $\varepsilon_{\vec{k}}$. It turns out that the matrix elements $v(\vec{G}-\vec{G}')$ multiplying $a_{\vec{k}-\vec{G}'}$ decay rapidly with increasing $\vec{G}-\vec{G}'$ in contrast to the NFE case. Thus, a relatively small set of OPWs is required for the band structure calculation.

5.9 Pseudopotential Method

We remarked in the discussion on NFE method that energy bands of several metals such as sodium, potassium, and aluminum were close to NFE bands.

This puzzled us because the electron–ion potential near the ions cannot be considered as weak. The answer to this puzzle is provided by the idea of pseudopotential, which we shall discuss now.

Our discussion is based on Phillips and Kleinman's work (1959), which starts with the OPW approach. We saw that an OPW can be written as

$$\left|\chi_{\vec{k}}\right\rangle = \left|\vec{k}\right\rangle - \sum_{j} \left\langle \vec{k}j \middle| \vec{k} \right\rangle \left| \vec{k}j \right\rangle \tag{5.64}$$

Let us define a projection operator P

$$P = \sum_{j} \left|\vec{k}j\right\rangle \left\langle \vec{k}j \right| \tag{5.65}$$

The operator P projects any function onto the core states. Thus, the OPW (Equation 5.64) can be written as

$$\begin{aligned} \left|\chi_{\vec{k}}\right\rangle &= \left|\vec{k}\right\rangle - P\left|\vec{k}\right\rangle \\ &= (1 - P)\left|\vec{k}\right\rangle \end{aligned} \tag{5.66}$$

Now, we expand $\psi_{\vec{k}}(\vec{r})$ in terms of OPWs

$$\psi_{\vec{k}}(\vec{r}) = \sum_{\vec{G}} a_{\vec{k}-\vec{G}} \left|\chi_{\vec{k}-\vec{G}}\right\rangle \tag{5.67}$$

$$\begin{aligned} &= \sum_{\vec{G}} a_{\vec{k}-\vec{G}} (1 - P) \left|\vec{k} - \vec{G}\right\rangle \\ &= (1 - P) \sum_{\vec{G}} a_{\vec{k}-\vec{G}} \left|\vec{k} - \vec{G}\right\rangle \\ &= (1 - P)\phi_{\vec{k}}(\vec{r}) \end{aligned} \tag{5.68}$$

where

$$\phi_{\vec{k}}(\vec{r}) = \sum_{\vec{G}} a_{\vec{k}-\vec{G}} \left|\vec{k} - \vec{G}\right\rangle \tag{5.69}$$

$\phi_{\vec{k}}(\vec{r})$ is called the pseudowave function. In contrast to the true wave function $\psi_{\vec{k}}(\vec{r})$, it is a smooth function because the expansion in Equation 5.69

contains only a few plane waves. Outside the core region, it is equal to the true wave function because the projection operation P in Equation 5.65 is zero outside the core region.

Let us now find the equation satisfied by the pseudowave function $\phi_{\vec{k}}(\vec{r})$. For this, we substitute $\psi_{\vec{k}}(\vec{r})$ from Equation 5.68 in the KS equation

$$\left[-\frac{\hbar^2}{2m_e}\nabla^2 + v(\vec{r})\right]\psi_{\vec{k}}(\vec{r}) = \varepsilon_{\vec{k}}\,\psi_{\vec{k}}(\vec{r})$$

and we get

$$-\frac{\hbar^2}{2m_e}\nabla^2\phi_{\vec{k}}(\vec{r}) + v(\vec{r})\phi_{\vec{k}}(\vec{r}) - \left[-\frac{\hbar^2}{2m_e}\nabla^2 + v(\vec{r})\right]P\,\phi_{\vec{k}}(\vec{r})$$

$$= \varepsilon_{\vec{k}}\phi_{\vec{k}}(\vec{r}) - \varepsilon_{\vec{k}}P\,\phi_{\vec{k}}(\vec{r})$$

or

$$-\frac{\hbar^2}{2m_e}\nabla^2\phi_{\vec{k}}(\vec{r}) + \left[v(\vec{r}) - \left\{-\frac{\hbar^2}{2m_e}\nabla^2 + v(\vec{r})\right\}P + \varepsilon_{\vec{k}}\,P\right]\phi_{\vec{k}}(\vec{r}) = \varepsilon_{\vec{k}}\,\phi_{\vec{k}}(\vec{r})$$

or

$$-\frac{\hbar^2}{2m_e}\nabla^2\phi_{\vec{k}}(\vec{r}) + W\,\phi_{\vec{k}}(\vec{r}) = \varepsilon_{\vec{k}}\,\phi_{\vec{k}}(\vec{r}) \tag{5.70}$$

where W is known as pseudopotential and is given by

$$W = v(\vec{r}) - \left[-\frac{\hbar^2}{2m_e}\nabla^2 + v(\vec{r})\right]P + \varepsilon_{\vec{k}}\,P \tag{5.71}$$

The second term in Equation 5.71 can be written as

$$\left[-\frac{\hbar^2}{2m_e}\nabla^2 + v(\vec{r})\right]P = \left[-\frac{\hbar^2}{2m_e}\nabla^2 + v(\vec{r})\right]\sum_j \left|\vec{k}j\right\rangle\left\langle\vec{k}j\right|$$

$$= \sum_j \varepsilon_{aj}\left|\vec{k}j\right\rangle\left\langle\vec{k}j\right| \tag{5.72}$$

where ε_{aj} are the core electrons' energies. Substituting Equation 5.72 in Equation 5.71, we get

$$W = v(\vec{r}) + \sum_j (\varepsilon_{\vec{k}} - \varepsilon_{aj}) \left| \vec{k}j \right\rangle \left\langle \vec{k}j \right| \tag{5.73}$$

We see from Equation 5.70 that $\phi_{\vec{k}}(\vec{r})$ satisfies the KS equation with pseudopotential W with the same eigenvalues of the original KS equation. Note that no approximation has been made in deriving Equation 5.70.

The pseudopotential W is a weak potential. This can be seen by looking at matrix element

$$\left\langle \phi_{\vec{k}} \right| W \left| \phi_{\vec{k}} \right\rangle = \left\langle \phi_{\vec{k}} \right| v \left| \phi_{\vec{k}} \right\rangle + \sum_j (\varepsilon_{\vec{k}} - \varepsilon_{aj}) \left| \left\langle \phi_{\vec{k}} \middle| \vec{k}j \right\rangle \right|^2 \tag{5.74}$$

Since valence electron energies always lie above the core energies (i.e., $\varepsilon_{\vec{k}} > \varepsilon_{aj}$), the second term in Equation 5.74 is always positive. Since $v(\vec{r})$ is attractive, the first term will be negative, which will be partially canceled by the second term. This statement is sometimes referred to as the cancellation theorem. Thus, the pseudopotential W is a weak potential in contrast to $v(\vec{r})$, which is quite strong near ion cores. This is an important result and explains why for some materials the NFE method works so well.

The pseudopotential W has some peculiar features that the true potential $v(\vec{r})$ does not have. First, the pseudopotential is a nonlocal potential, that is, it does not appear in Equation 5.70 as a simple multiplicative function of $\vec{r}$ only.

$$W \phi_{\vec{k}}(\vec{r}) = v(\vec{r})\phi_{\vec{k}}(\vec{r}) + \sum_j (\varepsilon_{\vec{k}} - \varepsilon_{aj}) \left| \vec{k}j \right\rangle \left\langle \vec{k}j \middle| \phi_{\vec{k}} \right\rangle$$

$$= v(\vec{r})\phi_k(\vec{r}) + \sum_j (\varepsilon_{\vec{k}} - \varepsilon_{aj}) \phi_{\vec{k}j}(\vec{r}) \int \phi^*_{\vec{k}j}(\vec{r}\,') \phi_{\vec{k}}(\vec{r}\,')\, d^3r'$$

To examine this further, let us write the matrix element in Equation 5.74 as

$$\left| \left\langle \phi_{\vec{k}} \middle| \vec{k}j \right\rangle \right|^2 = \iint \phi^*_{\vec{k}}(\vec{r}) \phi_{\vec{k}j}(\vec{r}) \phi^*_{\vec{k}j}(\vec{r}\,') \phi_{\vec{k}}(\vec{r}\,') d^3r\, d^3r'$$

so that $\langle \phi_{\vec{k}} | W | \phi_{\vec{k}} \rangle$ can be written as

$$\left\langle \phi_{\vec{k}} \right| W \left| \phi_{\vec{k}} \right\rangle = \iint \phi^*_{\vec{k}}(\vec{r}) W(\vec{r}, \vec{r}\,') \phi_{\vec{k}}(\vec{r}\,') d^3r\, d^3r' \tag{5.75}$$

where

$$W(\vec{r},\vec{r}') = v(\vec{r})\delta(\vec{r},\vec{r}') + \sum_j (\varepsilon_{\vec{k}} - \varepsilon_{aj})\phi_{\vec{k}j}(\vec{r})\phi^*_{\vec{k}j}(\vec{r}') \tag{5.76}$$

Thus, the pseudopotential has a local part (first term) and a nonlocal part (second term). Thus, Equation 5.73 can be interpreted as Equation 5.76, as the integration over $\vec{r}'$ is implicit in Equation 5.73. Also note that

$$\begin{aligned}\langle\phi_{\vec{k}}|W|\phi_{\vec{k}}\rangle^* &= \iint \phi_{\vec{k}}(\vec{r})W^*(\vec{r},\vec{r}')\,\phi^*_{\vec{k}}(\vec{r}')d^3r\,d^3r' \\ &= \iint \phi^*_{\vec{k}}(\vec{r}')W(\vec{r}',\vec{r})\,\phi_{\vec{k}}(\vec{r})d^3r'\,d^3r \\ &= \langle\phi_{\vec{k}}|W|\phi_{\vec{k}}\rangle \end{aligned} \tag{5.77}$$

However, $\langle\phi_{\vec{k}'}|W|\phi_{\vec{k}}\rangle^* \neq \langle\phi_{\vec{k}}|W|\phi_{\vec{k}'}\rangle$ as W depends on $\vec{k}$. Thus, W is not Hermitian. Note that the pseudopotential is an energy-dependent potential. Thus, one must be careful in applying many standard theorems that are derived for the energy-independent potentials.

The second peculiarity of the pseudopotential is that it is not unique. For example, in Equation 5.73, if one replaces $(\varepsilon_{\vec{k}} - \varepsilon_{aj})$ by any arbitrary function $F(\varepsilon_{\vec{k}}, \varepsilon_{aj})$, the resulting pseudopotential will give the same eigenvalue. To prove this let us write the KS Equation 5.70 for pseudowave function using the new pseudopotential.

$$\left[-\frac{\hbar^2}{2m_e}\nabla^2 + v(\vec{r})\right]\phi'_{\vec{k}}(\vec{r}) + \sum_j F(\varepsilon'_{\vec{k}}, \varepsilon_{aj})|\vec{k}j\rangle\langle\vec{k}j|\phi'_{\vec{k}}\rangle = \varepsilon'_{\vec{k}}\,\phi'_{\vec{k}}(\vec{r}) \tag{5.78}$$

Here, we have assumed that the new pseudopotential will give eigenvalue $\varepsilon'_{\vec{k}}$ and pseudowave function $\phi'_{\vec{k}}(\vec{r})$. By multiplying Equation 5.78 by $\psi_{\vec{k}}(\vec{r})$ on the left and integrating over the whole volume, we get

$$\left\langle\psi_{\vec{k}}\left|\left[-\frac{\hbar^2}{2m_e}\nabla^2 + v(\vec{r})\right]\right|\phi'_{\vec{k}}\right\rangle + \sum_j F(\varepsilon'_{\vec{k}}, \varepsilon_{aj})\langle\psi_{\vec{k}}|\vec{k}j\rangle\langle\vec{k}j|\phi'_{\vec{k}}\rangle = \varepsilon'_{\vec{k}}\langle\psi_{\vec{k}}|\phi'_{\vec{k}}\rangle \tag{5.79}$$

Since the Hamiltonian $-(\hbar^2/2m_e)\nabla^2 + v(\vec{r})$ is a Hermitian operator, in Equation 5.79, it can be applied to the left to give the first term equal to $\varepsilon_{\vec{k}}\langle\psi_{\vec{k}}|\phi'_{\vec{k}}\rangle$. The second term in Equation 5.79 is zero because $\langle\psi_{\vec{k}}|\vec{k}j\rangle = 0$. Therefore, we get

$$\varepsilon_{\vec{k}}\langle\psi_{\vec{k}}|\phi'_{\vec{k}}\rangle = \varepsilon'_{\vec{k}}\langle\psi_{\vec{k}}|\phi'_{\vec{k}}\rangle \tag{5.80}$$

From this, we conclude that either $\psi_{\vec{k}}$ is orthogonal to $\phi'_{\vec{k}}$ or $\varepsilon'_{\vec{k}} = \varepsilon_{\vec{k}}$. We see from Equation 5.68 that $\psi_{\vec{k}}$ cannot be orthogonal to $\phi'_{\vec{k}}$. Thus, for any choice of F, we obtain exactly the same eigenvalue. Because of the nonuniqueness of the pseudopotential, it is possible to generate several pseudopotentials that will give the same answer. This freedom allows construction of several different kinds of pseudopotentials. One approach, which was very popular in the 1960s was to adjust the pseudopotential to fit some experimental quantities. Such pseudopotentials are called empirical pseudopotentials. At present, very good pseudopotentials using first-principles can be constructed that do not have any adjustable parameters. We shall discuss how to construct such pseudopotentials in the next chapter. It turns out that the plane wave expansion method combined with the pseudopotential approach works very well for a large number of materials and gives accuracy comparable to more elaborate all electron methods such as KKR and APW, discussed in Chapter 7.

Once the pseudopotential is known, the energy bands can be calculated using the plane wave expansion method, which we discussed earlier. For simplicity, let us assume that the pseudopotential is local, that is, $W = W(\vec{r})$. Thus, the secular equation is

$$\det\left|(\varepsilon^{0}_{\vec{k}-\vec{G}} - \varepsilon_{\vec{k}})\delta_{\vec{G}\vec{G}'} + W(\vec{G} - \vec{G}')\right| = 0 \tag{5.81}$$

Since W is a weak potential, the convergence of the plane wave expansion is much better now. For the nonlocal pseudopotential, we shall derive the secular equation in the next chapter.

Atomic pseudopotential: The total pseudopotential W can be written as the sum of atomic pseudopotentials w, centered on lattice sites $\vec{R}_\ell$, that is

$$W(\vec{r}) = \sum_{\ell} w(\vec{r} - \vec{R}_\ell) \tag{5.82}$$

where we have assumed that the pseudopotential is local and we have one atom per unit cell. Equation 5.82 can easily be generalized to many atoms per unit cell (see next chapter). In fact, in the methods referred above, to construct pseudopotentials, it is w that is calculated.

Orthogonality hole: We know that the electron charge density can be written as

$$n(\vec{r}) = \sum_{\substack{\vec{k} \\ \text{occupied}}} \psi^{*}_{\vec{k}}(\vec{r})\psi_{\vec{k}}(\vec{r}) \tag{5.83}$$

where $\psi_{\vec{k}}(\vec{r})$ are the true wave functions and the sum is over the occupied states. We also know that $\psi_{\vec{k}}(\vec{r})$ can be expressed in terms of the pseudowave function $\phi_{\vec{k}}(\vec{r})$ using Equation 5.68, that is

$$\psi_{\vec{k}}(\vec{r}) = (1 - P)\phi_{\vec{k}}(\vec{r})$$

$$= \phi_{\vec{k}}(\vec{r}) - P\phi_{\vec{k}}(\vec{r}) \tag{5.84}$$

Thus

$$\langle \psi_{\vec{k}} | \psi_{\vec{k}} \rangle = \langle \phi_{\vec{k}} | \phi_{\vec{k}} \rangle - \langle \phi_{\vec{k}} | P\phi_{\vec{k}} \rangle - \langle P\phi_{\vec{k}} | \phi_{\vec{k}} \rangle + \langle P\phi_{\vec{k}} | P\phi_{\vec{k}} \rangle$$

$$\langle \phi_{\vec{k}} | P | \phi_{\vec{k}} \rangle = \langle \phi_{\vec{k}} | \sum_j |\vec{k}j\rangle\langle \vec{k}j | \phi_{\vec{k}} \rangle$$

$$= \sum_j \left| \langle \vec{k}j | \phi_{\vec{k}} \rangle \right|^2$$

Similarly

$$\langle P\phi_{\vec{k}} | \phi_{\vec{k}} \rangle = \langle \phi_{\vec{k}} | P | \phi_{\vec{k}} \rangle$$

$$= \sum_j \left| \langle \vec{k}j | \phi_{\vec{k}} \rangle \right|^2$$

$$\langle P\phi_{\vec{k}} | P\phi_{\vec{k}} \rangle = \langle \phi_{\vec{k}} | P^2 | \phi_{\vec{k}} \rangle$$

$$= \langle \phi_{\vec{k}} | P | \phi_{\vec{k}} \rangle$$

$$= \sum_j \left| \langle \vec{k}j | \phi_{\vec{k}} \rangle \right|^2$$

Thus

$$\langle \psi_{\vec{k}} | \psi_{\vec{k}} \rangle = \langle \phi_{\vec{k}} | \phi_{\vec{k}} \rangle - \sum_j \left| \langle \vec{k}j | \phi_{\vec{k}} \rangle \right|^2 \tag{5.85}$$

By substituting P from Equation 5.65, we see that $\psi_{\vec{k}}^* \psi_{\vec{k}}$ is less than $\phi_{\vec{k}}^* \phi_{\vec{k}}^*$. Thus, if we define pseudocharge density as

$$n_{ps}(\vec{r}) = \sum_{\substack{\vec{k} \\ \text{occupied}}} \phi_{\vec{k}}^*(\vec{r})\phi_{\vec{k}}(\vec{r}) \tag{5.86}$$

Then, Equation 5.85 implies that orthogonalization of $\phi_{\vec{k}}(\vec{r})$ to the core states punches a hole in the pseudocharge density around each ion. This positive charge density is called the orthogonality hole.

EXERCISES

5.1 Plot empty lattice bands in one dimension. It may be convenient to use k in units of π/a and energy in units of $\hbar^2\,\pi^2/2m_e\,a^2$. Compare your results with Figure 5.1.

5.2 Plot empty lattice bands for sc, bcc, and fcc structures along [100], [110], and [111] directions in the Brillouin zone. Take k in units of π/a and energy in units of $\hbar^2\,\pi^2/2m_e\,a^2$. Label symmetry points. Compare your results with Figures 5.2 through 5.4.

5.3 Show that the plane wave expansion (Equation 5.29) satisfies Bloch's theorem.

5.4 Using the plane wave expansion (Equation 5.29), show that the coefficients $C_{\vec{k}-\vec{G}}$ satisfy the equation

$$(\varepsilon^0_{\vec{k}-\vec{G}} - \varepsilon_{\vec{k}})C_{\vec{k}-\vec{G}} + \sum_{\vec{G}'} v(\vec{G}' - \vec{G})C_{\vec{k}-\vec{G}'} = 0.$$

5.5 Consider electrons moving in one-dimensional periodic potential

$$V(x) = V_1 \cos\frac{2\pi x}{a} + V_2 \cos\frac{4\pi x}{a}$$

where a is the lattice spacing.

a. Sketch the three lowest-energy bands in the first Brillouin zone.

b. Calculate the energy gap at $k = (\pi/a)$ between the first and second band and at $k = 0$ between the second and third band.

5.6 Consider a square lattice with lattice spacing a and a weak crystal potential

$$V(x,y) = -4V_o \sin\left(\frac{2\pi x}{a}\right)\sin\left(\frac{2\pi y}{a}\right)$$

where V_o is a positive constant.

a. What is the energy gap at $\mathbf{k} = (\pi/a,\ \pi/a)$?

b. What is the energy gap at $\mathbf{k} = (\pi/a,\ o)$?

c. If each site is occupied by a divalent atom, would NFE theory predict it to be an insulator for some value of V_o? If yes, what will be the value of V_o?

5.7 A two-dimensional solid has a rectangular lattice with primitive translational vectors $a\hat{i}$ and $b\hat{j}$ with $b > a$. There is one divalent atom per unit cell. Electrons experience a weak periodic potential

$$V(x,y) = -2V_1\left[\cos\left(\frac{2\pi x}{a}\right) + \cos\left(\frac{2\pi y}{b}\right)\right]$$

where V_1 is a positive number.

a. Using the NFE approximation, evaluate three lowest energies at $\vec{k} = (0,0)$ and $(\pi/a,0)$. What are the band gaps at $\vec{k} = (0,0)$ and $(\pi/a,0)$?

b. Is it possible to induce metal to insulator transition by changing V_1? Explain. If so, find the value of V_1 at which this will happen.

5.8 Show that the tight-binding wave function $\psi_{\vec{k}}(\vec{r})$ given by Equation 5.34 satisfies Bloch's theorem.

5.9 Using the tight-binding approximation, calculate the energy band dispersion arising from *s* atomic levels for (a) fcc and (b) bcc monoatomic crystals.

5.10 a. For a simple square lattice with lattice spacing a and overlap integral $-t$, find the expression for ε_k assuming *s*-type orbitals in the tight-binding approximation. Take $\varepsilon_o = 4t$.

b. Using the result of part (a), plot Fermi surface in the first BZ, when (i) the band is nearly empty, (ii) when the band is half-full, and (iii) when the band is nearly full.

5.11 For a two-dimensional simple hexagonal lattice with lattice spacing a and overlap integral t, find an expression for $\varepsilon_{\vec{k}}$, assuming *s*-type orbitals in tight-binding approximation. Show that $\varepsilon_{\vec{k}}$ has sixfold symmetry.

5.12 Calculate the DOS for a one-dimensional tight-binding model and plot it as a function of energy.

5.13 Calculate the DOS for a square lattice using tight-binding *s*-band model and plot it. (Hint: Express DOS in terms of the Bessel function and evaluate by numerical integration.)

5.14 Calculate the DOS for a cubic lattice using the *s*-band tight-binding model and plot it. (Hint: Express DOS in terms of the Bessel function and evaluate by numerical integration. Also see Jelitto (1969).)

5.15 The band structure and Fermi surface of Cu are shown in Figures 4.11 and 4.12. Note that the highest occupied band intersects the Fermi energy ε_F in the (100) direction (ΓX) but in the (111) direction (ΓL), it lies below ε_F in the vicinity of $(\frac{1}{2},\frac{1}{2},\frac{1}{2})$ point of the BZ shown by L point in Figure 4.6. The distances in the BZ are given in units of $2\pi/a$. The Fermi surface of Cu is not a sphere but develops necks around L point as shown in Figure 4.12. These necks are almost circular. Estimate radius k_N of these necks, assuming that the highest occupied band near point L can be described by

$$\varepsilon_k = \frac{h^2k^2}{2m_e} - \Delta$$

where Δ is 2 eV, and k is measured from the center of the zone. You may use the free electron approximation to estimate ε_F.

5.16 Show that the OPW (Equation 5.62) satisfies Bloch's theorem.

5.17 Show that the wave function of an electron can be expanded in the form of Equation 5.63.

Further Reading

Harrison W. 1970. *Solid State Theory*. New York: McGraw-Hill Book Company.

Harrison W. 1980. *Electronic Structure and Properties of Solids*. San Francisco: W H Freeman and Company.

Kaxiras E. 2003. *Atomic and Electronic Structure of Solids*. Cambridge: Cambridge University Press.

Martin R. M. 2004. *Electronic Structure, Basic Theory and Practical Methods*. Cambridge: Cambridge University Press.

Mihaly L. and Martin M. C. 1996. *Solid State Physics, Problems and Solutions*. New York: John Wiley & Sons.

Singh D. J. and Nordstrom L. 2006. *Planewaves, Pseudopotentials, and the LAPW Method* (Second Edition). New York: Springer.

Ziman J. M. 1969. *Principles of the Theory of Solids*. Cambridge: Cambridge University Press.

6

Methods of Electronic Structure Calculations II

6.1 Introduction

In Chapter 5, we introduced the concept of pseudopotential, which has emerged as a powerful tool for calculating the electronic structure of solids. A pseudopotential is a weak potential that gives the same eigenvalues as obtained by using the real potential as shown in Figure 6.1, which shows $\ell = 0$ pseudopotential and the Coulomb potential for silicon (here, ℓ denotes angular momentum quantum number). One can see that while the Coulomb potential is very strong and approaches infinity near the origin, the pseudopotential is close to zero.

As mentioned in Chapter 5, the pseudowave function is a very smooth function as compared to the true wave function as shown in Figure 6.2, which shows the 3*s* wave function and the pseudowave function for silicon.

Because of its smoothness, the pseudowave function requires a smaller number of plane waves in its expansion than needed for the true wave function, thus saving considerable computing costs. In the pseudopotential method, the core electrons are eliminated and one focuses on the valence electrons. In other words, it is assumed that all core levels remain fully occupied and unchanged in all atomic environments. This is known as the frozen core approximation and can be justified because the core electrons are hardly affected by changing the atomic environment. In solid state, the core electrons hardly take any part in properties like bonding or transport. Thus, one can focus only on the valence electrons, reducing the number of electrons in a band structure calculation. The error introduced by such an approximation is quite small but savings in the computational cost is enormous.

Although the idea of pseudopotential was very appealing, its construction posed challenging problems. Since the construction of the pseudopotential method by Phillips and Kleinman (1959) was complicated, initially in the sixties, empirical pseudopotentials were used, which had adjustable parameters obtained by fitting to some experimental results. Since then, the field has made huge progress and now one mostly uses the first-principles

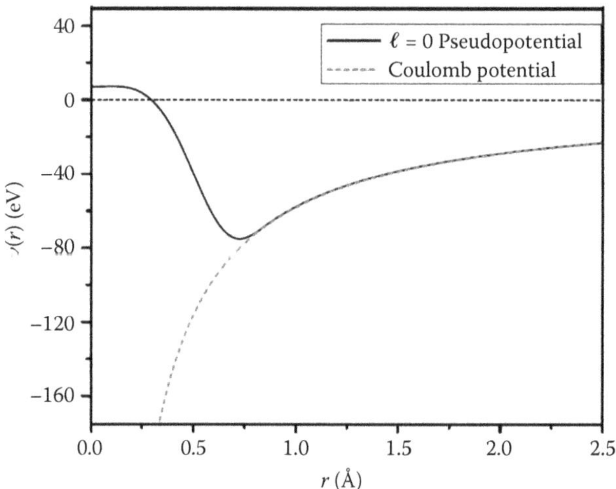

FIGURE 6.1
$\ell = 0$ Pseudopotential and Coulomb potential for silicon.

pseudopotentials, which have no adjustable parameters and yield results that are in good agreement with the full potential electronic structure calculations. The great advantage of the pseudopotential approach is that it is very efficient and saves a lot of computing cost. Therefore, in this chapter, we shall focus on this method. In Section 6.2, we shall take a scattering approach to understand the concept of pseudopotential, which will lead us to the basic principle for constructing an atomic pseudopotential. In Section 6.3, we shall

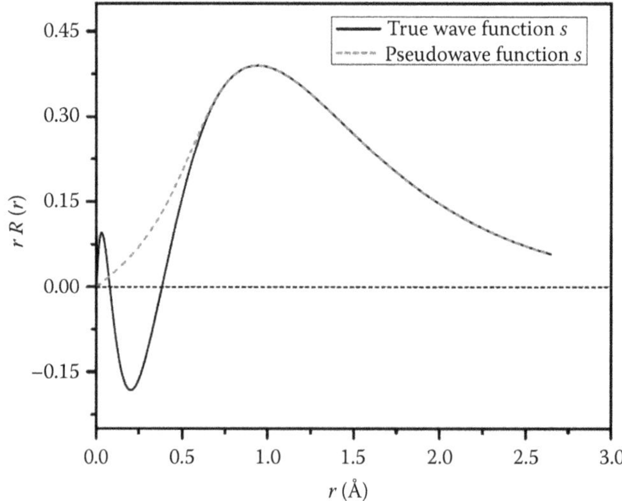

FIGURE 6.2
3*s* True wave function and pseudowave function for silicon atom.

briefly discuss how a first-principles atomic pseudopotential is constructed. In Section 6.4, we shall derive the secular equation to obtain energy eigenvalues. In Section 6.5, we shall discuss the calculation of the total energy. In Section 6.6, we very briefly discuss the ultrasoft pseudopotential and PAW methods. In Section 6.7, we shall discuss issues regarding the choice of energy cutoff and number of k-points. In Section 6.8, we shall discuss how to handle nonperiodic systems using supercell geometry.

6.2 Scattering Approach to Pseudopotential

In this section, we shall try to understand the concept of pseudopotential using the scattering theory. Let us consider a *single muffin-tin potential* $v(\vec{r})$, which is defined as

$$\begin{aligned} v(\vec{r}) &= v(r) \quad && r \le r_m \\ &= 0 && r \ge r_m \end{aligned} \tag{6.1}$$

where r_m is the muffin-tin radius, that is, the muffin-tin potential is spherically symmetric and has a value $v(r)$ within a sphere of radius r_m and zero outside the sphere. It is assumed to be attractive and is shown schematically in Figure 6.3.

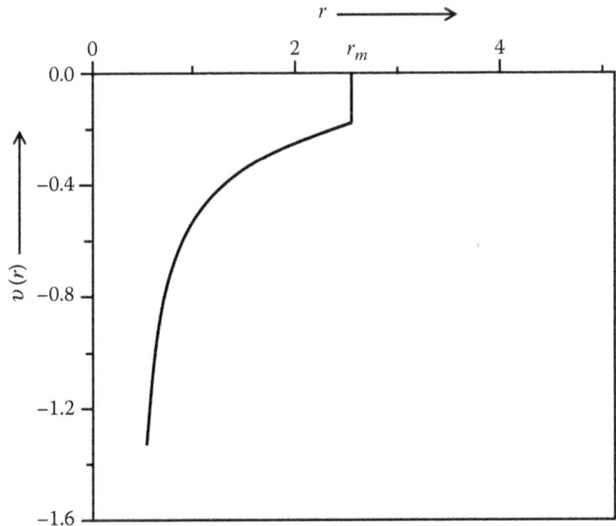

FIGURE 6.3
Schematic showing the muffin-tin potential $v(r)$ as a function r.

An electron moving in this potential will satisfy the Schrödinger equation

$$\left[-\frac{\hbar^2}{2m_e}\nabla^2 + v(r)\right]\psi(\vec{r}) = \varepsilon\psi(\vec{r}) \tag{6.2}$$

Since the potential $v(r)$ is spherically symmetric, the general solution of Equation 6.2 can be written as

$$\psi(\vec{r}) = \sum_L a_L R_\ell(r) Y_L(\vec{r}) \tag{6.3}$$

where $Y_L(\vec{r})$ is a spherical harmonic, a_L, expansion coefficient and as before, L is a composite angular momentum index for (ℓ, m). $R_\ell(r)$ is the solution of the radial Schrödinger equation

$$\left[-\frac{1}{2r}\frac{\hbar^2}{m_e}\frac{d^2}{dr^2}r + \frac{\hbar^2}{2m_e}\frac{\ell(\ell+1)}{r^2} + v(r)\right]R_\ell(r) = \varepsilon R_\ell(r) \tag{6.4}$$

The solutions of Equation 6.4 for energy $\varepsilon \geq 0$ correspond to the scattering states and for $\varepsilon < 0$, correspond to the bound states. Let us first consider the scattering case when $\varepsilon \geq 0$. We see from Equation 6.1 for $r \geq r_m$, the potential is zero. Therefore, Equation 6.4 for $r \geq r_m$ is the radial Schrödinger equation for free electrons. For $\varepsilon \geq 0$ and $r \geq r_m$, two linearly independent solutions of Equation 6.4 are spherical Bessel and Neumann functions $j_\ell\,(kr)$ and $n_\ell\,(kr)$, where $\varepsilon = (\hbar^2/2m_e)k^2$ (Merzbacher 1998). Thus, for $r \geq r_m$, $R_\ell(r)$ is a linear combination of j_ℓ and n_ℓ. Thus

$$R_\ell(r) = \cos\delta_\ell(\varepsilon) j_\ell(kr) - \sin\delta_\ell(\varepsilon) n_\ell(kr) \quad \text{for } r \geq r_m \tag{6.5}$$

where $\delta_\ell(\varepsilon)$ are energy-dependent parameters and are called phase-shifts.

Inside the muffin-tin radius, $R_\ell(r)$ is the solution that is regular at the origin. At $r = r_m$, its logarithmic derivative

$$\gamma_\ell(\varepsilon) = \frac{1}{R_\ell}\frac{d}{dr}R_\ell\bigg|_{r=r_m} \tag{6.6}$$

must be equal to the log derivative of $R_\ell\,(r)$ from the outside given by Equation 6.5. From this condition, we can express phase-shifts $\delta_\ell(\varepsilon)$

$$\delta_\ell(\varepsilon) = \cot^{-1}\left[\frac{k n'_\ell - \gamma_\ell n_\ell}{k j'_\ell - \gamma_\ell j_\ell}\right]_{r=r_m} \tag{6.7}$$

where prime denotes the derivative.

Let us now look at the asymptotic form (i.e., $r \to \infty$) of $R_\ell(r)$. The asymptotic forms of $j_\ell(kr)$ and $n_\ell(kr)$ are given by

$$\lim_{r\to\infty} j_\ell(kr) = \sin(kr - \ell\pi/2)/kr \tag{6.8}$$

$$\lim_{r\to\infty} n_\ell(kr) = -\cos(kr - \ell\pi/2)/kr \tag{6.9}$$

Substituting these in Equation 6.5, we get the asymptotic form for $R_\ell(r)$

$$\lim_{r\to\infty} R_\ell(r) = \frac{1}{kr}\sin(kr - \ell\pi/2 + \delta_\ell(\varepsilon)) \tag{6.10}$$

From Equation 6.10, we see that the asymptotic form of $R_\ell(r)$ differs from Equation 6.8, the asymptotic form of $v(r) = 0$ regular solution, by a shift of phase $\delta_\ell(\varepsilon)$. Thus, $\delta_\ell(\varepsilon)$ is called as phase-shift.

For the case, $\varepsilon < 0$, which corresponds to bound states, the solution of Equation 6.4 for $v(r) = 0$ must decay exponentially for $r > r_m$. The solution outside r_m is Hankel function of the first kind (Merzbacher 1998), that is

$$R_\ell(r) = A\,h_\ell^{(1)}(kr) \quad r > r_m \tag{6.11}$$

where $k = \sqrt{2m_e\varepsilon/\hbar^2}$ is now imaginary.

Let us now look at the solution of the free electron ($v(r) = 0$) problem with spherical boundary condition. We already know the solution of this problem with periodic boundary condition, which gives $\varepsilon = (\hbar^2/2m_e)k^2$ and $\psi = A\,e^{i\vec{k}\cdot\vec{r}}$. Let us solve the same problem with spherical boundary condition. We assume a spherical crystal of large radius R and take the origin at the center of the sphere. The potential is zero inside the sphere and is infinite outside the sphere. Therefore, the regular solution $j_\ell\,(kR)$ (Equation 6.8) must vanish at the surface of the sphere. This implies that

$$kR - \ell\pi/2 = n\pi \tag{6.12}$$

where n is an integer. This gives

$$k = (n\pi + \ell\pi/2)/R \tag{6.13}$$

and

$$\varepsilon = \frac{\hbar^2}{2m_e}k^2 = \frac{\hbar^2}{2m_e}(n\pi + \ell\pi/2)^2/R^2 \tag{6.14}$$

Now, let us introduce a single attractive muffin-tin potential $v(r)$ at the center of the spherical crystal of radius R as shown in Figure 6.4.

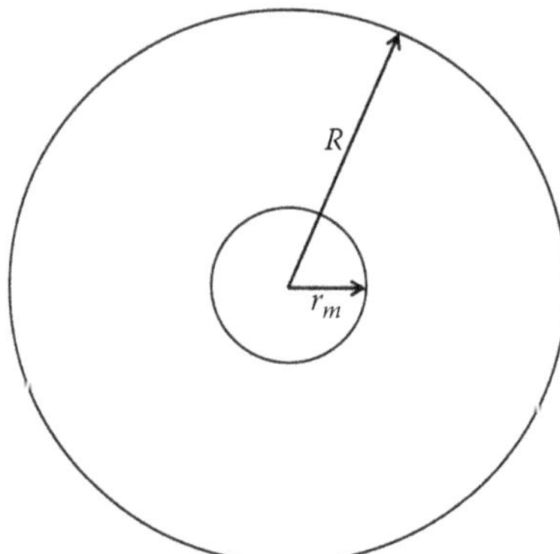

FIGURE 6.4
A single muffin-tin potential at the center of a large sphere of radius R.

Since the potential is infinite outside the crystal, $R_\ell(r)$ must vanish at the surface of the sphere. Using Equation 6.10, this gives

$$kR - \ell\,\pi/2 + \delta_\ell\,(\varepsilon) = \text{n}\pi \tag{6.15}$$

or

$$k = (n\pi + \ell\,\pi/2 - \delta_\ell)/R$$

and

$$\varepsilon = \frac{\hbar^2}{2m_e R^2}(n\pi + \ell\pi/2 - \delta_\ell)^2 \tag{6.16}$$

Equation 6.16 shows that the energy eigenvalues are determined by the phase-shifts or log derivatives at the muffin-tin radius. It further shows that by introducing a muffin-tin potential at the center of the big sphere, the energy levels get shifted and this shift depends on δ_ℓ.

If there is no perturbing potential at the center of the spherical crystal, that is, $v(r) = 0$, the phase-shift $\delta_\ell = 0$. Let us now introduce an attractive perturbing potential and slowly increase its strength. Note that for attractive potential, the phase-shifts are always positive (see, e.g., Roman 1965). As the strength of the potential increases from zero, the phase-shift grows and the solutions get modified. Let us see how it happens for s-state ($\ell = 0$) and $n = 1$. For this case, the boundary condition from Equation 6.15 is

$$kR + \delta_0(\varepsilon) = \pi \tag{6.17}$$

As the potential increases, the phase-shift δ_0 increases toward π and the wave number k approaches zero. At $\delta_0 = \pi$, $k = 0$, which implies $\varepsilon = 0$, that is, we have reached the bottom of the band. As the attractive potential is increased further, corresponding to a phase-shift greater than π, ε becomes negative and the nature of the state changes abruptly. For $\varepsilon < 0$, we get a bound state that decays exponentially outside the muffin-tin radius. Thus, we can say that at this strength of the potential, when δ_0 is slightly greater than π, a bound state is formed. We can generalize this result by saying that integral multiples of π in the phase-shift correspond to bound states, that is, in general, δ_ℓ can be written as

$$\delta_\ell = m\pi + \delta'_\ell \tag{6.18}$$

where m is an integer so that δ'_ℓ lies between 0 and π. Comparing free electron solution Equation 6.8 and asymptotic form of $R_\ell(r)$ in Equation 6.10, we see that $R_\ell(r)$ will have m additional nodes. To illustrate this, let us take an example. Assume that $v(r)$ in Equation 6.4 corresponds to a Si atom and let us consider 3s wave function, which corresponds to a valence electron. In Figure 6.5a, we have plotted $r\ R_0(r)$ for a free electron and in Figure 6.5b, $r\ R_0(r)$ for a 3s valence electron. We see that (i) 3s wave function is phase-shifted compared to the free electron wave function, (ii) it has two nodes (excluding $r = 0$), that is, $m = 2$ in Equation 6.18. This is because it has to be orthogonal to 1s (no node) and 2s (1 node) core electron wave functions. This means that m in Equation 6.18 also gives a number of core states of angular momentum ℓ below the valence electron state. From the phase-shift point of view, we can understand the oscillations in $R_\ell(r)$ as follows. We see that for $v(r) = 0$, there are no bound states and thus no oscillations in the core region. As the potential $v(r)$ is increased from zero, it pulls more and more of the wave function in the core region, increasing the phase-shift 0 to $m\pi + \delta'_\ell$. Whenever the phase-shift increases by π, one node is added in the core region. Thus, a stronger attractive potential will give rise to more nodes in the core region.

Let us now bring the pseudopotential into the picture. We know that a pseudopotential is a weak potential that gives the same eigenvalues as the real potential. We also know that pseudopotential is not unique. From the above discussion, it is clear that we may choose a pseudopotential in such a way that it simply eliminates integral multiples of π from the phase-shift (see Equation 6.18). This will imply that the pseudowave function will have no nodes in the core region ($r < r_m$). Also, outside the core region, the pseudowave functions will be equal to the true wave functions. This is illustrated in Figure 6.5c, which shows $r\ R_0(r)$ for the s pseudowave function for Si. Thus, the pseudowave function has no rapid oscillations in the core region so that its plane wave expansion is rapidly convergent. Note that such a pseudowave function will give the same logarithmic derivative as the true wave function, as is clear from Equations 6.7 and 6.18 and from the fact that cot $(m\pi + \delta_\ell) =$ cot

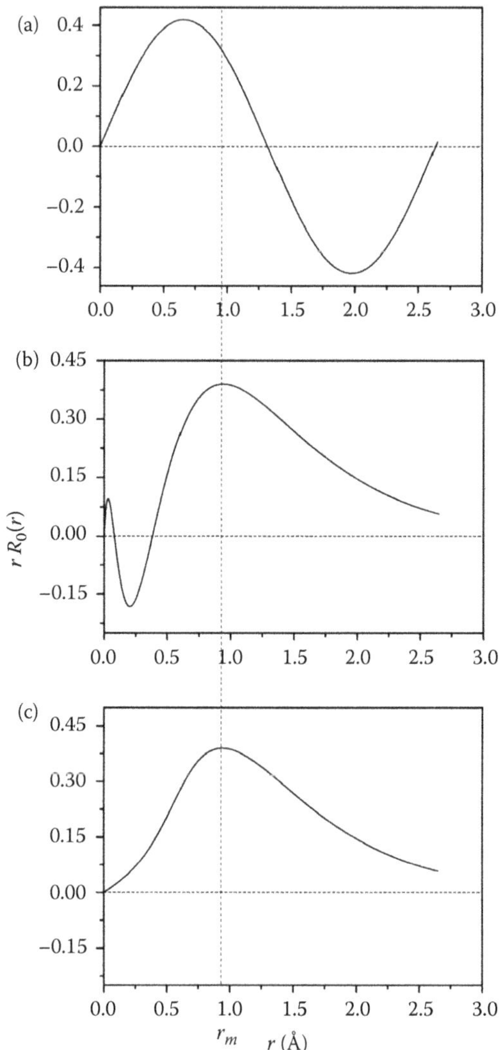

FIGURE 6.5
$r\,R_0(r)$ for (a) free electron, (b) 3*s* Si true wave function, and (c) *s* Si pseudowave function.

δ_ℓ. Thus, the integral multiples of π in Equation 6.18 have no effect on the log derivatives. Since eigenvalues are determined by the log derivatives, such a pseudopotential, which eliminates integral multiples of π in Equation 6.18, is guaranteed to give the same eigenvalues as the true potential. This suggests a way of constructing atomic pseudopotentials from first-principles. One first solves the Kohn–Sham equation for an atom and gets the true wave function. Then, one simply removes the nodes of the true wave function in the core region and obtains a pseudowave function. Now, an inverse problem is solved. We try to find a pseudopotential that will give this pseudowave function as its eigenfunction and the true eigenvalues as its eigenvalues. This is the basic principle of constructing first-principles pseudopotentials.

6.3 Construction of First-Principles Atomic Pseudopotentials

As we have discussed earlier, the pseudopotential is not unique. Thus, there is enormous choice in constructing the pseudopotential. As mentioned earlier, empirical pseudopotentials were used in the 1960s, which were determined by fitting experimental data. Such empirical pseudopotentials have several drawbacks, namely

1. The pseudopotential constructed for one environment of an atom will not work for another environment, that is, they lack transferability.
2. Although simple to use, they are not obtained from first-principles.

During the last two decades or so, there has been a revolution in the use of first-principles atomic pseudopotentials. All first-principles pseudopotentials are generated from all-electron atomic calculations. The basic principle of generating first-principles pseudopotential has already been discussed in Section 6.2. Here, we briefly discuss the main steps to construct pseduopotentials (Hamann et al. 1979, Bachelet et al. 1982):

1. First of all, true radial wave function $R_{n\ell}(r)$ of the atom is obtained by solving the radial Kohn–Sham equation

$$\left[-\frac{1}{2r}\frac{\hbar^2}{m_e}\frac{d^2}{dr^2}r + \frac{\hbar^2}{2m_e}\frac{\ell(\ell+1)}{r^2} + v(r)\right]R_{n\ell}(r) = \varepsilon_{n\ell}R_{n\ell}(r) \tag{6.19}$$

where

$$v(r) = -\frac{e^2}{4\pi\varepsilon_0}\frac{Z}{r} + v_H(r) + u_{xc}(r) \tag{6.20}$$

Here, $u_{xc}(r)$ is the exchange correlation potential and $v_H(r)$ is given by

$$v_H(r) = \int \frac{e^2}{4\pi\varepsilon_0}\frac{n(r')}{|\vec{r}-\vec{r}'|}d^3r' \tag{6.21}$$

where $n(r')$ is the total electron density

2. From true radial wave function $R_{n\ell}(r)$, a pseudowave function $\phi_\ell^{ps}(r)$ is constructed such that it fulfills the following conditions:
   i. The pseudowave function $\phi_\ell^{ps}(r)$ should contain no nodes. Since the pseudowave function is constructed only for valence electrons, we omit the principal quantum number n.
   ii. The normalized atomic radial pseudowave function $\phi_\ell^{ps}(r)$ is equal to normalized radial true wave function $R_\ell(r)$ beyond a suitably chosen cutoff radius r_c

$$\phi_\ell^{ps}(r) = R_\ell(r) \quad \text{for } r \geq r_c \tag{6.22}$$

   iii. The eigenvalues of pseudowave function must be equal to the true eigenvalues.
   iv. The charge enclosed within r_c for the pseudowave function and true wave function must be equal, that is

$$\int_0^{r_c} \left|\phi_\ell^{ps}(r)\right|^2 r^2 dr = \int_0^{r_c} \left|R_\ell(r)\right|^2 r^2 dr \tag{6.23}$$

   This is referred to as the norm-conservation condition and if a pseudopotential meets this condition, it is called a "norm-conserving pseudopotential" (NCPP). Note that a pseudowave function fulfilling these conditions can be constructed arbitrarily in many ways. This freedom is further exploited to produce a smooth pseudopotential.
3. Once the pseudowave function has been obtained, the radial Schrödinger equation can be inverted to yield a screened pseudopotential that has the pseudowave function as its eigenfunction at the correct eigenvalue.

$$\left[-\frac{1}{2r}\frac{\hbar^2}{m_e}\frac{d^2(r\phi_\ell^{ps}(r))}{dr^2} + \frac{\hbar^2}{2m_e}\frac{\ell(\ell+1)}{r^2}\phi_\ell^{ps}(r) + w_\ell^{sc}\,\phi_\ell^{ps}(r)\right] = \varepsilon_\ell\,\phi_\ell^{ps}(r) \tag{6.24}$$

Equation 6.24 implies that the pseudowave function must have continuous derivatives up to the second derivative if we want a continuous pseudopotential. Since the true wave function satisfies Equation 6.24 with true potential $v(r)$ and since for $r \geq r_c$, $\phi_\ell^{ps}(r) = R_\ell(r)$, the screened pseudopotential $w_\ell^{sc}(r)$ and the true potential are identical beyond r_c.

The screened pseudopotential w^{sc} thus obtained lacks transferability as the screening from the valence electrons depends strongly on the environment in which they are placed. Thus, the first step to improve transferability is to remove the screening effects of the valence electrons and generate an ionic pseudopotential w. This is done by subtracting $v_H(r)$ and $u_{xc}(r)$, calculated by using the valence pseudowave function, from w^{sc}, that is

$$w_\ell(r) = w_\ell^{sc}(r) - v_H(r) - u_{xc}(r) \tag{6.25}$$

where

$$v_H(r) = \int \frac{e^2}{4\pi\varepsilon_0} \frac{n^v(r')}{|\vec{r} - \vec{r}'|} d^3r' \tag{6.26}$$

and $u_{xc}(r)$ is also calculated from $n^v(r)$, where $n^v(r)$ is the valence electron density.

$$n^v(r) = \sum_{\ell,\text{ occupied}} |\phi_\ell^{ps}(r)|^2 \tag{6.27}$$

When such an ionic potential is placed in a different environment, $v_H(r)$ and $u_{xc}(r)$ are recalculated for that environment and again added to the potential.

Note that this does not guarantee that the pseudopotential is transferable to any environment. It can be used only in an environment in which eigenvalues do not change significantly from the eigenvalues used in its construction. The property that makes this energy range wider is loosely called smoothness, that is, the smoother the pseudopotential, the weaker the energy dependence.

In summary, the following conditions for the construction of pseudopotentials should be satisfied:

1. The pseudowave function $\phi^{ps}(r)$ is nodeless and is identical to the true wave function outside a suitably chosen cutoff radius r_c.
2. The first and second derivatives of the pseudowave function are continuous at r_c.
3. The eigenvalues of the pseudowave functions coincide with those of true wave functions.
4. The norm of the true and pseudowave functions inside r_c is the same

$$\int_0^{r_c} |r\phi^{ps}(r)|^2 \, dr = \int_0^{r_c} |r R_\ell(r)|^2 \, dr \tag{6.28}$$

5. Further conditions may be imposed to enhance the smoothness of the pseudopotential.

We had seen in Chapter 5 that the pseudopotential in general will be a nonlocal function, that is, it is not a simple multiplicative factor. The above procedure also shows that the pseudopotential is nonlocal, that is, each component of the wave functions will see a different potential. We see that the pseudopotential so generated is local in the radial coordinate but nonlocal in the angular coordinates. This potential can be written as

$$w = \sum_{\ell=0}^{\infty} \sum_{m=-\ell}^{\ell} |\ell m\rangle w^{\ell}(r) \langle \ell m| \tag{6.29}$$

$$= \sum_{\ell} w^{\ell}(r) \hat{P}_{\ell} \tag{6.30}$$

where $|\ell m\rangle$ are the spherical harmonics and $\hat{P}_{\ell} = \Sigma_{m=-\ell}^{\ell} |\ell m\rangle\langle \ell m|$ is the projection operator. Here, $w^{\ell}(r)$ is the pseudopotential for the angular component ℓ. When such a pseudopotential acts on a wave function, it decomposes it into its spherical harmonics components and then each of it is multiplied by the corresponding pseudopotential. Since $w^{\ell}(r)$ is a local operator in the radial coordinate, such a potential is called semilocal potential.

The summation on ℓ in Equation 6.30 is truncated at some value of $\ell = \ell_{max}$. The pseudopotential then can be written as

$$= w_{loc}(r) + \sum_{\ell=0}^{\ell_{max}} w_{nonlocal}^{\ell}(r) \hat{P}_{\ell} \tag{6.31}$$

where

$$w_{nonlocal}^{\ell}(r) = w^{\ell}(r) - w_{loc}(r) \tag{6.32}$$

Here, $w_{loc}(r)$ is the local part that is normally chosen as the pseudopotential for the first angular momentum that is not represented in the core, for example, $w^{d}(r)$ for Si. The idea is that it should represent those angular components of the pseudopotential adequately that are not corrected by the nonlocal part. The nonlocal part $w_{nonlocal}^{\ell}(r)$ as defined by Equation 6.32 is the difference between the l-dependent potential $w^{\ell}(r)$ and $w_{loc}(r)$. Since $w^{\ell}(r)$ and $w_{loc}(r)$ are identical outside the cutoff radius, $w_{nonlocal}^{\ell}(r)$ is zero outside the cutoff radius. Thus, $w_{nonlocal}^{\ell}(r)$ are short-ranged functions confined to the core region.

Other methods of pseudopotential construction that are quite popular are due to Troullier and Martins and due to Vanderbilt. Smoothest *norm-conserving* potentials are obtained by using the Troullier and Martins' construction (1991). On the other hand, Vanderbilt (1990) used a non-norm-conserving pseudopotential to obtain "ultrasoft" pseudopotential (see Section 6.6).

6.4 Secular Equation

We shall derive the secular equation for the NCPP. Here, we shall consider a more general case when we have M identical atoms in a unit cell at positions $\vec{d}_j$ and when the pseudopotential is nonlocal. The formulation can be easily generalized to many different types of atoms per unit cell. The total pseudopotential W can be written as

$$W(\vec{r},\vec{r}') = \sum_{i,j} w(\vec{r} - \vec{R}_i - \vec{d}_j, \vec{r}' - \vec{R}_i - \vec{d}_j) \tag{6.33}$$

where w is the atomic pseudopotential and $\vec{R}_i$ is a lattice vector. The general Hamiltonian can be written as

$$H = -\frac{\hbar^2}{2m_e}\nabla^2 + v_H(\vec{r}) + v_{xc}(\vec{r}) + W \tag{6.34}$$

where $v_{xc}(\vec{r})$ is the exchange correlation potential, $v_H(\vec{r})$ is the Hartree potential given by $v_H(\vec{r}) = \int(e^2/4\pi\varepsilon_0)n(\vec{r}')(d\vec{r}'/\vec{r} - \vec{r}')$. Here, $n(\vec{r})$ is the charge density of the valence electrons given by $n(\vec{r}) = \Sigma_{\vec{k}}^{occ}|\phi_{\vec{k}}^{ps}(\vec{r})|^2$. The pseudowave function $\phi^{PS}(\vec{r})$ is expanded in plane waves as

$$\begin{aligned}\phi_{\vec{k}}^{PS}(\vec{r}) &= \frac{1}{\sqrt{\Omega}}\sum_{\vec{G}} c_{\vec{k}+\vec{G}} \exp[i(\vec{k}+\vec{G})\cdot\vec{r}] \\ &= \sum_{\vec{G}} c_{\vec{k}+\vec{G}} \left|\vec{k}+\vec{G}\right\rangle \end{aligned} \tag{6.35}$$

We want to solve the equation

$$H\phi_{\vec{k}}^{PS}(\vec{r}) = \varepsilon_{\vec{k}}\,\phi_{\vec{k}}^{PS}(\vec{r}) \tag{6.36}$$

By multiplying Equation 6.36 by $\left\langle\vec{k}+\vec{G}\right|$, we get

$$\left\langle\vec{k}+\vec{G}\middle|H\middle|\phi_{\vec{k}}^{PS}(\vec{r})\right\rangle = \varepsilon_{\vec{k}}\left\langle\vec{k}+\vec{G}\middle|\phi_{\vec{k}}^{PS}(\vec{r})\right\rangle \tag{6.37}$$

Substitution of Equation 6.35 in Equation 6.37 gives

$$\sum_{G'} H_{\vec{k}+\vec{G},\vec{k}+\vec{G}'}\, c_{\vec{k}+\vec{G}'} = \varepsilon_{\vec{k}}\, c_{\vec{k}+\vec{G}} \tag{6.38}$$

where

$$H_{\vec{k}+\vec{G},\vec{k}+\vec{G}'} = \left\langle \vec{k}+\vec{G} \middle| H \middle| \vec{k}+\vec{G}' \right\rangle \tag{6.39}$$

Equation 6.38 is the secular equation by solving which energies $\varepsilon_{\vec{k}}$ can be found.

Let us now calculate matrix element (Equation 6.39). The matrix element of kinetic energy operator is very easy to evaluate and is given by

$$\left\langle \vec{k}+\vec{G} \middle| -\frac{\hbar^2}{2m_e}\nabla^2 \middle| \vec{k}+\vec{G}' \right\rangle = \frac{\hbar^2}{2m_e}\left|\vec{k}+\vec{G}\right|^2 \delta_{\vec{G},\vec{G}'} \tag{6.40}$$

The Hartree potential is given by

$$\begin{aligned} v_H(\vec{r}) &= \int \frac{e^2}{4\pi\varepsilon_0} \frac{n(\vec{r}')d^3r'}{|\vec{r}-\vec{r}'|} \\ &= \int d^3r' \frac{e^2}{4\pi\varepsilon_0} \sum_{\vec{G}} n(\vec{G}) e^{i\vec{G}\cdot\vec{r}'} \sum_{\vec{q}} \frac{4\pi}{q^2} e^{i\vec{q}\cdot(\vec{r}-\vec{r}')} \\ &= \sum_{\vec{G},\vec{q}} n(\vec{G}) \frac{e^2}{4\pi\varepsilon_0} \frac{4\pi}{q^2} e^{i\vec{q}\cdot\vec{r}} \delta_{\vec{q}\vec{G}} \\ &= \sum_{\vec{G}} n(\vec{G}) \frac{e^2}{4\pi\varepsilon_0} \frac{4\pi}{|\vec{G}|^2} e^{i\vec{G}\cdot\vec{r}} \end{aligned} \tag{6.41}$$

where $n(\vec{G})$ is the Fourier transform of the charge density $n(\vec{r})$. Now, the matrix element of $v_H(\vec{r})$ can be written as

$$\begin{aligned} \left\langle \vec{k}+\vec{G} \middle| v_H(\vec{r}) \middle| \vec{k}+\vec{G}' \right\rangle &= \frac{1}{\Omega}\int v_H(r) e^{-i(\vec{G}-\vec{G}')\cdot\vec{r}} d^3r = v_H(\vec{G}-\vec{G}') \\ &= n(\vec{G}-\vec{G}') \frac{e^2}{4\pi\varepsilon_0} \frac{4\pi}{|\vec{G}-\vec{G}'|^2} \end{aligned} \tag{6.42}$$

The pseudopotential W can be divided into two parts as

$$W(\vec{r},\vec{r}') = W^{loc}(\vec{r}) + W^{n\ell}(\vec{r},\vec{r}') \tag{6.43}$$

Let us first calculate the matrix element of the local part. We can write it as

$$\begin{aligned}
\left\langle \vec{k}+\vec{G}\middle| W^{loc} \middle| \vec{k}+\vec{G}' \right\rangle &= \frac{1}{\Omega}\int W^{loc}(\vec{r})e^{-i(\vec{G}-\vec{G}')\cdot\vec{r}}d^3r \\
&= \frac{1}{\Omega}\int \sum_{i,j} w^{loc}(\vec{r}-\vec{R}_i-\vec{d}_j)e^{-i(\vec{G}-\vec{G}')\cdot\vec{r}}d^3r \\
&= \frac{1}{\Omega}\sum_{i,j}\int w^{loc}(\vec{r})e^{-i(\vec{G}-\vec{G}')\cdot(\vec{r}+\vec{R}_i+\vec{d}_j)}d^3r \\
&= \frac{N}{\Omega}\sum_{j}\int w^{loc}(\vec{r})e^{-i(\vec{G}-\vec{G}')\cdot(\vec{r}+\vec{d}_j)}d^3r \\
&= \frac{N}{\Omega}\sum_{j} e^{-i(\vec{G}-\vec{G}')\cdot\vec{d}_j}\int w^{loc}(\vec{r})e^{-i(\vec{G}-\vec{G}')\cdot\vec{r}}d^3r \\
&= S(\vec{G}-\vec{G}')w^{loc}(\vec{G}-\vec{G}')
\end{aligned} \tag{6.44}$$

where $S(\vec{q})$ is the geometrical structure factor given by

$$S(\vec{q}) = \sum_{j} e^{-i\vec{q}\cdot\vec{d}_j} \tag{6.45}$$

and $w^{loc}(\vec{G})$ is the Fourier transform of the local atomic pseudopotential given by

$$w^{loc}(\vec{G}) = \frac{N}{\Omega}\int w^{loc}(\vec{r})e^{-i\vec{G}\cdot\vec{r}}d^3r = \frac{1}{\Omega_{cell}}\int w^{loc}(\vec{r})e^{-i\vec{G}\cdot\vec{r}}d^3r \tag{6.46}$$

where $\Omega_{cell} = \Omega/N$ is the unit cell volume. Note that the $\vec{G} = 0$ matrix element of $v_H(r)$ diverges as $1/G^2$. The $\vec{G} = 0$ matrix element of the bare ion–electron Coulomb interaction will also diverge as $1/G^2$. The local pseudopotential will have the same long-range behavior since at large distances the ion cores will look like point particles. The sum of the two contributions will diverge as $4\pi\, Q_T/G^2$, where Q_T is the total charge in the unit cell. Thus, to avoid $\vec{G} = 0$ divergence, such calculations can be done only for the charge-neutral systems.

The nonlocal part of the pseudopotential $W^{n\ell}$ can be expanded in terms of atomic contributions $w^{n\ell}$ as

$$W^{n\ell}(\vec{r},\vec{r}') = \sum_{i,j} w^{n\ell}(\vec{r}-\vec{R}_i-\vec{d}_j,\vec{r}'-\vec{R}_i-\vec{d}_j)$$

Thus

$$\left\langle \vec{k}+\vec{G}\right|W^{n\ell}\left|\vec{k}+\vec{G}'\right\rangle = \sum_{i,j}\left\langle \vec{k}+\vec{G}\right|w^{n\ell}(\vec{r}-\vec{R}_i-\vec{d}_j,\vec{r}'-\vec{R}_i-\vec{d}_j)\left|\vec{k}+\vec{G}'\right\rangle$$

$$= \frac{1}{\Omega}\iint\sum_{i,j} e^{-i(\vec{k}+\vec{G})\cdot\vec{r}}w^{n\ell}(\vec{r}-\vec{R}_i-\vec{d}_j,\vec{r}'-\vec{R}_i-\vec{d}_j)\,e^{i(\vec{k}+\vec{G}')\cdot\vec{r}'}d^3r d^3r'$$

$$= \frac{1}{\Omega}\iint\sum_{i,j} e^{-i(\vec{k}+\vec{G})\cdot(\vec{r}+\vec{R}_i+\vec{d}_j)}w^{n\ell}(\vec{r},\vec{r}')$$

$$\times\, e^{i(\vec{k}+\vec{G}')\cdot(\vec{r}'+\vec{R}_i+\vec{d}_j)}d^3r\, d^3r'$$

$$= N\sum_j e^{-i(\vec{G}-\vec{G}')\cdot\vec{d}_j}\left\langle \vec{k}+\vec{G}\right|w^{n\ell}\left|\vec{k}+\vec{G}'\right\rangle$$

$$= NS(\vec{G}-\vec{G}')\left\langle \vec{k}+\vec{G}\right|w^{n\ell}\left|\vec{k}+\vec{G}'\right\rangle \tag{6.47}$$

The matrix elements of $w^{n\ell}$ can be written as

$$\left\langle \vec{k}+\vec{G}\right|w^{n\ell}(\vec{r},\vec{r}')\left|\vec{k}+\vec{G}'\right\rangle = \sum_\ell \left\langle \vec{k}+\vec{G}\right|w^{n\ell}_\ell(r)\frac{1}{r^2}\delta(r-r')$$

$$\times\sum_m Y^*_{\ell m}(\vec{r})Y_{\ell m}(\vec{r}')\left|\vec{k}+\vec{G}'\right\rangle$$

$$= \frac{1}{\Omega}\sum_\ell \iint d^3r\,d^3r'\,e^{-i(\vec{k}+\vec{G})\cdot\vec{r}}w^{n\ell}_\ell(r)\frac{1}{r^2}\delta(r-r')$$

$$Y^*_{\ell m}(\vec{r})Y_{\ell m}(\vec{r}')e^{+i(\vec{k}+\vec{G}')\cdot\vec{r}'} \tag{6.48}$$

where we have assumed the pseudopotential to be of semilocal form (Equation 6.30). Let us evaluate Equation 6.48 by expanding the plane waves in terms of spherical harmonics with respect to the center of the atom as

$$e^{+i(\vec{k}+\vec{G}')\cdot\vec{r}'} = 4\pi\sum_{\ell=0}^{\infty}\sum_{m=-\ell}^{+\ell} i^\ell j_\ell(|\vec{k}+\vec{G}'|r)Y_{\ell m}(\vec{k}+\vec{G}')Y^*_{\ell m}(\vec{r}')$$

Substituting this in Equation 6.48, we get

$$w^{n\ell}_{\vec{k}+\vec{G},\vec{k}+\vec{G}'} = \frac{1}{\Omega}\sum_{\ell}\iint r^2\,dr\,d\Omega\,r'^2\,dr\,d\Omega'$$

$$\times (4\pi)^2 \sum_{L_1} i^{-\ell_1} j_{\ell_1}(|\vec{k}+\vec{G}|r) Y^*_{L_1}(\vec{k}+\vec{G}) Y_{L_1}(\vec{r})$$

$$\times w^{n\ell}_{\ell} \frac{1}{r^2}\delta(r-r') Y^*_L(\vec{r}) Y_L(\vec{r}')$$

$$\times \sum_{L_2} i^{\ell_2} j_{\ell_2}(|\vec{k}+\vec{G}'|r) Y_{L_2}(\vec{k}+\vec{G}') Y^*_{L_2}(\vec{r}') \tag{6.49}$$

where we have used a composite index $L \equiv (\ell, m)$.

Now, we use the addition theorem for spherical harmonics (see, e.g., *Quantum Mechanics* by Merzbacher (1998))

$$P_\ell(\cos\theta_{\vec{k}+\vec{G},\vec{k}+\vec{G}'}) = \frac{4\pi}{2\ell+1}\sum_{m,-\ell}^{+\ell} Y^*_L(\vec{k}+\vec{G}) Y_L(\vec{k}+\vec{G}') \tag{6.50}$$

and

$$\int Y^*_{L_1}(\vec{r}) Y_L(\vec{r})\,d\Omega = \delta_{L_1 L} \tag{6.51}$$

Substitution of Equations 6.50 and 6.51 in Equation 6.49 gives

$$w^{n\ell}_{\vec{k}+\vec{G},\vec{k}+\vec{G}'} = \frac{1}{\Omega}\sum_{\ell} 4\pi(2\ell+1) P_\ell(\cos\theta_{\vec{k}+\vec{G},\vec{k}+\vec{G}'})$$

$$\times \int r^2 dr\, j_\ell(|\vec{k}+\vec{G}|r)\, w^{n\ell}_{\ell}(r)\, j_\ell(|\vec{k}+\vec{G}'|r) \tag{6.52}$$

This form of the nonlocal pseudopotential is inconvenient for some applications. Therefore, Kleinman and Bylander (1982) proposed a form that is computationally more efficient. However, we shall not discuss it here and refer to the original paper. The exchange correlation potential is local in space. Thus

$$\left\langle \vec{k}+\vec{G} \middle| v_{xc}(\vec{r}) \middle| \vec{k}+\vec{G}' \right\rangle = v_{xc}(\vec{G}-\vec{G}')$$

Once all the matrix elements of H are known, $\varepsilon_{\vec{k}}$ can be calculated from Equation 6.38.

6.5 Calculation of the Total Energy

The total energy of a solid is a very important quantity, as a large number of its properties can be related to the total energy. For example, the structure and equilibrium lattice constant of the solid could be obtained by optimizing the total energy. Similarly, the bulk moduli can be related to the total energy. Thus, an accurate calculation of the total energy can predict several properties of solids. It is interesting to note that Wigner and Seitz (1955) had commented in 1955 that nothing much would be gained from such a calculation. However, the experience of the last three decades tells a different story. The calculation of the total energy now plays a central role in predicting unknown structures, designing new materials, *ab initio* molecular dynamics, calculations of phonon spectra, and so on as we shall see in subsequent chapters.

We shall now discuss how the total energy can be calculated using norm-conserving pseudopotentials. In the pseudopotential approximation, the core electrons are frozen and one only needs to calculate the wave functions and energy eigenvalues of the valence electrons. Since the core electrons do not play any role in bonding, this is a very good approximation. Also, it should be noted that in this approximation, the pseudowave function and the true wave function are identical outside the ionic cores. Thus, the total energy E can be written as

$$E = E_{kin} + E_{xc} + E_H + E_{ps}^{loc} + E_{ps}^{n\ell} + V_{II} \tag{6.53}$$

where E_{kin} is the kinetic energy, E_{xc} is the exchange correlation energy, E_{ps}^{loc} and $E_{ps}^{n\ell}$ are local and nonlocal pseudopotential energies, respectively, and V_{II} is the ion–ion repulsive energy.

The kinetic energy of the electrons, E_{kin}, can be written as

$$\begin{aligned} E_{kin} &= \sum_{\vec{k}} \int \phi_{\vec{k}}^{*}(\vec{r})\left(-\frac{\hbar^2}{2m_e}\nabla^2\right)\phi_{\vec{k}}(\vec{r})\,d^3r \\ &= \sum_{\vec{k},\vec{G}} \frac{\hbar^2}{2m_e}(\vec{k}+\vec{G})^2 c^{*}(\vec{k}+\vec{G})\,c(\vec{k}+\vec{G}) \end{aligned} \tag{6.54}$$

where $\phi_{\vec{k}}$ is the pseudowave function and for simplicity superscript "PS" over $\phi_{\vec{k}}$ has been dropped. The exchange-correlation energy E_{xc} is generally calculated in real space. In LDA, it can be calculated using

$$E_{xc} = \int \varepsilon_{xc}\left(n(\vec{r})\right) n(\vec{r})\,d^3r \tag{6.55}$$

where ε_{xc} is obtained from homogeneous electron gas calculation.

The Hartree energy E_H is given by

$$\begin{aligned}
E_H &= \frac{1}{2}\iint \frac{e^2}{4\pi\varepsilon_0}\frac{n(\vec{r})n(\vec{r}')}{|\vec{r}-\vec{r}'|}d^3r\,d^3r' \\
&= \frac{1}{2}\iint \sum_{\vec{q}} \frac{e^2}{4\pi\varepsilon_0}\frac{4\pi}{q^2}e^{i\vec{q}\cdot(\vec{r}-\vec{r}')}\sum_{\vec{G}\vec{G}'} n(\vec{G})e^{i\vec{G}\cdot\vec{r}} \\
&\quad \times n(\vec{G}')e^{i\vec{G}'\cdot\vec{r}'}\,d^3r\,d^3r' \\
&= \frac{1}{2}\iint \sum_{\vec{q}}\sum_{\vec{G}\vec{G}'} \frac{e^2}{4\pi\varepsilon_0}\frac{4\pi}{q^2}\delta_{\vec{q},-\vec{G}}\,\delta_{\vec{q},\vec{G}'}\,n(\vec{G})n(\vec{G}') \\
&= \frac{1}{2}\sum_{\vec{G}} \frac{e^2}{\varepsilon_0}\frac{1}{G^2}n(\vec{G})n(-\vec{G})
\end{aligned} \tag{6.56}$$

where $n(\vec{G})$ is the Fourier transform of $n(\vec{r})$.

E_{ps}^{loc} in Equation 6.53 denotes the local pseudopotential energy given by

$$E_{ps}^{loc} = \int W^{loc}(\vec{r})n(\vec{r})d^3r \tag{6.57}$$

where $W^{loc}(\vec{r})$ is the local part of the pseudopotential.

E_{ps}^{loc} can be evaluated as

$$\begin{aligned}
E_{ps}^{loc} &= \sum_{\vec{k}} \int \phi_{\vec{k}}^*(\vec{r})W^{loc}(\vec{r})\phi_{\vec{k}}(\vec{r})d^3r \\
&= \sum_{\vec{k},\vec{G},\vec{G}'} c^*(\vec{k}+\vec{G})\langle\vec{k}+\vec{G}|W^{loc}(\vec{r})|\vec{k}+\vec{G}'\rangle c(\vec{k}+\vec{G}') \\
&= \sum_{\vec{k},\vec{G},\vec{G}'} c^*(\vec{k}+\vec{G})S(\vec{G}-\vec{G}')w^{loc}(\vec{G}-\vec{G}')c(\vec{k}+\vec{G}')
\end{aligned} \tag{6.58}$$

$E_{ps}^{n\ell}$ is the nonlocal pseudopotential energy given by

$$E_{ps}^{n\ell} = \sum_{\vec{k}} \int \phi_{\vec{k}}^*(\vec{r})W^{n\ell}(\vec{r},\vec{r}')\phi_{\vec{k}}(\vec{r}')d^3r\,d^3r' \tag{6.59}$$

$$\begin{aligned}
&= \sum_{\vec{k},\vec{G},\vec{G}'} c^*(\vec{k}+\vec{G})\langle\vec{k}+\vec{G}|W^{n\ell}(\vec{r},\vec{r}')|\vec{k}+\vec{G}'\rangle c(\vec{k}+\vec{G}') \\
&= \sum_{\vec{k},\vec{G},\vec{G}'} NS(\vec{G}-\vec{G}')\langle\vec{k}+\vec{G}|\,w^{n\ell}\,|\,\vec{k}+\vec{G}'\rangle c^*(\vec{k}+\vec{G})c(\vec{k}+\vec{G}')
\end{aligned} \tag{6.60}$$

where we have used Equation 6.48. As we have already calculated $\langle \vec{k} + \vec{G}| w^{n\ell} | \vec{k} + \vec{G}' \rangle$ in Equation 6.52, $E_{ps}^{n\ell}$ can be evaluated.

V_{II} in Equation 6.53 is the contribution coming from ion–ion repulsion and is given by

$$V_{II} = \frac{1}{2} \sum_{\substack{I,J,i \\ I \neq J}} \frac{e^2}{4\pi\varepsilon_0} \frac{Z_I Z_J}{|\vec{R}_i + \vec{d}_I - \vec{d}_J|} \tag{6.61}$$

The evaluation of E_H, E_{ps}^{loc}, and V_{II} requires some care in computation because of the long-range nature of these interactions. For evaluating the lattice sum in Equation 6.58, Ewald's method is used. For details, we refer to the article by Bagno et al. (1991).

6.6 Ultrasoft Pseudopotential and Projector-Augmented Wave Method

It turns out that the norm-conserving potentials need a large number of plane waves or large energy cutoff (see the next section) for elements with strongly localized orbitals such as transition metal and rare-earth elements, thus requiring much more computational effort. Vanderbilt (1990) showed that this problem could be solved by relaxing the norm-conservation requirement. This greatly reduced the energy cutoff because a large value of r_c could be used. However, this comes at a price. Since the norm of the pseudowave function is not conserved, it results in the charge deficit. To compensate for this deficit, augmentation charges, which are defined as the charge density difference between true and pseudowave functions, are introduced in the core region. The core radius r_c can now be chosen quite large, about half the nearest-neighbor distance, making the pseudopotential ultrasoft. Using this method, one can obtain a pseudowave function as smooth as possible in the core region and reduce the number of plane waves needed in the plane wave expansion. The norm-conservation is taken into account afterward by solving a generalized eigenvalue equation. Vanderbilt's approach works really well for 3-*d* transition metals. It not only saves computational time but also improves accuracy. Thus, the ultrasoft pseudopotential (US-PP) method has been one of the important developments in the electronic structure methods.

Another very important development is due to Blöchl (1994), who developed the PAW method. The PAW method combines the ideas of the LAPW (see the next chapter) and ultrasoft pseudopotential methods, somewhat similar to the OPW method, which combines the plane wave and tight-binding methods. In this method, the all-electron wave function is decomposed in terms of a smooth pseudowave function and a rapidly varying contribution localized within the core region. The all-electron wave function ψ_{AE}^n, where n is a composite index for the band, $\vec{k}$, and spin, is related to the pseudowave function ψ_{PS}^n by a linear transformation

$$|\psi_{AE}^n\rangle = T|\psi_{PS}^n\rangle \tag{6.62}$$

The KS equation can be written as

$$H|\psi_{AE}^n\rangle = \epsilon_n|\psi_{AE}^n\rangle$$

$$HT|\psi_{PS}^n\rangle = \epsilon_n T|\psi_{PS}^n\rangle$$

or

$$T^{\dagger}HT|\psi_{PS}^n\rangle = T^{\dagger}T\epsilon_n|\psi_{PS}^n\rangle \tag{6.63}$$

Pseudowave functions are obtained by solving Equation 6.63 and then transformed back to true wave functions using Equation 6.62, which are then used to calculate the charge density and the total energy. Thus, the PAW method is an all-electron method and not a traditional pseudopotential method. It uses frozen core approximation but, in principle, it is not necessary. However, the method is cast in such a way that the machinery of the pseudopotential can be used with slight modification and some additional routines. The PAW method is one of the most powerful and accurate methods with the efficiency of the pseudopotential methods. For details, we refer to the original paper (Blöchl 1994).

6.7 Energy Cutoff and *k*-Point Convergence

There are two very important issues that must be settled before doing an electronic structure calculation:

1. How many plane waves must one use for the expansion of the wave function?
2. How many *k*-points must one use for an accurate determination of the total energy?

In principle, an infinite number of plane waves are required in the plane wave expansions of $\phi_{\vec{k}}^{PS}(\vec{r})$ in Equation 6.34. In practice, the series is terminated at some $\vec{G}_{max}$ corresponding to the cutoff energy $E_{cut} = (\hbar^2/2m_e)G_{max}^2$. A smoother and weaker pseudopotential will require a lower-energy cutoff. The truncation of the series will lead to an error in the calculated quantities, but this can be decreased by increasing energy cutoff. Therefore, it is important to vary E_{cut}

and examine how the total energy converges. The energy cutoff should be chosen such that one gets the total energy close to the converged value.

For Brillouin zone integration, one uses a discrete set of k-points. It is generally seen that a uniform mesh of k-points yields better results compared to a nonuniform mesh. Therefore, care should be taken that the k-points are uniformly distributed in the Brillouin zone. The total energy should be computed as a function of number of k-points and an appropriate number of k-points should be chosen so that one gets the total energy close to the converged value.

6.8 Nonperiodic Systems and Supercells

Until now we have been discussing methods of electronic structure calculations that are applicable to periodic systems only. However, nonperiodic systems can also be handled by these methods using a supercell. A supercell is a large unit cell containing many primitive unit cells, which is periodically repeated in all space. The supercell then becomes the unit cell for the calculation for a nonperiodic system. We shall give examples of supercells for some nonperiodic systems below:

a. *Impurity or defect in a solid*: An impurity or defect in a solid breaks the translational symmetry and, in principle, the methods we have discussed so far are not applicable. However, we can apply the above methods by choosing a large supercell and putting the impurity in the middle of the central cell as shown in Figure 6.6. The figure

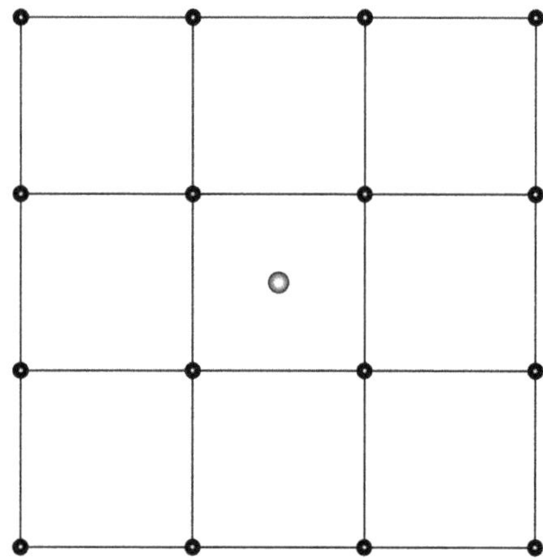

FIGURE 6.6
A 3×3 supercell for an impurity in a two dimensional solid. Impurity is shown in the central cell.

shows a 3×3 supercell in two dimensions. The unit cell of the periodic system is represented by a small square and the supercell consists of 3×3 such unit cells. The impurity is placed in the central cell. The supercell is chosen large enough so that there is almost no interaction between two impurities in the neighboring supercells. This supercell becomes the unit cell for the calculation involving the impurity. Note that impurities can also be handled by using the Green's function method as discussed in Chapter 8.

b. *Surface or interface*: In this case, there is periodicity in the plane of the surface but there is no periodicity in the direction perpendicular to the surface. For the case when there is a vacuum on one side and solid on the other, a supercell is shown in Figure 6.7. The supercell contains a crystal slab and vacuum region.

   This is also called slab geometry. To represent the isolated surface accurately, one must ensure that

   i. The width of the vacuum region must be wide enough so that the faces of the adjacent crystal slabs do not interact across the vacuum region.
   ii. The thickness of the crystal slab should be large enough so that the two surfaces of the crystal slab do not interact with each other. Similarly, an interface between the solids A and B can be handled. Now, instead of vacuum, one has a solid A or B.

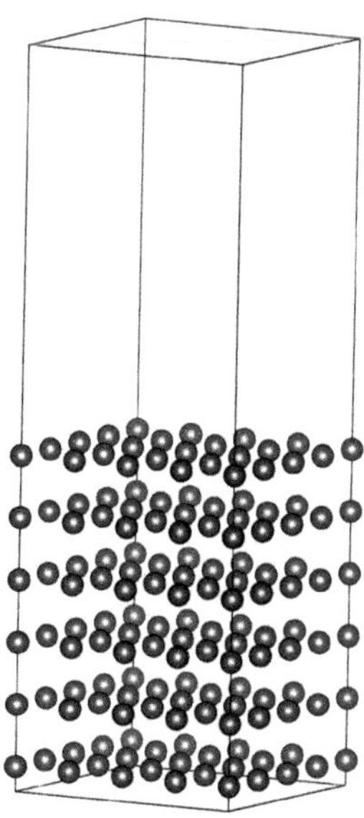

FIGURE 6.7
Supercell for a surface calculation. The upper part of the supercell is empty.

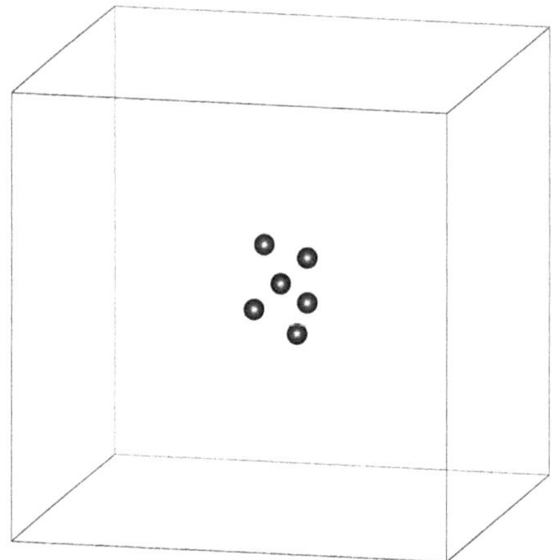

FIGURE 6.8
Supercell for a cluster calculation.

c. *Atomic cluster*: An isolated cluster of atoms can also be handled by using a large supercell as shown in Figure 6.8. The cluster is placed at the center of the supercell and the supercell is chosen to be large enough so that the two clusters in the neighboring supercells do not interact.

d. *Nanowire*: A nanowire has length that is very large compared to its diameter, which is of the order of nanometers. This can be approximated by assuming that there is a periodicity along the length of the wire but no periodicity in the direction perpendicular to it. In Figure 6.9, we show schematically the supercell geometry for the nanowire. The supercell is shown by solid lines. In directions perpendicular to the wire, there is vacuum. The width of this vacuum region should

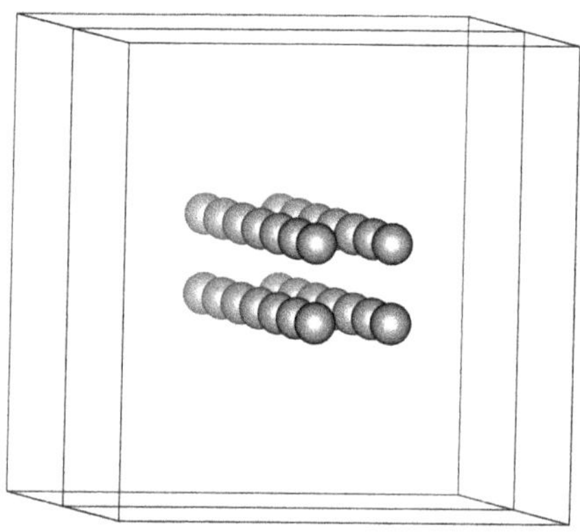

FIGURE 6.9
Supercell for a nanowire.

be large enough so that there is no interaction between the wires in the neighboring supercells.

We see that the size of the supercell is much larger than the primitive cell of the solid. As a result, there is a high increase in the computing cost, which goes as $\geq \Omega^3_{cell}$, where Ω_{cell} is the volume of the unit cell. For dealing with nonperiodic solids, one can develop specialized techniques that can handle nonperiodicity more efficiently (e.g., see Chapter 8). However, the pseudopotential technique is very efficient and at times it is more convenient to use the supercell technique rather than go for a specialized technique.

EXERCISES

6.1 Show that the phase-shift $\delta_\ell(\varepsilon)$ can be written in terms of log derivatives γ_ℓ as

$$\delta_\ell(\varepsilon) = \cot^{-1}\left[\frac{k n'_\ell - \gamma_\ell n_\ell}{k j'_\ell - \gamma_\ell j_\ell}\right]_{r=r_m}.$$

6.2 Show that the matrix element of the local part of the pseudopotential between plane waves $\langle \vec{k} + \vec{G}| W^{loc}(\vec{r}) |\vec{k} + \vec{G}\rangle$ is given by Equation 6.44.

6.3 Show that the matrix element of the atomic nonlocal potential between plane waves $\langle \vec{k} + \vec{G}| w^{n\ell}(\vec{r}, \vec{r}') |\vec{k} + \vec{G}'\rangle$ is given by Equation 6.52.

6.4 Show that the local pseudopotential energy, E^{loc}_{ps}, is given by Equation 6.58.

6.5 Show that the nonlocal pseudopotential energy, E^{nl}_{ps}, is given by Equation 6.60.

6.6 Generate the atomic pseudopotential for copper in atomic configuration $3d^{10}4s^1$ using one of the codes listed in Appendix A. Check the eigenvalues obtained by using the full potential and the pseudopotential. Compare (i) full potential and $\ell = 0$ and $\ell = 1$ components of the pseudopotential, and (ii) $4s$ wave function and $\ell = 0$ pseudowave function.

6.7 Generate the ultrasoft pseudopotential for copper using the codes listed in Appendix A and compare your results with the results obtained in Exercise 6.6.

6.8 Generate the atomic pseudopotential for silicon in its atomic ground state $3s^2 3p^2$ using one of the codes listed in Appendix A. Check the eigenvalues obtained by using full potential and the pseudopotential. Compare (i) full potential and $\ell = 0$ and $\ell = 1$ components of the pseudopotential and (ii) $3s$ wave function and $\ell = 0$ pseudowave function.

6.9 Using the pseudopotential generated in Exercise 6.6 for Cu, calculate band structure, density of states, and charge density in (110) plane for Cu in fcc structure using one of the plane wave codes listed in Appendix A. Compare your results with the results shown in Chapter 4.

6.10 Using the pseudopotential generated in Exercise 6.8 for silicon, calculate band structure, density of states, and charge density in (110) plane for silicon in the diamond structure using one of the plane wave codes listed in Appendix A. Compare your results with the results shown in Chapter 4.

Further Reading

Ashcroft N. and Mermin N. 1976. *Solid State Physics.* New York: W. B. Saunders Company.

Harrison W. 1970. *Solid State Theory.* New York: McGraw-Hill Book Company.

Kaxiras E. 2003. *Atomic and Electronic Structure of Solids.* Cambridge: Cambridge University Press.

Kohanoff J. 2006. *Electronic Structure Methods for Solids and Molecules.* Cambridge: Cambridge University Press.

Martin R. M. 2004. *Electronic Structure, Basic Theory and Practical Methods.* Cambridge: Cambridge University Press.

Singh D. J. and Nordstrom L. 2006. *Planewaves, Pseudopotentials, and the LAPW Method* (Second Edition). New York: Springer.

7

Methods of Electronic Structure Calculations III

7.1 Introduction

In this chapter, we shall discuss methods of electronic structure calculations which are based on muffin-tin potentials such as the KKR and APW methods. A muffin-tin potential is a spherically symmetric potential within a sphere and zero outside the sphere as shown in Figure 6.3. We begin with an introduction to the single particle Green's function in Section 7.2, which we need for discussion of the KKR method and theory of disordered alloys in the next chapter. In Section 7.3, we discuss how to handle perturbations using Green's function. In Section 7.4, we give an example of calculation of Green's function for the 3-dimensional electron gas. In Sections 7.5 and 7.6, we discuss the KKR and LMTO methods. In Sections 7.7 and 7.8, the APW and the LAPW methods are discussed. In Section 7.9, we briefly discuss the basic principles of linear scaling methods.

7.2 Green's Function

Green's functions are mathematical constructs that are used to obtain solutions of inhomogeneous differential equations and named after George Green who first proposed the concept in the eighteenth century. His work was concerned with solutions of Laplace's and Poisson's equations, but it contained the basic idea which found much wider application in classical as well as quantum physics. Green's functions have played an important role in the development of condensed matter physics and have found their widespread use in the study of many-body theory, electronic structure, magnetism, superconductivity, transport properties, disordered systems, and critical phenomena.

In many-body theory and condensed matter physics, several kinds of Green's function are defined, but here we focus only on the time-independent single particle Green's function which is designed to solve the time-independent Schrödinger or the KS equation, that is

$$H(\vec{r})\psi(\vec{r}) = \varepsilon\psi(\vec{r}) \tag{7.1}$$

where

$$H(\vec{r}) = -\frac{\hbar^2}{2m}\nabla^2 + v(\vec{r})$$

For calculating one particle properties such as charge density or density of states (DOS), Green's function $G(\vec{r},\vec{r}',z)$ is defined as

$$(z - H(\vec{r}))G(\vec{r},\vec{r}',z) = \delta(\vec{r} - \vec{r}') \tag{7.2}$$

Here, z is a complex variable with $\varepsilon = \text{Re}(z)$, $y = \text{Im}(z)$ and $H(\vec{r})$ is a time-independent Hermitian linear differential operator. We assume that $H(\vec{r})$ possesses a complete set of eigenfunctions ψ_i such that

$$H(\vec{r})\psi_i(\vec{r}) = \varepsilon_i\psi_i(\vec{r}) \tag{7.3}$$

and ψ_i are orthonormal, that is

$$\int \psi_i^*(\vec{r})\psi_j(\vec{r})d^3r = \delta_{ij} \tag{7.4}$$

Also the completeness of ψ_i implies that

$$\sum_i \psi_i(\vec{r})\psi_i^*(\vec{r}') = \delta(\vec{r} - \vec{r}') \tag{7.5}$$

Here, i may stand for a set of indices which may take discrete values for discrete part of the spectrum of $H(\vec{r})$ or may take continuous values for continuous part of the spectrum. Similarly, Σ_i will mean summation over the discrete part of the spectrum plus integration over the continuous spectrum.

We now introduce abstract vector space such that

$$\psi_i(\vec{r}) = \langle \vec{r} | \psi_i \rangle \tag{7.6}$$

$$\delta(\vec{r} - \vec{r}')H(\vec{r}) = \langle \vec{r} | H | \vec{r}' \rangle \tag{7.7}$$

$$\langle \vec{r} | \vec{r}' \rangle = \delta(\vec{r} - \vec{r}') \tag{7.8}$$

$$\int |\vec{r}\rangle\langle\vec{r}|\, d^3r = 1 \tag{7.9}$$

Thus, we can write Equations 7.2, 7.4, and 7.5 in matrix form in the Hilbert space as

$$(z - H)G(z) = 1 \tag{7.10}$$

$$\langle \psi_i | \psi_j \rangle = \delta_{ij} \tag{7.11}$$

$$\sum_i |\psi_i\rangle\langle\psi_i| = 1 \tag{7.12}$$

Our original Equation 7.2 is obtained by taking $\langle\vec{r}|\ldots|\vec{r}'\rangle$ matrix elements of Equation 7.10.

$$\begin{aligned}\langle \vec{r} |(z - H)G(z)| \vec{r}' \rangle &= \delta(\vec{r} - \vec{r}') \\ &= z\langle \vec{r} | G(z) | \vec{r}' \rangle - \langle \vec{r} | HG(z) | \vec{r}' \rangle = \delta(\vec{r} - \vec{r}')\end{aligned}$$

or,

$$= zG(\vec{r},\vec{r}',z) - \int \langle \vec{r} | H | \vec{r}'' \rangle \langle \vec{r}'' | G(z) | \vec{r}' \rangle d^3r'' = \delta(\vec{r} - \vec{r}'')$$

$$= (z - H(\vec{r}))G(\vec{r},\vec{r}',z) = \delta(\vec{r} - \vec{r}')$$

Similarly, we can obtain Equations 7.4 and 7.5 from Equations 7.11 and 7.12. We see the advantage of using the abstract vector space now. We can immediately write the formal solution of Equation 7.10 as

$$G(z) = (z - H)^{-1} \equiv \frac{1}{z - H} \tag{7.13}$$

$$\begin{aligned}G(z) &= \frac{1}{z - H}\sum_i |\psi_i\rangle\langle\psi_i| \\ &= \sum_i \frac{|\psi_i\rangle\langle\psi_i|}{z - \varepsilon_i}\end{aligned} \tag{7.14}$$

If H has discrete as well as continuous parts in its spectrum, then Equation 7.14 should be written as

$$G(z) = \sum_i{}' \frac{|\psi_i\rangle\langle\psi_i|}{z - \varepsilon_i} + \int \frac{|\psi_i\rangle\langle\psi_i|}{z - \varepsilon_i} di \tag{7.15}$$

where prime over the summation sign means that the summation is only over discrete part of the spectrum. Equation 7.15 can be written in $\vec{r}$-representation as

$$\begin{aligned} G(\vec{r},\vec{r}',z) &= \langle \vec{r}|G(z)|\vec{r}'\rangle \\ &= \sum_i{}' \frac{\psi_i(\vec{r})\psi_i^*(\vec{r}')}{z - \varepsilon_i} + \int \frac{\psi_i(\vec{r})\psi_i^*(\vec{r}')}{z - \varepsilon_i} di \end{aligned} \tag{7.16}$$

Thus, we have found the formal solution of Equation 7.2. Let us look at this solution more carefully. Note that H is a Hermitian operator, which implies that all of its eigenvalues ε_i are real. This further implies that for $y = \text{Im } z \neq 0$, $G(\vec{r},\vec{r}',z)$ is an analytic function of z in the entire complex plane. If the spectrum of H is discrete, $G(\vec{r},\vec{r}',z)$ has poles at $z = \varepsilon_i$. Since ε_i is real, these poles will lie on the real z-axis. This implies that if we know the poles of Green's function, we can find discrete eigenvalues of H. If spectrum of H is continuous, $G(\vec{r},\vec{r}',z)$ is not defined over that portion of the real z-axis corresponding to the continuous spectrum of H. If the eigenstates of H are extended, this produces a branch cut on the real z-axis as shown in Figure 7.1. This means that side limits of $G(\vec{r},\vec{r}',\varepsilon \pm iy)$ as $y \to 0^+$ exist but are different from each other.

Let us define these side limits as follows:

$$G^+(\vec{r},\vec{r}',\varepsilon) = \lim_{y\to 0^+} G(\vec{r},\vec{r}',\varepsilon + iy) \tag{7.17}$$

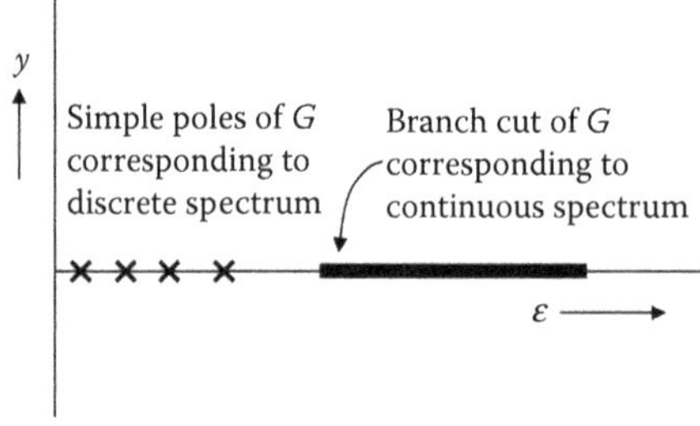

FIGURE 7.1
Poles and branch cut of Green's function.

$$G^{-}(\vec{r},\vec{r}',\varepsilon) = \lim_{y\to 0^{+}} G(\vec{r},\vec{r}',\varepsilon - iy) \tag{7.18}$$

where ε is in the continuous part of the spectrum. To show that these limits exist and that there is a discontinuity in $G(\vec{r},\vec{r}',z)$ as one crosses the branch cut, we shall use the following identity:

$$\lim_{y\to 0} \frac{1}{\varepsilon \pm iy - a} = P\frac{1}{\varepsilon - a} \mp i\pi\delta(\varepsilon - a) \tag{7.19}$$

where P stands for the principal value, the meaning of which will be clear in the proof given below.

To prove Equation 7.19, let us consider the integral

$$\lim_{y\to 0} \int_{-\infty}^{\infty} \frac{f(\varepsilon)}{\varepsilon + iy - a} d\varepsilon \tag{7.20}$$

where $f(\varepsilon)$ is a smooth function without any singularities.

We see that in evaluating Equation 7.20, the problem arises close to $\varepsilon = a$, as there is a pole at $\varepsilon = a$. This means that except for a small interval from $a - \delta$ to $a + \delta$, where δ is small, we can integrate Equation 7.20 by putting $y = 0$ but near $\varepsilon = a$ we have to be careful. Thus, near $\varepsilon = a$, we slightly deform the path of integration as shown in Figure 7.2.

The integral over the straight line which avoids the range $a - \delta$ to $a + \delta$ is called the principal value, denoted by sign P. To find the contribution from the semicircular path, let us write

$$z = a + \delta \exp(i\theta)$$

The semicircle is infinitesimally small (as δ is very small) but gives a finite contribution to the integral. This is equal to

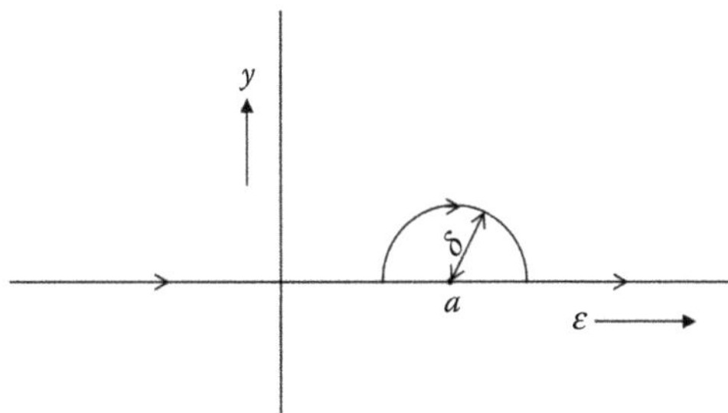

FIGURE 7.2
Deformed path of integration near the pole.

$$\int_{\text{semicircle}} \frac{f(z)}{z-a}dz = \int_{\pi}^{0} \frac{f(a)}{\delta\, e^{i\theta}} i\delta\, e^{i\theta} d\theta = -i\pi\, f(a)$$

Thus, the integral in Equation 7.20 can be written as

$$\ell im_{y\to 0} \int_{-\infty}^{\infty} \frac{f(\varepsilon)}{\varepsilon + iy - a} d\varepsilon = P\int_{-\infty}^{\infty} \frac{f(\varepsilon)}{\varepsilon - a} d\varepsilon - i\pi f(a) \tag{7.21}$$

or we can write it symbolically as

$$\ell im_{y\to 0} \frac{1}{\varepsilon + iy - a} = P\frac{1}{\varepsilon - a} - i\pi\delta(\varepsilon - a) \tag{7.22}$$

We can see that if we consider $\varepsilon - iy - a$ instead of $\varepsilon + iy - a$ in Equation 7.22, the only difference we will get is +ve sign in the last term, which proves identity (7.19).

Let us apply identity (7.22) to evaluate Equations 7.17 and 7.18. Thus, from Equations 7.16 and 7.17

$$G^{+}(\vec{r},\vec{r}',\varepsilon) = \ell im_{y\to 0} \sum_i \frac{\psi_i(\vec{r})\psi_i^*(\vec{r}')}{\varepsilon + iy - \varepsilon_i} \tag{7.23}$$

$$= P\sum_i \frac{\psi_i(\vec{r})\psi_i^*(\vec{r}')}{\varepsilon - \varepsilon_i} - i\pi\sum_i \delta(\varepsilon - \varepsilon_i)\psi_i(\vec{r})\psi_i^*(\vec{r}') \tag{7.24}$$

Here, summation over i implies summation over discrete as well as continuous portions of the spectrum. Let us put $\vec{r} = \vec{r}'$ in Equation 7.24. Then,

$$G^{+}(\vec{r},\vec{r},\varepsilon) = P\sum_i \frac{|\psi_i(\vec{r})|^2}{\varepsilon - \varepsilon_i} - i\pi\sum_i \delta(\varepsilon - \varepsilon_i)|\psi_i(\vec{r})|^2 \tag{7.25}$$

We see that the first term is real while the last term is imaginary. Taking the imaginary part and integrating it over $\vec{r}$, we get

$$\int \mathrm{Im}\, G^{+}(\vec{r},\vec{r},\varepsilon) d^3r = -\pi\sum_i \delta(\varepsilon - \varepsilon_i)\int |\psi_i(\vec{r})|^2 d^3r$$

$$= -\pi\sum_i \delta(\varepsilon - \varepsilon_i) = -\pi N\rho(\varepsilon) \tag{7.26}$$

where $\rho(\varepsilon) = (1/N)\sum_i \delta(\varepsilon - \varepsilon_i)$ is the DOS and N is the number of cells in the solid. Thus,

$$\rho(\varepsilon) = -\frac{1}{\pi N}\int \mathrm{Im}G^{+}(\vec{r},\vec{r},\varepsilon)d^3r \tag{7.27}$$

This is an important result and is frequently used to calculate the DOS if Green's function is known. Note that the integration in Equation 7.27 is overall the volume of the solid. However, in periodic solids, the integration can be limited to one unit cell as each unit cell, contributes the same due to the periodicity. In that case, the factor of N in Equation 7.27 cancels out. This result can also be written as

$$\rho(\varepsilon) = -\frac{1}{\pi N}\mathrm{Im}TrG^{+}(\varepsilon) \tag{7.28}$$

where Tr stands for trace. Similarly, if we take an imaginary part of $G^{+}(\vec{r},\vec{r},\varepsilon)$ in Equation 7.25 and integrate over energy ε, we get

$$\int_{-\infty}^{\varepsilon_F} \mathrm{Im}\,G^{+}(\vec{r},\vec{r},\varepsilon)d\varepsilon = -\pi\sum_i |\psi_i(\vec{r})|^2 \int_{-\infty}^{\varepsilon_F} \delta(\varepsilon-\varepsilon_i)d\varepsilon$$

$$= -\pi \sum_{\substack{i \\ \text{occupied}}} |\psi_i(\vec{r})|^2 = -\pi\, n(\vec{r})$$

or

$$n(\vec{r}) = -\frac{1}{\pi}\int_{-\infty}^{\varepsilon_F} \mathrm{Im}\,G^{+}(\vec{r},\vec{r},\varepsilon)d\varepsilon \tag{7.29}$$

Here, $n(\vec{r})$ is the electron density and we have assumed that all the levels up to the Fermi energy, ε_F, are filled ($T = 0\,\mathrm{K}$ case). Equation 7.29 is another important result that relates electron density to Green's function. Another useful result can be derived as follows:

$$G(\vec{r},\vec{r}',z) = \sum_i \frac{\psi_i(\vec{r})\psi_i^{*}(\vec{r}')}{z-\varepsilon_i}$$

$$= \sum_i \int_{-\infty}^{\infty} \delta(\varepsilon-\varepsilon_i)\frac{\psi_i(\vec{r})\psi_i^{*}(\vec{r}')}{z-\varepsilon}d\varepsilon$$

$$= \int_{-\infty}^{\infty} \frac{1}{z-\varepsilon}\sum_i \delta(\varepsilon-\varepsilon_i)\psi_i(\vec{r})\psi_i^{*}(\vec{r}')d\varepsilon$$

$$= -\frac{1}{\pi}\int_{-\infty}^{\infty} \frac{\mathrm{Im}\,G^{+}(\vec{r},\vec{r}',\varepsilon)}{z-\varepsilon}d\varepsilon$$

By taking trace of both sides, we get

$$Tr\ G(z) = -\frac{1}{\pi}\int_{-\infty}^{\infty}\frac{\operatorname{Im} TrG^{+}(\varepsilon)}{z-\varepsilon}\,d\varepsilon \;=\; \int_{-\infty}^{\infty}\frac{N\rho(\varepsilon)}{z-\varepsilon}\,d\varepsilon \tag{7.30}$$

where we have used Equation 7.28.

Let us now get back to Equation 7.24. If we now evaluate $G^{-}(\vec{r},\vec{r},\varepsilon)$ by replacing $\varepsilon + iy$ to $\varepsilon - iy$ in Equation 7.23, the only change we will get in Equation 7.24 will be the + sign of the second term. Thus, in general, we get two different limits of Green's function $G(\vec{r},\vec{r},\varepsilon + iy)$ depending on whether we go from below or above in the complex energy plane. If the spectrum is discrete and $\varepsilon \neq \varepsilon_i$, that is, we avoid the poles, then these limits are the same. However, if ε lies in the continuous spectrum, there is no way of crossing the real axis without avoiding this discontinuity. As we have seen above, this continuity is related to the DOS.

From Equation 7.16, we see that

$$G^{*}(\vec{r},\vec{r}',z) = G(\vec{r}',\vec{r},z^{*}) \tag{7.31}$$

or in matrix form, we can write

$$G^{\dagger}(z) = G(z^{*}) \tag{7.32}$$

where † denotes the Hermitian conjugate. Green's function $G(z)$ satisfies the criteria of the Herglotz function. A function $F(z)$ is Herglotz if it satisfies the following properties.

1. $F(z)$ is a analytic function of z in the entire complex plane except at the real axis, that is, for $\operatorname{Im} z \neq 0$. Its spectrum is bounded and $F(z)$ behaves as $1/z$ as $z \to \pm\infty$.
2. $F^{\dagger}(z) = F(z^{*})$ (7.33)
3. $\operatorname{Im} F(z) \leq 0$ for $\operatorname{Im} z > 0$

Clearly $G(z)$ satisfies all these properties.

Now, we shall show that the formal solution of the following inhomogeneous differential equation can be written in terms of Green's function:

$$(z - H(\vec{r}))\psi(\vec{r}) = f(\vec{r}) \tag{7.34}$$

We know that $G(\vec{r},\vec{r}',z)$ satisfies the equation

$$(z - H(\vec{r}))G(\vec{r},\vec{r}',z) = \delta(\vec{r} - \vec{r}') \tag{7.35}$$

Then $\psi(\vec{r})$ can be written as

$$\psi(\vec{r}) = \varphi(\vec{r}) + \int G(\vec{r},\vec{r}',z) f(\vec{r}') d^3r' \quad \text{for } z \neq \varepsilon_i \tag{7.36}$$

where $\phi(\vec{r})$ is the solution of the homogeneous equation (i.e., without $f(\vec{r})$ in Equation 7.34) at energy ε. Equation 7.36 is indeed a solution of Equation 7.34, which can be checked by substituting for $\psi(\vec{r})$ into Equation 7.34 and then using Equation 7.35. Note that both $\phi(\vec{r})$ and $\psi(\vec{r})$ should satisfy the same boundary conditions. We see that Green's function provides a method to solve an inhomogeneous differential equation like Equation 7.34, provided we know Green's function.

In summary, we have shown that Green's function $G(\vec{r},\vec{r}',z)$ is an analytic function in the whole complex plane except on the real axis. Its poles give the discrete eigenvalues. Its branch cut along the real axis corresponds to the continuous spectrum of H and the discontinuity in $G(\vec{r},\vec{r}',z)$ across this branch cut is related to the DOS and the electron density.

7.3 Perturbation Theory Using Green's Function

One advantage of using Green's function is that it is relatively easier to handle perturbations. Another advantage is that there is no need to assume the perturbation to be small as is the case with usual perturbation theory. This makes it a much more powerful tool.

Let us divide our one electron Hamiltonian H into two parts, unperturbed Hamiltonian H_0 and a perturbation H_1

$$H = H_0 + H_1 \tag{7.37}$$

We assume that the eigenvalues ε_i^0 and the eigenfunction $|\psi_i^0\rangle$ of H_0 are known so that Green's function G_0 (z)

$$G_0(z) = (z - H_0)^{-1} = \sum_i \frac{|\psi_i^0\rangle\langle\psi_i^0|}{z - \varepsilon_i^0} \tag{7.38}$$

is also known. Our aim is to find Green's function $G(z)$ for the Hamiltonian H, given by

$$G(z) = (z - H)^{-1} \tag{7.39}$$
$$= (z - H_0 - H_1)^{-1} = [(z - H_0)(1 - (z - H_0)^{-1} H_1)]^{-1}$$

$$= \left[G_0^{-1}(1 - G_0 H_1)\right]^{-1}$$

$$= (1 - G_0\, H_1)^{-1}\, G_0 \tag{7.40}$$

or

$$(1 - G_0\, H_1) G = G_0$$

$$G = G_0 + G_0\, H_1\, G \tag{7.41}$$

Equation 7.41 is known as Dyson's equation. This can be rewritten as

$$G = G_0 + G_0\, H_1\, (G_0 + G_0\, H_1\, G)$$

$$= G_0 + G_0\, H_1\, G_0 + G_0\, H_1\, G_0\, H_1\, G_0 + \cdots \tag{7.42}$$

$$= G_0 + G_0[H_1 + H_1\, G_0\, H_1 + \cdots]G_0$$

$$= G_0 + G_0\, T\, G_0 \tag{7.43}$$

where we have introduced an operator T, which is called t-matrix and is given by

$$T = H_1 + H_1\, G_0\, H_1 + \cdots \tag{7.44}$$

$$= H_1 + H_1\, G_0(H_1 + H_1\, G_0\, H_1 + \cdots)$$

$$= H_1 + H_1\, G_0 T \tag{7.45}$$

$$= H_1 + (H_1 + H_1\, G_0\, H_1 + \cdots)G_0\, H_1$$

$$= H_1 + TG_0\, H_1 \tag{7.46}$$

It is clear from Equation 7.43 that if T is known, G can be found. Equation 7.44 can be rewritten as

$$T = H_1[1 + G_0\, H_1 + G_0\, H_1\, G_0\, H_1 + \cdots]$$

$$= H_1[G_0 + G_0 H_1 G_0 + \cdots]G_0^{-1}$$

$$= H_1 G G_0^{-1} = G_0^{-1} G H_1 \tag{7.47}$$

$$= H_1\, G(z - H_0) \tag{7.48}$$

where we have used Equation 7.38. From Equation 7.48, it is clear that $T(z)$ will have analytic properties similar to $G(z)$, that is, it will be analytic in the whole complex z-plane except on the real axis. Various techniques such as diagrammatic techniques have been used to evaluate $T(z)$, but we shall not discuss them here.

The above equations are written in matrix form in Hilbert space but can be easily written in real or $\vec{k}$-space. For example, Equation 7.41 can be written in real space as

$$G(\vec{r},\vec{r}\,',z) = G_0(\vec{r},\vec{r}\,',z) + \iint G_0(\vec{r},\vec{r}_1,z)H_1(\vec{r}_1,\vec{r}_2)\, G(\vec{r}_2,\vec{r}\,',z)d^3r_1d^3r_2 \quad (7.49)$$

or in $\vec{k}$-space

$$G(\vec{k},\vec{k}',z) = G_0(\vec{k},\vec{k}',z) + \sum_{\vec{k}_1}\sum_{\vec{k}_2} G_0(\vec{k},\vec{k}_1,z)H_1(\vec{k}_1,\vec{k}_2)\, G(\vec{k}_2,\vec{k}',z) \quad (7.50)$$

We now discuss how the eigenstates $|\psi\rangle$ associated with H can be found. Let $|\phi\rangle$ be the eigenstates of H_0, that is

$$(\varepsilon - H_0)|\phi\rangle = 0 \quad (7.51)$$

and

$$(\varepsilon - H)\ |\psi\rangle = 0 \quad (7.52)$$

or

$$(\varepsilon - H_0 - H_1)|\psi\rangle = 0$$

or

$$(\varepsilon - H_0)|\psi\rangle = H_1\ |\psi\rangle \quad (7.53)$$

Thus, $|\psi\rangle$ can be written as

$$|\psi\rangle = |\phi\rangle + G_0\,(\varepsilon)H_1\ |\psi\rangle \quad (7.54)$$

This can be checked by substituting Equation 7.54 into Equation 7.53. If ε does not belong to the spectrum of H_0, then

$$|\psi\rangle = G_0\,(\varepsilon)H_1\ |\psi\rangle \quad (7.55)$$

or

$$\psi(\vec{r}) = \int G_0(\vec{r},\vec{r}',\varepsilon)H_1(\vec{r}')\psi(\vec{r}')d^3r' \tag{7.56}$$

Equation 7.56 can be used to find eigenfunctions corresponding to the discrete spectrum assuming that none of the discrete eigenvalues of H and H_0 coincide. If H_0 is taken to be the free electron Hamiltonian, it has no discrete spectrum and we can use Equation 7.56 without any problem to find $\psi(\vec{r})$ corresponding to the discrete spectrum of H. If ε belongs to the continuous part of the spectrum, then the limiting procedure as discussed before has to be used, that is, we have to use $G_0^{\pm}(\varepsilon)$ in Equation 7.54

$$\left|\psi^{\pm}\right\rangle = \left|\phi\right\rangle + G_0^{\pm}(\varepsilon)H_1\left|\psi^{\pm}\right\rangle \tag{7.57}$$

or

$$\psi^{\pm}(\vec{r}) = \phi(\vec{r}) + \int G_0^{\pm}(\vec{r},\vec{r}',\varepsilon)H_1(\vec{r}',\vec{r}'')\psi^{\pm}(\vec{r}'')d^3r'd^3r'' \tag{7.58}$$

Generally $H_1(\vec{r}',\vec{r}'') = \delta(\vec{r}' - \vec{r}'')v(\vec{r}')$, thus

$$\psi^{\pm}(\vec{r}) = \phi(\vec{r}) + \int G_0^{\pm}(\vec{r},\vec{r}',\varepsilon)v(\vec{r}')\psi^{\pm}(\vec{r}')d^3r' \tag{7.59}$$

Equation 7.58 or Equation 7.59 is called the Lippman–Schwinger equation. Note that if $\phi(\vec{r})$ does not satisfy the same boundary condition as $\psi^{\pm}(\vec{r})$, it should be dropped in these equations.

Let us now express Equation 7.57 in terms of T. This can be done by substituting for $|\psi^{\pm}\rangle$ on the right-hand side of Equation 7.57, that is

$$\begin{aligned}
\left|\psi^{\pm}\right\rangle &= \left|\phi\right\rangle + G_0^{\pm}(\varepsilon)H_1\left|\phi\right\rangle + G_0^{\pm}(\varepsilon)H_1G_0^{\pm}(\varepsilon)H_1\left|\phi\right\rangle + \cdots \\
&= \left|\phi\right\rangle + G_0^{\pm}(H_1 + H_1G_0^{\pm}H_1 + \cdots)\left|\phi\right\rangle \\
&= \left|\phi\right\rangle + G_0^{\pm}\, T^{\pm}\left|\phi\right\rangle
\end{aligned} \tag{7.60}$$

Comparing Equations 7.57 and 7.60, we get

$$H_1\ |\psi^{\pm}\rangle = T^{\pm}\ |\phi\rangle \tag{7.61}$$

Equation 7.47 can also be written as

$$T = G_0^{-1}GH_1$$

or

$$G_0 T = GH_1$$

Substituting this into Equation 7.60 we get,

$$|\psi^{\pm}\rangle = |\phi\rangle + G^{\pm}H_1|\phi\rangle \tag{7.62}$$

Equations 7.60 and 7.62 are sometimes very useful as these express $|\psi^{\pm}\rangle$ in terms of $|\phi\rangle$ and not in terms of $|\psi^{\pm}\rangle$ as is the case with Equation 7.57.

7.4 Free Electron Green's Function in Three Dimensions

As an example, we shall find Green's function for a free electron gas in three dimensions (3D). For simplicity, we shall use the atomic units (a.u.) in this section (see Appendix C).

For the free electron gas, the Hamiltonian, its eigenvalues, and eigenfunctions are given by

$$H(\vec{r}) = -\frac{\hbar^2}{2m_e}\nabla^2 = -\frac{1}{2}\nabla^2 \quad \text{(in a.u.)} \tag{7.63}$$

$$\varepsilon_{\vec{k}} = \frac{\hbar^2 k^2}{2m_e} = \frac{k^2}{2} \quad \text{(in a.u.)} \tag{7.64}$$

$$\left\langle \vec{r} \middle| \vec{k} \right\rangle = \frac{1}{\sqrt{\Omega}} e^{i\vec{k}\cdot\vec{r}} \tag{7.65}$$

Green's function satisfies the equation

$$\left(z + \frac{\nabla^2}{2}\right) G(\vec{r},\vec{r}',z) = \delta(\vec{r} - \vec{r}') \tag{7.66}$$

Using Equation 7.14, the solution of Equation 7.66 can be written as

$$G(\vec{r},\vec{r}',z) = \sum_{\vec{k}} \frac{\left\langle \vec{r} \middle| \vec{k} \right\rangle \left\langle \vec{k} \middle| \vec{r}' \right\rangle}{z - k^2/2} = \int \frac{e^{i\vec{k}\cdot(\vec{r}-\vec{r}')}}{z - k^2/2} \frac{d^3k}{(2\pi)^3} \tag{7.67}$$

To evaluate the integral in Equation 7.67, let us write

$$\vec{d} = (\vec{r} - \vec{r}')$$

Also we choose spherical polar coordinates such that k_z axis coincides with $\vec{d}$. Thus,

$$e^{i\vec{k}\cdot(\vec{r}-\vec{r}')} = e^{i\vec{k}\cdot\vec{d}} = e^{ikd\cos\theta}$$

Therefore, Equation 7.67 can be written as

$$G(\vec{r},\vec{r}',z) = \int \frac{e^{ikd\cos\theta}}{z - k^2/2} \frac{d^3k}{(2\pi)^3}$$

$$= \int \frac{2\pi}{z - k^2/2} \frac{k^2 dk}{8\pi^3} \int_0^{\pi} e^{ikd\cos\theta} \sin\theta \, d\theta$$

$$= \frac{1}{4\pi^2} \int \frac{k^2 dk}{z - k^2/2} \left(-\frac{e^{ikdt}}{ikd} \right) \Bigg|_1^{-1}$$

$$G(\vec{r},\vec{r}',z) = \frac{1}{4\pi^2} \int_0^{\infty} \frac{1}{ikd} (e^{ikd} - e^{-ikd}) \frac{k^2 dk}{z - k^2/2}$$

$$= \frac{1}{4\pi^2 id} \int_0^{\infty} (e^{ikd} - e^{-ikd}) \frac{k\, dk}{z - k^2/2}$$

$$= \frac{1}{4\pi^2 id} \int_{-\infty}^{\infty} \frac{k\, e^{ikd}}{z - k^2/2} dk \tag{7.68}$$

The integral in Equation 7.68 can be easily evaluated in a complex plane. We first note that the integrand has poles at

$$k^{\pm} = \pm\sqrt{2}\, z^{1/2} \tag{7.69}$$

One of these poles will have positive imaginary part unless z is a real and positive number. Thus, the integration in Equation 7.68 can be done by closing the path by an infinite semicircle in the upper half-plane as shown in Figure 7.3 and then using the residue theorem.

Here, we are interested in finding $G^{\pm}(\vec{r},\vec{r}',\varepsilon)$. Let us first calculate $G^{+}(\vec{r},\vec{r}',\varepsilon)$ and write $z = \varepsilon + iy$ where $y \to 0$. Thus,

$$k^{\pm} = \pm\sqrt{2}(\varepsilon + iy)^{1/2}$$

$$= \pm\sqrt{2}\sqrt{\varepsilon}\left(1 + \frac{iy}{\varepsilon}\right)^{1/2} = \pm\sqrt{2\varepsilon}\left(1 + \frac{iy}{2\varepsilon}\right) \tag{7.70}$$

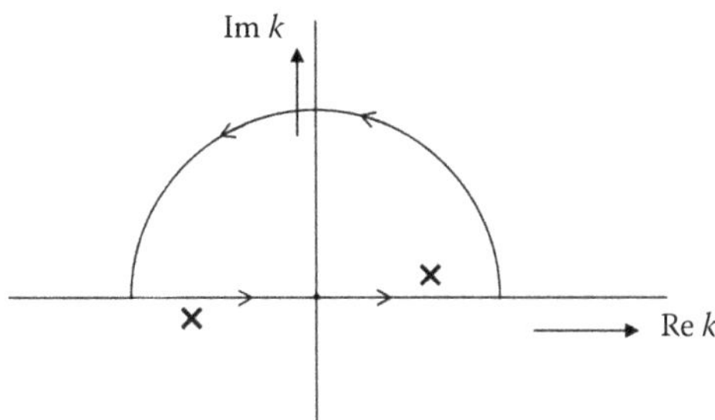

FIGURE 7.3
Path of integration. The poles are indicated by crosses.

Thus, the integral I in Equation 7.68 is

$$\lim_{k\to k^+} + 2\pi i \frac{2\sqrt{2\varepsilon}\, e^{i\sqrt{2\varepsilon}d}}{(\sqrt{2z}-k)(\sqrt{2z}+k)}(k-k^+)$$

$$= -2\pi\, i e^{i\sqrt{2\varepsilon}d}$$

Thus,

$$G^+(\vec{r},\vec{r}\,',\varepsilon) = -\frac{1}{4\pi^2 id} 2\pi\, i\, e^{i\sqrt{2\varepsilon}d}$$

$$= -\frac{1}{2\pi}\frac{e^{i\sqrt{2\varepsilon}|\vec{r}-\vec{r}'|}}{|\vec{r}-\vec{r}'|} \quad \text{if } \varepsilon > 0 \tag{7.71}$$

To calculate $G^-(\vec{r},\vec{r}\,',\varepsilon)$, we will have to replace z by $\varepsilon - iy$ and then take $y \to 0$ limit. Now the pole lying in the upper half of the complex plane is $k^- = \sqrt{2\varepsilon}(1-(iy/2\varepsilon))$. This will give

$$G^-(\vec{r},\vec{r}',\varepsilon) = -\frac{1}{2\pi}\frac{e^{-i\sqrt{2\varepsilon}|\vec{r}-\vec{r}'|}}{|\vec{r}-\vec{r}'|} \quad \varepsilon > 0 \tag{7.72}$$

Note that for $\varepsilon > 0$, G^+ and G^- have discontinuity as expected. For $\varepsilon < 0$, there is no singularity in the integrand in Equation 7.68 on the real axis. Now, the poles are on the imaginary axis.

$$k^{\pm} = \pm\sqrt{2}(-|\varepsilon|)^{1/2} = \pm i\sqrt{2|\varepsilon|}$$

Green's function, then, is given by

$$G(\vec{r},\vec{r}',\varepsilon) = -\frac{1}{2\pi}\frac{e^{-\sqrt{2|\varepsilon|}|\vec{r}-\vec{r}'|}}{|\vec{r}-\vec{r}'|} \quad \varepsilon < 0 \tag{7.73}$$

From Equation 7.71

$$\mathrm{Im}\, G^{+}(\vec{r},\vec{r}\,',\varepsilon) = -\frac{1}{2\pi}\frac{\sin(\sqrt{2\varepsilon}\,|\vec{r}-\vec{r}\,'|)}{|\vec{r}-\vec{r}\,'|}$$

$$\lim_{\vec{r}\to\vec{r}\,'} \mathrm{Im}\, G^{+}(\vec{r},\vec{r}\,',\varepsilon) = -\frac{\sqrt{2\varepsilon}}{2\pi}$$

The DOS $\rho(\varepsilon)$ is, thus, given by

$$\rho(\varepsilon) = -\frac{1}{\pi N}\int \mathrm{Im}\, G^{+}(\vec{r},\vec{r},\varepsilon)d^3r$$

$$= -\frac{1}{\pi N}\frac{\sqrt{2\varepsilon}}{2\pi}\Omega = -\frac{\Omega}{2N\pi^2}\sqrt{2}\varepsilon^{1/2} \tag{7.74}$$

which goes as $\varepsilon^{1/2}$ as expected.

7.5 Korringa–Kohn–Rostoker (KKR) Method

The KKR method was first developed by Korringa (1947) and then by Kohn and Rostoker (1954) to calculate the energy bands in solids and is known as the KKR method. It can be generalized to disordered alloys as we shall see in the next chapter. The KKR method is also known as the Green's function method as it is formulated in terms of Green's function. The details of the method are quite involved so here I shall only give a brief outline.

Let us consider an electron moving in a periodic potential $v(\vec{r})$. Our aim is to find energy bands, which are obtained by solving the Kohn–Sham equation,

$$\left(\frac{-\hbar^2}{2m_e}\nabla^2 + v(\vec{r})\right)\psi_{\vec{k}}(\vec{r}) = \varepsilon_{\vec{k}}\psi_{k}(\vec{r})$$

or in atomic units

$$\left(\varepsilon_{\vec{k}} + \frac{\nabla^2}{2}\right)\psi_{\vec{k}}(\vec{r}) = v(\vec{r})\psi_{\vec{k}}(\vec{r}) \tag{7.75}$$

We already know the solution of Equation 7.75 without $v(\vec{r})$, which are plane waves given by Equation 7.65 and the corresponding Green's function (7.71). We can use Equation 7.59 to write the solution of Equation 7.75 as

$$\psi_{\vec{k}}(\vec{r}) = \int G_0^+(\vec{r},\vec{r}',\varepsilon_{\vec{k}})v(\vec{r}')\psi_{\vec{k}}(\vec{r}')d^3r' \tag{7.76}$$

There are two points that are worth noting regarding Equation 7.76. First, we have chosen G_0^+ to build up our solution, which is the normal practice in the literature, but the whole analysis can be carried out using G_0^- as well. Second, we have not added in Equation 7.76, the solution of homogenous equation, which is a plane wave. This is related to the boundary condition. Since $v(\vec{r})$ is periodic, $\psi_{\vec{k}}(\vec{r})$ satisfies Bloch's theorem

$$\psi_{\vec{k}}(\vec{r} + \vec{R}_i) = e^{i\vec{k}\cdot\vec{R}_i}\,\psi_{\vec{k}}(\vec{r}) \tag{7.77}$$

where $\vec{R}_i$ is a lattice vector. The solution of Equation 7.75 with $v(\vec{r}) = 0$ is

$$\phi = \frac{1}{\sqrt{\Omega}}e^{i\vec{k}'\cdot\vec{r}} \tag{7.78}$$

where

$$k' = \sqrt{2\varepsilon_{\vec{k}}} \tag{7.79}$$

From this, it is clear that in general $k' \neq k$. Thus, $\psi_{\vec{k}}(\vec{r})$ and $\phi(\vec{r})$ cannot satisfy Bloch's theorem with the same $\vec{k}$ and the only way $\psi_{\vec{k}}(\vec{r})$ can satisfy Bloch's theorem is to drop $\phi(\vec{r})$.

The periodic potential $v(\vec{r})$ can be written as

$$v(\vec{r}) = \sum_i v_a(\vec{r} - \vec{R}_i) \tag{7.80}$$

where $v_a(\vec{r})$ is the contribution from each atom. Here, we are assuming that there is only one atom per primitive cell. The potential $v_a(\vec{r})$ is assumed to be of the muffin-tin form, that is

$$v_a(\vec{r}) = \begin{cases} v_a(r) & \text{if } r \le r_m \\ 0 & \text{if } r > r_m \end{cases} \tag{7.81}$$

That is, the potential $v_a(\vec{r})$ is assumed to be spherically symmetric within a radius, r_m, called the muffin-tin radius, and is assumed to be zero in the remaining part of the primitive cell. The muffin-tin potentials are nonoverlapping potentials as shown schematically in Figure 7.4 for a two-dimensional solid. The solid is assumed to have a square lattice. The figure shows few unit cells (Wigner–Seitz unit cell) and each cell is divided into two

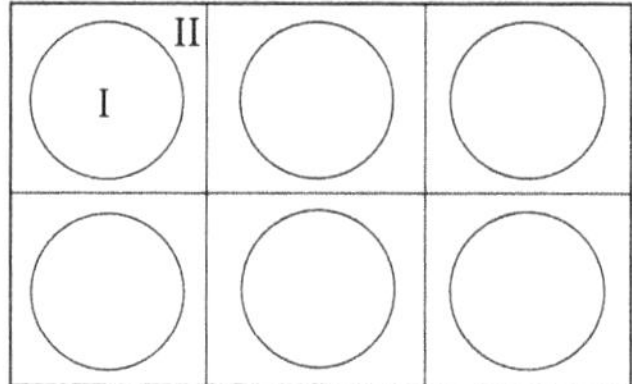

FIGURE 7.4
Muffin-tin potentials are nonoverlapping potentials shown schematically for a two-dimensional solid.

regions, I and II. Inside the circles (region I), the potential is attractive and deep (see Figure 6.3) but outside the circle (region II) it is flat. We see that Figure 7.4 looks like a muffin-tin and thus such a potential is called a muffin-tin potential. This kind of potential was first proposed by Slater while developing the APW method (see Section 7.7).

Substituting Equation 7.71 for G_0^+ and Equation 7.81 for $v(\vec{r})$ into Equation 7.76, we get

$$\psi_{\vec{k}}(\vec{r}) = -\frac{1}{2\pi}\sum_i \int \frac{e^{i\kappa|\vec{r}-\vec{r}'|}}{|\vec{r}-\vec{r}'|} v_a(\vec{r}'-\vec{R}_i)\psi_{\vec{k}}(\vec{r}')d^3r' \tag{7.82}$$

where

$$\kappa = \sqrt{2\varepsilon} \tag{7.83}$$

Replacing $\vec{r}'$ by $\vec{r}' + \vec{R}_i$ and using Equation 7.77, we get

$$\begin{aligned}\psi_{\vec{k}}(\vec{r}) &= -\frac{1}{2\pi}\sum_i \int \frac{e^{i\kappa|\vec{r}-\vec{r}'-\vec{R}_i|}}{|\vec{r}-\vec{r}'-\vec{R}_i|} v_a(\vec{r}')e^{i\vec{k}\cdot\vec{R}_i}\psi_{\vec{k}}(\vec{r}')d^3r' \\ &= -\frac{1}{2\pi}\int g(\vec{r},\vec{r}',\kappa,\vec{k})v_a(\vec{r}')\psi_{\vec{k}}(\vec{r}')d^3r' \end{aligned} \tag{7.84}$$

where

$$g(\vec{r},\vec{r}',\kappa,\vec{k}) = \sum_i \frac{e^{i\kappa|\vec{r}-\vec{r}'-\vec{R}_i|}}{|\vec{r}-\vec{r}'-\vec{R}_i|} e^{i\vec{k}\cdot\vec{R}_i} \tag{7.85}$$

and is called structural Green's function as it depends only on the structure of the lattice and not on the potential. Since $v_a(r')$ is zero outside the radius r_m, the integration in Equation 7.84 is limited only over a single muffin-tin sphere.

Now a functional Λ is constructed, which upon variation gives Equation 7.84. This can easily be seen to be

$$\Lambda = \int \psi_{\vec{k}}^{*}(\vec{r}) v_a(r) \psi_{\vec{k}}(\vec{r}) d^3r + \frac{1}{2\pi} \int \psi_{\vec{k}}^{*}(\vec{r}) v_a(r) g(\vec{r}, \vec{r}', \kappa, \vec{k}) v_a(r') \psi_{\vec{k}}(\vec{r}') \, d^3r \, d^3r' \tag{7.86}$$

This can be seen by taking the functional derivative $\delta\Lambda/\delta\psi_{\vec{k}}^{*}(r)$ and setting it to be zero. Note that in addition to $\psi_{\vec{k}}^{*}(\vec{r})$, $v_a(r)$ was also introduced in Equation 7.86. This has the advantage that all the integrals in Equation 7.86 are over muffin-tin spheres. These integrals can further be simplified by using Green's identity, which reduces them to surface integrals over the spheres.

The use of muffin-tin potential has more advantages. As it is spherically symmetric, we can choose a trial function of the following form inside each cell:

$$\psi_{\vec{k}}(\vec{r}) = \sum_{\ell m} C_{\ell m} \phi_\ell(r) Y_{\ell m}(\theta, \phi) = \sum_{L} C_L \phi_\ell(r) Y_L(\theta, \phi) \tag{7.87}$$

where $\phi_\ell(r)$ is the solution of the radial Schrödinger equation for $v_a(r)$ and $Y_{\ell m}(\theta,\phi)$ are the spherical harmonics. In Equation 7.87, we have used a composite index $L \equiv (\ell, m)$. The substitution of Equation 7.87 into Equation 7.86 leads to a function

$$\Lambda = \sum_{L} \sum_{L'} \Lambda_{LL'} C_L C_{L'}^{*} \tag{7.88}$$

where $\Lambda_{LL'}$ are coefficients and here we shall not discuss how to evaluate them. For details, the paper by Kohn and Rostoker (1954) should be consulted. However, the final form of $\Lambda_{LL'}$ is quite interesting, therefore, I shall just quote the final result, which is

$$\Lambda_{LL'} = A_{LL'} + \kappa \delta_{LL'} \cot \delta_\ell \tag{7.89}$$

Here, the coefficients $A_{LL'}$ depend on $\vec{k}, \kappa$ and the structure of the solid and are called KKR structure constants. In Equation 7.89, δ_ℓ are phase shifts which depend only on the potential (see Chapter 6). Thus, the form of Equation 7.89 is particularly suited for efficient computation. For the same structure, one need not calculate $A_{LL'}$ again and again.

The variational functional Λ in Equation 7.88 has to be minimized with respect to $C_{L'}^{*}$, and this leads to a set of linear algebraic equations in C_L, that is

$$\sum_{L} \Lambda_{LL'} C_L = 0 \tag{7.90}$$

These equations have nontrivial solution only if

$$\det|\Lambda_{LL'}| = 0$$

or from Equation 7.89

$$\det\left|A_{LL'} + \kappa\delta_{LL'}\cot\delta_{\ell}\right| = 0 \tag{7.91}$$

This is the KKR secular equation. This determinant is a function of $\vec{k}$ and energy ε, and its roots give the band structure, that is, $\varepsilon_{\vec{k}}$ vs. $\vec{k}$. For calculating band structure, one fixes $\vec{k}$ and varies ε to find the roots of the determinant. In principle, the size of the determinant in Equation 7.91 is infinite, but in practical calculation, $\ell_{\max}$ is chosen at about 3. This makes the size of the determinant 16×16, which is quite small. Using this method, Moruzzi et al. (1978) have calculated band structures of a large number of elemental solids and were able to explain various trends seen in properties such as bulk modulus, cohesive energy, and Wigner–Seitz radius. This proved to be a turning point that established the density function theory as the foundation on which most electronic structure calculations are based.

The muffin-tin approximation to the potential is the main approximation in the KKR method. It is because of this approximation that the calculation becomes quite simple. For metals, this approximation is particularly good, because the charge density is nearly spherically symmetric in a large region around the atom. But in semiconductors and insulators, the charge density is highly directional and the muffin-tin approximation is not a good approximation for these materials. It turned out to be difficult to extend the scheme for the full potential, although several attempts have been made to develop a KKR-type scheme by using the full-potential rather than muffin-tin potential. Such methods are known as full-potential KKR methods or FP-KKR methods. For details, see for example, the paper by Asato et al. (1999), who have done electronic structure calculations using this method for several metals and semiconductors and have found this method quite satisfactory.

7.6 Linear Muffin-Tin Orbital (LMTO) Method

We saw that the energy eigenvalues $\varepsilon_{\vec{k}}$ in methods like the plane wave expansions method were obtained by solving an equation like

$$det\left|(\varepsilon^{0}_{\vec{k}-\vec{G}} - \varepsilon_{\vec{k}})\delta_{\vec{G}\vec{G}'} + v(\vec{G}' - \vec{G})\right| = 0 \tag{7.92}$$

Thus by solving this equation, one could get all the eigenvalues $\varepsilon_{\vec{k}}$ in a single diagonalization. In the KKR method, we get the eigenvalues by finding the roots of the equation

$$det\left|A_{LL'} + \kappa\delta_{LL'}\cot\delta_{\ell}\right| = 0 \tag{7.93}$$

Finding roots of Equation 7.93 is a much slower process than getting $\varepsilon_{\vec{k}}$ from Equation 7.92. The roots are found one by one by changing energy while in the plane wave expansion method, one gets all the eigenvalues in one go. If somehow Equation 7.93 can be cast in the form of Equation 7.92, the method can become very fast.

The reason why the KKR method is slow is that the basis set is energy dependent in contrast to the plane wave method. The plane waves form an energy-independent basis set, which results in a simpler secular equation like Equation 7.92. In the LMTO method, by using a suitable approximation, it was possible to cast the secular equation in the form like Equation 7.92, thus speeding up the calculation.

Andersen (1975) first constructed a muffin-tin orbital (MTO) $\phi_L(\varepsilon,\vec{r})$ for a single muffin-tin potential and then expanded it around reference energy ε_v, using Taylor's expansion retaining terms up to linear in energy, that is

$$\phi_L(\varepsilon,\vec{r}) = \phi_L(\varepsilon_v,\vec{r}) + (\varepsilon - \varepsilon_v)\dot{\phi}_L(\varepsilon_v,\vec{r}) \tag{7.94}$$

where $\dot{\phi}_L$ is the energy derivative of ϕ_L evaluated at $\varepsilon = \varepsilon_v$. The orbital ϕ_L given by Equation 7.94 is called an LMTO. Now, the wave function of electron $\psi_{\vec{k}}(\vec{r})$ can be expanded in terms of LMTOs as

$$\psi_{\vec{k}}(\vec{r}) = \sum_{\vec{R},L} a_L^{\vec{k}} e^{i\vec{k}\cdot\vec{R}}\phi_L(\varepsilon,\vec{r} - \vec{R}) \tag{7.95}$$

where the sum is over all the lattice vector $\vec{R}$ and angular momentum index L. Here, we have assumed one atom per primitive unit cell. This leads to an equation for $\varepsilon_{\vec{k}}$ as

$$det\left|H_{LL'}(\vec{k}) - \varepsilon_{\vec{k}}O_{LL'}(\vec{k})\right| = 0 \tag{7.96}$$

where $H_{LL'}(\vec{k})$ and $O_{LL'}(\vec{k})$ are energy-independent matrix elements. Equation 7.96 has the same form as Equation 7.92 and thus gives all the eigenvalues in a single diagonalization and saves enormous computing time. Andersen also gave the tight-binding version of the LMTO known as the TB-LMTO. More recently, a new NMTO method has been presented, which goes beyond the linear approximation and gives more accurate results. For details, we refer to the original papers (Andersen 1975, Andersen and Saha-Dasgupta 2000).

7.7 Augmented Plane Wave (APW) Method

The APW method was given by J.C. Slater (1937), one of the pioneers in the field of electronic structure. We assume that in each cell the potential is of muffin-tin form, that is, it is spherically symmetric within a radius r_m and zero outside.

$$\begin{aligned} v(\vec{r}) &= v(r) \quad \text{for } r < r_m \\ &= 0 \quad \text{for } r > r_m \end{aligned}$$

For simplicity, we assume that we have only one atom per unit cell. Pictorially, it is shown in Figure 7.4 for a square lattice in two dimensions. We can divide the whole space in two parts: region I corresponding to muffin-tin spheres and region II, the interstitial region as indicated in Figure 7.4. In region II the potential is zero, so the wave function of the electron in this region is likely to be of extended nature, while in region I it is likely to be of more localized character. Thus, if we could construct a basis which is more localized in region I and has extended character like a plane wave in region II, we are likely to have a very small number of basis functions in the expansion of the wave function of the electron. This is the idea behind the APW method.

In region I, the solution of the Schrödinger equation is of the form

$$\chi(\vec{r}) = \sum_{\ell m} a_{\ell m}\phi_\ell(\varepsilon, r)Y_{\ell m}(\hat{r}) \tag{7.97}$$

where $Y_{\ell m}(\hat{r})$ are spherical harmonics for direction $\hat{r}$, $a_{\ell m}$, the expansion coefficients and $\phi_\ell(\varepsilon, r)$ satisfies the radial equation (in atomic units)

$$\left[\frac{1}{2}\left(-\frac{d^2}{dr^2} + \frac{\ell(\ell+1)}{r^2}\right) + v(r)\right] r\phi_\ell(\varepsilon, r) = \varepsilon r\phi_\ell(\varepsilon, r) \tag{7.98}$$

In the interstitial region the potential is zero and the solutions of the Schrödinger equation are plane waves. Therefore, $\chi(\vec{r})$ given by Equation 7.97 is exactly matched to plane wave $e^{i\vec{k}\cdot\vec{r}}$ at $r = r_m$. This condition gives $a_{\ell m}$ as follows:

Suppose $\vec{r}$ lies in nth cell in the interstitial region, we can write

$$\vec{r} = \vec{r}_n + \vec{\rho} \tag{7.99}$$

where $\vec{\rho}$ is measured with respect to the center of the nth sphere as shown in Figure 7.5.

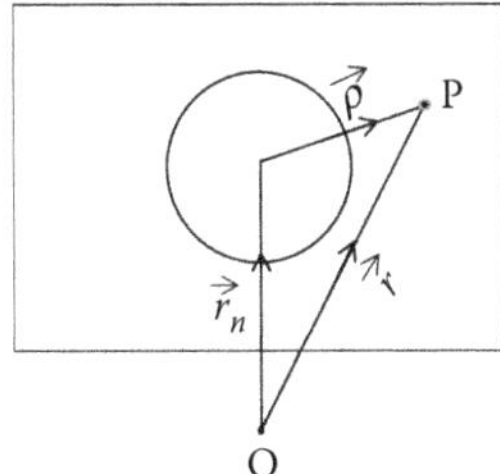

FIGURE 7.5
The figure shows how $\vec{r}$ can be expanded about the center of the nth cell.

We can write

$$\begin{aligned}\chi(\vec{r}) &= e^{i\vec{k}\cdot\vec{r}} = e^{i\vec{k}\cdot(\vec{r}_n+\vec{\rho})} \\ &= e^{i\vec{k}\cdot\vec{r}_n} e^{i\vec{k}\cdot\vec{\rho}}\end{aligned} \tag{7.100}$$

we can expand $e^{i\vec{k}\cdot\vec{\rho}}$ as

$$e^{i\vec{k}\cdot\vec{\rho}} = 4\pi \sum_{\ell m} i^{\ell} j_{\ell}(k\rho) Y_{\ell m}(\vec{\rho}) Y^{*}_{\ell m}(\vec{k}) \tag{7.101}$$

so that

$$e^{i\vec{k}\cdot\vec{r}} = 4\pi\, e^{i\vec{k}\cdot\vec{r}_n} \sum_{\ell m} i^{\ell} j_{\ell}(k\rho) Y_{\ell m}(\vec{\rho}) Y^{*}_{\ell m}(\vec{k}) \tag{7.102}$$

Now matching $e^{i\vec{k}\cdot\vec{r}}$ at the surface of nth sphere ($\rho = r_m$) to Equation 7.97 gives $a_{\ell m}$

$$a_{\ell m} = 4\pi\, e^{i\vec{k}\cdot\vec{r}_n} i^{\ell} j_{\ell}(kr_m) Y^{*}_{\ell m}(\vec{k}) / \phi_{\ell m}(\varepsilon, r_m) \tag{7.103}$$

Thus, we define an APW as

$$\chi_{\vec{k}}(\vec{r}) = \begin{cases} e^{i\vec{k}\cdot\vec{r}} \text{ outside sphere} \\ 4\pi e^{i\vec{k}\cdot\vec{r}_n} \sum_{\ell m} i^{\ell} j_{\ell}(kr_m) Y^{*}_{\ell m}(\vec{k}) (\phi_{\ell m}(\varepsilon,\rho)) / \phi_{\ell m}(\varepsilon, r_m) \text{ inside } n\text{th sphere} \end{cases} \tag{7.104}$$

Note that $\chi_{\vec{k}}(\vec{r})$ satisfies Bloch's theorem.

$$\chi_{k}(\vec{r} + \vec{R}) = e^{i\vec{k}\cdot\vec{R}} \chi_{\vec{k}}(\vec{r}) \tag{7.105}$$

and is discontinuous in slope at the surface of the muffin-tin sphere. Now, just like in the plane wave or OPW methods, we expand the wave function of the electron in terms of APWs as

$$\psi_{\vec{k}}(\vec{r}) = \sum_{\vec{G}} C_{\vec{k}+\vec{G}}\chi_{\vec{k}+\vec{G}}(\vec{r}) \tag{7.106}$$

where $C_{\vec{k}+\vec{G}}$ are unknown coefficients, which are determined by a variational procedure. The wave function (7.106) is used as a trial wave function and coefficients are adjusted so as to give minimum energy. We shall not go into details of this calculation here and refer to the book by Loucks (1967). The final result has a very interesting form, very much similar to the one we got for the plane wave expansion method,

$$\sum_{\vec{G}} \left\{ \left[\frac{1}{2}(\vec{k}+\vec{G})^2 - \varepsilon \right] \delta_{\vec{G}'\vec{G}} + V_{\vec{G}'\vec{G}}(\varepsilon,\vec{k}) \right\} C_{\vec{k}+\vec{G}} = 0 \tag{7.107}$$

where

$$\begin{aligned} V_{\vec{G}'\vec{G}}(\varepsilon,\vec{k}) = {} & \frac{4\pi r_m^2}{\Omega_{cell}} \left(\frac{1}{2}\left|\vec{k}+\vec{G}\right|^2 - \varepsilon \right) \frac{j_\ell\left(\left|\vec{G}-\vec{G}'\right| r_m\right)}{\left|\vec{G}-\vec{G}'\right|} \\ & + \frac{2\pi r_m}{\Omega_{cell}} \sum_{\ell} \left\{ (2\ell+1) P_\ell(\cos\theta_{\vec{G}\vec{G}'}) j_\ell\left(\left|\vec{k}+\vec{G}'\right| r_m\right) j_\ell\left(\left|\vec{k}+\vec{G}\right| r_m\right) \right\} \\ & \times \left[r\frac{d}{dr}\ln\phi_\ell(\varepsilon,r) - r\frac{d}{dr}\ln j_\ell\left(\left|\vec{k}+\vec{G}\right| r\right) \right]_{r=r_m} \end{aligned} \tag{7.108}$$

where $\theta_{\vec{G}\vec{G}'}$ is the angle between $\vec{k}+\vec{G}$ and $\vec{k}+\vec{G}'$. The eigenvalues $\varepsilon_{\vec{k}}$ are obtained by seeking the solution of linear equations in $C_{\vec{k}+\vec{G}'}$, which can exist only if the determinant of the coefficients given in curly brackets in Equation 7.107 vanishes. These coefficients depend on ε. Thus, one finds the determinant for a certain energy ε and then varies it until the root is found as in the KKR method. By repeating this procedure for each $\vec{k}$, one can find the band structure $\varepsilon_{\vec{k}}$ vs. $\vec{k}$.

7.8 Linear Augmented Plane Wave (LAPW) Method

One big disadvantage of the APW method is that the basis functions are energy dependent. Thus, the energy bands cannot be determined by a single diagonalization. As we saw above that to find $\varepsilon_{\vec{k}}$, it is necessary to find the secular determinant as a function of ε and then find its roots. This is computationally very demanding and a nonlinear procedure to find $\varepsilon_{\vec{k}}$. What we would like is to obtain all eigenvalues at one $\vec{k}$ by a single diagonalization as is the case in the plane wave method, where the basis functions are energy independent. In such cases, we have linear equations for eigenvalues. In the LAPW method, this is achieved by defining basis functions as the linear combination of radial function $\phi_\ell(\varepsilon_v, r)$ and its energy derivative $\dot{\phi}_\ell(\varepsilon_v, r)$ evaluated at chosen fixed energy ε_v. The energy derivative satisfies the equation (in atomic units)

$$\left[\frac{1}{2}\left\{-\frac{d^2}{dr^2} + \frac{\ell(\ell+1)}{r^2}\right\} + v(r) - \varepsilon\right] r\,\dot{\phi}_\ell(\varepsilon_v, r) = r\,\phi_\ell(\varepsilon_v, r) \tag{7.109}$$

where $\phi_\ell(\varepsilon_v, r)$ satisfies Equation 7.98, same as in the APW method.

The LAPW basis is defined as

$$\chi_{\vec{k}}^{LAPW} = \begin{cases} e^{i\vec{k}\cdot\vec{r}}, & \text{in the interstitial region} \\ \sum_{\ell m} \{a_{\ell m}\phi_\ell(\varepsilon_v, \rho) + b_{\ell m}\dot{\phi}_\ell(\varepsilon_v, \rho)\}\, Y_{\ell m}(\vec{\rho}), & \rho < r_m \end{cases}$$

where $a_{\ell m}$ and $b_{\ell m}$ are coefficients, analogous to the APW method and ρ is measured from the center of the nth cell (see Figure 7.5). Thus, the LAPW are just plane waves in the interstitial region but have more variational freedom inside the spheres compared to the APWs. This is because if ε_v differs slightly from the band energy, a linear combination will reproduce the APW radial function as

$$\phi_\ell(\varepsilon, r) = \phi_\ell(\varepsilon_v, r) + (\varepsilon - \varepsilon_v)\dot{\phi}_\ell(\varepsilon_v, r) + O(\varepsilon - \varepsilon_v)^2$$

where $O(\varepsilon - \varepsilon_v)^2$ denotes errors that are quadratic in this energy difference. Thus, the LAPW method introduces error of the order of $(\varepsilon - \varepsilon_v)^2$ in the wavefunction and error of the order of $(\varepsilon - \varepsilon_v)^4$ in the band energy. Because of the high order of this error, the LAPWs form a good basis set over a relatively large energy region, so that all valence bands can be treated with a single set of ε_v.

The resulting equations have the same form as Equation 7.107 with the simplification that $V_{\vec{G}'\vec{G}}$ is independent of energy. This is a big advantage because in the LAPW method, one can get all $\varepsilon_{\vec{k}}$ at a $\vec{k}$ by a single diagonalization, thus saving much computational cost compared to the APW method. The LAPW method has also been generalized to full potential and is known as the full-potential LAPW (FP-LAPW) method. For details, we refer to the book by Singh and Nordstrom (2006).

7.9 Linear Scaling Methods

The computational effort involved in electronic structure calculations using the methods discussed so far scales at least as N^3, where N is the number of electrons in the unit cell. This means that if N is doubled or the cell is doubled, the computational time will increase at least eight times. Thus, treating large unit cells with atoms more than 100 becomes very difficult using these methods. To handle such large systems using the first-principles techniques, several methods, for which computational effort scales linearly with N, have been developed during the last two decades. Such methods are known as linear scaling methods or $O(N)$ methods. Here, we shall briefly discuss the principle involved to achieve linear scaling. It turns out that linear scaling is not achieved simply by some mathematical trick but involves deep understanding of physics, namely, the concept of locality in quantum mechanics.

The locality in quantum mechanics means that properties of a region comprising of one or a few atoms are mainly dependent on the local environment and weakly influenced by regions which are far from it. For example, semiconductors are often described in terms of covalent bonds which are mainly dependent on the immediate neighborhood. In metals, the total charge density can be obtained within a good accuracy by the superposition of atomic charge densities. Since the atomic charge density decays rapidly with distance, this implies that the charge density in a metal is mainly determined by its local environment. Kohn (1996) describes such quantities which depend mainly on its local environment using the term "nearsightedness." The "nearsightedness" is a consequence of quantum mechanical destructive interference of electrons.

The "nearsightedness" or locality is not reflected in standard electronic structure calculations because such calculations are formulated in terms of wave functions which extend over the entire volume. Due to this, much computational effort is wasted when dealing with large systems as these methods do not take advantage of the "nearsightedness." During the last two decades, many linear scaling approaches have been developed and all of them take advantage of the "nearsightedness" in one way or another. Most of these approaches are formulated in terms of localized orbitals, density matrix, or Green's functions. In most of the orbital and density matrix approaches a "divide-and-conquer" strategy is used for taking advantage of the "nearsightedness." In the "divide-and-conquer" strategy, the full system is divided into subsystems that are overlapping portions of the full system. These subsystems, also called "localization regions," have the same physical and chemical properties as the full system but do not depend on the volume of the full system. The electron density is calculated for each subsystem using conventional methods. The Hamiltonian for each subsystem contains potential energy of other neighboring subsystems, so that the electron density and energy in the original subsystem converge. Thus, the need for diagonalizing the Hamiltonian of the full system is avoided. In these methods, the computational cost will scale as $(N_{at})^3$ where N_{at} is the number of atoms in the original subsystem + neighboring subsystems. For large systems with atoms $N > N_{at}$, the computational cost will scale linearly with N. A different approach is based on energy functional minimization with respect to either localized functions or localized density matrices. The Green's function approach uses a different strategy. The Hamiltonian is expressed in terms of a orthogonal localized basis in which it is tridiagonal and symmetric. Green's function is expressed as a continued fraction which is terminated at some level by using the so-called "terminator." The accuracy of the method depends on what level the continued fraction is terminated and the choice of the terminator. Once Green's function is obtained, the DOS and energy can be calculated. For details, we refer to the book of Martin (2004) and the review article by Goedecker (1999).

EXERCISES

7.1 Show that

$$H_1 G_0 T = TG_0 H_1$$

7.2 Find Green's function for the free electron gas in one dimension. Using this, calculate the DOS and compare your result with that of Exercise 2.8.

7.3 Calculate the DOS for the free electron gas in two dimensions using the Green's function method and compare your result with that of Exercise 2.8.

7.4 Take the functional derivative of Equation 7.86 with respect to $\psi^*_{\vec{k}}(\vec{r})$ and by setting it to zero show that one gets Equation 7.84.

7.5 Show that the expansion coefficients in Equation 7.97 are given by

$$a_{\ell m} = 4\pi e^{i\vec{k}\cdot\vec{r}_n} i^\ell j_\ell(kr_m) Y^*_{\ell m}(\vec{k}) / \phi_{\ell m}(\varepsilon, r_m).$$

7.6 Using the KKR code mentioned in Appendix A, calculate band structure, DOS, and charge density of copper. Compare your results with those shown in Chapter 4.

7.7 Using the LAPW code mentioned in Appendix A, calculate band structure, Fermi surface, DOS, and charge density of copper. Compare your results with those shown in Chapter 4.

7.8 Using the LMTO code mentioned in Appendix A, calculate the band structure, DOS, and charge density of copper. Compare your results with those shown in Chapter 4.

Further Reading

Ashcroft N. and Mermin N. 1976. *Solid State Physics*. New York, NY: W. B. Saunders Company.

Economou E. N. 1979. *Green's Function in Quantum Physics*. Berlin: Springer.

Singh D. J. and Nordstrom L. 2006. *Planewaves, Pseudopotentials, and the LAPW Method*, Second Edition. New York, NY: Springer.

Ziman J. M. 1969. *Principles of the Theory of Solids*. Cambridge: Cambridge University Press.

8

Disordered Alloys

8.1 Introduction

In this chapter, we shall discuss how to calculate the electronic structure of disordered alloys. Disordered alloys such as bronze (CuZn) have been known to mankind for ages and form an important class of materials. These materials do not have translational symmetry; therefore, Bloch's theorem is not applicable to them. Thus, the methods of electronic structure calculations that were developed for ordered materials are, in general, not applicable to such materials. In a limited way, one can use supercell geometry to handle nonperiodic systems, as discussed in Chapter 6, but it is very expensive. The methods that we shall discuss are more cost-effective and general.

Disordered materials can be broadly classified in two categories:

1. Substitutional disordered alloys: In such materials, there is an underlying lattice but each site of the lattice can be occupied by any constituents of the alloy as shown schematically in Figure 8.1.
2. Amorphous systems: Such systems do not have any underlying lattice as shown schematically in Figure 8.2. Amorphous Si, ordinary glass, and metallic glasses are some examples of such systems.

Theoretically, amorphous systems are more difficult to handle and will be discussed later in Chapter 11. Here, we shall first try to develop methods for calculating the electronic structure of substitutional disordered alloys. This problem is best formulated in terms of Green's function, which we discussed in Chapter 7. For simplicity, we shall focus our discussion on random binary alloys. A random binary alloy has two components and each lattice site can be randomly occupied by any component with some probability. We shall denote an alloy of A and B atoms by A_xB_y, where x and y denote the atomic concentrations of A and B atoms in the alloy. We shall start with the discussion of short and long-range order (LRO) in alloys in Section 8.2. Then, we discuss Green's function for an ordered solid and for a substitutional impurity embedded in the ordered solid in Section 8.3. In Section 8.4, we shall come to the problem of disordered alloys and develop approximations such as CPA. In Section 8.5, we

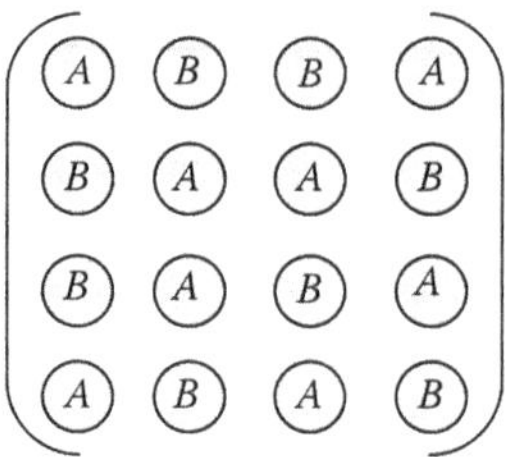

FIGURE 8.1
Schematic representation of two-dimensional substitutional alloy of A and B atoms that lie on a periodic lattice.

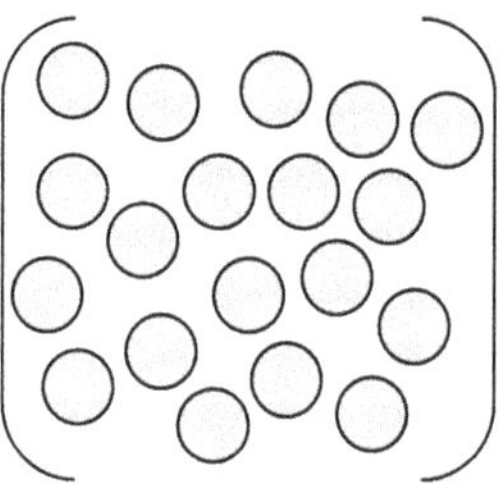

FIGURE 8.2
Schematic representation of an amorphous solid in two dimensions.

apply the theory to a tight-binding model. Since such a simple tight-binding model cannot be applied to realistic systems, we shall develop the formulation in the KKR framework using the muffin-tin model in Section 8.6. This theory is known as the KKR-CPA, which has been very successful in determining the electronic structure of realistic alloy systems and in making contact with experiments such as ARPES, 2D-ACAR, Compton, and optical experiments. Some examples of the KKR-CPA results for some alloys and comparison with experimental results will be given in Section 8.7. Finally, we shall discuss some theoretical attempts to go beyond the CPA and other recent developments in Section 8.8.

8.2 Short- and Long-Range Order

A random alloy, in which each site is occupied randomly by its constituent atoms, is an idealization. In general, there will be a deviation from a perfect random alloy and a real alloy may show some order. Depending on the length scales over which this order exists, we talk of short-range order (SRO) and long-range order (LRO). The SRO corresponds to the order that may exist at length scales of the order of nearest-neighbor distances whereas the LRO corresponds to the order that may exist over length scales approaching infinity. For example, in a crystalline solid, the atoms are arranged periodically and thus both LRO and SRO exist. In contrast, in an ideal gas, the atomic arrangement is totally random and therefore, both SRO and LRO are absent. In disordered alloys and amorphous materials, the LRO is absent but there may be some SRO. Similarly, in ferromagnets and antiferromagnets, the magnetic moments are aligned periodically, and, therefore, there is magnetic LRO.

The SRO can be characterized by an SRO parameter. For a disordered A_xB_y alloy, let $p(A,B)$ be the probability of finding a B atom as a nearest neighbor of A atom. Warren–Cowely short-range parameter α for a nearest-neighbor pair is defined as

$$\alpha = 1 - p(A,B)/y$$

where y is the concentration of B atoms in the alloy. It can be seen from the above equation that $\alpha = 0$ represents a perfect random alloy as $p(A,B) = y$ in this case. Similarly, $\alpha < 0$ when $p(A,B) > y$, which means AB- and BA-type ordering is favored. Using similar logic, we can see that $\alpha > 0$ represents a system in which AA- and BB-type clustering is favored.

8.3 An Impurity in an Ordered Solid

Before we discuss Green's function for an impurity in an ordered solid, we shall discuss how to obtain Green's function for an ordered solid in the tight-binding representation.

a. *Green's function for an ordered solid* Using the procedure outlined in Chapter 7, we will construct Green's function for a perfect solid, using a simple tight-binding model. In this case, the system is periodic and each site is identical. Let us assume a simple tight-binding Hamiltonian

$$H = \sum_i |i\rangle \varepsilon_i \langle i| + \sum_{ij} |i\rangle h_{ij} \langle j| \tag{8.1}$$

where $|i\rangle$ and $\langle j|$ denote Wannier functions centered at sites i and j, ε_i are site energies, and h_{ij} are hopping matrix elements. Since each site is identical, $\varepsilon_i = \varepsilon_0$. For simplicity, let us assume that

$$\begin{aligned} h_{ij} &= h \quad \text{where } i \text{ and } j \text{ are nearest neighbors} \\ &= 0 \quad \text{otherwise} \end{aligned} \tag{8.2}$$

Since $|i\rangle$ and $\langle j|$ are orthonormal

$$\langle i|j\rangle = \delta_{ij} \tag{8.3}$$

The Wannier functions are related to Bloch functions $|\vec{k}\rangle$ by (see Equation 5.54)

$$|\vec{k}\rangle = \frac{1}{\sqrt{N}}\sum_i e^{i\vec{k}\cdot\vec{R}_i}|i\rangle \tag{8.4}$$

where $\vec{R}_i$ is the position vector of site i. The eigenvalues are given by (see Equation 5.48)

$$\varepsilon_{\vec{k}} = \varepsilon_0 + h{\sum_i}' e^{i\vec{k}\cdot\vec{R}_i} \tag{8.5}$$

where the prime over summation indicates that $i = 0$ term is excluded. In Equation 8.4, N is the total number of sites in the solid.

From Equation 8.4, we have

$$\langle i|\vec{k}\rangle = \frac{1}{\sqrt{N}}e^{i\vec{k}\cdot\vec{R}_i} \tag{8.6}$$

Using Equation 7.14, we can write Green's function as

$$G_0(z) = \sum_{\vec{k}}\frac{|\vec{k}\rangle\langle\vec{k}|}{z - \varepsilon_{\vec{k}}}$$

Let us define

$$\begin{aligned} G_0(i,j,z) &= \langle i|G(z)j\rangle \\ &= \sum_{\vec{k}}\frac{\langle i|\vec{k}\rangle\langle\vec{k}|j\rangle}{z-\varepsilon_{\vec{k}}} = \frac{1}{N}\sum_{\vec{k}}\frac{e^{i\vec{k}\cdot(\vec{R}_i-\vec{R}_j)}}{z-\varepsilon_{\vec{k}}} \end{aligned} \tag{8.7}$$

If we put $i = j$, we get

$$\begin{aligned} G_0(i,i,z) &= \frac{1}{N}\sum_{\vec{k}}\frac{1}{z-\varepsilon_{\vec{k}}} \\ &= \frac{1}{N}\int\sum_{\vec{k}}\frac{1}{z-\varepsilon}\delta(\varepsilon-\varepsilon_{\vec{k}})d\varepsilon \end{aligned} \tag{8.8}$$

$$= \int \frac{\rho(\varepsilon)}{z - \varepsilon} d\varepsilon \tag{8.9}$$

Thus, if the DOS $\rho(\varepsilon)$ is known, $G_0(i,i,z)$ can be constructed using Equation 8.9. Alternatively, if $G_0(i,i,z)$ is known, $\rho(\varepsilon)$ can be calculated by taking the imaginary part of Equation 8.9 as we had seen in Chapter 7

$$\rho(\varepsilon) = -\frac{1}{\pi} Im\, G_0(i,i,\varepsilon^+) \tag{8.10}$$

b. *Green's function for a single impurity in a solid* Let us consider an impurity that is kept on a site i of an otherwise perfect solid. We assume that site energy at site i is

$$\epsilon_i = \epsilon_0 + \delta$$

where ϵ_0 is the site energy at all other sites. The hopping elements between the nearest-neighboring sites are assumed to be h. Let us write the Hamiltonian as

$$H = H_0 + H_1 \tag{8.11}$$

where

$$H_0 = \sum_i |i\rangle \varepsilon_0 \langle i| + \sum_{ij}{}' |i\rangle h \langle j| \tag{8.12}$$

and

$$H_1 = |i\rangle \delta \langle i| \tag{8.13}$$

A prime over Σ in Equation 8.12 indicates that $i = j$ is excluded and the sum is only over the nearest neighbors.

For Hamiltonian 8.12, we already know Green's function G_0 (see Equation 8.7). The Green function G for Hamiltonian 8.11 can be obtained by using Dyson's equation (see Equation 7.41)

$$G = G_0 + G_0 H_1 G$$

$$G = G_0 + G_0 H_1 (G_0 + G_0 H_1 G)$$

$$= G_0 + G_0 H_1 G_0 + G_0 H_1 G_0 H_1 G_0 + \cdots$$

$$G = G_0 + G_0 t_i G_0 \tag{8.14}$$

where

$$
\begin{aligned}
t_i &= H_1 + H_1 G_0 H_1 + \cdots \\
&= |i\rangle\delta\langle i| + |i\rangle\delta\langle i|\, G_0\, |i\rangle\delta\langle i| + \cdots \\
&= |i\rangle\, [\delta + \delta^2 G_0(i,i,z) + \delta^3 G_0^2(i,i,z) + \cdots]\langle i| \\
&= |i\rangle \frac{\delta}{1 - \delta G_0(i,i,z)} \langle i| \\
&= |i\rangle \frac{\delta}{1 - \delta F(z)} \langle i|
\end{aligned} \tag{8.15}
$$

where

$$F(z) = G_0(i,i,z) \tag{8.16}$$

Taking the matrix element of Equation 8.14, we get

$$
\begin{aligned}
G(i,i,\varepsilon) &= \langle i|G|i\rangle \\
&= G_0(i,i,\varepsilon) + \frac{\delta G_0^2(i,i,\varepsilon)}{1 - \delta G_0(i,i,\varepsilon)}
\end{aligned} \tag{8.17}
$$

Thus, we have found Green's function for an impurity in a solid, from which we can find the DOS at site i.

8.4 Disordered Alloy: General Theory

We assume a random binary alloy of say A and B atoms with concentration x and $1 - x$, respectively. This means each site can be occupied by A with probability x and by B with probability $1 - x$. As we have mentioned earlier, in a disordered alloy, there is no translational symmetry and as a result, Bloch's theorem is not applicable. Thus, all the methods we have developed so far could not be applied to obtain the electronic structure of a disordered alloy. Since we are considering a substitutional disordered alloy, which has an underlying periodic lattice, we can make a simple approximation, that is, we replace the alloy by a periodic system of effective atoms as shown in Figure 8.3. The effective atom can be chosen in many different ways:

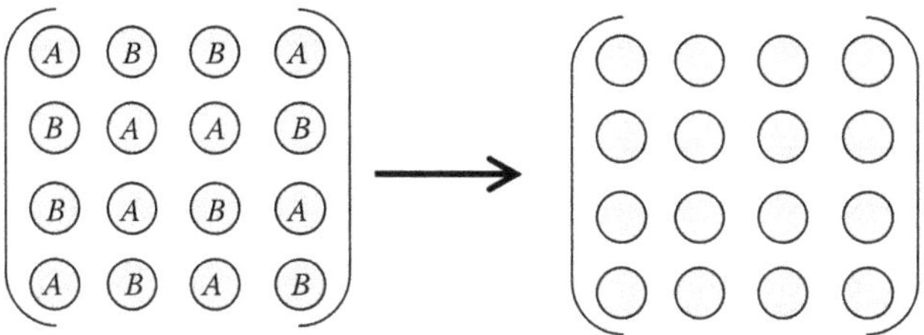

FIGURE 8.3
A disordered alloy is replaced by an ordered system of effective atoms.

1. *Virtual Crystal Approximation (VCA)* We can assume that the effective atom has a potential, which is an average of the potentials of A and B atoms, that is

$$v_{eff} = xv_A + (1 - x)v_B$$

   where v_A and v_B are potentials of A and B atoms, respectively. This approximation is known as the VCA.

2. *Average t-Matrix Approximation (ATA)* In this approximation, we assume that the t-matrix of the effective atom is an average of t_A and t_B, the t-matrix of A and B atoms, respectively, that is

$$\langle t \rangle = xt_A + yt_B$$

   where $y = 1 - x$.

3. *Coherent Potential Approximation (CPA)* In this approximation, the effective atom is chosen so that the average scattering from the effective atom is zero, that is, the average t-matrix from A or B atoms embedded in the effective medium is zero

$$\langle t \rangle = xt_A + yt_B = 0$$

   where t_A and t_B are the t-matrixes of A and B atoms embedded in the effective medium. This is known as CPA and is the best of the three approximations.

In the VCA, since the potential is real, Bloch's states are still the solution of the problem, with infinite lifetimes. This is unphysical because due to the disorder in the system, Bloch's states will scatter and will have a finite lifetime. However, in spite of this problem, it is sometimes used due to its simplicity. Now, we shall develop a theory of alloys using the ATA and CPA and derive the condition for finding the effective atom.

We assume that the one-electron Hamiltonian for the alloy A_xB_{1-x} can be written as

$$H = H_0 + V \tag{8.18}$$

where H_0 is some suitably chosen periodic unperturbed Hamiltonian and V is the deviation from this periodic part. We assume that V can be written as the sum over site contributions

$$V = \sum_i v_i \tag{8.19}$$

where i is the site index. For the binary A_xB_{1-x}, v_i will take values v_A or v_B depending on whether i is occupied by A or B. Since for any value of x, there will be a large number of possible configurations, H will be configuration dependent. However, we will be interested in configurational averages of physical quantities, such as charge density or DOS. For this purpose, we shall develop a theory in terms of the average Green's function that can be directly related to these properties.

The average Green's function $G_{av}(z)$ can be written as

$$G_{av}(z) = \langle G_c(z) \rangle \tag{8.20}$$

where $\langle\ \rangle$ denotes the configurational average and $G_c(z)$ is the configuration-dependent Green's function

$$G_c(z) = (z - H)^{-1} \tag{8.21}$$

The average Green's function can also be written in terms of self-energy $\Sigma(z)$, which is periodic, that is

$$\begin{aligned} G_{av}(z) &= \langle (z - H)^{-1} \rangle \\ &= \left(z - H_0 - \Sigma(z)\right)^{-1} \end{aligned} \tag{8.22}$$

We shall develop the approximation to find $G_{av}(z)$ or alternatively $\Sigma(z)$.

We first rewrite the Hamiltonian H by adding and subtracting a periodic potential

$$\widetilde{V}(z) = \sum_i \tilde{v}_i(z) \tag{8.23}$$

Thus,

$$H = H_0 + V = H_0 + \tilde{V} - \tilde{V} + V$$
$$= \tilde{H} + U(z) \tag{8.24}$$

where

$$\tilde{H} = H_0 + \tilde{V}$$

and

$$U(z) = V - \tilde{V} = \sum_i (v_i - \tilde{v}_i(z)) = \sum_i u_i(z) \tag{8.25}$$

As we had seen in Section 7.3, $G_c(z)$ can be expanded using Dyson's equation

$$G_c(z) = (z - H_0 - V)^{-1}$$
$$= (z - \tilde{H} - U)^{-1}$$
$$= \tilde{G} + \tilde{G}UG_c \tag{8.26}$$

where

$$\tilde{G} = (z - \tilde{H})^{-1} = (z - H_0 - \tilde{V})^{-1} \tag{8.27}$$

By repeatedly substituting Equation 8.26 for G_c on the right-hand side of Equation 8.26, we get an infinite series as

$$G_c(z) = \tilde{G} + \tilde{G}\,U\,\tilde{G} + \tilde{G}\,U\,\tilde{G}\,U\,\tilde{G} + \cdots$$
$$= \tilde{G} + \tilde{G}\,(U + U\,\tilde{G}\,U + \cdots)\,\tilde{G}$$
$$= \tilde{G} + \tilde{G}\,T\,\tilde{G} \tag{8.28}$$

where t-matrix T is

$$T = U + U\tilde{G}U + U\tilde{G}U\tilde{G}U + \cdots$$
$$= \sum_i u_i + \sum_i u_i\tilde{G}\sum_j u_j + \cdots \tag{8.29}$$

Suppose, we have only one impurity at site i, then $u_j = 0$ for $j \neq i$ and $T = t_i$, which can be written by using Equation 8.29 as

$$
\begin{aligned}
t_i &= u_i + u_i \tilde{G} u_i + \cdots \\
&= u_i + u_i \tilde{G}(u_i + u_i \tilde{G} u_i + \cdots) \qquad (8.30) \\
&= u_i + u_i \tilde{G}\, t_i \qquad (8.31)
\end{aligned}
$$

where we used Equation 8.30 in writing Equation 8.31.

From Equation 8.31, we can get t_i in terms of u_i and $\tilde{G}$ as

$$t_i = (1 - u_i \tilde{G})^{-1} u_i \qquad (8.32)$$

Now, let us take the configurational average of Equation 8.28

$$G_{av} = \langle G_c(z) \rangle = \tilde{G} + \tilde{G}\langle T \rangle \tilde{G}$$

This is because $\tilde{G}$ is a periodic function and is similar for each configuration. This can be rewritten as

$$G_{av} = (1 + \tilde{G}\langle T \rangle)\tilde{G} \qquad (8.33)$$

or

$$
\begin{aligned}
G_{av}^{-1} &= \tilde{G}^{-1}(1 + \tilde{G}\langle T \rangle)^{-1} \\
&= \tilde{G}^{-1}(1 + \tilde{G}\langle T \rangle - \tilde{G}\langle T \rangle)(1 + \tilde{G}\langle T \rangle)^{-1} \\
&= \tilde{G}^{-1}(1 - \tilde{G}\langle T \rangle(1 + \tilde{G}\langle T \rangle)^{-1}) \qquad (8.34)
\end{aligned}
$$

Using Equation 8.22, the self-energy can be written as

$$
\begin{aligned}
\Sigma(z) &= z - H_0 - G_{av}^{-1} = z - H_0 - \tilde{G}^{-1}(1 - \tilde{G}\langle T \rangle)(1 + \tilde{G}\langle T \rangle)^{-1} \\
&= z - H_0 - \tilde{G}^{-1} + \langle T \rangle(1 + G\langle T \rangle)^{-1} \\
&= \tilde{V} + \langle T \rangle(1 + G\langle T \rangle)^{-1} \qquad (8.35)
\end{aligned}
$$

where we have used Equations 8.34 and 8.27.

Now, we return to Equation 8.29 and write it as

$$T = \sum_i T_i \qquad (8.36)$$

where

$$
\begin{aligned}
T_i &= u_i + u_i \tilde{G} \sum_j u_j + u_i \tilde{G} \sum_j u_j \tilde{G} \sum_k u_k + \cdots \\
&= u_i + u_i \tilde{G} \left(\sum_j u_j + \sum_j u_j \tilde{G} \sum_k u_k + \cdots \right) \\
&= u_i + u_i \tilde{G} T \\
&= u_i (1 + \tilde{G} T)
\end{aligned}
\tag{8.37}
$$

where we have used Equation 8.29. Equation 8.37 can be further rewritten using Equation 8.36 as

$$
\begin{aligned}
T_i &= u_i + u_i \tilde{G} \sum_j T_j \\
&= u_i + u_i \tilde{G} T_i + u_i \tilde{G} \sum_{j \neq i} T_j
\end{aligned}
$$

or

$$
(1 - u_i \tilde{G}) T_i = u_i + u_i \tilde{G} \sum_{j \neq i} T_j
$$

or

$$
\begin{aligned}
T_i &= (1 - u_i \tilde{G})^{-1} u_i + (1 - u_i \tilde{G})^{-1} u_i \tilde{G} \sum_{j \neq i} T_j \\
&= t_i + t_i \tilde{G} \sum_{j \neq i} T_j
\end{aligned}
\tag{8.38}
$$

where we have used Equation 8.32. This can be rewritten by the repeated use of Equation 8.38 on the right-hand side

$$
T_i = t_i + t_i \tilde{G} \sum_{j \neq i} t_j + t_i \tilde{G} \sum_{j \neq i} t_j \sum_{k \neq j} t_k + \cdots \tag{8.39}
$$

Substituting into Equation 8.36, we get

$$T = \sum_i t_i + \sum_i t_i \tilde{G} \sum_{j \neq i} t_j + \sum_i t_i \tilde{G} \sum_{j \neq i} t_j \tilde{G} \sum_{k \neq j} t_k + \cdots \tag{8.40}$$

The form of this expansion is interesting. We can see that the total scattering operator can be written as the sum of contributions coming from single-site scattering (first term), two-site scattering (second term), and so on. This is a multiple scattering expansion of T and, therefore, this theory is sometimes referred to as the multiple scattering theory.

Now, let us take the configurational average of Equation 8.36, that is

$$\langle T \rangle = \sum_i \langle T_i \rangle \tag{8.41}$$

where using Equation 8.38, we get

$$\begin{aligned} \langle T_i \rangle &= \left\langle t_i \left(1 + \tilde{G} \sum_{j \neq i} T_j \right) \right\rangle \\ &= \langle t_i \rangle \left(1 + \tilde{G} \sum_{j \neq i} \langle T_j \rangle \right) \\ &\quad + \langle (t_i - \langle t_i \rangle) \tilde{G} \sum_{j \neq i} (T_j - \langle T_j \rangle) \rangle \end{aligned} \tag{8.42}$$

Now, we neglect the second term, that is, we neglect the correlations between sites j and i. This is known as the single-site approximation (SSA). Such an approximation is not valid if there is short-range ordering or clustering in the system.

Now, using Equation 8.41, we write

$$\langle T \rangle = \langle T_i \rangle + \sum_{j \neq i} \langle T_j \rangle$$

or

$$\sum_{j \neq i} \langle T_j \rangle = \langle T \rangle - \langle T_i \rangle \tag{8.43}$$

Now, substituting this into Equation 8.42 and neglecting the second term

$$\langle T_i \rangle = \langle t_i \rangle (1 + \tilde{G}(\langle T \rangle - \langle T_i \rangle))$$

or

$$(1 + \langle t_i \rangle \tilde{G})\langle T_i \rangle = \langle t_i \rangle (1 + \tilde{G}\langle T \rangle)$$

or

$$\langle T_i \rangle = (1 + \langle t_i \rangle \tilde{G})^{-1} \langle t_i \rangle (1 + \tilde{G}\langle T \rangle) \tag{8.44}$$

Now, we return to Equation 8.35 and write it as

$$\begin{aligned} \Sigma(z) &= \sum_i \tilde{v}_i + \sum_i \langle T_i \rangle (1 + \tilde{G}\langle T \rangle)^{-1} \\ &= \sum_i \sigma_i(z) \end{aligned} \tag{8.45}$$

where

$$\begin{aligned} \sigma_i(z) &= \tilde{v}_i + \langle T_i \rangle (1 + \tilde{G}\langle T \rangle)^{-1} \\ &= \tilde{v}_i + (1 + \langle t_i \rangle \tilde{G})^{-1} \langle t_i \rangle \end{aligned} \tag{8.46}$$

where we have used Equation 8.44.

Equation 8.46 suggests two ways of calculating self-energy Σ.

1. We can assume some periodic potential $\tilde{V}$ and calculate $\tilde{G}$ using Equation 8.27, so that $\langle t_i \rangle$ can be calculated as

$$\langle t_i \rangle = x t_A + y t_B \tag{8.47}$$

   where t_A and t_B can be calculated using Equation 8.32. This approximation is known as the ATA and is not unique as it depends on the choice of $\tilde{V}$.
2. We can choose $\tilde{V}$ such that $\langle t_i \rangle = 0$ so that the second term in Equation 8.46 is zero and

$$\sigma_i(z) = \tilde{v}_i(z)$$

   or

$$\Sigma(z) = \tilde{V}(z) \tag{8.48}$$

That is, we choose the effective medium such that the average scattering from each site with respect to the effective medium is zero, that is

$$\langle t_i \rangle = x t_A + y t_B = 0 \tag{8.49}$$

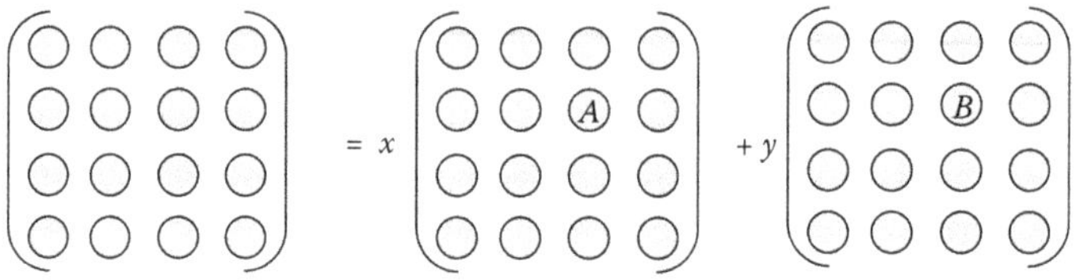

FIGURE 8.4
In the CPA, the effective medium is determined by requiring that the average scattering from an A or B site embedded in the effective medium is zero.

In other words, if we embed an A or B atom in the effective medium, the average scattering from this site is zero in accordance with Equation 8.49. This is known as CPA and is shown in Figure 8.4. This is the best single-site approximation.

Let us write the CPA condition in terms of Green's function. For this purpose, we fix the $i = 0$ site by putting an A atom and taking the configurational average of G_c (Equation 8.28) over all other sites

$$\langle G_c \rangle_{0=A} = \tilde{G} + \tilde{G} \langle T \rangle_{0=A} \tilde{G} \tag{8.50}$$

where the symbol $\langle\ \rangle_{0=A}$ represents restricted site averaging and means that configurational averaging has been taken with respect to all sites except $i = 0$ site.

Using Equation 8.36, we write

$$\begin{aligned} \langle T \rangle_{0=A} &= \sum_i \langle T_i \rangle_{0=A} \\ &= \langle T_0 \rangle_{0=A} + \sum_{i \neq 0} \langle T_i \rangle_{0=A} \end{aligned} \tag{8.51}$$

In the SSA, each site is surrounded by the effective medium; as a result, all fluctuations of the local environment are neglected. This implies that

$$\langle T_i \rangle_{0=A} = \langle T_i \rangle \quad \text{for } i \neq 0 \tag{8.52}$$

Also, by using Equation 8.38

$$\begin{aligned} \langle T_0 \rangle_{0=A} &= t_A + t_A \tilde{G} \sum_{i \neq 0} \langle T_i \rangle_{0=A} \\ &= t_A \left(1 + \tilde{G} \sum_{i \neq 0} \langle T_i \rangle \right) \end{aligned} \tag{8.53}$$

where we have used Equation 8.52. Now, by using Equation 8.42 (neglecting the last term of Equation 8.42), this can be written as

$$\langle T_0 \rangle_{0=A} = t_A \langle t_0 \rangle^{-1} \langle T_0 \rangle \tag{8.54}$$

Thus, we can write Equation 8.51 as

$$\begin{aligned}
\langle T_0 \rangle_{0=A} &= t_A \langle t_0 \rangle^{-1} \langle T_0 \rangle + \sum_{i \neq 0} \langle T_i \rangle \\
&= t_A \langle t_0 \rangle^{-1} \langle T_0 \rangle + \langle T \rangle - \langle T_0 \rangle \\
&= \langle T \rangle + (t_A \langle t_0 \rangle^{-1} - 1) \langle T_0 \rangle \\
&= \langle T \rangle + (t_A - \langle t_0 \rangle) \langle t_0 \rangle^{-1} \langle T_0 \rangle \\
&= \langle T \rangle + (t_A - x t_A - y t_B) \langle t_0 \rangle^{-1} \langle T_0 \rangle \\
&= \langle T \rangle + y (t_A - t_B) \langle t_0 \rangle^{-1} \langle T_0 \rangle
\end{aligned} \tag{8.55}$$

where $y = 1 - x$.

Similarly, it can be shown that

$$\langle T \rangle_{0=B} = \langle T \rangle + x (t_B - t_A) \langle t_0 \rangle^{-1} \langle T_0 \rangle \tag{8.56}$$

Equations 8.55 and 8.56 imply that

$$x \langle T \rangle_{0=A} + y \langle T \rangle_{0=B} = \langle T \rangle \tag{8.57}$$

As a result from Equation 8.50

$$\begin{aligned}
x \langle G_c \rangle_{0=A} + y \langle G_c \rangle_{0=B} &= \tilde{G} + \tilde{G} (x \langle T \rangle_{0=A} + y \langle T \rangle_{0=B}) \tilde{G} \\
&= \tilde{G} + \tilde{G} \langle T \rangle \tilde{G} \\
&= G_{av}
\end{aligned} \tag{8.58}$$

Equation 8.58 is another way of writing CPA conditions that we shall use later.

8.5 Application to the Single Band Tight-Binding Model of Disordered Alloy

We assume a tight-binding model for the disordered alloy A_xB_{1-x}, described by the Hamiltonian H

$$H = \sum_i |i\rangle \varepsilon_i \langle i| + \sum_{ij}{}' |i\rangle h \langle j| \tag{8.59}$$

Here, all ε_i's are not equal but take values, ε_A or ε_B depending on whether the site is occupied by atom A or B. In Equation 8.59, h is the hopping matrix element between the nearest-neighbor sites and is assumed to be similar irrespective of the disorder. In other words, we ignore the off-diagonal disorder in H.

For convenience, we shall express ε_A and ε_B in units of ω and define the zero of energy such that

$$\varepsilon_A = \frac{1}{2}\omega\delta \quad \text{and} \quad \varepsilon_B = -\frac{1}{2}\omega\delta \tag{8.60}$$

where

$$\delta = \frac{\varepsilon_A - \varepsilon_B}{\omega} \tag{8.61}$$

ω is a scaling factor, which for convenience will be chosen equal to unity.

We choose

$$\tilde{H} = H_o + \sum_i |i\rangle \tilde{V} \langle i|$$

and

$$H_{eff} = H_o + \sum_i |i\rangle \Sigma \langle i| \tag{8.62}$$

where H_o is the Hamiltonian corresponding to a simple tight-binding model of a simple cubic lattice, $\tilde{H}$, the Hamiltonian of a periodic system with potential $\tilde{V}$ at each site, and H_{eff}, effective Hamiltonian for the alloy with self-energy Σ at each site. Note that $\tilde{V}$ and Σ are simply numbers and not operators as in the previous section.

Now, we shall use ATA and CPA conditions (Equations 8.47 and 8.48), to obtain self-energy $\Sigma(z)$. For this, we have to know $\langle t_i \rangle$, which we can find using Equation 8.15.

$$\begin{aligned}\langle t_i \rangle &= xt_A + yt_B \\ &= |i\rangle\left[x\frac{(\varepsilon_A - \tilde{V})}{1-(\varepsilon_A - \tilde{V})\tilde{F}} + y\frac{(\varepsilon_B - \tilde{V})}{1-(\varepsilon_B - \tilde{V})\tilde{F}}\right]\langle j| \end{aligned} \tag{8.63}$$

where

$$\begin{aligned}\tilde{F} &= \tilde{G}(i,i,z) \\ &= G_o(i,i,z-\tilde{V}) \end{aligned} \tag{8.64}$$

This is because $\tilde{H}$ differs from H_0 only by a constant $\tilde{V}$. Substituting Equation 8.63 into Equation 8.46 of the previous section, we get the self-energy $\Sigma(z)$ in ATA.

The self-energy in the CPA is obtained by substituting $\tilde{V} = \Sigma$ into Equation 8.63 and setting $\langle t_i \rangle = 0$. This gives

$$\Sigma = x\varepsilon_A + (1-x)\varepsilon_B - (\varepsilon_A - \Sigma)\tilde{F}(\varepsilon_B - \Sigma) \tag{8.65}$$

Using this equation, Σ can be obtained by using an iterative procedure. Once Σ is known, one obtains the DOS

$$\begin{aligned}\rho(\varepsilon) &= -\frac{1}{\pi} Im\, G(i,i,\varepsilon) \\ &= -\frac{1}{\pi} Im\, G_0(i,i,\varepsilon - \Sigma) \end{aligned} \tag{8.66}$$

The DOS for this simple cubic tight-binding model is schematically shown in Figure 8.5 for various values of δ and $x = 0.15$ (Ehrenreich and Schwartz 1976). We see that for $\delta = 0.4$, the DOS gets somewhat distorted but looks similar to the DOS of the ordered material (see Figure 5.9) except that there are no sharp features corresponding to van Hove singularities. For $\delta = 0.75$, there is appearance of a structure in the upper part of the DOS. For $\delta = 1.5$, something very drastic happens and the DOS splits into two, with an energy gap in between. Thus, by increasing δ beyond 1.0, the single s band splits into two bands that is an important effect of the disorder. This range of $\delta > 1$ is called the split band regime. This is similar to what happens in the Hubbard model for large electron–electron repulsion U. This is not surprising because it is possible to map the Hubbard model onto the tight-binding alloy model for $x = y = (1/2)$ (Velický et al. 1968).

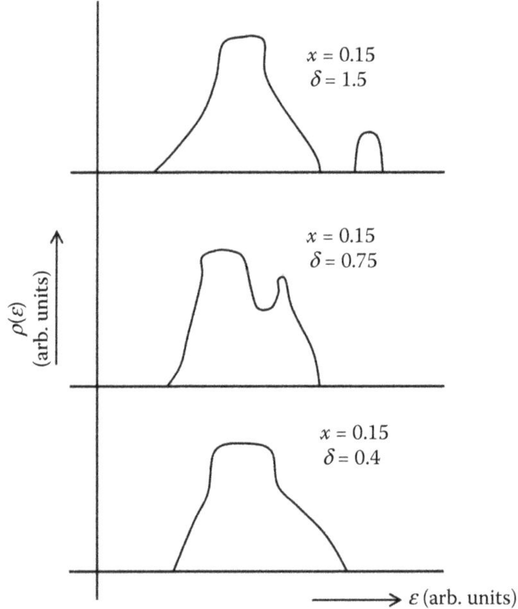

FIGURE 8.5
Schematic diagram showing the DOS of a disordered cubic tight-binding alloy model using the CPA for $x = 0.15$ and various values of δ.

8.6 Muffin-Tin Model: KKR-CPA

The single band tight-binding model of a disordered alloy discussed in Section 8.5 is not applicable to real systems such as Cu–Zn alloy because a real system has many bands. One can generalize this model to many bands, but then it has too many parameters. What we would like to develop is a parameter-free theory of the electronic structure of a disordered alloy similar to the one we have already discussed for the periodic solids. We shall see that this is possible if we use muffin-tin model of the alloy and apply the idea of the CPA developed in Section 8.4. In Chapter 7, we developed the KKR method for periodic solids using the muffin-tin model and Green's function. For disordered alloys, we shall develop the KKR-CPA method that combines the ideas of the KKR method and the CPA. Within the DFT, this gives rise to a first-principles theory of the electronic structure of disordered alloys. As the algebra involved in developing the theory is quite complicated, we shall highlight the important points and give only the outline of the derivation.

As we had seen in Chapter 7, a muffin-tin potential is spherically symmetric and finite within a radius r_m called muffin-tin radius and is zero outside it. Thus, for a disordered A_xB_{1-y} alloy, there will be two types of muffin-tin potentials, $v_A(r)$ and $v_B(r)$ representing A and B atoms. We assume that the potentials centered on different sites do not overlap. For a particular configuration of the alloy, one-electron Hamiltonian can be written as

$$H(\vec{r}) = H_0(\vec{r}) + V(\vec{r}) \tag{8.67}$$

where $H_0(\vec{r})$ is the free-electron Hamiltonian and $V(\vec{r})$ can be decomposed as

$$V(\vec{r}) = \sum_i v_i(r_i) \tag{8.68}$$

where the sum is over all sites and $\vec{r}_i = \vec{r} - \vec{R}_i$, $\vec{R}_i$ being the position vector of the ith site. $v_i(r_i)$ is the muffin-tin potential centered at $\vec{R}_i$ and is equal to $v_A(r_i)$ or $v_B(r_i)$ depending on whether i is occupied by an A or B atom.

As in Chapter 7, we shall use the abstract vector space and use the convention that

$$H(\vec{r}) = \langle \vec{r} | H | \vec{r} \rangle \tag{8.69}$$

where

$$H = H_0 + V \tag{8.70}$$

Green's function for this configuration in the abstract vector space can be written as

$$G_c(\varepsilon) = (\varepsilon + i0 - H)^{-1} \tag{8.71}$$

We are interested in $G_{av}(\varepsilon)$, the average of G_c over all configurations of the alloy, that is

$$G_{av} = \langle G_c(\varepsilon) \rangle = \langle (\varepsilon + i0 - H)^{-1} \rangle \tag{8.72}$$

As we had seen in Section 8.4, $G_c(\varepsilon)$ can be expanded as

$$G_c(\varepsilon) = G_0(\varepsilon) + G_0 T G_0 \tag{8.73}$$

T is given by

$$T = \sum_t t_i + \sum_t t_i G_0 \sum_{j \neq i} t_j + \cdots \tag{8.74}$$

where t_i is atomic t-matrix given by

$$t_i = v_i(1 - G_0 v_i)^{-1} \tag{8.75}$$

Equation 8.74 is an infinite series, which can be written in a closed form if we decompose T as

$$T = \sum_{ij} T_{ij} \tag{8.76}$$

Then T_{ij} satisfies the equation

$$T_{ij} = t_i\delta_{ij} + t_i G_0 \sum_{k \neq i} T_{kj} \tag{8.77}$$

This can be checked by substituting Equation 8.77 into Equation 8.76 repeatedly that gives us Equation 8.74. T_{ij} are called path operators and describe all possible scattering processes beginning at site i and ending at site j (Gyorffy 1972). Equation 8.77 is exact and represents infinite coupled operator equations for T_{ij}, which in general cannot be solved. The use of nonoverlapping muffin-tin potentials greatly simplifies the algebra making these equations solvable. Because of the spherical symmetry of the muffin-tin potential, the use of angular momentum representation is particularly convenient and simplifies the algebra. The matrix elements of t_i and T_{ij} in the coordinate representation can be expanded as

$$\begin{aligned}\langle \vec{r} | t_i | \vec{r}' \rangle &= t_i(\vec{r}_i, \vec{r}_i') \\ &= \sum_L Y_L(\vec{r}_i) t_i^{\ell}(r_i, r_i') Y_L(\vec{r}_j')\end{aligned} \tag{8.78}$$

$$\begin{aligned}\langle \vec{r} | T_{ij} | \vec{r}' \rangle &= T_{ij}(\vec{r}_i, \vec{r}_j') \\ &= \sum_{LL'} Y_L(\vec{r}_i) T_{ij}^{LL'}(\vec{r}_i, \vec{r}_j') Y_{L'}(\vec{r}_j')\end{aligned} \tag{8.79}$$

where L is a composite index (ℓ, m) and Y_L are spherical harmonics. The Fourier transform of $t_i(\vec{r}_i, \vec{r}_i^{\,\prime})$ can be written as

$$t_i(\vec{k}, \vec{k}') = \iint e^{-i\vec{k}\cdot\vec{r}_i} t_i(\vec{r}_i, \vec{r}_i') e^{i\vec{k}\cdot\vec{r}_i'} d^3r_i d^3r_i' \tag{8.80}$$

The $k = k' = \kappa = \sqrt{2\varepsilon}$ (in a.u.) matrix element, called the energy shell matrix element, can be easily evaluated in terms of the phase shift $\delta_i^\ell(\kappa)$ of potential $v_i(r)$

$$t_i^\ell(\kappa,\kappa) \equiv t_i(\kappa) = -\frac{1}{\kappa}e^{i\delta_i^\ell}\sin\delta_i^\ell \tag{8.81}$$

It turns out that most of the properties of interest such as the DOS can be expressed in terms of only the energy shell matrix elements of t_i and T_{ij} operators. Thus, Equation 8.77 reduces to the following system of linear algebraic equations in terms of energy shell matrix elements $t_i^\ell(\kappa)$ and $T_{ij}^{LL'}(\kappa)$

$$T_{ij}^{LL'}(\kappa) = t_i^\ell(\kappa)\delta_{ij}\delta_{LL'} + t_i^\ell(\kappa)\sum_{pL_1} B_{ip}^{LL_1}(\kappa)T_{pj}^{L_1L'}(\kappa) \tag{8.82}$$

where $B_{ip}^{LL'}(\kappa)$ is off-diagonal, that is, zero for $i = p$. It is related to the Fourier transform of the usual KKR structure constants and depends only on the lattice structure (Ehrenreich and Schwartz 1976). The solution of Equation 8.82 is

$$T_{ij}^{LL'}(\kappa) = \left[(t^{-1}(\kappa) - B)^{-1}\right]_{ij}^{LL'} \tag{8.83}$$

Now, let us find the configuration-dependent Green's function G_c in the coordinate representation. It is more convenient to express $G_c(\vec{r},\vec{r}',\varepsilon)$ in terms of regular and irregular solutions $Z_i^L(\vec{r},\varepsilon)$ and $J_i^L(\vec{r},\varepsilon)$ of the Schrödinger equation for an isolated potential $v_i(r)$ (Faulkner and Stocks 1980). $Z_i^L(\vec{r},\varepsilon)$ and $J_i^L(\vec{r},\varepsilon)$ are normalized such that for $r \geq r_m$

$$Z_i^L(\vec{r},\varepsilon) = \kappa Y_L(\vec{r}_i)\left[n_\ell(\kappa r_i) - \cot\delta_i^\ell j_\ell(\kappa r_i)\right] \tag{8.84}$$

and

$$J_i^L(\vec{r},E) = Y_L(\vec{r}_i)j_\ell(\kappa r_i) \tag{8.85}$$

where j_ℓ and n_ℓ are spherical Bessel and Neumann functions, respectively. In terms of these functions, $G_c(\vec{r},\vec{r}',\varepsilon)$ can be expressed as

$$\begin{aligned} G_c(\vec{r}_i,\vec{r}_i',\varepsilon) = &\sum_{LL'} Z_i^L(\vec{r}_i,\varepsilon)T_{ij}^{LL'}(\kappa)Z_i^{L'}(\vec{r}_i',\varepsilon) \\ &- \sum_{L} Z_i^L(\vec{r}_i,\varepsilon)J_i^L(\vec{r}_i',\varepsilon) \end{aligned} \tag{8.86}$$

Equation 8.86 is exact within the muffin-tin approximation and can be used to evaluate Green's function for a finite system such as a cluster of atoms. For infinite systems, such as the disordered alloy, this is not possible as it requires inversion of a matrix of infinite size in Equation 8.83 and then averages over all configurations. Therefore, to make further progress, we resort to the CPA, which in this context is known as the KKR-CPA. As we had seen in Section 8.4, in this approximation, the alloy A_xB_{1-x} is replaced by a periodic effective medium, which is determined by the requirement that if a particular site say $i = 0$ site is replaced by an A or B atom, the average scattering from this site with respect to the medium is zero, that is

$$x\langle G_c\rangle_{0=A} + y\langle G_c\rangle_{0=B} = \langle G_c\rangle \tag{8.87}$$

If t_A and t_B represent the energy shell matrix element of the atomic t-matrix for A and B atoms, respectively and t_c represents the energy shell matrix element of the effective atom, then Equation 8.87 gives the KKR-CPA condition as

$$t_c^{-1} = xt_A^{-1} + (1-x)t_B^{-1} + (t_A^{-1} - t_c^{-1})T_{00}^c(t_B^{-1} - t_c^{-1}) \tag{8.88}$$

where

$$T_{00}^c = \frac{1}{N}\sum_{\vec{k}}\left[t_c^{-1} - B(\vec{k},\varepsilon)\right]^{-1} \tag{8.89}$$

where N is the number of unit cells in the alloy. The energy shell matrix element t_c can be obtained by solving Equation 8.88 self-consistently. This is a nontrivial task as it involves Brillouin zone integration at each step through Equation 8.89. This was a big hurdle in the early stage that was overcome by using a simple method of integration called special direction method (Bansil 1975, Prasad and Bansil 1980). Although this method is computationally very efficient, it is not as accurate as the tetrahedron method that was later developed for disordered alloys (Kaprzyk and Mijnarends 1986).

For the charge self-consistent calculation, we need charge densities in A and B cells. These are obtained from the restricted site averages of Green's function in Equation 8.86 that we shall denote by $\langle G(\vec{r},\vec{r}',\varepsilon)\rangle_{A(B)}$. Let $\rho_{A(B)}(\vec{r})$ denote the charge density associated with an $A(B)$ cell that can be written as

$$\rho_{A(B)}(\vec{r}) = -\frac{1}{\pi}\int_{-\infty}^{\varepsilon_F} Im\langle G(\vec{r},\vec{r},\varepsilon)\rangle_{A(B)}d\varepsilon \tag{8.90}$$

where ε_F is the Fermi energy. Note that we have assumed the system to be at absolute zero temperature. The integration in Equation 8.90 is along the real energy that takes much computer time. This problem was addressed by Zeller et al. (1982), who used the complex energy method to speed up the calculation. This also makes the k-space integration faster because the integrand becomes smoother for the complex energies.

The Fermi energy ε_F is calculated by requiring that the integrated density of states (IDOS) at the Fermi energy gives the average number of electrons in the cell. Generally, the IDOS is calculated from Lloyd's formula given by

$$N(\varepsilon) = N_0(\varepsilon) + \frac{2}{\pi} Im[x\, ln\, ||t_C^{-1} - t_B^{-1}|| + y\, ln\, ||t_C^{-1} - t_A^{-1}||]$$
$$- \frac{2}{\pi N} \sum_{\vec{k}} Im\, ln\, ||t_C^{-1} - B(\vec{k}, \varepsilon)|| \tag{8.91}$$

where $N(\varepsilon)$ is the IDOS for the alloy and $N_0(\varepsilon)$ is the IDOS for the free electrons. This formula is very convenient to use but sometimes gives unphysical jumps. This problem was pursued by Kaprzyk and Bansil (1990) who derived a generalized Lloyd formula that does not give unphysical jumps.

The average DOS can be calculated by differentiating IDOS $N(\varepsilon)$ from Equation 8.91. But the component DOS is calculated from the restricted site-averaged Green's function $\langle G(\vec{r}, \vec{r}', \varepsilon)\rangle_{A(B)}$ as

$$\rho_{A(B)}(\varepsilon) = -\frac{1}{\pi} \int Im \langle G(\vec{r}, \vec{r}, \varepsilon)\rangle_{A(B)} d^3 r \tag{8.92}$$

where the integration is over the unit cell.

The charge self-consistent KKR-CPA calculation is carried out in the following steps:

1. We start with some initial guess for the potentials v_A and v_B and solve the KKR-CPA Equation 8.88.
2. The new charge densities in A and B cells are then obtained using Equation 8.90.
3. New potentials $v_{A(B)}$ in A and B cells are obtained from these charge densities.
4. The whole procedure is iterated until the self-consistency is achieved. This means that, at the final step, the input and output potentials are the same within some tolerance. We note that the converged potentials do not depend on the initial choice of the potentials v_A and v_B and hence, the final results for the band structure, DOS, and so on are also independent of the initial choice of the potentials.

We see that the KKR-CPA theory does not involve any adjustable parameter and therefore is a first-principles theory of the random alloys. It shows correct limiting behavior, that is, in $x \to 0$ limit, it reduces to the standard

KKR band theory discussed in Chapter 7. It also describes the impurity limit correctly. Thus, it is a very general theory that can treat pure metals, single impurity, and concentrated alloys at the same level. In the following section, we discuss some applications of the KKR-CPA.

8.7 Application of the KKR-CPA: Some Examples

The KKR-CPA method has been applied to a large number of alloy systems. To illustrate some salient features of the KKR-CPA, we present some representative results. For more applications to many other systems, we refer to the review articles given at the end of this chapter.

8.7.1 Density of States

As an example of the average DOS of a disordered alloy, we show in Figure 8.6 the KKR-CPA average DOS for $Cu_{0.9}Ge_{0.1}$ alloy (Prasad and Bansil 1982). The DOS for pure Cu is also shown in the figure along with the experimental results by Norris and Williams (1978). We note that the experimental results are in reasonably good agreement with the theoretical results. We see that there is a large broadening of the structure due to alloying at the top and

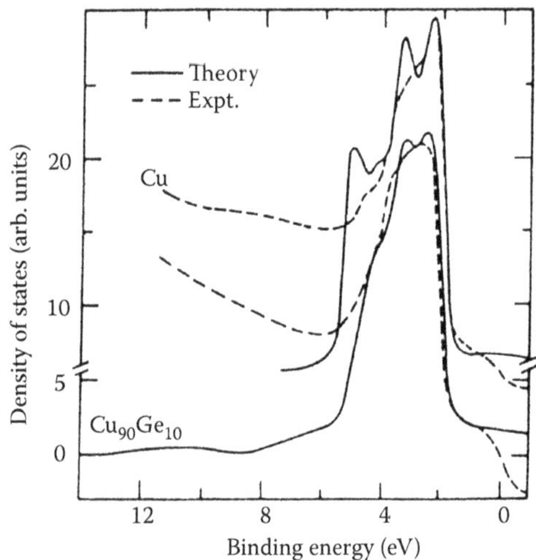

FIGURE 8.6
Average DOS for $Cu_{0.9}Ge_{0.1}$ and Cu. The experimental results of Norris and Williams are also shown. (From Prasad R. and Bansil A. 1982. *Phys. Rev. Lett.* 48: 113–116.) The energy zero is placed at the Fermi energy.

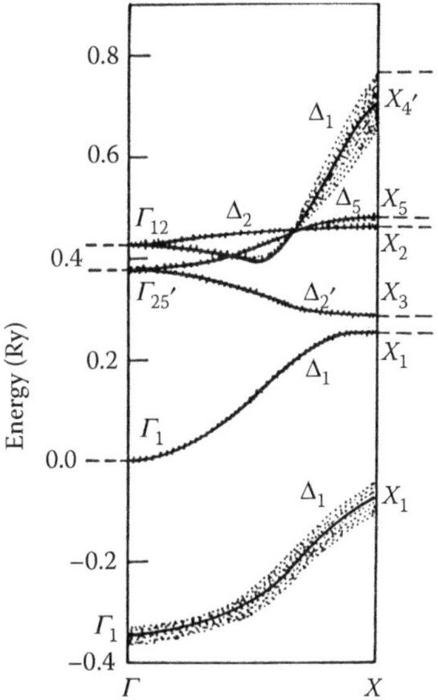

FIGURE 8.7
Complex energy bands of $Cu_{0.9}Ge_{0.1}$ along ΓX direction. The shading around bands represents disorder smearing. (From Prasad R. and Bansil A. 1982. *Phys. Rev. Lett.* 48: 113–116.)

bottom of the Cu-*d*-band (i.e., around 2 and 5 eV below 0) that is correctly predicted by the theory. Further, there is a small bump in the energy range of 10–12 eV below 0 in the KKR-CPA results for the alloy. This bump arises due to a new band in the alloy band structure that can be attributed to Ge in the alloy. This is similar to what we saw in the tight-binding model of the disordered alloy when we increased δ beyond 1 in Section 8.5. This means that the new band arises because *s–p* states of Cu and Ge experience a large disorder. We see this new band in Figure 8.7 as discussed below.

8.7.2 Complex Energy Bands

The concept of energy bands is an important concept in the theory of ordered solids that is very useful in interpreting various experimental results. This concept can be generalized to disordered alloys but now the energy bands become complex. To see this, recall that in the KKR theory of ordered solids that we discussed in Chapter 7, the energy bands are given by the KKR equation that can be written as (Exercise 8.4)

$$||t^{-1} - B(\vec{k},\varepsilon)|| = 0 \tag{8.93}$$

It turns out that for disordered alloys, one gets a similar equation for the energy bands (Bansil 1978, 1979a,b, 1987)

$$\|t_C^{-1} - B(\vec{k},\varepsilon)\| = 0 \tag{8.94}$$

This equation reduces to Equation 8.94 in the limit of pure $A(B)$ material, that is, when $t_c = t_{A(B)}$. This equation can be solved in two ways: (i) first fix $\vec{k}$ and then look for the root of the determinant to obtain $\varepsilon_{\vec{k}}$ and (ii) first fix ε and then find the roots of the determinant to find constant energy surfaces in $\vec{k}$ space. To obtain energy bands, one uses the first method whereas to obtain Fermi surface, one uses the second method. First, let us focus on $\varepsilon_{\vec{k}}$ versus $\vec{k}$ curve. We know that $\varepsilon_{\vec{k}}$ are real for the ordered material but interestingly turn out to be complex in alloys. This is because of disorder scattering that makes determinant 8.94 to be complex; as a result, it has roots at complex energies if we assume $\vec{k}$ to be real. Thus, the imaginary part of $\varepsilon_{\vec{k}}$ can be associated with the disorder scattering and is an important effect of disorder. It makes the alloy bands fuzzy and therefore is also known as disorder smearing.

It is interesting to note that in Chapters 4 through 7, the concept of energy bands and energy gaps emerged from translational symmetry. The theory of disordered alloys tells us that it may not be necessarily the case. The disordered alloys can have well-defined bands, provided the separation between them is much larger than their disorder smearing. To see that this is indeed the case, we shall show energy bands of disordered $Cu_{0.9}Ge_{0.1}$ and $Cu_{0.9}Al_{0.1}$ alloys. In Figure 8.7, we show the complex energy bands for $Cu_{0.9}Ge_{0.1}$ alloy along ΓX direction (Prasad and Bansil 1982). The dark line represents the real part of the energy and the shading around the bands represents the disorder smearing. We clearly see that there are well-defined bands and the disorder smearing is much smaller than the band separations. We see that, in general, the disorder smearing changes from band to band and could be $\vec{k}$ dependent. Most importantly, we see that as a result of alloying a new band around −0.2 Ry due to Ge impurities appears below Cu band structure. As we had discussed above, this band appears because of the large disorder in Cu and Ge s–p states. The band structure of Cu–Ge alloys was measured by ARPES experiment and was found to be in good agreement with the KKR-CPA results (Bansil et al. 1984). In Figure 8.8, we show the complex energy bands for disordered $Cu_{0.9}Al_{0.1}$ alloy (Asonen et al. 1982). In the figure, the disorder smearing of a band is shown by the vertical width of the band. We clearly see that there are well-defined bands and the disorder smearing is much smaller than the band separations as was the case with Cu–Ge alloy. We see that the disorder smearing changes from band to band and is $\vec{k}$ dependent. We further note that the band structure of the $Cu_{0.9}Al_{0.1}$ alloy looks similar to the band structure of copper shown in Figure 4.11 except that the bands in the alloy are smeared. In Figure 8.8, the d bands lie in the range of 4–7 eV whereas the band near the Fermi energy largely has s–p character. We also see that the disorder smearing in s–p bands is larger than that of the d bands. This is because the d states see much less disorder as there are no d

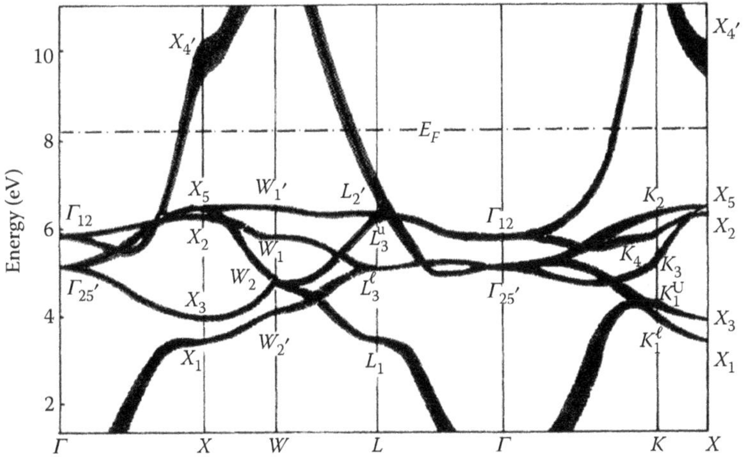

FIGURE 8.8
Complex energy bands of $Cu_{0.9}Al_{0.1}$ disordered alloy along some symmetry directions in the Brillouin zone. The vertical width of the shading of a band is 2× the imaginary part of the complex energy. (From Asonen H. et al. *Phys. Rev.* B 25, 7075–7085.)

electrons in Al. The band structure of Figure 8.8 was found to be in good agreement with ARPES results.

The energy bands and band gaps in disordered alloys can be probed by a variety of experimental techniques such as ARPES, optical absorption, differential reflectivity, and piezoreflectance techniques. These experiments have confirmed the existence of energy bands and energy gaps in disordered alloys. For example, using ARPES, the alloy energy bands have been measured for many alloys and are found in good agreement with the KKR-CPA results (see, e.g., Bansil and Pessa 1983, Bansil et al. 1984, Asonen et al. 1982). Rao et al. (1983) calculated the composition dependence of various gaps using the KKR-CPA for many Cu-based alloy systems and found it in good agreement with the experimental results. As was the case for ordered materials, the complex energy band structure of a disordered alloy has been found to be very useful in interpreting various experiments. It gives deeper insight, helps understand the formation of new bands, and how the bands of a disordered alloy evolve as a function of alloying.

8.7.3 Fermi Surface

We saw in Section 8.7.2 that the disordered alloys can have well-defined bands. If the band intersecting the Fermi energy is well defined, the disordered alloy should have a well-defined Fermi surface. Fermi surface is obtained by fixing ε and looking for the roots of determinant 8.94. Now, the energy ε is real; therefore, the roots of Equation 8.94 occur at complex values of k, whose imaginary part represents disorder scattering and is called disorder smearing. Note that

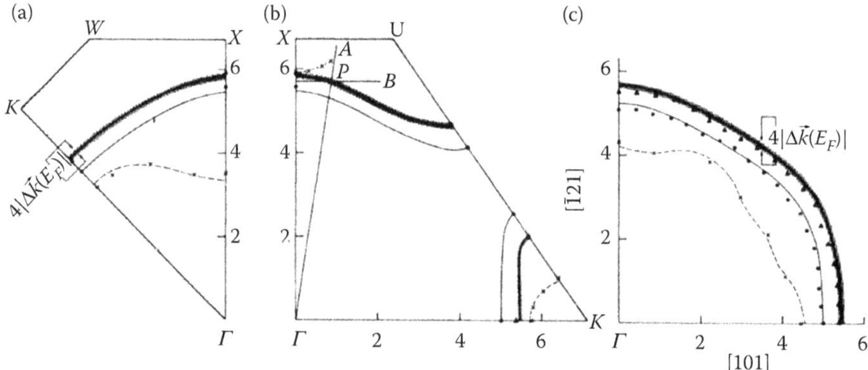

FIGURE 8.9
Fermi surface cross-section in various planes of Brillouin zone for $Cu_{0.7}Zn_{0.3}$ and Cu. The curve with smearing corresponds to the disordered alloy. The experimental points for Cu and the alloy are shown by dots and triangles, respectively. (From Prasad R., Papadopoulos S. C., and Bansil A. 1981. *Phys. Rev.* B 23: 2607–2613.)

the disorder smearing of the Fermi surface depends on the direction of $\vec{k}$. As a result of disorder smearing, the Fermi surface of a disordered alloy is fuzzy and is not sharp as is the case for pure metals. As an example of the alloy Fermi surface in Figure 8.9, we show few typical cross-sections of the Fermi surface of $Cu_{0.7}Zn_{0.3}$ alloy obtained by using the KKR-CPA (Prasad et al. 1981). We see that the alloy Fermi surface is fuzzy as opposed to the Cu Fermi surface that is also shown in Figure 8.9 and is sharp. In the figure, the theoretical results have been compared with the results obtained from two-dimensional angular correlation of positron annihilation radiation (2D-ACAR) experiment and are found to be in good accord. The 2D-ACAR technique has been used to obtain Fermi surface dimensions for several disordered alloys such as Cu–Zn, Cu–Pd, Nb–Mo, and Li–Mg (see, e.g., Prasad et al. 1981, Matsumoto et al. 2001, Rajput et al. 1993, 1996). It has been found that the KKR-CPA does extremely well in predicting the Fermi surfaces of disordered alloys. Also, experiments have established that, in general, concentrated metallic alloys do have well-defined Fermi surfaces.

8.8 Beyond CPA

As we have already mentioned, the CPA is a single-site approximation and, therefore, neglects correlations between the two atoms in the alloy. In many alloy systems, such as CuPd and CuNi that show short-range ordering or clustering, such correlations cannot be neglected and thus cannot be handled by using the CPA. To deal with such local environmental effects, one has to go beyond such single-site schemes such as the CPA and develop theories using multisite or cluster approximations. However, this problem turned out to be extremely difficult. Here, briefly we shall review some attempts to go beyond the CPA.

It turns out that the CPA has many good features such as it always yields positive DOS. In other words, the CPA Green's function is always analytic in the upper half of the complex energy plane. Also, it preserves the translational invariance and the point-group symmetry of the underlying lattice. It was found that a simple cluster generalization of the CPA does not preserve these properties leading to unphysical results such as negative DOS (Butler 1973, Nickel and Butler 1973). One of the first successful attempts that preserves the above-mentioned properties was due to Mills and Ratanavararaksa (1978) who developed a traveling cluster approximation using a diagrammatic technique. Mookerjee and coworkers (Mookerjee 1973, Bishop and Mookerjee 1974, Razee et al. 1991) developed a CCPA using the augmented space formalism that also preserves the analytic properties of the averaged Green's function. This method was generalized to the muffin-tin model and the KKR-CCPA scheme was developed (Razee et al. 1990, Razee and Prasad 1992, 1993a,b, Mookerjee and Prasad 1993). Recently, this method has been applied to CuZn alloys by Rahaman and Mookerjee (2009). Mookerjee and coworkers also developed a method that incorporates environmental effects using the recursion scheme in the augmented space (Saha et al. 1994). More recently, Rowland and coworkers have developed an NLCPA, which also preserves the analytical properties and translational invariance of the underlying lattice. Rowland and coworkers have also applied it to the muffin-tin model and developed KKR–NLCPA (Rowlands 2009). Also, Skriver and Ruban have developed an LSGF method that is also analytic and computationally very efficient (Ruban and Abrikosov 2008). For details, we refer to the original papers and review articles.

EXERCISES

8.1 Derive the CPA condition 8.65 for a single band tight-binding model of a disordered alloy.

8.2 Calculate the DOS for a binary random alloy with a cubic lattice using the CPA and the single band tight-binding model. Compare your results with the results of Figure 8.5.

8.3 Calculate the DOS for a binary random alloy when the DOS corresponding to the Hamiltonian H_0 in Equation 8.62 is given in the form

$$\rho_0(\epsilon) = \left(\frac{2}{\pi w^2}\right)(w^2 - \varepsilon^2) \quad \text{for } |\varepsilon| \le w$$

$$\rho_0(\epsilon) = 0 \quad \text{for } |\varepsilon| \ge w.$$

Do you get the same trend as in Exercise 8.2 by changing δ and x? (Hint: see the paper by Velický et al. (1968).)

8.4 Show that the KKR Equation 7.91 can be written in the form of

$$\| t^{-1} - B(\vec{k}, E) \| = 0$$

where t is the energy shell matrix element of the atomic t-matrix given by Equation 8.81.

8.5 Calculate the DOS for disordered $Cu_{0.9}Ge_{0.1}$ alloy using the KKR-CPA method. Compare your result with Figure 8.6. You may use the KKR-CPA code mentioned in Appendix A.

8.6 Calculate the DOS for the disordered $Cu_{0.7}Zn_{0.3}$ alloy using the KKR-CPA method. You may use the KKR-CPA code mentioned in Appendix A.

8.7 Calculate the Fermi surface for disordered $Cu_{0.7}Zn_{0.3}$ alloy using the KKR-CPA method. Compare your results with Figure 8.9. You may use the KKR-CPA code mentioned in Appendix A.

Further Reading

Bansil A. 1987. Modern band theory of disordered alloys. *Lect. Notes Phys.* 283: 273–298. Berlin: Springer.

Faulkner J.S. 1982. Modern theory of alloys. *Progress in Material Science*, 27:1–187.

Prasad R. 1994. KKR approach to random alloys. In *Methods of Electronic Structure Calculations* eds., Andersen O. K., Kumar V., and Mookerjee A. pp. 211–230. Singapore: World Scientific.

9

First-Principles Molecular Dynamics

9.1 Introduction

Until now we had assumed that the ions were fixed at their ionic positions and focused on the electronic structure problem, that is we know how to calculate electronic band structure and the total electronic energy using the first-principles approach for a fixed ionic configuration. Now, we will relax the condition that the ions are fixed. We shall see that it is possible to study ion dynamics from the first principles, which, in the most general form, is known as the first-principles molecular dynamics (MD).

MD is a computer simulation technique in which one looks at the time evolution of a system of interacting particles by integrating their equations of motion. In this way, one obtains trajectory of each particle, that is, their coordinates as a function of time. Thus, the technique allows one to view each particle as it moves; therefore, it is like a "theoretical microscope" by which one can see very fine details of a process occurring in a system as a function of time. The technique is also called "computer experiment." The physical properties are obtained by taking their time averages over the trajectories. Thus, MD is basically a statistical mechanics method.

MD can be of two types (i) classical MD and (ii) first-principles MD. In the classical MD, one uses empirical potentials, while in the first-principles MD the potential is obtained quantum mechanically without using any empirical parameter. Classical MD, although first introduced by Alder and Wainwright (1959) for hard spheres, was developed by Rahman (1964) using empirical potentials. The first-principles MD was first proposed by Bendt and Zunger (1983) and is known as Born–Oppenheimer MD. However, they were not able to implement it on real systems. Car and Parrinello (1985) proposed another way of first-principles MD, which is known as Car–Parrinello MD and also implemented it on a real system (crystalline silicon) to calculate its static and dynamical properties. The development of Car–Parrinello MD has been a major development in the field of electronic structure.

The chapter is organized as follows. In Section 9.2, we will discuss the classical MD and in Section 9.3, we show how one can calculate physical properties. In Section 9.4, we derive basic equations leading to the Born–Oppenheimer

MD. In Section 9.5, Car–Parrinello MD is discussed. In Section 9.6, we give a brief discussion on comparison of the Born–Oppenheimer and Car–Parrinello MD. In Sections 9.7 and 9.8, we discuss the method of steepest descent (SD) and simulated annealing to obtain the ground state. In Section 9.9, we discuss the Hellmann–Feynman theorem and in Section 9.10 the calculation of forces. In Section 9.11, we mention some applications of the first-principles MD.

9.2 Classical MD

Suppose we have N atoms in a fixed volume Ω interacting via some interparticle potential $V(\vec{R}_1, \vec{R}_2, \ldots, \vec{R}_N)$, where $\vec{R}_i$ are atomic coordinates. We denote velocities by $\dot{\vec{R}}_i$, accelerations by $\ddot{\vec{R}}_i$, and assume that the atoms are classical particles and follow Newton's laws. Force $\vec{F}_i$ on ith atom is given by

$$\vec{F}_i = -\vec{\nabla}_i V \tag{9.1}$$

Once the force $\vec{F}_i$ is known, one can write Newton's equation of motion

$$M\ddot{\vec{R}}_i = -\vec{\nabla}_i V \tag{9.2}$$

where we have assumed that each atom has the same mass. In MD, one solves the equations of motion (9.2) numerically and studies the time evolution of the system. In classical MD, V is a phenomenological potential, which is generally obtained by assuming a parametric form and then fitting it to some experimental results. For example, for inert gases, it could be expressed as a sum over Lennard–Jones potentials (see, e.g., Rahman 1964).

The equations of motion (9.2) can be integrated numerically by using, for example, Verlet algorithm (Verlet 1967). By using Taylor expansion, one can write

$$\vec{R}_i(t + \Delta t) = \vec{R}_i(t) + \Delta t\, \dot{\vec{R}}_i(t) + \frac{(\Delta t)^2}{2} \ddot{\vec{R}}_i(t) + \frac{(\Delta t)^3}{6} \dddot{\vec{R}}_i(t) + O[\Delta t^4] \tag{9.3}$$

where Δt is a small interval of time of the order of 10^{-14} s. Similarly,

$$\vec{R}_i(t - \Delta t) = \vec{R}_i(t) - \Delta t\, \dot{\vec{R}}_i(t) + \frac{(\Delta t)^2}{2} \ddot{\vec{R}}_i(t) - \frac{(\Delta t)^3}{6} \dddot{\vec{R}}_i(t) + O[\Delta t^4] \tag{9.4}$$

Adding Equations 9.3 and 9.4, we get

$$\vec{R}_i(t + \Delta t) = 2\vec{R}_i(t) - \vec{R}_i(t - \Delta t) + (\Delta t)^2 \ddot{\vec{R}}_i(t) + O[\Delta t^4] \tag{9.5}$$

In Equation 9.5, one can substitute for $\ddot{\vec{R}}_i(t)$ from Equation 9.2. Thus, if one knows $\vec{R}_i(t)$ and $\vec{R}_i(t - \Delta t)$, one can find the positions of atoms at subsequent time intervals. Equation 9.5 is known as the Verlet algorithm. Note that in the Verlet algorithm the local truncation error varies as $O[\Delta t^4]$ and therefore, the global error, the total error over the time of interest is of the order of Δt^3. Thus, the Verlet algorithm is more efficient than the algorithm based on Equation 9.3 in which one stops at second derivative of positions. However, it is not a self-starting algorithm, that is, to calculate $\vec{R}_i(t + \Delta t)$ one has to know both $\vec{R}_i(t)$ and $\vec{R}_i(t - \Delta t)$. Therefore, to start it one needs another algorithm such as Equation 9.3 to obtain first few steps.

Since our system that has a fixed number of particles, N and volume Ω, is isolated, the total energy is a constant of motion, that is

$$E = \frac{1}{2}\sum_i M\dot{\vec{R}}_i^2 + V = \text{constant} \tag{9.6}$$

Here, the velocities at time t are obtained from

$$\dot{\vec{R}}_i(t) = \frac{\vec{R}_i(t + \Delta t) - \vec{R}_i(t - \Delta t)}{2\Delta t} \tag{9.7}$$

The time step Δt should be so chosen that the constant of motion (total energy) comes out to be a constant within the desired accuracy.

Using Equations 9.5 and 9.7, the phase space trajectories $\{\vec{R}_i(t), \dot{\vec{R}}_i(t)\}$ can be calculated for each atom for several thousand time steps. From this, time average of a physical quantity can be calculated as

$$\langle A \rangle = \lim_{\tau \to \infty} \frac{1}{\tau} \int_{t_0}^{t_0 + \tau} A(t)\,dt \tag{9.8}$$

Note that at equilibrium, the average cannot depend on initial time t_0. According to the ergodic hypothesis, the time average (9.8) provided by MD is the same as the ensemble average. Thus, using Equation 9.8, one can calculate the temperature of the system as

$$\frac{3}{2}k_B T = \left\langle \frac{1}{2}\sum_{i=1}^{N} M\dot{\vec{R}}_i^2 \right\rangle = \lim_{\tau \to \infty} \frac{1}{\tau}\int_0^{\tau} dt \frac{1}{2}\sum_{i=1}^{N} M\dot{\vec{R}}_i^2(t) \tag{9.9}$$

where we have assumed that each atom has only 3 degrees of freedom.

It is sometimes useful to define "instantaneous" temperature as

$$T(t) = \frac{1}{3k_B}\sum_{i=1}^{N} M\dot{\vec{R}}_i^2(t) \tag{9.10}$$

Note that this is not a physical quantity as it is not a time-averaged quantity over the trajectories. But it could be used as a helpful device in doing simulations. For more details of classical MD, we refer to the books by Haile (1997), Rapaport (2004), and Allen and Tildesley (1989).

9.3 Calculation of Physical Properties

Once the phase space trajectories have been obtained, physical properties of a system of particles can be calculated by taking the time average over all the trajectories. Here, we shall discuss mean-squared displacement, diffusion constant, velocity auto-correlation function and phonon density of states. The formulae for these quantities are also valid for the first-principles MD.

We define the mean-squared displacement $R^2(t)$as

$$R^2(t) = \frac{1}{N}\left\langle \sum_{i=1}^{N} [\vec{R}_i(t) - \vec{R}_i(0)]^2 \right\rangle \tag{9.11}$$

This quantity in the limit of t tending to infinity can be shown as

$$R^2(t) = 6Dt \quad t \to \infty \ (3d) \tag{9.12}$$

$R^2(t)$ as a function of t is shown schematically in Figure 9.1. We see that for short times it goes as t^2. This is because for short times we can assume that the particle moves with almost constant velocity v and does not suffer any collision. Therefore,

$$R = vt$$

or

$$R^2 = v^2\, t^2$$

For long times, $R^2(t)$ goes as t. This is because for large t we can assume the particle to be a random walker for which $R^2(t)$ varies linearly with t. Thus, by

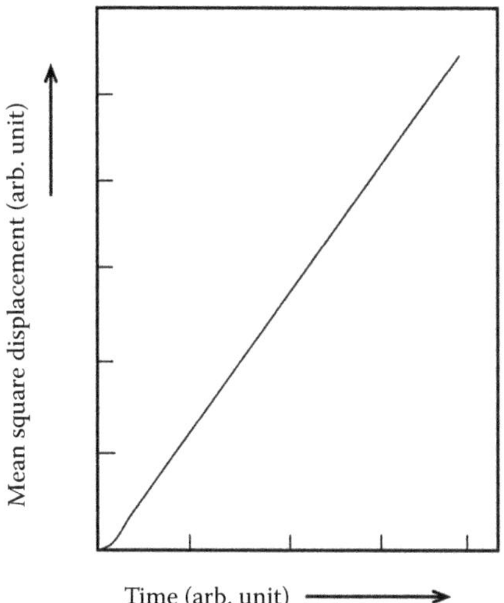

FIGURE 9.1
Schematic plot of the mean-squared displacement as a function of time.

calculating the slope of $R^2(t)$ at large t, the diffusion constant can be calculated from Equation 9.12.

Another important quantity is velocity autocorrelation function $Z(t)$, which tells how the velocity $v(t)$ of a particle at time t is correlated with its velocity at time $t = 0$ on average, that is

$$Z(t) = \left\langle \frac{1}{N} \sum_{i=1}^{N} \vec{v}_i(0) \cdot \vec{v}_i(t) \right\rangle \tag{9.13}$$

This is related to the diffusion constant by

$$D = \frac{1}{3} \int_0^\infty Z(t)\, dt \tag{9.14}$$

Another important quantity, which one can get from the $Z(t)$, is the phonon density of states, $D(\omega)$, which can be obtained by taking the Fourier transform of $Z(t)$ (Dickey and Paskin 1969)

$$D(\omega) = c \int_0^\infty Z(t) \cos(\omega t)\, dt \tag{9.15}$$

where c is a constant.

9.4 First-Principles MD: Born–Oppenheimer Molecular Dynamics (BOMD)

To study ion dynamics using classical MD, one needs to know the phenomenological potential. For a given material, it is difficult to find a good potential. These are often obtained by hit and trial method and by fitting parameters to experimental data. Even if a good potential is known for a material, it has limited use. For example, it cannot be used in situations where electrons may play an important role.

In first-principles or *ab initio* MD, no such phenomenological potential is used and the forces on ions are calculated using quantum mechanical calculations. These calculations have no adjustable parameters. The only information one needs to know is the atomic numbers and masses of the atoms constituting the system.

Recall that in Chapter 2, the total many-body wave function ϕ could be decomposed by using the Born–Oppenheimer approximation as (see Equation 2.3)

$$\phi = \chi_I(\vec{R})\,\psi_E(\vec{r},\vec{R}) \tag{9.16}$$

where χ_I is the wave function for ions and ψ_E is the wave function for electrons with ions at frozen positions $\vec{R}$. χ_I and ψ_E satisfy the following equations:

$$\left[-\frac{\hbar^2}{2m_e}\sum_i \nabla_i^2 + V_{EE}(\vec{r}) + V_{IE}(\vec{r},\vec{R})\right]\psi_E(\vec{r},\vec{R}) = E_E(\vec{R})\,\psi_E(\vec{r},\vec{R}) \tag{9.17}$$

and

$$\left[-\frac{\hbar^2}{2M}\sum_\ell \nabla_\ell^2 + V_{II}(\vec{R}) + E_E(\vec{R})\right]\chi_I(\vec{R}) = E_t\chi_I(\vec{R}) \tag{9.18}$$

where m_e is the mass of the electron and M is the mass of the ion. In the previous chapters, we have focused on solving Equation 9.17 using the DFT. Note that E_E is the total ground state electronic energy for the fixed position of ions and $\psi_E(\vec{r},\vec{R})$ is the many-electron wave function. This problem was reduced to a one-electron problem using the DFT. We shall now focus on the solution of Equation 9.18 in which $V_{II} + E_E$ provides the potential for ions.

We start with Ehrenfest's theorem, which gives the following equations for the evolution of mean values of position and momentum operators for the ions (see, e.g., quantum mechanics by Merzbacher 1998).

$$M\frac{d\langle\vec{R}\rangle}{dt} = \langle\vec{P}\rangle \tag{9.19}$$

and

$$\frac{d}{dt}\langle\vec{P}\rangle = -\langle\vec{\nabla}(V_{II}(\vec{R}) + E_E(\vec{R}))\rangle \tag{9.20}$$

Substitution of Equation 9.19 into Equation 9.20 gives

$$M\frac{d^2}{dt^2}\langle\vec{R}\rangle = -\langle\vec{\nabla}(V_{II}(\vec{R}) + E_E(\vec{R}))\rangle \tag{9.21}$$

Note that Equation 9.21 is valid only for the mean value of the position operator. However, we make an approximation here and identify the mean values with the positions of classical particles. This approximation is sometimes referred to as classical nuclei approximation. This is quite reasonable since the atoms have large mass and their de Broglie wavelength at room temperature is ~0.1 Å, much smaller than the interatomic separations (of the order of few ångstroms). Thus, Equation 9.21 can be rewritten as

$$M\frac{d^2\vec{R}}{dt^2} = -\vec{\nabla}(V_{II}(\vec{R}) + E_E(\vec{R})) \tag{9.22}$$

We note that Equation 9.22 is very similar to Equation 9.2 of the classical MD, but now the potential is obtained from the first principles. Such MD in which the potential is obtained from the first principles is called first-principles or *ab initio* MD and the dynamics based on Equation 9.22 is known as the Born–Oppenheimer MD (BOMD). Such a dynamics can be based on the Hartree–Fock method or DFT, but here we shall assume that we are using DFT. In BOMD, the potential for ions is $V_{II}(\vec{R}) + E_E(\vec{R})$ and thus at each time step KS equations have to be solved and $E_E(\vec{R})$ has to be computed. Thus, in this dynamics, the electrons are always in the ground state and the dynamics are only due to the ions. In BOMD, the total energy of the system is a constant of motion which is

$$E_{tot} = \frac{1}{2}\sum_{\ell} M\left|\dot{\vec{R}}_{\ell}\right|^2 + V_{II}(\vec{R}) + E_E(\vec{R}) \tag{9.23}$$

9.5 First-Principles MD: Car–Parrinello Molecular Dynamics (CPMD)

As mentioned above, a first-principles MD can be based on Equation 9.22. To do this, we will have to calculate $E_E(\vec{R})$, which is the total energy of the electrons in the ground state corresponding to the fixed positions of the atoms. This means that at each step we will have to calculate the ground state of the system self-consistently, which could be computationally very time consuming. Car and Parrinello (1985) proposed an alternative which did not involve electronic self-consistency at each step and thus made the first-principles MD computationally very efficient.

To do this, they proposed a Lagrangian in which the single particle electronic orbitals $\psi(\vec{r},t)$ are also dynamical variables along with nuclear coordinates $\vec{R}(t)$. The dynamics of nuclei is real while the dynamics of the orbitals is fictitious. This was a great idea which led to considerable savings in computational cost and made the first-principles MD possible as we shall see. Thus, Car and Parrinello (1985) proposed the following Lagrangian:

$$L = \frac{1}{2}\sum_{\ell} M|\dot{\vec{R}}_\ell|^2 + \sum_{i,occ}\int \mu|\dot{\psi}_i(\vec{r},t)|^2\, d^3r - E[\{\psi_i\}\{\vec{R}_\ell\}]$$

$$+\sum_{ij}\Lambda_{ij}\left(\int \psi_i^*(\vec{r},t)\psi_j(\vec{r},t)\, d^3r - \delta_{ij}\right) \tag{9.24}$$

where

$$E[\{\psi_i\}\{\vec{R}_\ell\}] = E_E(\vec{R}) + V_{II}(\vec{R})$$

$$= \sum_{i=1}^{N}\int \psi_i^*(\vec{r},t)\left(-\frac{\hbar^2}{2m_e}\nabla_i^2\right)\psi_i(\vec{r},t)\, d^3r$$

$$+\int v(\vec{r})n(\vec{r})\,d^3r + \frac{1}{2}\iint \frac{e^2}{4\pi\varepsilon_0}\frac{n(\vec{r})n(\vec{r}\,')}{|\vec{r}-\vec{r}\,'|}\,d^3r\,d^3r'$$

$$+ E_{xc}[n] + V_{II}(\vec{R}) \tag{9.25}$$

where

$$n(\vec{r}) = \sum_{i,\text{ occupied}} |\psi_i(\vec{r},t)|^2 \tag{9.26}$$

The first term in Equation 9.24 is the total kinetic energy of the atoms and the last term arises because of the constraint that the orbitals be orthonormal. In this term, Λ_{ij} are the Lagrange multipliers. Since the Lagrangian is real, the matrix Λ is Hermitian, that is $\Lambda_{ij} = \Lambda^*_{ji}$. The second term was introduced by Car and Parrinello (1985) as a computational device and as such has no physical meaning. It represents a fictitious kinetic energy associated with the orbitals which are frequently referred to as electronic degrees of freedom. μ is fictitious mass such that, in the limit $\mu \to 0$, we get the true Lagrangian of the system.

Using Lagrangian (9.24), equations of motions for the dynamical variables $\{\vec{R}_\ell\}$ and $\{\psi_i\}$ can be written as

$$\frac{d}{dt}\left(\frac{\partial L}{\partial \dot{\vec{R}}_\ell}\right) = \frac{\partial L}{\partial \vec{R}_\ell} \tag{9.27}$$

$$\frac{d}{dt}\left(\frac{\delta L}{\delta \dot{\psi}^*_i(\vec{r},t)}\right) = \frac{\delta L}{\delta \psi^*_i(\vec{r},t)} \tag{9.28}$$

Note that Equation 9.28 involves functional derivatives because the orbitals are not simple variables but are continuous scalar fields. From Equations 9.27 and 9.28, we get

$$M\ddot{\vec{R}}_\ell = -\frac{\partial E[\{\psi_i\}\{\vec{R}_\ell\}]}{\partial \vec{R}_\ell} \tag{9.29}$$

and

$$\mu\ddot{\psi}_i(\vec{r},t) = -\frac{\delta E[\{\psi_i\}\{\vec{R}_\ell\}]}{\delta \psi^*_i(\vec{r},t)} + \sum_j \Lambda_{ij}\psi_j(\vec{r},t) \tag{9.30}$$

$$= -H_{KS}\,\psi_i(\vec{r},t) + \sum_j \Lambda_{ij}\psi_j(\vec{r},t) \tag{9.31}$$

where

$$H_{KS} = -\frac{\hbar^2}{2m_e}\nabla^2 + v(\vec{r}) + v_{xc}(\vec{r}) + \int \frac{e^2}{4\pi\varepsilon_0}\frac{n(\vec{r}\,')}{|\vec{r} - \vec{r}\,'|}d^3r' \tag{9.32}$$

Equations 9.29 and 9.31 are the desired equations of motion for $\{\vec{R}_\ell\}$ and $\{\psi_i(\vec{r},t)\}$. Note that while the dynamics of the ions is real, the dynamics of orbitals is fictitious and is introduced as a convenient computational tool to avoid costly solution of the KS equations at each time step as is the case in

BOMD. Equation 9.30 tries to simulate the adiabatic evolution of KS orbitals on the Born–Oppenheimer (B–O) surface which is given by

$$H_{KS}(t)\psi_i^{KS}(\vec{r},t) = \varepsilon_i(t)\psi_i^{KS}(\vec{r},t) \tag{9.33}$$

where time dependence of H_{KS} (t) comes from time dependence of the ionic coordinates. The dynamics generated by Equation 9.30 for the KS orbitals will be, in general, different from the adiabatic evolution of $\psi_i^{KS}(\vec{r},t)$ given by Equation 9.33. Thus, this scheme will work only if the wave functions obtained by Equation 9.31 remain close to the solutions of Equation 9.33. If one starts from an initial condition, when all the electrons are in the ground state (i.e., the system is on the B–O surface), from Equations 9.29 and 9.31, one can find the solution for $\{\vec{R}_\ell\}$ and $\{\psi_i(\vec{r},t)\}$ at any later time by using algorithms such as the Verlet algorithm. Care should be taken so that the orbitals $\{\psi_i(\vec{r},t)\}$ are orthonormal to each other at each time step. This can be achieved by using the SHAKE algorithm discussed in Appendix E. Note that at subsequent time steps one need not solve the electronic structure problem (Equation 9.33), but one simply updates the orbitals using Equation 9.31. Thus, the Car–Parrinello strategy results in tremendous savings in computational cost.

Now we shall prove that the eigenvalues of the matrix Λ are the single-particle energies of the KS equations. Assume that the left-hand side of Equation 9.31 is zero, that is, $\ddot{\psi}_i = 0$. This means that the forces on the orbitals are zero which would be the case when the electrons are in the ground state. Then Equation 9.31 reads as

$$H_{KS}\,\psi_i = \sum_j \Lambda_{ij}\psi_j$$

or in matrix form

$$\hat{H}_{KS}\,\psi = \Lambda\,\psi \tag{9.34}$$

Since matrix Λ is Hermitian, it can be diagonalized by a unitary matrix. Let U be the unitary matrix which diagonalizes it. Then,

$$U^{-1}\hat{H}_{KS}U(U^{-1}\psi) = (U^{-1}\Lambda U)(U^{-1}\psi)$$

or

$$\hat{H}'_{KS}\,\psi' = \varepsilon'\,\psi' \tag{9.35}$$

where ε' is a diagonal matrix consisting of eigenvalues ε_i' and $\hat{H}_{KS}'$ and ψ' are transformed matrices under the unitary transformation. Now since $\hat{H}_{KS}$ and $\hat{H}_{KS}'$ are related by a unitary transformation, they have the same eigenvalues. Therefore, ε_i', that is, eigenvalues of matrix Λ can be interpreted as the KS eigenvalues.

An explicit expression for Λ_{ij} can be obtained by multiplying Equation 9.31 by ψ_j^* and then integrating,

$$\Lambda_{ij} = H_{KS}^{ji} + \mu \int \psi_j^*(\vec{r},t)\ddot{\psi}_i(\vec{r},t)d^3r \tag{9.36}$$

where

$$H_{KS}^{ji} = \int \psi_j^*(\vec{r},t) H_{KS}\, \psi_i(\vec{r},t)d^3r$$

Equation 9.36 can be simplified by noting that H_{KS} and Λ are Hermitian, that is

$$\Lambda_{ij} = \frac{1}{2}(\Lambda_{ij} + \Lambda_{ji}^*)$$

$$= H_{KS}^{ji} + \frac{1}{2}\mu \int (\psi_j^*\, \ddot{\psi}_i + \ddot{\psi}_j^*\psi_i)\, d^3r \tag{9.37}$$

Let us differentiate the following orthonormality condition twice with respect to time, that is

$$\int \psi_j^*\, \psi_i\, d^3r = \delta_{ij}$$

This gives

$$\int (\psi_j^*\ddot{\psi}_i + \ddot{\psi}_j^*\psi_i)d^3r = -2\int \dot{\psi}_j^*\dot{\psi}_i\, d^3r$$

Thus, we get

$$\Lambda_{ij} = H_{KS}^{ji} - \mu \int \dot{\psi}_j^*\dot{\psi}_i\, d^3r \tag{9.38}$$

Although this equation is not useful for numerical calculation, this is the only form of the Lagrange multiplier matrix compatible with the equation of motion (9.31).

The conserved quantity for Lagrangian (9.24) is the total energy E_{tot} which is

$$E_{tot} = \frac{1}{2}\sum_{\ell} M|\dot{\vec{R}}_{\ell}|^2 + \sum_{i,\,\text{occupied}} \mu\int|\dot{\psi}_i|^2\, d^3r + E[\{\psi_i\}\{\vec{R}_{\ell}\}] \tag{9.39}$$

Note that the fictitious kinetic energy is included in E_{tot}.

9.6 Comparison of the BOMD and CPMD

As mentioned earlier in BOMD, the dynamics are only due to the ions and the electrons are always in the ground state. Thus in BOMD, one can use a relatively large time step for integration of motion compared to CPMD where the time step has to be chosen small as the electrons are also involved in the dynamics. However, in CPMD, the KS equations have to be solved only in the beginning while in BOMD one has to do it at each time step which may be very time consuming.

In BOMD, the electrons are always in the ground state as KS equations are solved at each time step. The CPMD avoids the need of solving the KS equations at each time step by using a fictitious dynamics of electrons. Thus, the Car–Parrinello scheme will work well if the electrons always stay very close to the instantaneous ground state, that is, the system is close to the B–O surface. This will happen if the exchange of energy between nuclear and electronic degrees of freedom is small. Thus, if there is a large energy gap between the highest occupied level (HOMO) and the lowest unoccupied level (LUMO), the scheme will generally work well. This is because if the gap is much larger than $\hbar\omega$, where ω is the typical vibrational frequency of atoms, there will be very small exchange of energy between nuclear and electronic degrees of freedom. Therefore, the electrons will remain close to the ground state assuming that the system was started from the ground state. If the gap is comparable to $\hbar\omega$ or there is no gap, there will be an exchange of energy between nuclear and electronic degrees of freedom and the scheme will not work well. If the gap opens and closes periodically during the simulations, the scheme will also not work. The BOMD does not face this kind of problem.

9.7 Method of Steepest Descent

If one is not interested in the dynamics but only wants to find the energy minimum or the ground state geometry, the simplest method to use is the method of SD. In this method, one solves the following first-order differential equations instead of Equations 9.29 and 9.30 of CPMD.

$$M\dot{\vec{R}}_\ell = -\frac{\partial E}{\partial \vec{R}_\ell} \tag{9.40}$$

$$\mu\dot{\psi}_i = -\frac{\delta E}{\delta \psi_i^*} + \sum_j \Lambda_{ij}\psi_j \tag{9.41}$$

From Equation 9.40, one can see that once the energy minimum is attained, $\dot{\vec{R}}_\ell = 0$, that is, the nuclear coordinates do not change any further. Similarly, $\dot{\psi} = 0$ when the ground state of the system is attained. Solving Equations 9.40 and 9.41 together using the SD method is referred to as combined electron–ion minimization. Thus, the SD method is a very efficient way to find the energy minimum. Note that in the SD method, only the end product matters and time t does not have any physical significance. It is just a parameter or tool to achieve the end. The main drawback of the SD method is that one can get stuck in a local minimum and once one is stuck, there is no way of coming out of it.

The SD method can be used to solve Equation 9.41, that is, to solve the KS problem. This is frequently referred to as electron minimization. There are various other schemes in use, but we shall not discuss them here.

9.8 Simulated Annealing

The main drawback of the SD method to find the ground state configuration is that one can get trapped in a local minimum. The energy function $E(\{\vec{R}_\ell\}\{\psi_i\})$ can have several local minima, but we are interested in the global minimum. The situation is schematically shown in Figure 9.2, which shows a hypothetical energy function $E(x)$ as a function of some variable x. This function has local minima at x_1, x_2, x_4 and has the global minimum at x_3. We are interested in finding the global minimum.

The problem of finding the global minimum of the energy function is the problem of nonlinear optimization encountered in several fields. In the current case, we have to minimize energy, which depends on several variables in a nonlinear fashion, and the minimization has to be done subject to certain constraints. One very interesting and powerful method to solve this kind of problem was suggested by Kirkpatrick et al. (1983). This method is known as simulated annealing, as it tries to mimic the natural process of solidification of liquids by slow cooling. The idea can be understood as follows. Suppose we consider a metal and heat it so that it melts and forms a liquid. At high

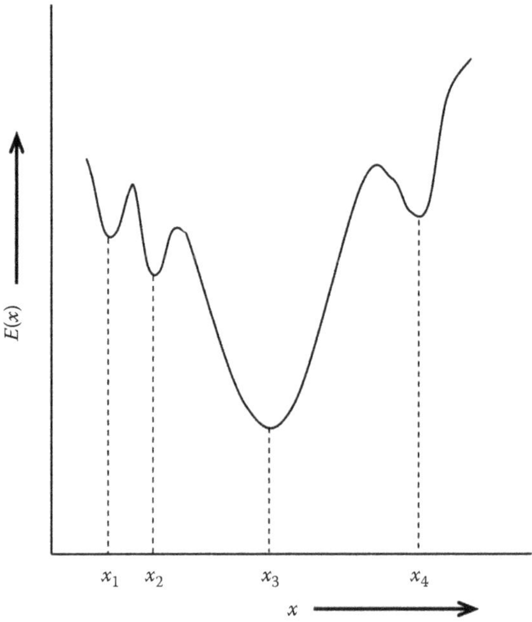

FIGURE 9.2
Schematic diagram of energy function $E(x)$.

temperatures in the liquid state, its atoms will move rapidly with respect to each other. As the liquid is slowly cooled, the atoms will move slower and slower. At certain temperature the liquid will solidify and if the cooling rate is slow enough, it may solidify in a crystalline state, which is the lowest state of energy. If one cools the liquid quickly or "quenches" it, the system may solidify in an amorphous or disordered state. Thus, by slow cooling or "annealing," we can find the minimum energy state. This is the idea behind the simulated annealing.

Simulated annealing can be implemented either using the Metropolis algorithm or MD. When implemented using the MD, it is also referred to as dynamical simulated annealing. In the current case, when we want to study dynamics of ions, it is natural to implement simulated annealing using MD. It is implemented as follows:

1. Heat the system so that the system is in liquid state. This can be easily done by rescaling the velocities of atoms so that the temperature increases above the melting temperature.
2. Equilibrate the system for several thousand time steps. The idea is that all the atoms should move sufficiently far from their original positions.
3. Cool the system very slowly to temperature $T = 0$.

4. Note the position of atoms and the energy at $T = 0$.
5. Repeat steps 1 through 4 several times.

One hopes that by following the above steps one obtains the state of minimum energy. Note that in the simulated annealing method, the system can get out of the local minimum if it has sufficient kinetic energy. This was not possible in SD method, because as soon as the system reaches the minimum, it loses its kinetic energy.

9.9 Hellmann–Feynman Theorem

The Hellmann–Feynman theorem is used to calculate forces on the ions. From Equation 9.29, the force on the ℓth ion is

$$\vec{F}_\ell = -\frac{\partial E}{\partial \vec{R}_\ell} \tag{9.42}$$

where

$$E = E_E + V_{II}$$

is given by Equation 9.25.

Thus,

$$\vec{F}_\ell = -\left[\frac{\partial E_E}{\partial \vec{R}_\ell} + \frac{\partial V_{II}}{\partial \vec{R}_\ell}\right] = \vec{F}_\ell^E + \vec{F}_\ell^I \tag{9.43}$$

E_E can be written as

$$E_E = \langle \psi | H | \psi \rangle \tag{9.44}$$

where ψ is the many-electron wave function and H is the Hamiltonian of the system.

$$\begin{aligned}\frac{\partial E_E}{\partial \vec{R}_\ell} &= \frac{\partial}{\partial \vec{R}_\ell}\left(\langle \psi | H | \psi \rangle\right) \\ &= \left\langle \psi \left| \frac{\partial H}{\partial \vec{R}_\ell} \right| \psi \right\rangle + \left\langle \frac{\partial \psi}{\partial \vec{R}_\ell} \middle| H \middle| \psi \right\rangle + \left\langle \psi \middle| H \middle| \frac{\partial \psi}{\partial \vec{R}_\ell} \right\rangle\end{aligned} \tag{9.45}$$

If ψ is the eigenfunction of H, then the last two terms of Equation 9.45 can be written as

$$E_E\left\langle \frac{\partial\psi}{\partial\vec{R}_\ell}\middle|\psi\right\rangle + E_E\left\langle\psi\middle|\frac{\partial\psi}{\partial\vec{R}_\ell}\right\rangle = E_E\frac{\partial}{\partial\vec{R}_\ell}\langle\psi|\psi\rangle = 0$$

Thus,

$$\vec{F}_\ell^E = -\frac{\partial E_E}{\partial\vec{R}_\ell} = -\langle\psi|\frac{\partial H}{\partial\vec{R}_\ell}|\psi\rangle \tag{9.46}$$

Thus, if ψ is the normalized eigenfunction of H, $\partial E_E/\partial\vec{R}_\ell$ is given by Equation 9.46. This statement is known as the Hellmann–Feynman theorem.

If ψ is not normalized, Equations 9.44 and 9.45 can be expressed as

$$E_E = \frac{\langle\psi|H|\psi\rangle}{\langle\psi|\psi\rangle} \tag{9.47}$$

and

$$\vec{F}_\ell^E = -\frac{\partial E_E}{\partial\vec{R}_\ell} = \frac{-\langle\psi|(\partial H/\partial\vec{R}_\ell)|\psi\rangle}{\langle\psi|\psi\rangle} \tag{9.48}$$

This is because $\partial/\partial\vec{R}_\ell\langle\psi|\psi\rangle = 0$.

Let us look at the change in energy δE_E and the change in force $\delta\vec{F}_\ell^E$ due to a slight change in wave function by an amount $\delta\psi$.

$$\begin{aligned}
\delta E_E &= \frac{\langle\psi+\delta\psi|H|\psi+\delta\psi\rangle}{\langle\psi+\delta\psi|\psi+\delta\psi\rangle} - \frac{\langle\psi|H|\psi\rangle}{\langle\psi|\psi\rangle} \\
&= \frac{\langle\psi|H|\delta\psi\rangle + \langle\delta\psi|H|\psi\rangle + \langle\delta\psi|H|\delta\psi\rangle + \langle\psi|H|\psi\rangle}{\langle\psi|\psi\rangle + O(\delta\psi^2)} - \frac{\langle\psi|H|\psi\rangle}{\langle\psi|\psi\rangle} \\
&= \frac{\langle\psi|H|\delta\psi\rangle + \langle\delta\psi|H|\psi\rangle}{\langle\psi|\psi\rangle} + O(\delta\psi^2) \\
&= \frac{E_E\delta\langle\psi|\psi\rangle}{\langle\psi|\psi\rangle} + O(\delta\psi^2) \\
&= O(\delta\psi^2)
\end{aligned}$$

or

$$E_E = E_E^0 + O(\delta\psi^2) \tag{9.49}$$

Thus, the change in energy is second order with respect to the change in wave function.

If the change in $\delta\psi$ is of the same order as the change in $\vec{R}_\ell$, then the change in the force is of first order with respect to the change in the wave function. Thus, the forces calculated using the Hellmann–Feynman theorem are more sensitive to the errors in the wave functions. Thus, it is easier to calculate an accurate total energy than to calculate an accurate force (see, e.g., Payne et al. 1992).

9.10 Calculation of Forces

The force acting on ℓth ion is given by Equation 9.29, which is

$$\vec{F}_\ell = -\frac{\partial E[\{\psi_i\}\{\vec{R}_\ell\}]}{\partial \vec{R}_\ell} \tag{9.50}$$

where E is given by

$$E = \sum_{i=1}^{N}\int \psi_i^*(\vec{r})\left(-\frac{\hbar^2}{2m_e}\nabla^2\right)\psi_i(\vec{r})\,d^3r + \int v(\vec{r})n(\vec{r})d^3r + \frac{1}{2}\iint \frac{e^2}{4\pi\varepsilon_0}\frac{n(\vec{r})n(\vec{r}\,')}{|\vec{r}-\vec{r}\,'|}d^3r\,d^3r' + E_{xc}[n] + V_{II}(\vec{R}) \tag{9.51}$$

The calculation of forces is the most time-consuming part of the MD calculation. This calculation can be speeded up by directly calculating forces on ions rather than first calculating E and then numerically differentiating it. The Hellmann–Feynman theorem in the form of Equation 9.46 is not very useful, as ψ appearing in this equation is the many-electron wave function. What we would like is an expression for the force in terms of single-particle orbitals ψ_i or one electron charge density $n(\vec{r})$.

To accomplish this, let us substitute Equation 9.51 into Equation 9.50 to get

$$\vec{F}_\ell = -\frac{\partial}{\partial \vec{R}_\ell}\left[E_{kin} + E_{xc} + E_H + \int v(\vec{r})n(\vec{r})d^3r\right] - \frac{\partial}{\partial \vec{R}_\ell}V_{II}(\vec{R}) \tag{9.52}$$

We also know that $\psi_i(\vec{r})$ satisfies the KS equations

$$\left[-\frac{\hbar^2}{2m_e}\nabla^2 + v(\vec{r}) + \int \frac{e^2}{4\pi\varepsilon_0}\frac{n(\vec{r}')}{|\vec{r}-\vec{r}'|}d^3r' + \frac{\delta E_{xc}}{\delta n(\vec{r})}\right]\psi_i(\vec{r}) = \varepsilon_i\psi_i(\vec{r}) \tag{9.53}$$

Note that $\psi_i(\vec{r})$ and ε_i depend on atomic positions $\vec{R}_\ell$.

Let us now evaluate various terms in Equation 9.52. The first term gives

$$-\frac{\partial E_{kin}}{\partial \vec{R}_\ell} = -\sum_i \int\left[\frac{\partial \psi_i^*}{\partial \vec{R}_\ell}\left(-\frac{\hbar^2}{2m_e}\nabla^2\right)\psi_i + \psi_i^*\left(-\frac{\hbar^2}{2m_e}\nabla^2\right)\frac{\partial \psi_i}{\partial \vec{R}_\ell}\right]d^3r \tag{9.54}$$

The second term gives

$$-\frac{\partial E_{xc}[n]}{\partial \vec{R}_\ell} = -\int \frac{\partial E_{xc}[n]}{\partial n(\vec{r})}\frac{\partial n(\vec{r})}{\partial \vec{R}_\ell}d^3r \tag{9.55}$$

The third term gives

$$\begin{aligned}-\frac{\partial E_H}{\partial \vec{R}_\ell} &= -\left[\frac{1}{2}\iint \frac{e^2}{4\pi\varepsilon_0}\frac{n(\vec{r})}{|\vec{r}-\vec{r}'|}\frac{\partial n(\vec{r}')}{\partial \vec{R}_\ell}d^3r\,d^3r'\right.\\ &\quad\left.+\frac{1}{2}\iint \frac{e^2}{4\pi\varepsilon_0}\frac{\partial n(\vec{r})}{\partial \vec{R}_\ell}\frac{n(\vec{r}')}{|\vec{r}-\vec{r}'|}d^3r\,d^3r'\right]\\ &= -\iint \frac{e^2}{4\pi\varepsilon_0}\frac{n(\vec{r}')}{|\vec{r}-\vec{r}'|}\frac{\partial n(\vec{r})}{\partial \vec{R}_\ell}d^3r\,d^3r'\end{aligned} \tag{9.56}$$

The fourth term in Equation 9.52 gives

$$-\int \frac{\partial v(\vec{r})}{\partial \vec{R}_\ell}n(\vec{r})\,d^3r - \int v(\vec{r})\frac{\partial n(\vec{r})}{\partial \vec{R}_\ell}d^3r \tag{9.57}$$

The last term gives

$$-\frac{\partial}{\partial \vec{R}_\ell}V_{II}(\vec{R}) = -\frac{\partial}{\partial \vec{R}_\ell}\frac{1}{2}\sum_{I\neq J}\frac{1}{4\pi\varepsilon_0}\frac{Z_IZ_Je^2}{|\vec{R}_I-\vec{R}_J|}$$

$$= -\frac{1}{2}\sum_{I\neq\ell}\frac{1}{4\pi\varepsilon_0}\frac{Z_I Z_\ell e^2}{|\vec{R}_I - \vec{R}_\ell|^3}(\vec{R}_I - \vec{R}_\ell)$$

$$+\frac{1}{2}\sum_{J\neq\ell}\frac{1}{4\pi\varepsilon_0}\frac{Z_\ell Z_J e^2}{|\vec{R}_\ell - \vec{R}_J|^3}(\vec{R}_\ell - \vec{R}_J)$$

$$= \sum_{J\neq\ell}\frac{1}{4\pi\varepsilon_0}\frac{Z_\ell Z_J e^2}{|\vec{R}_\ell - \vec{R}_J|^3}(\vec{R}_\ell - \vec{R}_J) \tag{9.58}$$

Also we know that

$$n(\vec{r}) = \sum_i \psi_i^*(\vec{r})\,\psi_i(\vec{r}) \tag{9.59}$$

so that

$$\frac{\partial n(\vec{r})}{\partial \vec{R}_\ell} = \sum_i \left[\frac{\partial \psi_i^*(\vec{r})}{\partial \vec{R}_\ell}\psi_i(\vec{r}) + \psi_i^*(\vec{r})\frac{\partial \psi_i(\vec{r})}{\partial \vec{R}_\ell}\right] \tag{9.60}$$

Substituting Equations 9.53 through 9.60 into Equation 9.52, we get

$$\vec{F}_\ell = \sum_{j\neq\ell}\frac{1}{4\pi\varepsilon_0}\frac{Z_\ell Z_J e^2(\vec{R}_\ell - \vec{R}_J)}{|\vec{R}_\ell - \vec{R}_J|^3} - \int\frac{\partial v(\vec{r})}{\partial \vec{R}_\ell}n(\vec{r})d^3r$$

$$-\left\{\sum_i\int\frac{\partial \psi_i^*}{\partial \vec{R}_\ell}\left[-\frac{\hbar^2}{2m_e}\nabla^2 + \frac{\delta E_{xc}}{\delta n(\vec{r})} + \int\frac{e^2}{4\pi\varepsilon_0}\frac{n(\vec{r}')}{|\vec{r} - \vec{r}'|}d^3r' + v(\vec{r})\right]\psi_i(r)d^3r\right.$$

$$\left.+\sum_i\int\psi_i^*\left[-\frac{\hbar^2}{2m_e}\nabla^2 + \frac{\delta E_{xc}}{\delta n(\vec{r})} + \int\frac{e^2}{4\pi\varepsilon_0}\frac{n(\vec{r}')}{|\vec{r} - \vec{r}'|}d^3r' + v(\vec{r})\right]\frac{\partial \psi_i(\vec{r})}{\partial \vec{R}_\ell}d^3r\right\} \tag{9.61}$$

Using Equation 9.53 the last two terms of Equation 9.61 (terms in the curly brackets) can be written as

$$-\sum_i\int\left[\frac{\partial \psi_i^*}{\partial \vec{R}_\ell}\varepsilon_i\psi_i + \psi_i^*\varepsilon_i\frac{\partial \psi_i}{\partial \vec{R}_\ell}\right]d^3r = -\sum_i\varepsilon_i\frac{\partial}{\partial \vec{R}_\ell}\int\psi_i^*\psi_i d^3r = 0 \tag{9.62}$$

Thus, the force on ℓth atom can be given as

$$\vec{F}_\ell = \sum_{j\neq\ell}\frac{1}{4\pi\varepsilon_0}\frac{Z_\ell Z_J e^2(\vec{R}_\ell - \vec{R}_J)}{|\vec{R}_\ell - \vec{R}_J|^3} - \int\frac{\partial v(\vec{r})}{\partial \vec{R}_\ell}n(\vec{r})d^3r \tag{9.63}$$

This is the Hellmann–Feynman theorem in terms of single particle charge density or single particle orbitals. This theorem greatly simplifies the calculation of forces on ions. It should be emphasized that this theorem is applicable only if the electronic wave functions are eigenstates of the KS Hamiltonian as we have assumed in Equation 9.62. If this is not so, then the terms in curly brackets in Equation 9.61 are not zero. Thus, in general, we should write the total force on ions as a sum of two terms as

$$\vec{F}_\ell = \vec{F}_\ell^{HF} + \vec{F}_\ell^{v} \tag{9.64}$$

where $\vec{F}_\ell^{HF}$ is the Hellmann–Feynman force given by Equation 9.63 and $\vec{F}_\ell^{v}$ is the term in curly brackets in Equation 9.61 and we refer to it as variational force. This force will be present if the electronic wave functions are not the eigenstates of the KS Hamiltonian or if the wave function has been expanded using an incomplete basis set as we shall see below.

Since the two terms in the curly brackets in Equation 9.61 are complex conjugates, we can write $\vec{F}_\ell^{v}$ as

$$\vec{F}_\ell^{v} = -2\,\mathrm{Re}\sum_i \int \frac{\partial \psi_i^*}{\partial \vec{R}_\ell}\left[-\frac{\hbar^2}{2m_e}\nabla^2 + v_{eff}(\vec{r})\right]\psi_i\, d^3r \tag{9.65}$$

where Re denotes the real part and

$$v_{eff}[n] = \frac{\delta E_{xc}}{\delta n} + \int \frac{e^2}{4\pi\varepsilon_0}\frac{n(r')}{|\vec{r} - \vec{r}'|}d^3r' + v(\vec{r}) \tag{9.66}$$

Equation 9.65 can be written as

$$\vec{F}_\ell^{v} = -2\,\mathrm{Re}\sum_i \int \frac{\partial \psi_i^*}{\partial \vec{R}_\ell}\left[-\frac{\hbar^2}{2m_e}\nabla^2 + v_{eff}[n] - \varepsilon_i\right]\psi_i\, d^3r \tag{9.67}$$

where we have used

$$\frac{\partial}{\partial \vec{R}_\ell}\left(\varepsilon_i \int \psi_i^* \psi_i\, d^3r\right) = 0$$

Recall that the KS equations are solved self-consistently. That is, one starts with a charge density $\tilde{n}(r)$, constructs potential $v_{eff}[\tilde{n}]$, and then obtains charge density by solving KS equations. The process is repeated until self-consistency is achieved, that is, output potentials and input potentials are the same within some tolerance. Although for most purposes, this may yield the desired accuracy, yet the self-consistency is never exact. Thus, if the calculation is not fully self-consistent, the wave functions are calculated for different potential $v_{eff}[\tilde{n}]$ related to the previous iteration. We add and subtract this potential in Equation 9.67, which gives

$$\vec{F}_\ell^v = -2\,\mathrm{Re}\sum_i \int \frac{\partial \psi_i^*}{\partial \vec{R}_\ell}\left[-\frac{\hbar^2}{2m_e}\nabla^2 + v_{eff}[\tilde{n}] - \varepsilon_i\right]\psi_i\, d^3r$$

$$-2\,\mathrm{Re}\sum_i \int \frac{\partial \psi_i^*}{\partial \vec{R}_\ell}\left[v_{eff}[n] - v_{eff}[\tilde{n}]\right]\psi_i\, d^3r \tag{9.68}$$

$$= \vec{F}_\ell^{IBS} + \vec{F}_\ell^{NSC} \tag{9.69}$$

The second term in Equation 9.69, $\vec{F}_\ell^{NSC}$ is a measure of nonself-consistency in the solution of the KS equation. To make this term small, it is important to bring the calculation very close to self-consistency. If this is achieved, $\vec{F}_\ell^{NSC}$ can be made arbitrarily small.

The first term $\vec{F}_\ell^{IBS}$ in Equation 9.69 arises due to the use of incomplete basis set and is usually referred to as the Pulay force (Pulay 1969, Srivastava and Weaire 1987, Martin 2004). This term includes the single particle equations, which should have been solved in the last but one iteration. But such a solution is usually not exact, because a finite, N_b dimensional basis set has been used in expanding the wave function

$$\psi_i(\vec{r},\vec{R}) = \sum_{j=1}^{N_b} C_{ij}(\vec{R})\phi_j(\vec{r},\vec{R}) \tag{9.70}$$

where $\vec{R}$ denotes the ionic coordinates $\{\vec{R}_\ell\}$ and $\phi_j(\vec{r},\vec{R})$, the basis functions.

Substituting this in the first term of Equation 9.68, we get

$$\vec{F}_\ell^{IBS} = -2\,\mathrm{Re}\sum_{i=1}^{N}\sum_{j=1}^{N_b}\sum_{j'=1}^{N_b} \int \frac{\partial C_{ij}^*}{\partial \vec{R}_\ell}\phi_j^*\left[-\frac{\hbar^2}{2m_e}\nabla^2 + v_{eff}[\tilde{n}] - \varepsilon_i\right]C_{ij'}\phi_{j'}\, d^3r$$

$$-\,2\,\mathrm{Re}\sum_{i=1}^{N}\sum_{j=1}^{N_b}\sum_{j'=1}^{N_b} \int C_{ij}^*\frac{\partial \phi_j^*}{\partial \vec{R}_\ell}\left[-\frac{\hbar^2}{2m_e}\nabla^2 + v_{eff}[\tilde{n}] - \varepsilon_i\right]C_{ij'}\phi_{j'}\, d^3r \tag{9.71}$$

Note that the actual secular equation, which one solves, is

$$\sum_{j'=1}^{N_b} \int \phi_j^*\left[-\frac{\hbar^2}{2m_e}\nabla^2 + v_{eff}[\tilde{n}] - \varepsilon_i\right]C_{ij'}\phi_{j'}\, d^3r = 0 \tag{9.72}$$

Because of this, the first term in Equation 9.71 becomes zero but the second term will be usually different from zero. Thus,

$$\vec{F}_{\ell}^{IBS} = -2\,\mathrm{Re}\sum_{i=1}^{N}\sum_{j=1}^{N_b}\sum_{j'=1}^{N_b}\int C_{ij}^{*}C_{ij'}\frac{\partial\phi_j^{*}}{\partial\vec{R}_{\ell}}\left[-\frac{\hbar^2}{2m_e}\nabla^2 + v_{eff}[\tilde{n}] - \varepsilon_i\right]\phi_{j'}d^3r \tag{9.73}$$

Thus, if the basis set depends on the ionic position as is the case with LMTO or LCAO basis sets, the Pulay forces will be finite. However, for the plane wave basis set, which does not depend on ionic coordinates, we see from Equation 9.73 that the Pulay forces are zero. This is a great advantage of using the plane wave basis set.

Note that we have proved Equation 9.73 using the DFT and using all electron potential $v(\vec{r})$. This proof can also be extended for pseudopotential.

9.11 Applications of the First-Principles MD

The first-principles MD has been applied to study various systems such as atomic clusters, nanowires, amorphous systems, surfaces and interfaces, biomolecules, water and nonaqueous liquids, chemical reactions, catalysis, and so on. Some of these applications will be discussed in subsequent chapters of the book. For more information, we refer to the excellent review articles by Car and Parrinello (1989), Payne et al. (1992), and Remler and Madden (1990) and the books by Kohanoff (2006) and Marx and Hutter (2009).

EXERCISES

9.1 Show that the mean-squared displacement $R^2(t)$ in three dimensions, in the limit of t tending to infinity, can be shown as

$$R^2(t) = 6Dt \quad t \to \infty \text{ (3 dimensions)}$$

9.2 Show that the diffusion constant D can be written as

$$D = \frac{1}{3}\int_0^{\infty} Z(t)\,dt$$

9.3 Show that the phonon density of states $D(\omega)$ can be obtained by taking Fourier transform of $Z(t)$

$$D(\omega) = c\int_0^{\infty} Z(t)\cos(\omega t)\,dt$$

where c is a constant.

9.4 Calculate the mean-squared displacement $R^2(t)$ for a system of particles interacting via Lennard–Jones potential given by

$$V(r) = 4\varepsilon\left[\left(\frac{\sigma}{r}\right)^{12} - \left(\frac{\sigma}{r}\right)^{6}\right]$$

where r is the interatomic distance and ε and σ are parameters. Take the values of the parameters corresponding to argon (see Rahman 1964) and check that the mean-squared displacement $R^2(t)$ has correct behavior as time t tends to zero and infinity.

9.5 Prove that

$$\Lambda_{ij} = H_{KS}^{ji} + \mu\int \psi_j^*(\vec{r},t)\ddot{\psi}_i(\vec{r},t)d^3r$$
$$= H_{KS}^{ji} - \mu\int \dot{\psi}_j^*\dot{\psi}_i\, d^3r$$

9.6 Show that for a plane wave basis set the Pulay force is zero.

9.7 Find bond length and vibrational frequency of a silicon dimer using the CPMD and BOMD. Use one of the MD codes listed in Appendix A. Take a large cubic supercell of length about 10 Å and use the method of SD for finding bond length. For finding the vibrational frequency, displace one of the atoms by a small amount ~0.01 Å and calculate time period of one oscillation using CPMD or BOMD. Compare the results obtained by CPMD and BOMD.

9.8 Calculate the structure of a silicon trimer using the simulated annealing. Use a large supercell and one of the MD codes listed in Appendix A. Heat the system upto 1500 K and then slowly cool it to 300 K. Below 300 K, you may use the method of SD to get the structure.

Further Reading

Haile J. M. 1997. *Molecular Dynamics Simulation: Elementary Methods.* New York: John Wiley & Sons.

Kohanoff J. 2006. *Electronic Structure Methods for Solids and Molecules.* Cambridge: Cambridge University Press.

10

Materials Design Using Electronic Structure Tools

10.1 Introduction

In the last few chapters, we discussed some important methods to obtain the electronic structure of materials using the DFT. These methods are first-principles, that is, they do not use any adjustable parameters. The only inputs to these calculations are electronic mass, electronic charge, atomic numbers, and masses of the atoms constituting the material. These calculations can predict fairly accurately the structure, lattice constants, charge densities, and various electronic, magnetic, optical, transport properties, etc. The question then arises is, can we use the first-principles methods to design new materials with specific properties? With the advances made during the last two decades in developing fast algorithms and computer capabilities, the answer to this question is "yes." Designing new materials and predicting their properties from first-principles is one of the fastest growing and most exciting areas of theoretical research. Search for new materials using computers is fast compared to the experimental search and has great cost advantage too. This is because the experimental search would take time for preparation of materials in the laboratory and then check whether such materials have desired properties. This exercise would be very expensive and time consuming. Therefore, a better approach would be to theoretically design new materials having desired properties and finally check them experimentally.

In this chapter, we shall discuss how electronic structure methods can be used to design new materials. In Section 10.2, we discuss the structure–property relationship. In Section 10.3, we point out the limitations of the first-principles calculations regarding time and length scales. In Section 10.4, we briefly discuss the multiscale approach to bridge the gap in time and length scales. In Section 10.5, we discuss the capabilities of the present-day electronic structure calculations and what kinds of questions can be easily answered by such calculations.

10.2 Structure–Property Relationship

The properties of a material depend on how its atoms are arranged. For example, carbon exists in various forms namely, diamond, graphite, graphene, C_{60}, and carbon nanotubes. Diamond is very hard, while graphite is very soft. Recently, its two-dimensional form graphene has been discovered, which has very different properties compared to graphite or diamond. It can also exist in the form of a C_{60} cluster which has entirely different properties. Similarly carbon nanotubes show very different properties from those of diamond, graphite, graphene, or C_{60}. Thus, properties of a material can be changed by changing the atomic arrangement, its size, or dimensionality. A material will have different properties in a nanowire form or in cluster form. Thus, understanding the structure–property relationship is very important for materials design. A material's properties can also be changed by doping. For example, conductivity of silicon can be changed by doping with boron or phosphorus. Alloying with another element has also been used in materials design. For example, in $Ga_{1-x}In_xAs$, the band gap can be adjusted by changing x.

10.3 First-Principles Approaches and Their Limitations

As mentioned earlier, first-principles approaches are based on quantum mechanics and do not have any adjustable parameters. The only inputs to the calculations are fundamental constants like electronic charge, electronic mass, atomic numbers, and atomic masses. In first-principles approaches, one starts by considering a material as a collection of interacting electrons and nuclei. The electronic and ionic problems are separated using the Born–Oppenheimer approximation. The electronic problem is generally solved using DFT, which reduces it to solving KS equations. The solution of these equations for large systems is difficult and requires huge computer memory. The computational effort involved in such calculations scales at least as N^3 where N is the number of atoms in the system. Even with the best computers available at present, we can handle systems with only about few thousand atoms. Typically, for Si system, for example, this requires a cell size of approximately few nanometers. This is one limitation of the first-principles methods that we can deal with length scales only up to nanometers. The other limitation is the time scale, when we want to study a time-dependent phenomenon. We read in Chapter 9 that the time step in *ab initio* MD is of the order of 10^{-15} s. Even with several thousand steps, we can access a time scale of the order of 10^{-11} s.

The other limitation of the first-principles approaches is due to handling of strong electron correlations. In systems containing d and f electrons, such correlations can be handled approximately by using LDA+U approach. A more sophisticated approach like DMFT is computationally very expensive and is not used commonly. Also, there are difficulties in calculations of properties which involve excited states of electrons such as optical properties. The GW method has been applied in some cases, but it is computationally very expensive. The TD-DFT approach has been mainly applied to small systems like clusters and is very difficult to implement for large systems. Also, there is difficulty in describing van der Waals interaction. A lot of research work is being done to solve these problems.

10.4 Problem of Length and Time Scales: Multiscale Approach

We read in the previous section that time and length scales which one can handle with the first-principles approaches are of the order of picoseconds (10^{-12} s) and few nanometers. But the time and length scales involved in most laboratory experiments and applications are of the order of seconds (or beyond) and centimeters (or more). Thus, there is a huge gap between the scales accessible by the first-principles approaches and the experiments. To bridge this gap, several approaches have been developed which are called multiscale approaches.

Before we proceed further, let us divide the length and time scales in four regimes as shown in Figure 10.1. In the first regime which we call first-principles regime, the time scale ranges from 10^{-15} to 10^{-11} s and the length scale from approximately 1 to 100 Å. In this regime, the main role is played by the electrons and this is the regime accessible by the first-principles approaches. In the second regime which we call atomistic, the length scale ranges from 100 Å to 1 μm and the time scale from 10^{-12} to 10^{-9} s. This is the regime in which the main role is played by the atoms which can be handled by an approach like classical MD. The third regime in Figure 10.1 is denoted as the mesoscopic regime. In this regime, the length scale ranges from 1 μm to 10^{-3} m, and the main role is played by defects, dislocations, and grain boundaries. The fourth regime is the macroscopic (continuum) regime in which time scale ranges from milliseconds to minutes (and beyond) and the length scale ranges from 10^{-3} m to centimeters and beyond and the material can be viewed as a continuous medium. In this regime, the continuous fields such as density, velocity, temperature, displacement, and stress play the main role and finite element approaches have been developed to examine large-scale properties of materials.

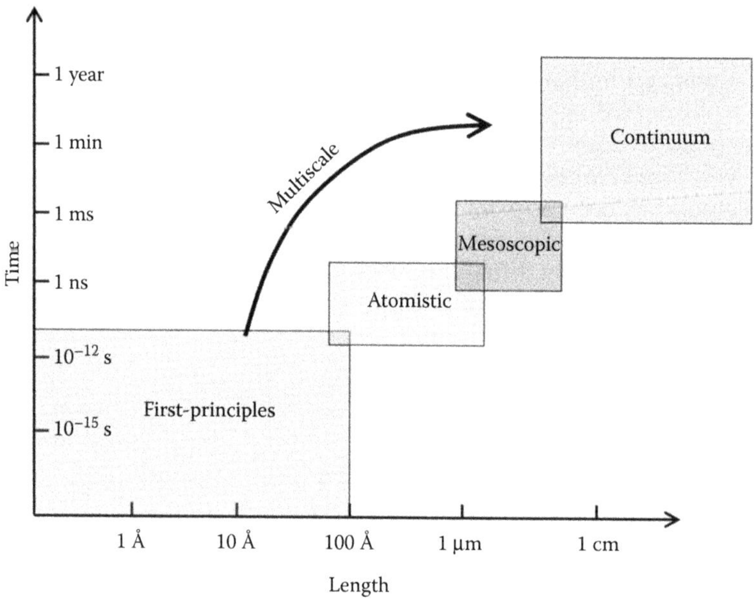

FIGURE 10.1
Length and time scales associated with first-principles, atomistic, mesoscopic, and continuum calculations.

We saw that different approaches have been developed to handle different regimes in Figure 10.1. Thus, the challenge is to develop approaches which combine different methods specialized at different scales so that computational power is effectively used. For example, in the problem of crack propagation in silicon, the electron plays a role only near the tip of the crack where bonds are being broken but away from the tip, electrons do not play an important role. Thus, the region near the tip can be treated using the first-principles methods and the region away from the tip can be handled using the classical MD. This approach has been successfully implemented and is a good example of a multiscale modeling (Lu and Kaxiras 2005).

Several other strategies have been used for multiscale modeling. For example, phase diagram, short-range order of disordered phases, and so on can be obtained by combining first-principles approaches with cluster expansion and Monte Carlo methods. Similarly, formation energies of intermetallic compounds calculated using DFT have been used as input in the CALPHHAD models for thermodynamics of materials (Pettifor 2003). Another interesting multiscale modeling approach is the integration of the DFT with phase-field models. For details, we refer to the review articles on the subject (Ceder 2010, Hafner 2008, Hafner et al. 2007, Lu and Kaxiras 2005, Nieminen 2002).

10.5 Applications of the First-Principles Methods to Materials Design

With the availability of fast computers and efficient codes for the first-principles calculations, it has become possible to apply these methods to materials design. However, there are several checks one should make before undertaking such a problem. One must clearly define the role of the first-principles calculation. For example, one may have a particular application in mind. Then the problem has to be formulated in such a way so that the first-principles calculation can solve it. One must also check whether the length and time scales fall within the domain of the first-principles approach. One must realize that most of the first-principles codes only give basic quantities such as band structure, density of states, charge densities, total energies, lattice constant, etc. So one must ask the question whether the knowledge of these quantities will suffice for this purpose or one needs to calculate other properties such as transport or optical properties. In that case, one must get (or write) a suitable code to calculate these quantities using the wave functions and charge densities provided by the first-principles codes. Many such codes have been developed and there is much code development in this direction. These post-processing tools have substantially increased the capabilities of the first-principles methods. Here, we list some of the quantities which can be calculated using the first-principles "tools" and some materials which have been studied.

1. *Crystal structure, phase stability, and phase diagrams*: For crystalline materials, the calculations of crystal structure and phase stability are quite routine now. For alloys, phase diagrams can be calculated using cluster expansion method (Adjaoud et al. 2009).
2. *Magnetic properties*: Magnetic structures including noncollinear structures can be calculated. Magnetic moments, exchange coupling, magnetic anisotropy, magneto-optical effects, etc. have also been calculated.
3. *Polarization, ferroelectric, and multiferroic materials*: Polarization can be calculated using the Berry phase approach in several ferroelectric and multiferroic materials (King-Smith and Vanderbilt 1993, Roy et al. 2010, 2011).
4. *Mechanical properties*: Elastic constants, tensile strengths, etc. can be calculated using the first-principles approach. Elastic property maps have been calculated for austenitic stainless steels (Vitos et al. 2002).
5. *Phonon spectrum*: the phonon spectra can be calculated either via force constant methods or by linear response method.

6. *Transport properties*: Transport properties can be calculated using the first-principles approach. There has been a lot of activity in the calculation of transport properties of nanowires and nanostructured materials.
7. *Optical properties*: Optical properties such as the real and imaginary part of the dielectric function, refractive index, absorption and reflection coefficients, and so on can be calculated using the first-principles codes.
8. *Surfaces and interfaces*: First-principles methods have been used to calculate surfaces and interface energies, surface states, surface reconstruction, and so on.
9. *Catalysis and adsorption and reactions*: First-principles methods have been applied to study catalysis, adsorption, and reactions.
10. *Diffusion constant*: Diffusion is an important phenomenon in materials and plays an important role in understanding properties such as segregation, phase transformation, corrosion, and so on. The diffusion constant can now be calculated using first-principles methods (see, e.g., Wimmer et al. 2008).
11. *Disordered alloys*: The electronic structure of disordered alloys can be calculated by using methods such as the KKR-CPA. Ordering energy and effective cluster interactions can be calculated (Ruban and Abrikosov 2008).
12. *Amorphous solids*: First-principles methods have been used to study amorphous systems using *ab initio* MD.
13. *Atomic clusters and nanostructured materials*: First-principles approaches are ideal tools to study these systems. The structure, stability, and melting of clusters have been studied. Nanowires and multilayers have been studied using the first-principles methods.
14. *Materials at extreme conditions*: One great advantage that the first-principles approaches offer is that one can study matter at extreme thermodynamic conditions, such as very high temperatures and pressures that cannot be realized easily in the laboratory. Many such calculations have been done.
15. *Biomaterials*: First-principles methods can now be applied to study systems which contain a large number of atoms such as biomaterials. These methods have been applied to understand complex processes in biological materials such as the visual process in the eye (Sugihara et al. 2004).
16. *Geomaterials*: The first-principles methods are being applied to understand various materials present in the Earth's core and to understand various processes taking place inside the core. For example, Oganov and Ono (2004) applied *ab initio* methods to understand the Earth's

lower mantle. Alfe et al. (2002) used these methods to understand composition and temperature of the Earth's core.

17. *Battery materials*: There is an intense activity in finding suitable materials for rechargeable batteries. First-principles methods are being used to find new materials for anode, cathode, and electrolyte and understand various processes undergoing in a battery (see, e.g., Ceder 2010, Prasad et al. 2003, Prasad et al. 2005).

Some applications of these methods will be discussed in subsequent chapters. For more details, we refer to review articles given under "References" at the end of the book.

11

Amorphous Materials

11.1 Introduction

Amorphous materials form a class of disordered materials as was mentioned in Chapter 8, as they do not possess translational symmetry, the characteristic of a crystalline material. In amorphous materials, the disorder is positional or topological that is, positions of the atoms are not on a regular lattice as shown schematically in Figure 8.2. This is very different from the substitutional disorder, in which case there is an underlying periodic lattice but the lattice sites are occupied by different atoms as shown schematically in Figure 8.1.

Some examples of amorphous materials are amorphous Si, SiO_2, $Pd_{0.4}Ni_{0.4}P_{0.2}$, As_2S_3, $Au_{0.8}Si_{0.2}$, etc. One way of producing an amorphous structure is to quench it from the molten state. In the molten state, atoms are randomly distributed and continuously moving. By cooling the system, one slows down the motion of the atoms. If one quenches the system to low temperatures, that is, cools it at a fast rate, it is possible to arrest the disorder in the liquid state and obtain an amorphous material. If the cooling rate is fast, the atoms in the material do not get enough time to organize themselves into the crystalline form and consequently one gets the amorphous structure. This is the usual method to obtain metallic glasses.

Theoretical progress in understanding amorphous solids has been slow mainly due to the absence of translational symmetry in these solids. This implies that Bloch's theorem and the methods based on it cannot be applied to these materials. However, a large amount of experimental data exists on many systems giving information regarding structure and electronic structure. The information regarding the structure can be used in modeling the structure of the material, which can further provide a deeper understanding of these systems. We shall start by defining the pair correlation and radial distribution functions in Section 11.2, then briefly discuss the structural modeling that is needed to calculate the electronic structure in Section 11.3. In Section 11.4, we shall discuss the phenomenon of electron localization due to disorder for which P.W. Anderson was awarded the Nobel Prize in 1977. In Section 11.5, we shall discuss the application of *ab initio* molecular dynamics to obtain structure and electronic structure of amorphous Si and hydrogenated amorphous silicon. At present, this is the only way of obtaining *ab initio* electronic structure of an amorphous material.

11.2 Pair Correlation and Radial Distribution Functions

Consider a homogeneous and isotropic distribution of N atoms of the same kind in volume Ω. Let the number of atoms between two spherical shells of radii r and $r + dr$ be $dn(r)$. We define pair correlation function or pair distribution function (PDF), $g(r)$, by

$$dn(r) = \frac{N}{\Omega} g(r) 4\pi r^2 dr \tag{11.1}$$

or

$$g(r) = \frac{\Omega}{N} \frac{1}{4\pi r^2} \frac{dn(r)}{dr} \tag{11.2}$$

In other words, $g(r)$ is proportional to the probability of finding another atom between r and $r + dr$, provided there is already an atom at $r = 0$. Since atoms have finite size, $g(r) \to 0$ as $r \to 0$. Also, at a large distance from the central atom, one is sure to find another atom implying $g(r) \to 1$ as $r \to \infty$. For amorphous silicon, $g(r)$ is shown in Figure 11.1 (for details, see Section 11.5). The first peak corresponds to the nearest neighbors and so on.

If we have more than one species of atoms present, we can define partial pair correlation function, $g_{\alpha\beta}(r)$. Suppose we have atom α at the origin and $dn_{\alpha\beta}(\vec{r})$ is the number of β atoms between spherical shells of radii r and $r + dr$, then $g_{\alpha\beta}(r)$ is defined as

$$dn_{\alpha\beta}(r) = \frac{N_\beta}{\Omega} g_{\alpha\beta}(r) 4\pi r^2 dr \tag{11.3}$$

where N_β is the number of β atoms in volume Ω.

There is some ambiguity regarding the definition of the radial distribution function (RDF) in the literature. In some books, it is the same as the pair correlation function, but Elliot (1984) defines it as

$$J(r) = 4\pi r^2 g(r) \tag{11.4}$$

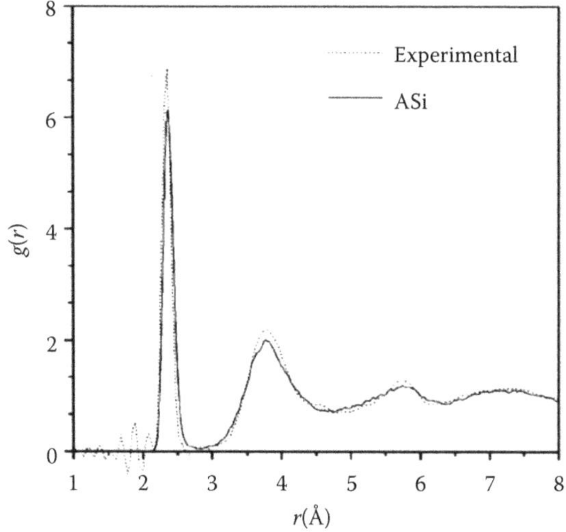

FIGURE 11.1
Radial distribution function of amorphous silicon. (From Singh R. et al. 2004. *Phys. Rev.* B70: 115213.)

and this differs from $g(r)$ by a factor of $4\pi r^2$. One advantage of this definition is that the area under a given peak of $J(r)$ gives the effective coordination number for that particular shell of atoms. Information about the RDF can be obtained from x-ray or neutron diffraction experiments.

11.3 Structural Modeling

To calculate the electronic structure, one needs to know the positions of the atoms constituting the material. In crystalline solids, this information is readily provided by the experiments. In amorphous solids, the experiments provide only partial information about the structure such as RDF, bond lengths, bond angle distributions, and coordination number. Various attempts have been made to model the amorphous structure in accordance with the experimental data. We shall briefly discuss some of these models.

Amorphous materials that have nondirectional bonding such as amorphous metals can be modeled by dense random packing (DRP) of hard spheres. Here, we assume that the atoms in the material can be considered to be "hard spheres," which would like to get packed as close as possible. We know that if the atoms are given sufficient time to adjust, they will arrange themselves in the fcc or hcp structure. However, if the material is obtained from rapid quench of the liquid state, the atoms do not have sufficient time to come to the perfect fcc or hcp structures and get stuck to some arbitrary configurations but come as close to the perfect structure as possible. For further details, we refer the reader to Elliot (1984) or Zallen (1998).

The continuous random network (CRN) model is used for modeling the structure of amorphous materials that have covalent bonding such as amorphous silicon. In the CRN model, the same coordination number is maintained as in the crystalline counterpart, but bond lengths and bond angles can deviate from their crystalline counterpart. For example, in the CRN model of amorphous Si, the coordination number of 4 is maintained. The idea is that because of the covalent bonding, the local environment around each atom is approximately maintained. But because of the slight deviations in bond angles and bond lengths, the long-range order is not obtained and one gets the amorphous structure. It should be pointed out that the CRN model is not a totally random network as there is some degree of local order present due to the presence of directed bonds.

Computer simulations have played an important role in modeling the structure of amorphous materials. Both Monte Carlo and molecular dynamics simulations have been used. In the Monte Carlo approach, one starts from some initial configurations and then calculates its RDF. Now, an atom is chosen at random and moved in an arbitrary direction. If it results in the improvement of the fit of the calculated RDF with the experimental curve, this is accepted, otherwise rejected. This process is continued until the root mean square deviation of the RDF from the experimental curve is less than a certain amount. This method has been used to generate structural models of several amorphous materials. Using a somewhat similar approach, Wooten et al. (1985) generated a structural model for amorphous Si and Ge known as the WWW model. They started with the diamond structure and repeatedly rearranged the structure by switching the bonds randomly. After each switch, the structure was partially relaxed with Keating potential. After a sufficiently large number of operations, an amorphous structure was obtained, which had no residual crystalline order. This model agreed well with experimental scattering data for amorphous Si.

Molecular dynamics (MD), particularly the first-principles MD simulations, is more frequently used now to generate amorphous structure. In this method, one prepares the liquid state of the material in a computer and then rapidly quenches it to obtain the amorphous structure. One starts with a crystalline sample and then heats it to a high temperature to obtain the liquid state. The sample is equilibriated for sufficiently long time and then rapidly cooled to obtain the amorphous structure. The method has been used to generate samples of amorphous Si, hydrogenated amorphous Si, amorphous Ge, etc. and one generally obtains samples with RDF in good agreement with experiments. Using the first-principles MD, one can also obtain the electronic structure and vibrational spectrum.

11.4 Anderson Localization

We saw earlier that an impurity in a crystalline solid can introduce an impurity state. An electron in this impurity state remains localized around this impurity and the wave function decays exponentially as one moves far away from the impurity as shown schematically in Figure 11.2.

Note that in band theory, all wave functions in a crystalline solid are extended (delocalized) as shown schematically in Figure 11.3.

Strong electron–electron interaction can also introduce localization, giving rise to an insulating state known as a Mott insulator. Anderson (1958) showed that strong disorder can also localize electrons in a solid, resulting in an insulating state. This was an important discovery for which Anderson (Figure 11.4) was awarded the Nobel Prize in 1977. It turns out that in one

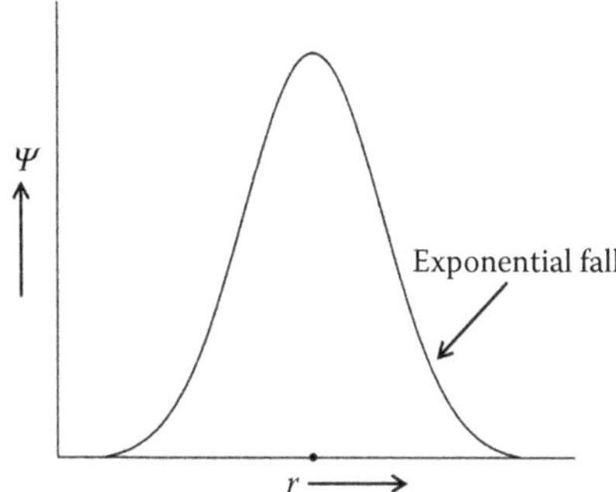

FIGURE 11.2
Schematic showing a localized state centered around an atom shown by a dot on the x-axis.

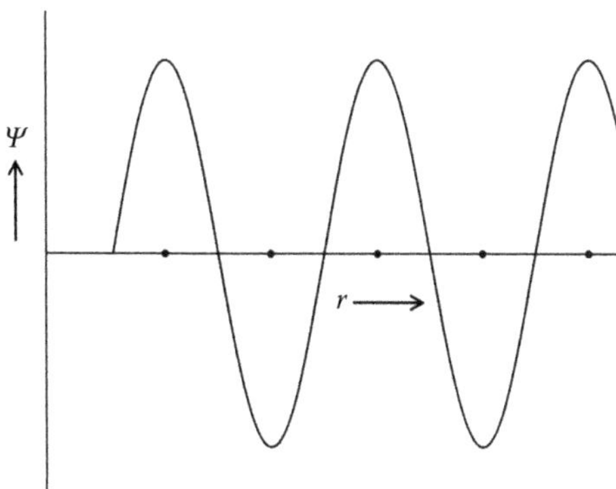

FIGURE 11.3
A schematic showing an extended state. The dots on x-axis indicate locations of atoms.

FIGURE 11.4
P. W. Anderson (1923–) discovered the concept of disorder-induced localization of electrons known as Anderson localization, for which he was awarded the Nobel Prize in 1977. (Courtesy of Princeton University.)

and two dimensions, even a small disorder can localize all states. Only in three dimensions, it is possible to have extended to localized state transition (Anderson transition) by tuning the strength of disorder.

To show that disorder can localize electrons, we shall use an *s*-band tight-binding model. Recall that for a crystalline solid, the electronic energies for a one *s*-band tight-binding model are given by

$$\varepsilon_{\vec{k}} = \varepsilon_0 + \gamma \sum_{\substack{j \neq 0 \\ n \cdot n}} e^{i\vec{k}\cdot\vec{R}_j} \tag{11.5}$$

where sum is over the nearest neighbors and γ is the overlap integral. ε_0 is the site energy, which, for a crystalline solid, is the same for all the sites. This would produce a band of bandwidth $B = 2z\gamma$, where z is the coordination number (see Section 5.5). To represent an amorphous solid in the tight-binding model, we assume that the site energy ε_i changes from site to site (i is the site label) randomly and is drawn from a uniform probability distribution

$$P(\varepsilon) = \frac{1}{W} \quad \text{for} -\frac{1}{2}W \le \varepsilon \le \frac{1}{2}W \tag{11.6}$$

that is, the site energies are distributed randomly in a width W with a uniform distribution. W will be zero for a crystalline solid, and as the disorder increases, W will increase. Note that for simplicity, we are assuming that the underlying lattice is periodic and the fact that the site energy changes from site to site randomly with distribution (Equation 11.6) takes care of the amorphous structure.

Recall that when $\gamma = 0$, there is no overlap between the atomic wave functions of the neighboring sites and the electrons are localized at their sites. On the other hand, for a large γ, there will be a band of bandwidth $2z\gamma$ and the electrons would be in extended states. Now, if we start with $\gamma = 0$ and then increase γ slowly, keeping W constant, and ask the question at what value of γ the wave functions are likely to be delocalized. To answer this question, we treat γ as perturbation and use the perturbation theory to look at the wave function ψ_i at the impurity site i, which we have assumed to be localized to begin with.

$$\psi_i = \psi_i^o + \sum_{\substack{j \neq i \\ n \cdot n}} \frac{\gamma}{\varepsilon_i - \varepsilon_j} \psi_j^o + \cdots \tag{11.7}$$

where ψ_i^o is the unperturbed localized wave function at site i and the sum in Equation 11.7 is assumed to be only over the nearest neighbors. There will be higher-order terms in Equation 11.7, but let us focus on the first-order term. The largest contribution to this first-order term will come from the term involving the smallest value of $\varepsilon_i - \varepsilon_j$. This is difficult to evaluate exactly but some probabilistic arguments can be given. Since ε_i and ε_j are distributed in the energy range W, the average value of $\varepsilon_i - \varepsilon_j$ is $W/2$. But we are interested in the smallest value of $\varepsilon_i - \varepsilon_j$, which can be obtained as follows. Let us first place ε_i at the center of the distribution of width W ranging from energy $-W/2$ to $W/2$. We also assume that ε_j are distributed uniformly at the interval of W/z in this energy interval. Thus, the smallest value of $|\varepsilon_i - \varepsilon_j|$ in this situation will be equal to W/z. This is the upper estimate because, in general, ε_i will not be distributed uniformly. Since the smallest value of $\varepsilon_i - \varepsilon_j$ could approach zero, the average value will be $W/2z$. Thus, the largest terms in the perturbation expansion (11.7) will involve factors of the order of

$$\frac{\gamma}{W/2z} = \frac{2z\gamma}{W} = B/W \tag{11.8}$$

Thus, if the perturbation expansion (11.7) has to converge

$$\frac{B}{W} < 1 \tag{11.9}$$

Since ψ_i is a localized function to start with, for $B < W$ or $W > B$, the successive terms in Equation 11.7 will get smaller and smaller and the resulting wave function will still be localized around the site i.

For $W < B$, the expansion (11.7) will not converge, as the successive term in Equation 11.7 will give larger and larger contribution, giving ψ_i an extended

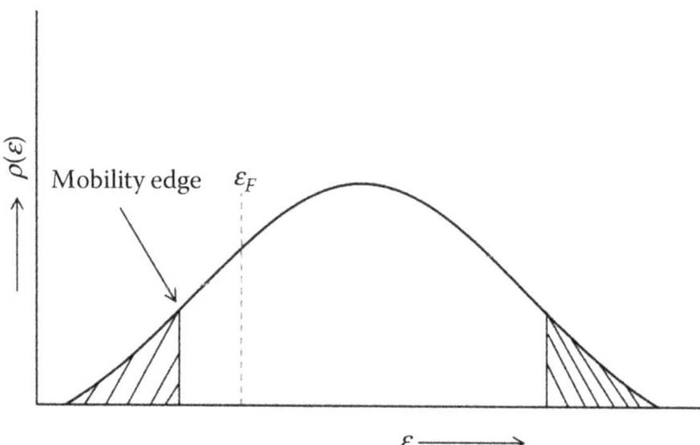

FIGURE 11.5
Schematic density of states of an amorphous solid. The shaded regions in the DOS correspond to the localized states and the unshaded region to the extended states.

character. Thus, we see that there is competition between W (disorder) and B (overlap). When $W > B$, the disorder dominates, giving rise to the localization. When W is small or zero, we get an extended (delocalized) state. Thus, by changing W, there will be a metal-to-insulator transition known as the Anderson transition.

Although the argument given above is rather crude, it shows that localization indeed can occur when the disorder exceeds some critical value. Later studies have shown that not all the states get localized, but it also depends on other parameters such as energy ε. In Figure 11.5, we show a schematic DOS, which shows two regions, shaded and unshaded. In the tail region shown by the shaded region, all the states are localized while in the unshaded region, all the states are extended. The line separating the shaded and unshaded region is known as mobility edge. As the disorder increases, the tail region increases and the unshaded region shrinks. The metal-to-insulator transition occurs when the Fermi energy ε_F, originally in the unshaded region, falls in the shaded region. For further details, we refer to many excellent books (Elliot 1984, Ziman 1982, Zallen 1998) and a review article (Lee and Ramakrishnan 1985).

11.5 Structural Modeling of Amorphous Silicon and Hydrogenated Amorphous Silicon

As an example of the structural modeling of an amorphous structure, we shall briefly discuss the modeling of amorphous silicon (*a*-Si) and hydrogenated amorphous silicon (*a*-Si:H). In addition to intrinsic interest, these materials are important materials from the application point of view. For example, these materials are used in solar cells and thin-film transistors for use in large-area electronics. In these materials, hydrogen plays an important role and passivates the defect levels. At the same time, hydrogen also causes the Staebler–Wronski effect, a degradation of material when exposed to light (see, e.g., Prasad and Shenoy 1996, Staebler and Wronski 1977). Therefore, understanding the role of hydrogen to find a solution to such problems is of paramount importance. The first-principles simulations could be used to shed light on this problem.

The WWW model of Wooten et al. (1985), as discussed in Section 11.3, gives a reasonably good representation of the structure of amorphous silicon. A more general method of obtaining amorphous structure is quenching from the melt and using this method. Several models have been generated. Since the amorphous structure is in a metastable state, it is not unique like a crystalline structure and thus a large number of amorphous structures can exist. Therefore, the calculated properties of an amorphous material depend on the sample. Here, we shall briefly discuss the structure and properties of some samples that have been prepared using the simulated annealing by Singh et al. (2004) who employed density functional-based tight-binding (DFTB) approach (Elstner et al. 1998). For *a*-Si, we shall consider two samples, one obtained by quenching from the melt, and another, the WWW model obtained by Wooten et al. (1985). The one obtained by quenching from the melt is denoted by ASi and contains 216 Si atoms. The sample was generated by starting with the crystalline sample and then raising the temperature to 4000 K. The sample is then equilibriated for some time so that memories of the initial structure are destroyed and then it is cooled down to 1800 K and then further equilibriated. The sample is then quenched to 300 K where it is again equilibriated for some time.

Similarly, four samples of *a*-Si:H were generated, which contained 216 Si and 24 H atoms. Samples ASiH and BSiH were generated with different cooling rates and sample CSiH was obtained by the hydrogenation of amorphous sample ASi. This was done by dividing the cell into 27 equal cubes and placing H at arbitrary positions in 24 of the outer 26 cubes. The resulting structure was equilibriated for some time at 300 K. To increase the diffusion of H atoms, the temperature was raised to 1200 K and further equilibriated. It was then quenched to 300 K and again equilibriated for some time. Sample WSiH was similarly obtained by hydrogenation of the WWW model of *a*-Si.

In Figure 11.1, we show the pair correlation function, $g(r)$, for amorphous silicon sample ASi together with the experimental data. There is good agreement between the calculated and experimental pair correlation curves. The first peak in the pair correlation curve, which occurs at 2.37 Å corresponds to the first nearest neighbors while the second and third peaks correspond to the second and third nearest neighbors. The positions of these peaks are in good agreement with the peaks in the RDFs for the WWW model.

In Figure 11.6, we show the partial correlation functions of Si–Si, Si–H, and H–H correlations in the four *a*-Si:H samples, ASiH, BSiH, CSiH, and WSiH.

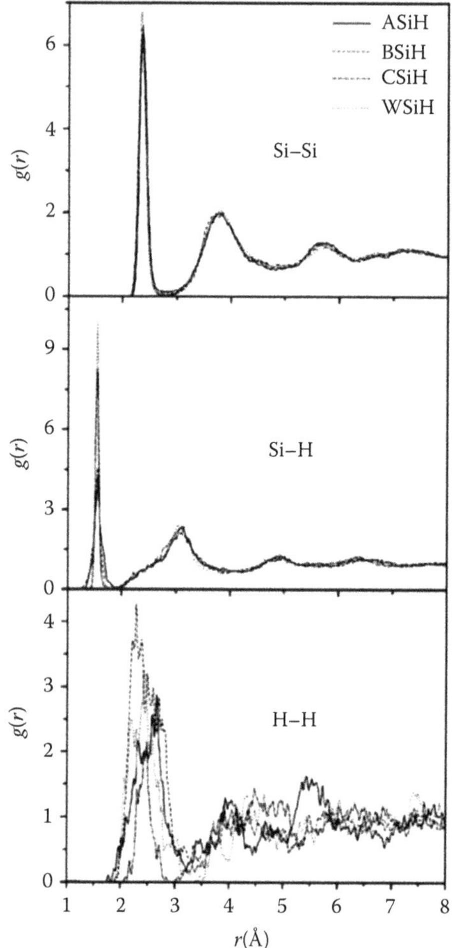

FIGURE 11.6
Partial correlation functions for Si–Si, Si–H, and H–H in four *a*-Si:H samples. (From Singh R. et al. 2004. *Phys. Rev.* B70: 115213.)

While the Si–Si and Si–H pair correlation functions for all the four samples are in close agreement, we see that H–H correlation functions for the four samples are sample dependent. Also, we note that the first peak in the H–H correlation function occurs above 2.0 Å, which indicates nonexistence of molecular hydrogen that has a bond length ~1.0 Å. The peaks in these correlation functions are in good agreement with those obtained experimentally (Bellisent et al. 1989).

In these amorphous samples, there are three- and fivefold coordinated atoms in addition to the fourfold coordinated Si atoms. In Table 11.1, we show the average percentage of three-, four-, and fivefold coordinated Si atoms and

TABLE 11.1

Average Percentages of *n*-Fold (3, 4, 5) Si Atoms and Mean Coordination Numbers Z_{Si-Si}, Z_{Si-H}, Z_{H-Si}, and $Z_{tots} = Z_{Si-Si} + Z_{Si-H}$

Samples	Threefold	Fourfold	Fivefold	Z_{Si-Si}	Z_{Si-H}	Z_{H-Si}	Z_{tots}
WWW	0.00	99.10	0.90	4.01	—	—	4.01
ASi	3.08	90.85	5.91	4.03	—	—	4.03
ASiH	4.47	88.20	7.33	3.92	0.12	1.04	4.04
BSiH	3.58	90.81	5.61	3.91	0.11	1.00	4.02
CSiH	2.41	95.52	2.07	3.89	0.11	1.00	4.00
WSiH	1.84	95.85	2.29	3.89	0.12	1.04	4.01

Source: Singh R. et al. 2004. *Phys. Rev.* B70: 115213.

the mean coordination numbers. The WWW model has the least number of defects while the ASi sample has a much higher percentage of defects. The ASiH and BSiH samples, generated from liquid quench, have more coordination defects than those of CSiH and WSiH, generated by hydrogenation of amorphous silicon samples. Thus, the samples generated from the liquid quench are found to be more disordered as compared to those generated by hydrogenation of amorphous Si samples. The table also gives the mean coordination numbers Z_{Si-Si}, Z_{Si-H}, Z_{H-Si}, and $Z_{tot} = Z_{Si-Si} + Z_{Si-H}$. We see that the total coordination number Z_{tot} of a Si atom in all samples is nearly equal to 4. The Z_{Si-Si} of Si atom in ASi sample is 4.03, which is quite close to 4.01 of the WWW model, but it reduces to 3.89 upon hydrogenation in CSiH and WSiH. The Z_{Si-Si} and Z_{tot} of a Si atom in ASiH and BSiH samples are larger than that of CSiH and WSiH samples indicating that a Si atom in ASiH and BSiH is overcoordinated as compared to that in CSiH and BSiH samples. The Z_{H-Si} of a hydrogen atom in all the *a*-Si:H samples is close to 1.0, implying that almost all the hydrogen atoms are bonded as Si–H monohydrides, in agreement with experimental results.

In Figure 11.7, we show the electronic DOS for amorphous Si samples, ASi and WWW, which shows a clear formation of band gap around the Fermi level that has been set at $\varepsilon = 0$ eV. The general shape of the curve for both the samples is almost the same except that the band gap for the WWW sample is slightly larger than that of the ASi sample. This is because there are more defects in the ASi sample compared to the WWW sample, as we have already seen in Table 11.1. The shape of the band tails near the gap region is exponential in both these samples, which is a universal feature of the amorphous semiconductors caused by the short-range disorder in these materials. Another interesting property is the short-ranged potential fluctuation observed by Agarwal (1996), which has been explained by Balamurugan and Prasad (2001) by using cluster calculations. Amorphous materials show many interesting properties that are still not understood. For more details, we refer to various books given in Further Reading.

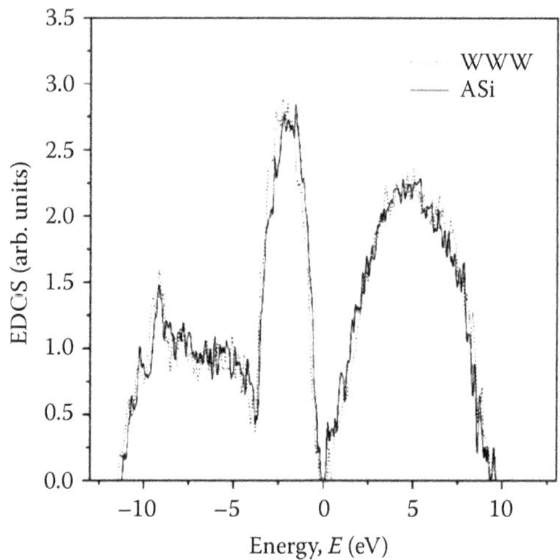

FIGURE 11.7
Electronic density of states for amorphous silicon. (From Singh R. et al. 2004. *Phys. Rev.* B70: 115213.)

EXERCISES

11.1 Calculate the radial distribution function for an amorphous solid obtained by rapid quenching of an originally liquid sample using classical molecular dynamics. Assume that the atoms are interacting via Lennard–Jones potential given by

$$V(r) = 4\varepsilon\left[\left(\frac{\sigma}{r}\right)^{12} - \left(\frac{\sigma}{r}\right)^{6}\right]$$

where r is the interatomic distance and ε and σ are parameters. You may take parameters from the paper of Rahman et al. (1976).

11.2 Calculate the velocity autocorrelation function and the phonon density of states of the amorphous Lennard–Jones solid given in Exercise 11.1.

11.3 Generate a sample of amorphous silicon using the classical molecular dynamics and potential due to Stillinger and Weber (1985). Calculate the radial distribution function and the phonon density of states for the amorphous silicon.

11.4 Generate a sample of amorphous silicon using a first-principles method or DFTB as mentioned in Appendix A. You may take a sample of 216 atoms. Calculate the radial distribution function, phonon and electron DOS for this sample and compare your results with that of Singh et al. (2004) and of Exercise 11.3.

Further Reading

Kaxiras E. 2003. *Atomic and Electronic Structure of Solids*. Cambridge: Cambridge University Press.

Lee P. A. and Ramakrishnan T. V. 1985. Disordered electronic systems. *Rev. Mod. Phys.* 57: 287–337.

Mardar M. 2010. *Condensed Matter Physics*. New York: John Wiley and Sons.

Street R. A. 1991. *Hydrogenated Amorphous Silicon*. Cambridge: Cambridge University Press.

12

Atomic Clusters and Nanowires

12.1 Introduction

In this chapter and the next, we shall be focusing on the systems of reduced dimensionality. The dimensionality of a system plays an important role in determining its electronic properties. For example, we saw in Chapter 2 how strongly the density of states (DOS) changes by changing the dimensionality. In this chapter, we shall discuss atomic clusters and nanowires, which can be considered systems with zero and one dimensionality, respectively. In the next chapter, we shall discuss two-dimensional systems such as surfaces and interfaces.

Atomic clusters are finite aggregates of atoms ranging from a few to several thousand atoms. Their properties are quite different from isolated atoms or bulk and strongly depend on the size of the cluster. Although clusters have been around for quite a long time, they came into prominence only after Richard Feynman (1959) gave his famous talk "There's plenty of room at the bottom." He recognized their technological potential and challenged scientists to develop a field where devices and machines could be constructed from components consisting of a small number of atoms.

The study of physical and chemical properties of clusters is one of the most active fields in physics, chemistry, and materials science. Clusters of varying sizes can now be generated using a variety of techniques. Their electronic, magnetic, optical, and chemical properties are found to be very different from their bulk form and are dependent on their size, shape, and composition (see e.g., Khanna and Castleman 2003; Chacko et al. 2004). A large fraction of their atoms reside on their surfaces, which makes them very different from bulk matter. Another feature that distinguishes them from bulk matter is the discreteness of electronic levels, which is due to the confinement of electrons in a small volume. Highly stable clusters have been designed that can act as building blocks or a basis to form a cluster-assembled solid. This offers new avenues for designing materials with desirable properties by assembling suitably chosen clusters. During the last two decades, there have been many important developments in this field. However, it will not be possible to discuss every important development and hence we shall focus only on a few developments that might provide a glimpse of the field.

One very important development was regarding magic clusters. It was found in the mass distribution of clusters that some clusters are found more abundant than others. These clusters were found to be very stable and are called magic clusters. The magic clusters of simple metals such as sodium are found when the number of valence electrons is 8, 20, 40, 58, and so on. The stability and magic behavior of these clusters can be understood by using the jellium model, which we shall discuss in Section 12.2. It turns out that the magic cluster is like a big atom that has 1*s*, 1*p*, 1*d*, 2*s* shells similar to the shell model of nuclei. Shell closure occurs for clusters having 8, 20, 40, 58, ... electrons explaining the stability and magic behavior of sodium clusters.

The jellium model is a very simple model that ignores the ionic structure. For more accurate results, one has to include ions when solving the KS equations. A large number of first-principles calculations have been done on a variety of clusters. The first-principles calculations play an important role in elucidating the geometry and structure of a cluster, which cannot be determined directly from experiments. Thus, in Section 12.3, we discuss the application of first-principles calculations to clusters. In particular, we shall discuss the application of the Car–Parrinello method to determine the geometry and structure of silicon and hydrogenated silicon clusters in Section 12.3.1. In Section 12.3.2, we shall discuss how the photoabsorption cross section is calculated using the time-dependent density functional theory and show its application to hydrogenated silicon clusters.

In Section 12.3.3, we shall discuss some carbon clusters. Among clusters, C_{60} is the most famous example, for which Kroto, Smalley, and Curl were awarded the Nobel Prize in 1996. C_{60} has a highly symmetric "football"-like structure in which 60 carbon atoms are arranged in 12 pentagons and 20 hexagons. It is interesting to know that this kind of high symmetry has long been used by artists and architects. Richard Buckminster Fuller was an architect who used this kind of symmetry for geodesic domes and thus C_{60} is known as "buckyball" or fullerene.

Section 12.4 is devoted to nanowires that are essentially one-dimensional (1D) objects. Because of its 1D nature, nanowires can exhibit very interesting phenomena such as the Peierls distortion and quantization of conductance (inverse of resistance). In Section 12.4.1, we discuss Peierls distortion. In Section 12.4.2, we apply the jellium model to obtain the band structure and DOS of a nanowire. In this section, we shall also discuss the quantization of conductance, which is a remarkable phenomenon observed in some nanowires. It has been found that in some nanowires the measured conductance is in integral units of $2e^2/\hbar$. These quantized values of conductance are manifestation of the restraint on the number of electrons that can travel through the wire at the nanometer scale. In Section 12.4.3, we briefly discuss some work on nanowires using the first-principles electronic structure calculations.

12.2 Jellium Model of Atomic Clusters

It was observed in mass spectra of sodium clusters that certain clusters were more abundant than others. These clusters were very stable and were called magic clusters. To explain the occurrence of magic clusters, a spherical jellium model was proposed by Knight et al. (1984). In this model, the ions are replaced by a constant positive charge background or jellium of spherical symmetry. The model is characterized by single parameter r_s, which is the radius of the volume per electron in the bulk metal.

Before we discuss the jellium model proposed by Knight et al., let us discuss a very simple model, in which an electron is confined in a spherically symmetric potential $v(r)$ such that

$$\begin{aligned} v(r) &= 0 \quad \text{for } r < R \\ &= \infty \quad \text{for } r \geq R \end{aligned} \tag{12.1}$$

that is, the electron is confined in a sphere of radius R. We solve the Schrödinger equation for $r < R$

$$-\frac{\hbar^2}{2m_e}\nabla^2\psi_i(\vec{r}) = \varepsilon_i\psi_i(\vec{r}) \tag{12.2}$$

where m_e is the mass of the electron. Because of the spherical symmetry, the radial and angular parts of $\psi_i(\vec{r})$ can be separated as

$$\psi_{n\ell m}(\vec{r}) = R_{n\ell}(r)Y_{\ell m}(\theta,\phi) \tag{12.3}$$

where n,ℓ,m, *respectively* are the principal quantum number, orbital quantum number, and magnetic quantum numbers. The radial solution is

$$R_{n\ell}(r) = j_\ell(k\,r) \tag{12.4}$$

where $k = 2m_e\varepsilon/\hbar^2$ and $j_\ell(k\,r)$ is the spherical Bessel function. The boundary condition is that $R_{n\ell}(r)$ vanishes at $r = R$, that is

$$j_\ell(k\,R) = 0$$

Thus, the root of $j_\ell(k\,R)$ gives the energy eigenvalues, which can be labeled as $\varepsilon_{n\ell}$, where $n = 1, 2, 3, \ldots$ and denotes the nth root of $j_\ell(k\,R)$. Therefore, one gets discrete eigenvalues for this model, which are in the order of 1*s*, 1*p*, 1*d*, 2*s*, 1*f*, 2*p*, 1*g*, 2*d*, 3*s*, … (Exercise 12.1). The magic cluster appears when some shell is just filled and another higher shell begins to fill, just like in inert gas atoms.

Although the simple model discussed above explains the occurrence of magic clusters, it is too crude. Now, we shall discuss the model proposed by Knight et al. (1984), which is more realistic and explains the mass spectrum of sodium clusters shown in Figure 12.1a. The one electron potential inside the cluster was simulated by assuming a spherically symmetric rounded potential well $v(r)$

$$v(r) = -\frac{v_0}{\exp[(r - r_0)/\sigma] + 1} \tag{12.5}$$

where v_0 is the sum of Fermi energy (3.23 eV) and the work function (2.7 eV) of the bulk. The parameter σ determines the variation of the potential at the edge of the sphere and was chosen to be 1.5 a.u., and r_0 is the effective radius of the clusters that was assumed to be $r_s N^{1/3}$, where N is the number of atoms in the cluster.

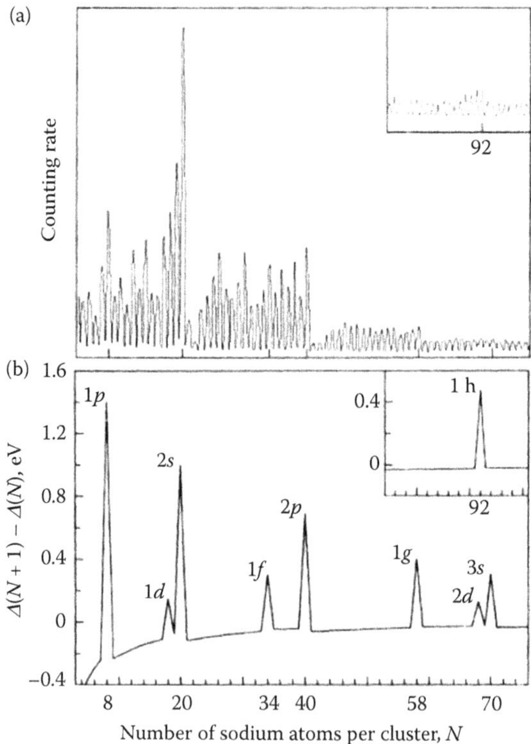

FIGURE 12.1
(a) Mass spectrum of sodium clusters. (b) $\Delta(N)$ as a function of N. (Reprinted with permission from Knight et al. *Phys. Rev. Lett.* 52: 2141–2143. 1984. Copyright 2001, by the American Physical Society.)

With potential given by Equation 12.5, the Schrödinger equation is

$$\left(-\frac{\hbar^2}{2m_e}\nabla^2 + v(r)\right)\psi_i(\vec{r}) = \varepsilon_i\psi_i(\vec{r}) \tag{12.6}$$

Since $v(r)$ is spherically symmetric, $\psi_i(\vec{r})$ can be separated in radial and angular parts, as

$$\psi_{n\ell m}(\vec{r}) = R_{n\ell}(\vec{r})Y_{\ell m}(\theta,\phi)$$

where n,ℓ,m are the principal quantum number, the orbital angular momentum number, and magnetic quantum numbers. In this model, the energy level order is 1*s*, 1*p*, 1*d*, 2*s*, 1*f*, 2*p*, 1*g*, 2*d*, 3*s*, ... The total energy $E(N)$ of the cluster was obtained by summing over the eigenvalues of the occupied states. The difference in total energy between adjacent clusters $\Delta(N)$ is defined as

$$\Delta(N) = E(N) - E(N-1) \tag{12.7}$$

To determine the stability, one plots $\Delta_2(N)$, the change in $\Delta(N)$

$$\Delta_2(N) = \Delta(N+1) - \Delta(N) = E(N+1) + E(N-1) - 2E(N) \tag{12.8}$$

Thus, $\Delta_2(N)$ is the relative binding energy of a cluster with N atoms with respect to those with $N+1$ and $N-1$ atoms. Therefore, peaks in $\Delta_2(N)$ correspond to relatively more stable clusters (see, e.g., Harbola 1992). We show a plot of $\Delta_2(N)$ as a function of N in Figure 12.1b. Peaks in the curve result when $\Delta(N+1)$ increases discontinuously because an orbital such as 1*p* is just filled and the next orbital such as 1*d* starts filling for the next $N+1$ atom cluster. These peaks match quite well with the peaks in the experimental result of mass spectrum for sodium clusters, which correspond to the magic clusters.

The jellium model totally ignores the ionic structure and explains the magic numbers mainly by quantum confinements of electrons. However, this could be somewhat misleading. The jellium model is rather an oversimplified model that totally ignores ionic cores and does not work for systems beyond simple systems such as clusters of alkali metals. In general, the ionic cores and their arrangement play an important role in determining the properties of clusters, as we shall see in the next section.

12.3 First-Principles Calculations of Atomic Clusters

First-principles calculations play an important role in the study of atomic clusters. Many properties such as geometry and structure of a cluster are not directly obtainable from experiments. In such cases, first-principles calculations provide very useful information about the geometry and structure of a cluster, which can then be verified by indirect means. The first-principles calculations can predict stable structures and hence can be used to design clusters with desired properties. In addition to geometry and structure determinations, such calculations are useful in understanding the properties such as magnetic, optical, and vibrational spectra. A variety of first-principles methods have been employed including CI, Hartree–Fock, MP, and plane wave methods such as Car–Parrinello, Gaussian-based methods, etc. The quantum chemical methods, such as CI, are very expensive and can be done only on clusters containing a few atoms. Therefore, the DFT-based first-principles methods have become very popular as they take less time and can handle fairly large clusters. Here, we shall mainly focus on the results obtained by using plane wave methods on some selected systems.

12.3.1 Ground-State Structures of Silicon and Hydrogenated Silicon Clusters

There has been a lot of interest in silicon clusters perhaps due to their importance in electronics, as silicon is the material of choice for chips and electronics devices. We shall see that in Si clusters the band gap increases compared to the bulk. Thus, with appropriate choice of a silicon cluster, it is possible to emit light in the optical range, as the bulk Si gap corresponds to the infrared region. Also in clusters, there is strong emission of light, which is due to quantum confinement of electrons and breaking of bulk selection rules. Ground-state structures of small Si_n clusters for $n \leq 10$ have been studied extensively by several workers such as Raghavachari and Logovinsky (1985), Tomanek and Schluter (1987), Andreoni (1991), Vasiliev, et al. (1997), and so on. Larger Si_n clusters ($n > 10 \leq 45$) have been studied by Kaxiras and Jackson (1993), Ho et al. (1998), and other workers. One important question is, when do the bulk-like features such as directional bonding start to appear as the cluster size increases? It has been found that for $n \leq 45$, there is no bulk-like directional bonding.

The ground-state structures for small silicon clusters for $n \leq 10$ are shown in Figure 12.2 (Balamurugan and Prasad 2001, Thingna et al. 2011). These structures were obtained using the CPMD and simulated annealing as explained in Chapter 9. The norm-conserving Bachelet–Hamann–Schlüter (1982) pseudopotential was used along with the LDA. A large fcc supercell with a side length of 35 a.u. was used to avoid interaction between clusters. These structures of clusters are in good agreement with the structures obtained by earlier workers. We see that Si_3 has a triangular structure while Si_4 is a flat rhombus. Si_5 is a "squashed" trigonal bipyramid and Si_6 is a distorted octahedron. Si_7, Si_8, and other clusters have more complicated geometry as shown in Figure 12.2.

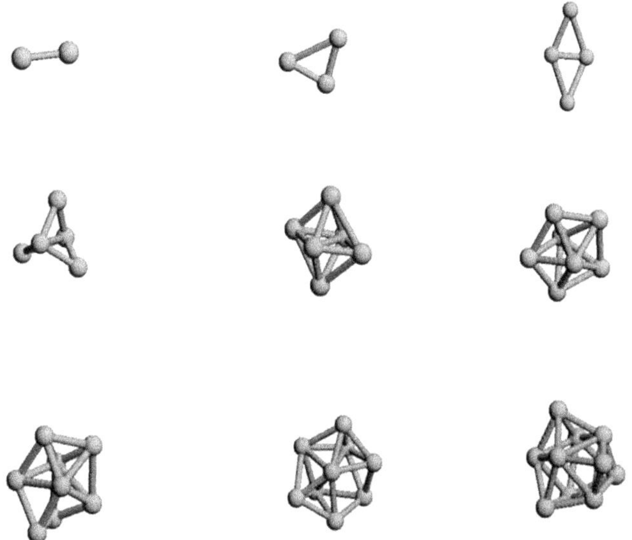

FIGURE 12.2
Ground-state structure of Si_2, Si_3, Si_4, Si_5, Si_6, Si_7, Si_8, Si_9, and Si_{10} clusters. (From Thingna J., Prasad R., and Auluck S. 2011. *J. Phys. Chem. Solids* 72: 1096–1100.)

To study the effect of the addition of a single hydrogen atom on Si_n clusters, we shall discuss ground-state structures of Si_nH cluster with $2 \leq n \leq 10$, which were obtained using the CPMD (Balamurugan and Prasad, 2001). In Figure 12.3, we show the ground-state structures of these clusters. Si_2H has a triangular structure in which two silicon atoms are bonded to each other not only via Si–Si bond but also via Si–H–Si bridge-type bond. Si_3H has a planar geometry with twofold symmetry and has some resemblance with the Si_4 cluster, shown in Figure 12.2. Also, in this cluster, there is a bridge-type Si–H–Si bond between two Si atoms numbered 1 and 2. In Si_4H, four Si atoms form a flat rhombus as in Si_4 and the hydrogen atom is above the plane and bonded with one of the Si atoms. In Si_5H, there are two lowest-energy structures that are very close in energy differing by only 0.06 eV. The structure (d) is higher in energy compared to (e). In the lowest-energy structure (e), there is a bridge-type bond connecting two Si atoms number 1 and 4. From Si_6H onward, we do not get this bridge-type bond and hydrogen simply attaches to the Si_n cluster causing very small distortions. Another interesting feature in these clusters, which have cage-like structure, is that the H atom attaches to the cage from outside and does not like to be inside the cage.

By comparing the structures in Figures 12.2 and 12.3, we see that a single H atom does not distort the Si_n cluster too much and the distortion becomes less and less as n increases. However, when the number of hydrogen atoms increases, there could be big changes in the structure. In Figure 12.4, we show the effect of further hydrogenation on the structure of Si_5 (Balamurugan,

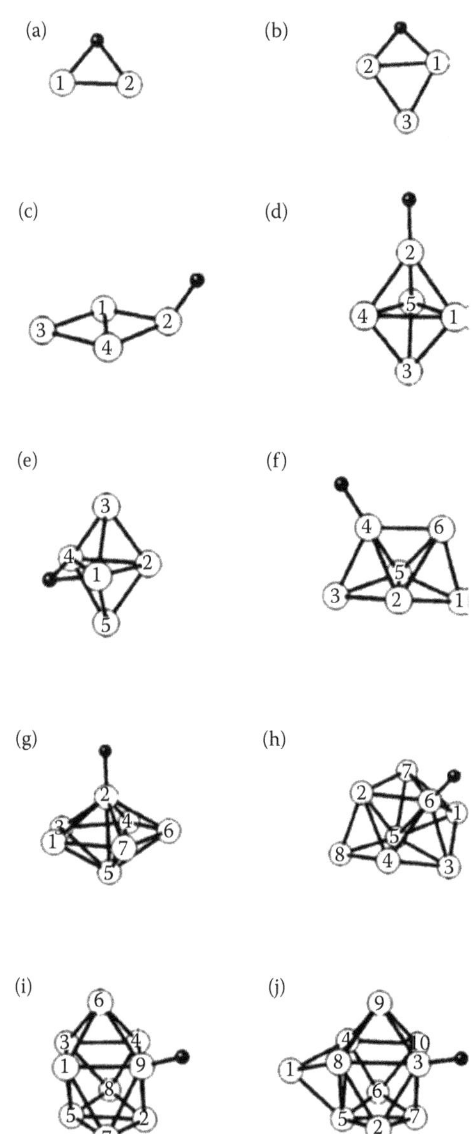

FIGURE 12.3
Ground-state structure of (a) Si_2H, (b) Si_3H, (c) Si_4H. (d) Higher energy structure of Si_5H. Ground-state structure of (e) Si_5H, (f) Si_6H, (g) Si_7H, (h) Si_8H, (i) Si_9H, and (j) $Si_{10}H$. Silicon atoms are numbered and hydrogen atoms are shown by dark circles. (From Balamurugan D. and Prasad R. 2001. *Phys. Rev. B* 64: 205406.)

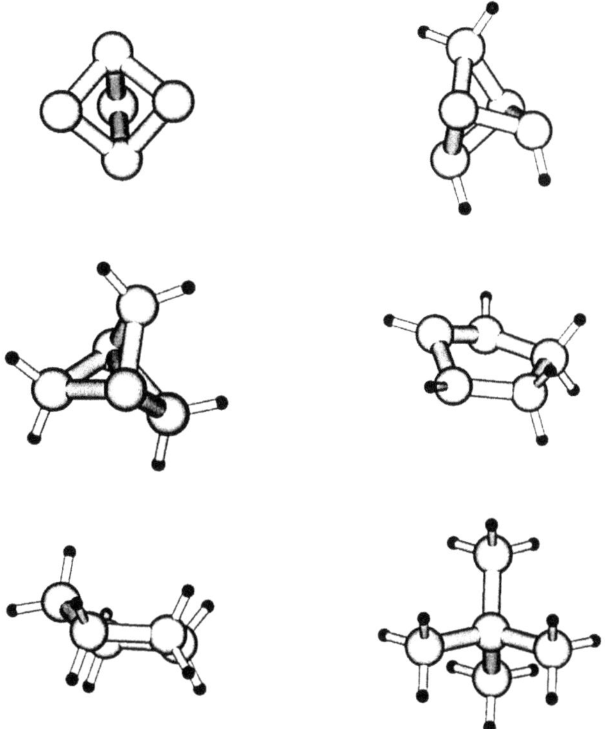

FIGURE 12.4
Ground-state structures of Si_5H_n clusters for $n = 0, 4, 6, 7, 10$, and 12. Empty sphere represents Si atom and black sphere (smaller) represents H atom. (From Balamurugan D. 2004. Ground and excited state properties of small hydrogenated silicon clusters: A first principles electronic structure study. PhD thesis, IIT Kanpur.)

2004). We see that as the cluster grows, it undergoes significant rearrangement and reconstruction. The structure of Si_5 does not change much up to the addition of three hydrogen atoms, but with the addition of the fourth hydrogen atom, the Si_5 cage begins to distort, as seen in Figure 12.4. With the addition of six hydrogen atoms, the structure is fully saturated as each Si has four saturated bonds. With the addition of one more hydrogen, this structure changes drastically to a ring-like structure. With 10 hydrogen atoms, this pentagonal ring gets fully saturated. With 12 hydrogen atoms, this structure changes into a more open structure and results in a symmetric tetrahedral structure. Thus, we see that the hydrogenation of a Si cluster can cause very large changes in the geometry and structure of the cluster.

As was mentioned in Section 12.1, properties of a cluster depend on its size. This is schematically shown in Figure 12.5. When the size of a cluster is small, say up to 10 or 20 atoms, the value of a physical property fluctuates with size due to the quantum size effect, but as the size of the cluster grows,

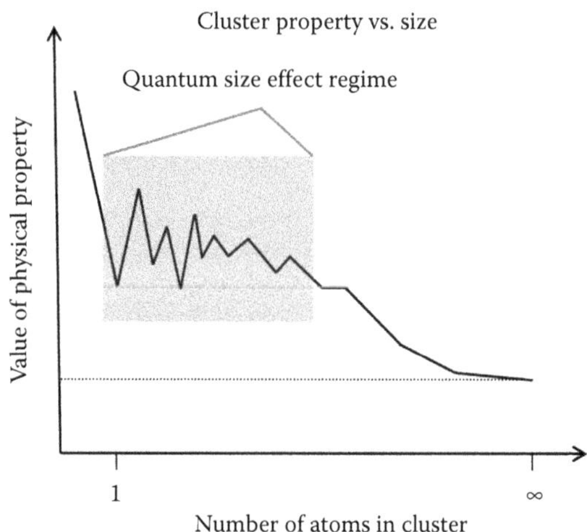

FIGURE 12.5
Schematic showing variation of a property with the size of a cluster. The dashed line indicates the bulk value.

the fluctuations get smaller. Eventually, when the number of atoms is large, say ~100, it approaches the bulk value. Thus, by tuning its size, a cluster having the desired value of the property can be designed. As an example, we show how the energy gap in a silicon cluster varies with its size. In Figure 12.6, we plot the energy gap between the first excited state and the highest occupied state for Si_n and Si_nH clusters as a function of size when the number

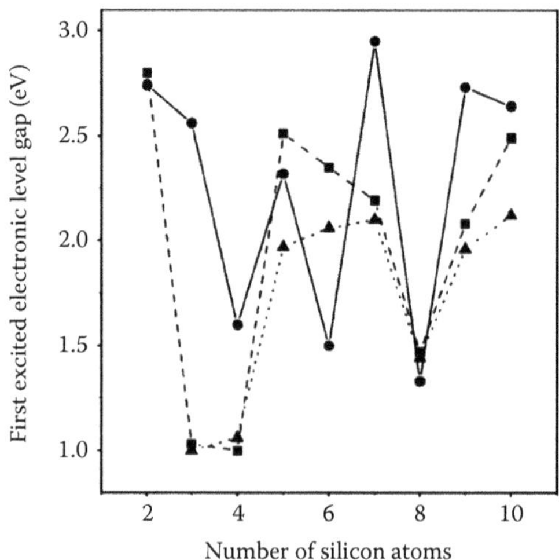

FIGURE 12.6
First excited-state energy gap in Si_n and Si_nH clusters. The circles correspond to Si_nH and triangles and squares to Si_n clusters. (From Balamurugan D. and Prasad R. 2001. *Phys. Rev. B* 64: 205406.)

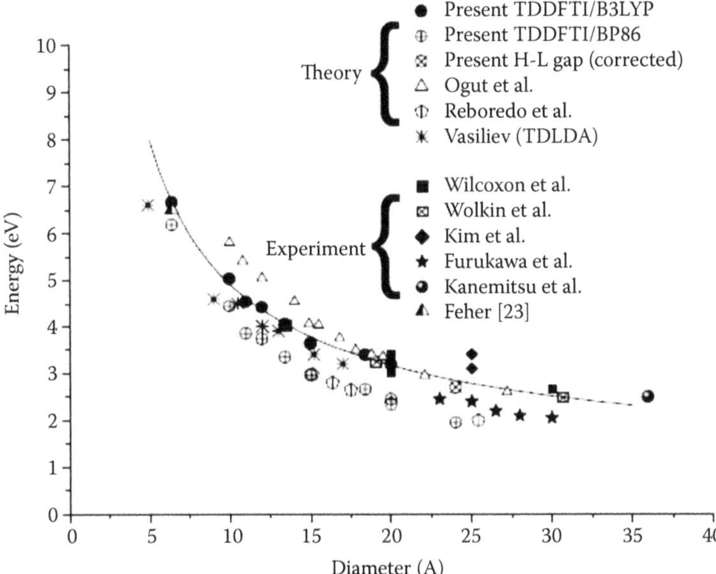

FIGURE 12.7
Energy gap of hydrogenated silicon clusters as a function of size. (Reprinted with permission from Garoufalis, C.S., et al. *Phys. Rev. Lett.* 87: 276402. Copyright 2001, by the American Physical Society.)

of atoms is small (Balamurugan and Prasad 2001, Prasad 2005). We see that the gap fluctuates. But as the number of atoms in the cluster becomes larger, the fluctuations become smaller and the gap tends to approach the bulk value as shown in Figure 12.7 (Garoufalis et al. 2001). Note that the methods and approximations used in these calculations are different, but the qualitative nature of the results conforms to the trend shown in Figure 12.5.

Clusters can show a very large change in properties by a slight manipulation of its structure or its charge state (Balamurugan et al. 2004, Balamurugan and Prasad 2006, Prasad 2005). For example, neutral SiH_4 has a very symmetric tetrahedral structure as shown in Figure 12.8. When it is charged, it shows a

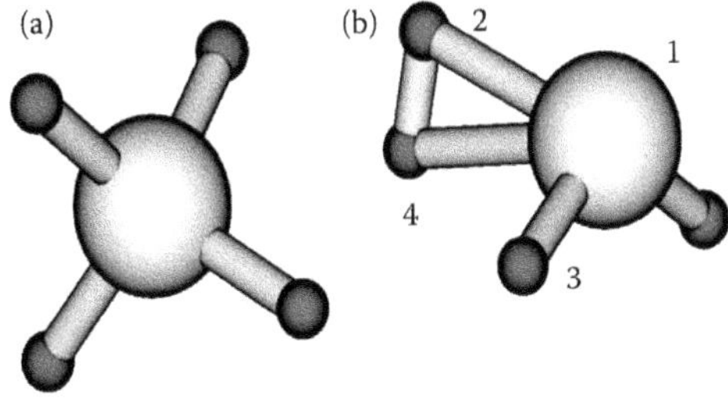

FIGURE 12.8
Ground-state structure of (a) SiH_4 and (b) SiH_4^+. (From Balamurugan D., Harbola M. K., and Prasad R. 2004. *Phys. Rev. A* 69: 033201.)

very large change in its geometry and no longer has the tetrahedral symmetry. Thus, we see that changing the cluster breaks the symmetry and is referred to as symmetry breaking. This symmetry breaking is due to the fact that in the distorted structure, the system lowers its energy by lifting the degeneracy of electronic levels in the symmetric structure. This phenomenon of lifting the degeneracy by lowering the symmetry is known as Jahn–Teller effect.

12.3.2 Photoabsorption Spectra

Now, we shall briefly discuss the photoabsorption spectra of silicon and hydrogenated silicon clusters. Peaks in the photoabsorption spectrum give information about electronic excited states of a cluster. The quantity of interest here is photoabsorption cross section that is defined as (Merzbacher 1998, Sakurai 1999)

$$\sigma(\omega)d\omega = \frac{\text{Energy absorbed in frequency interval } d\omega}{\text{Energy flux}}$$

As mentioned in Chapter 3, the DFT methods are basically developed for the ground state and therefore they do not give a good description of the excited states. To deal with excited states, one must use the TDDFT (for a review, see, e.g., Dreizler and Gross 1990, Chelikowsky et al. 2003). Theoretical formulation for the calculation of photoabsorption cross section using TDDFT has been presented by Yabana and Bertsch (1996, 1999). Here, we shall present the results for silicon and hydrogenated silicon clusters using this formulation (Thingna et al. 2011).

In Figure 12.9, we show the photoabsorption (PA) spectra for SiH_4 and Si_2H_6 (Thingna et al. 2011) along with the results of Marques et al. (2001). We get reasonably good agreement with the results of Marques et al. For SiH_4, we get peaks at 8.2, 9.2, and 9.8 eV, which are in good agreement with experimental results of 8.8, 9.7, and 10.7 eV (Itoh et al. 1986), respectively. The peaks in Si_2H_6 spectrum at 7.3, 8.6, 9.5 eV, and 10.6 eV are in good agreement with the experimental values of 7.6, 8.4, 9.5, and 9.9 eV.

In Figure 12.10, the photoabsorption spectra of Si_n clusters are shown for $n = 1$–10 along with their optimized structures (Thingna et al. 2011). For $n \leq 7$, we see that the photoabsorption spectrum is a combination of many peaks and looks like that of isolated atoms. However, for $n > 7$, it looks bulk-like. We also show the absorption spectra for Si_nH clusters in Figure 12.10. We find a similar trend in these spectra, which we saw in the case of Si_n clusters. Comparing the PA spectra of Si_nH and Si_n, we see that there is a slight shift of the Si_nH spectra to the left, that is, it is red-shifted. This is because the optical gap gets smaller by introducing a single hydrogen (Thingna et al. 2011).

To see the effect of further hydrogenation, we show the PA spectra of intermediate as well as fully hydrogenated clusters in Figure 12.11 (Thingna et al. 2011). The PA spectra are shown for Si_3H_3, Si_4H_4, Si_5H_6, Si_5H_{12}, Si_6H_7, and Si_6H_{14}

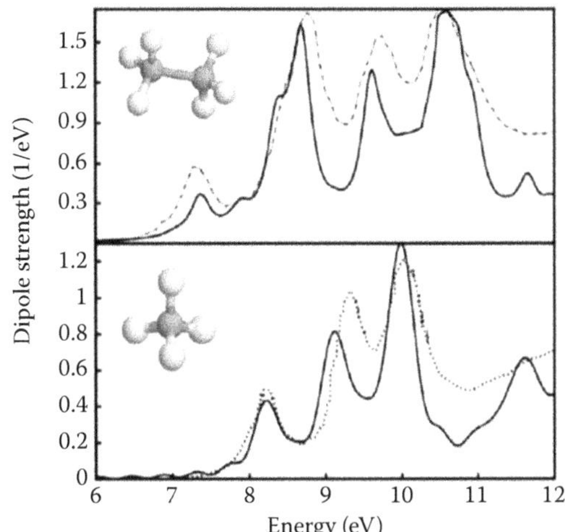

FIGURE 12.9
Photoabsorption spectra of S_2H_6 (top) and SiH_4. The dashed lines show the results of Marques et al. (2001). (From Thingna J., Prasad R., and Auluck S. 2011. *J. Phys. Chem. Solids* 72: 1096–1100.)

clusters. In all these clusters, a similar trend is found, that is, atomic-like behavior to a bulk-like transition in the PA spectra. However, in these clusters, the bulk-like behavior is observed much earlier at $n = 4$ compared to $n = 7$ for the singly hydrogenated clusters. The effect is attributed to the presence of a larger number of hydrogen atoms.

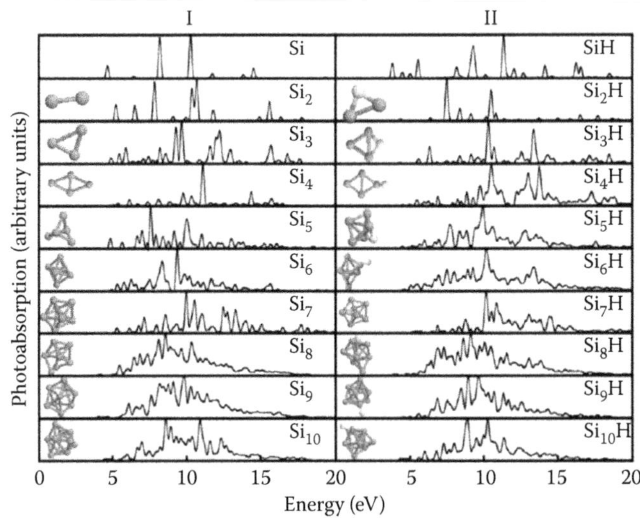

FIGURE 12.10
Photoabsorption spectra of Si_n and Si_nH clusters. (From Thingna J., Prasad R., and Auluck S. 2011. *J. Phys. Chem. Solids* 72: 1096–1100.)

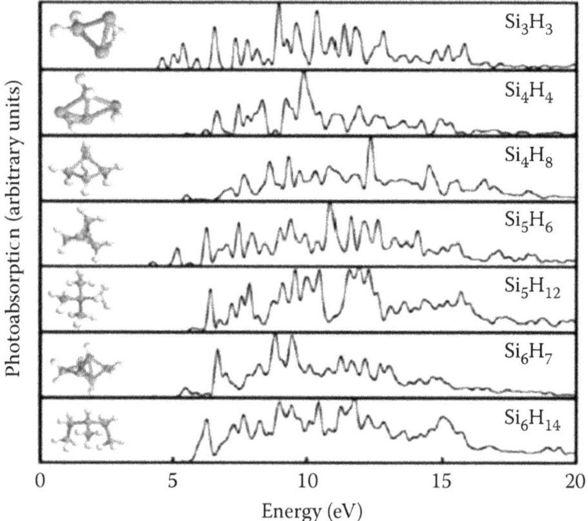

FIGURE 12.11
Photoabsorption spectra of Si_nH_m clusters. (From Thingna J., Prasad R., and Auluck S. 2011. *J. Phys. Chem. Solids* 72: 1096–1100.)

12.3.3 Carbon Clusters

Although carbon clusters have attracted much interest for many decades, it was the discovery of C_{60} that brought the field into the limelight. Carbon clusters are very common and can be found in hydrocarbon flames and other soot-forming systems. An understanding of the properties of these clusters is important for a variety of systems such as carbon stars, comets, and interstellar molecular clouds. These clusters are also important for understanding the chemical vapor deposition process for making diamond and silicon carbide films (Van Orden and Saykally 1998).

Several first-principles studies of ground-state structure of C_n clusters have appeared in the literature (e.g., Raghavachari and Binkley 1987, Tomanek and Schluter 1991, Jones 1999, Kosimov et al. 2010). According to the CI calculation of Raghavachari and Binkley (1987) for $2 \leq n \leq 10$, C_n clusters have structures that alternate between linear and ring. For even n, the clusters prefer to form rings, while for odd n, they form chains. However, photoelectron spectroscopy on C_n^- indicated linear structures up to $n = 9$ and monocyclic rings from $n = 10$–29 (Yang et al. 1988). Also, for C_{20}, there has been some controversy regarding its structure. The LDA predicts a cage-like structure; the H–F gives a ring structure while the quantum Monte Carlo gives a bowl-type structure (Grossman et al. 1995). The disagreement between theory and experiment can perhaps be attributed to the difficulty in performing the experiments on highly reactive clusters, practical necessity of studying charged species, and inherent approximations used in the calculations.

The time-of-flight mass spectra show that when $n > 30$, even-numbered C_n are more stable (e.g., Handschuh et al. 1995). These clusters have fullerene structure and the most stable fullerenes are C_{60} and C_{70}. C_{60} has a football-like structure with 12 pentagons and 20 hexagons as shown in Figure 12.12. As mentioned earlier, the discovery of C_{60} ushered in a new era in this field; therefore, we shall mainly focus on this cluster. For this discovery, Kroto, Curl, and Smalley (Kroto et al. 1985) were awarded the Nobel Prize in Chemistry in 1996. The easiest way of producing C_{60} carbon is by the arc method. This can be purified with good yield. It is stable at temperatures well above room temperature. In C_{60}, carbon molecules reside on a sphere of about 7 Å diameter. There are two different bond lengths in C_{60}, 1.40 and 1.46 Å. It is one of the most symmetrical clusters (Forro and Mihaly, 2001).

Several electronic structure calculations exist on C_{60} (e.g., Haddon et al. 1986, Feuston et al. 1991, Scuseria 1991). Haddon et al. (1986) used the Huckel molecular orbital theory to calculate electronic structure and bonding in C_{60} and predicted the cluster to be stable. Scuseria (1991) studied C_{60} at Hartree–Fock level and calculated equilibrium structure and ionization potential. They predicted the bond lengths to be 1.37 and 1.45 Å in reasonable agreement with the experimental value. Feuston et al. (1991) used *ab initio* CPMD to study the electronic and vibrational properties of C_{60}. They obtained the geometry and structure with bond lengths of 1.39 and 1.45 Å in good agreement with experimental results. It is interesting that the *ab initio* methods do a pretty good job in predicting the structure of C_{60}, although there were problems in predicting the structure of small carbon clusters.

C_{60} is a highly stable structure. It crystallizes in a crystalline form, which has an fcc lattice. The nearest-neighbor C_{60}–C_{60} distance is 10.02 Å at room temperature and the size of the cubic unit cell is 14.17 Å. Saito and Oshiyama (1991) studied cohesive mechanism and energy bands of C_{60} using the LDA

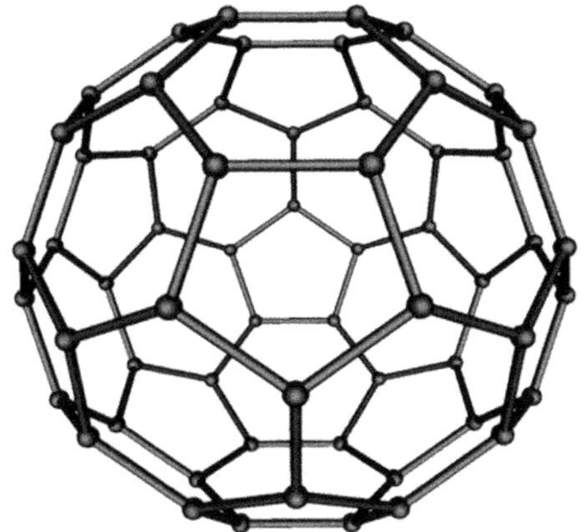

FIGURE 12.12
Structure of C_{60}. (Courtesy of Michael Ströck, Wikimedia Commons.)

and norm-conserving pseudopotential and an expansion in terms of Gaussian orbital basis set. They found that C_{60} clusters condense due to van der Waals force and the resulting C_{60} solid is a semiconductor with a direct energy gap of 1.5 eV. Zhang et al. (1991) studied the structure and dynamics of C_{60} solid using the *ab initio* CPMD simulations. They found that the C_{60} fullerene structure is very well preserved in the solid. They also found C_{60}–C_{60} interaction to be weak due to which there are only minor changes in the bond lengths and bond angles on solidification. They found the calculated electronic structure in excellent agreement with the photoemission data. In their constant-temperature MD simulations, C_{60} molecule started to rotate at low temperatures, in good agreement with NMR data. At high temperatures, the vibrational amplitude was found to be large but the cage structure was preserved.

As mentioned earlier, the C_{60}–C_{60} distance is very large ~10 Å, which is several times larger than a typical atom. Thus, there are large empty spaces between C_{60} clusters in solid C_{60} in which an atom can easily be inserted. Thus, solids such as A_nC_{60}, where A stands for Na, K, Rb, Cs, and so on, have been prepared, which are nearly stoichiometric. These solids are often called "doped fullerenes." There was much excitement in the field when it was found that when C_{60} solid is doped with Rb, it becomes superconducting with T_c approaching 28 K (Rosseinsky et al. 1991). It is believed that the pairing between the superconducting electrons arises due to the electron–phonon coupling. The T_c of doped fullerene is much higher compared to that of doped graphite and is thought to be so due to the curvature of C_{60}. This has led to a lot of interest in smaller fullerenes, which have larger curvature than C_{60}. It is indeed found that the electron–phonon interaction gets enhanced in C_{36} fullerene. A vast amount of work has been done on C_{60} and other fullerenes, which we cannot cover here. We refer to some excellent review articles (e.g., Forro and Mihaly, 2001, Van Orden and Saykally 1998). We shall discuss graphene and carbon nanotubes in Chapter 14.

After the discovery of C_{60}, there have been several attempts to seek fullerene-like forms of silicon. It turns out that silicon does not form a fullerene-like structure. However, metal-encapsulated silicon cage structures can be formed and have been predicted by the density functional *ab initio* calculations (Kumar and Kawazoe 2001).

12.4 Nanowires

During the last decade, there has been a lot of interest in the study of nanowires of atomic dimensions. Typically, the diameter of a nanowire is of the order of 1 nm while its length is unconstrained. Thus, the electrons in a nanowire are confined in two dimensions perpendicular to its length but are unconfined along its length and thus it is essentially a one-dimensional (1D) object. Because of this confinement and its 1D nature, it can show very different electronic, magnetic, and optical properties compared to bulk material, such as the Peierls distortion and quantization of conductance (inverse of resistance). Thus, there is much interesting physics in these 1D systems.

Another reason for the great interest in nanowires is due to their potential for use in nanotechnology. Because metallic nanowires have very small dimensions, they could be used for electrical conductions in integrated nanoscale systems and to connect nanoparticles. With modern techniques such as the scanning tunneling microscope (STM), it is now possible to precisely control the formation of nanowires on substrates. Freely standing metallic nanowires can be obtained by using the mechanically controllable break junction (MCBJ) method. The principle of the method is quite simple: two pieces of the material initially in contact are pulled away from each other over atomic distances. By using this method, it is possible to obtain very thin nanowires consisting of only a single atom strand. For example, gold nanowires have been obtained in this way (e.g., Ohnishi et al. 1998). Nanowires can also be obtained by electron beam irradiation of thin films (Kondo and Takayanagi 1997).

12.4.1 Peierls Distortion

Before we proceed further, we would like to discuss a well-known result due to Peierls regarding one-dimensional metals. He showed that a one-dimensional metal can become unstable with respect to a periodic distortion known as Peierls distortion. To understand this, let us consider a one-dimensional metal of monovalent atoms with lattice spacing a. Assuming the nearly free electron model, its electronic band structure can be obtained and is shown schematically in Figure 12.13. Since we have one electron per unit cell, the valence band is half-filled. Now, suppose that there is a periodic distortion

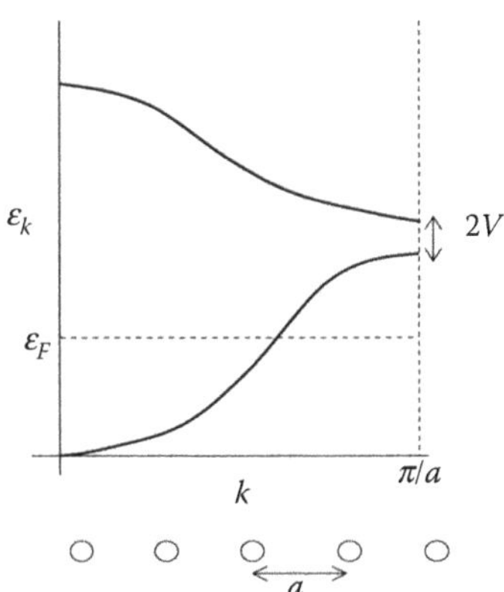

FIGURE 12.13
Schematic showing band structure of a one-dimensional solid with lattice constant a.

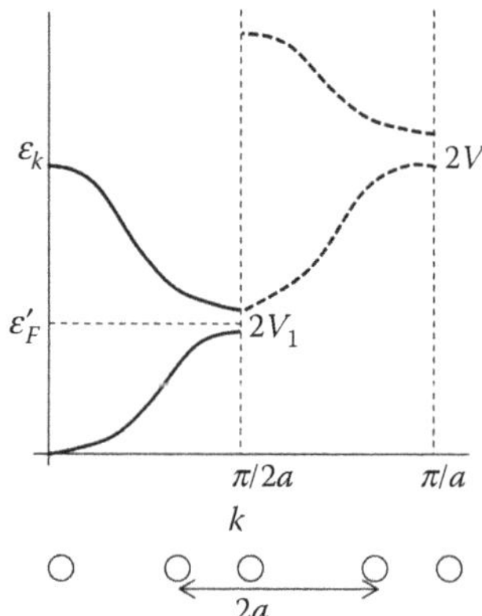

FIGURE 12.14
Schematic showing the band structure of the distorted 1D solid with lattice constant $2a$.

such that alternate pairs of atoms are closer and the periodicity of the lattice is $2a$ as shown in Figure 12.14. Since the periodicity is $2a$, the new BZ boundary will be at $\pi/2a$ and the new band structure will appear as shown in Figure 12.14. Now, there is a band gap at $k = \pi/2a$ equal to $2V_1$ depending on the electron–ion potential of the system. In Figure 12.14, we have also shown the bands in the original BZ of lattice constant a (Figure 12.13) by dashed lines. We note that the new Fermi level ε'_F is in the band gap and the system is now an insulator. The highest filled level is lower by V_1 compared to the old Fermi level ε_F and as a result, the total electronic energy of the new system would be lower compared to the old one. If the net gain in electronic energy is not overcome by an increase in elastic energy, the system would like to be in the distorted state and it becomes an insulator. Thus, there is always a possibility that a one-dimensional metal would be unstable with respect to the Peierls distortion and then there would be a transition from metal to an insulator. However, a nanowire is truly not a one-dimensional system. It has a diameter and there are going to be other effects such as surface effects, which may suppress such a distortion. A Peierls insulator is expected to show a giant Kohn anomaly in the phonon spectrum at some finite temperature (Renker et al. 1973, Prasad et al. 1975).

12.4.2 Jellium Model of Nanowire

Let us consider a very simple jellium model to understand the electronic structure and quantization of conductance in metallic nanowires. We model the nanowire as a long cylinder of radius R and length L as shown in Figure 12.15. Let (r, ϕ, z) denote the coordinates of an electron in the cylindrical coordinate system. We assume that the noninteracting electrons in the wire are confined in the radial direction by an infinite square well potential $V(r)$ such that

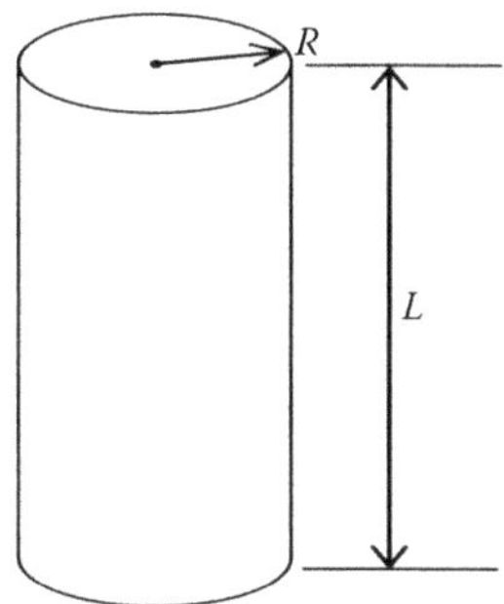

FIGURE 12.15
A nanowire of radius R and length L.

$$V(r) = 0 \quad \text{for } r < R$$
$$= \infty \quad \text{for } r \geq R$$

The Schrödinger equation for an electron moving in the nanowire in cylindrical coordinates is

$$-\frac{\hbar^2}{2m_e}\left(\frac{1}{r}\frac{\partial}{\partial r} + \frac{\partial^2}{\partial r^2} + \frac{1}{r^2}\frac{\partial^2}{\partial \phi^2} + \frac{\partial^2}{\partial z^2}\right)\psi(r,\phi,z) = \varepsilon\psi(r,\phi,z) \tag{12.9}$$

where m_e denotes the mass of the electron. Owing to the cylindrical symmetry, the equation is separable into r, ϕ, and z. The solution along the z axis will have a plane wave-like solution $e^{ik_z z}$ with k_z as wave vector and angular solution will be like $e^{im\phi}$, where m is an integer ($m = 0, \pm 1, \pm 2, \ldots$). Thus, the solution of Equation 12.9 can be written as

$$\psi_{mnk_z}(r,\phi,z) = \frac{e^{ik_z z}}{\sqrt{L}}\frac{e^{im\phi}}{\sqrt{2\pi}}R_{mn}(r) \tag{12.10}$$

where $R_{mn}(r)$ is the radial wave function and n is the radial quantum number. Substituting Equation 12.10 in Equation 12.9, we get

$$-\frac{\hbar^2}{2m_e}\left(\frac{1}{r}\frac{\partial}{\partial r} + \frac{\partial^2}{\partial r^2} - \frac{m^2}{r^2} - k_z^2\right)R_{mn}(r) = \varepsilon_{mn}R_{mn}(r) \tag{12.11}$$

where we have labeled the eigenvalues of $R_{mn}(r)$ as ε_{mn}. Equation 12.11 can be rewritten as

$$\left(\frac{1}{r}\frac{\partial}{\partial r} + \frac{\partial^2}{\partial r^2} - \frac{m^2}{r^2} - k_z^2\right)R_{mn}(r) = -\frac{2m_e\varepsilon_{mn}}{\hbar^2}R_{mn}$$

or

$$\frac{1}{r}\frac{\partial}{\partial r}R_{mn} + \frac{\partial^2 R_{mn}}{\partial r^2} + \left(\frac{2m_e\varepsilon_{mn}}{\hbar^2} - \frac{m^2}{r^2} - k_z^2\right)R_{mn} = 0$$

or

$$r^2\frac{\partial^2 R_{mn}}{\partial r^2} + r\frac{\partial R_{mn}}{\partial r} + (k^2r^2 - m^2)R_{mn} = 0 \tag{12.12}$$

where

$$k = \left(\frac{2m_e\varepsilon_{mn}}{\hbar^2} - k_z^2\right)^{1/2} \tag{12.13}$$

Equation 12.12 is Bessel's equation, whose general solution can be written as

$$R_{mn}(r) = A\,J_m(kr) + B\,N_m(kr) \tag{12.14}$$

where $J_m(kr)$ and $N_m(kr)$ are, respectively, the Bessel and Neumann functions of order m, and A and B are constants that can be determined from the boundary conditions. As the solution must be finite at $r = 0$ and since $N_m(kr) \to \infty$ as $r \to 0$, B must be zero.

Thus

$$R_{mn} = A\,J_m(kr) \tag{12.15}$$

The boundary condition requires that $R_{mn} = 0$ at $r = R$. This implies that $J_m(kR) = 0$ or

$$kR = \alpha_{mn}$$

where α_{mn} is the nth zero of the mth-order Bessel function. Using Equation 12.13, we can express ε_{mn} in terms of α_{mn}, that is

$$\frac{2m_e\varepsilon_{mn}}{\hbar^2} - k_z^2 = k^2 = \frac{\alpha_{mn}^2}{R^2}$$

or

$$\varepsilon_{mn} = \frac{\hbar^2k_z^2}{2m_e} + \frac{\hbar^2\alpha_{mn}{}^2}{2m_eR^2} \tag{12.16}$$

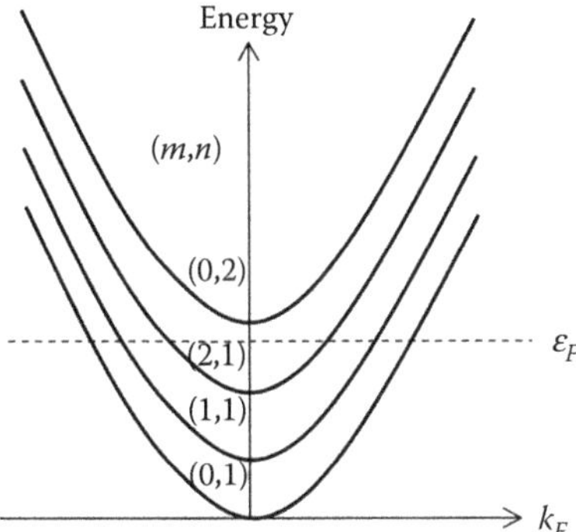

FIGURE 12.16
Schematic energy bands of a nanowire.

The energy bands obtained by Equation 12.16 are shown schematically in Figure 12.16, which shows parabolic subbands whose bottoms are located at energy

$$\varepsilon_{mn}^{0} = \frac{\hbar^2 \alpha_{mn}^2}{2m_e R^2} \tag{12.17}$$

The DOS for the nanowire can be obtained from Equation 12.16, similar to the calculation of DOS for the jellium model in Chapter 2 (Exercise 12.3). The DOS per unit length can be written as

$$\rho(\varepsilon) = \frac{1}{\pi}\frac{m_e}{\hbar^2}\sum_{mn}\frac{1}{\sqrt{\frac{2m_e}{\hbar^2}(\varepsilon - \varepsilon_{mn}^{0})}} \tag{12.18}$$

In Figure 12.17, we show a schematic plot of $\rho(\varepsilon)$, which diverges at the bottom of the parabolic subbands $\varepsilon = \varepsilon_{mn}^{0}$. This is very similar to the case of

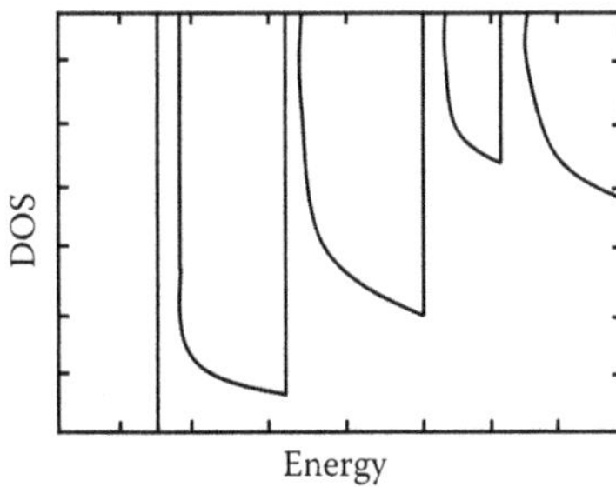

FIGURE 12.17
Schematic plot of the DOS of a nanowire.

1D free electron gas where the DOS goes as $\varepsilon^{-1/2}$. This behavior is responsible for many interesting properties of nanowires such as the quantization of conductance.

Using this model, Stafford et al. (1997) calculated conductance (inverse of resistance) of a nanowire. They found that the conductance of the nanowire could be written as

$$G = \frac{2e^2}{\hbar} \sum_{mn} T_{mn} \tag{12.19}$$

where T_{mn} is the transmission probability of each subband. Note that $G_0 = 2e^2/\hbar$ is the quantum of conductance. The conductance G is schematically plotted in Figure 12.18 as a function of the elongation of the wire, which shows plateaus in agreement with the observed behavior.

This is a remarkable result that can be understood in terms of energy band structure (Figure 12.16). For occupied bands, we can approximate $T_{mn} \sim 1$ and zero for unoccupied bands. Then, G can be written as

$$G = \frac{2e^2}{\hbar} v \tag{12.20}$$

where v is the number of occupied subbands. We see from Equation 12.17 that the spacing between the subbands depends on R. As the wire elongates, R also changes, and, therefore, the spacing. At some particular elongation of the wire, the topmost subband moves above the Fermi energy causing a drop in G according to Equation 12.20. As the wire is elongated further, nothing happens, until another subband moves above ε_F. This explains the behavior plotted in Figure 12.18.

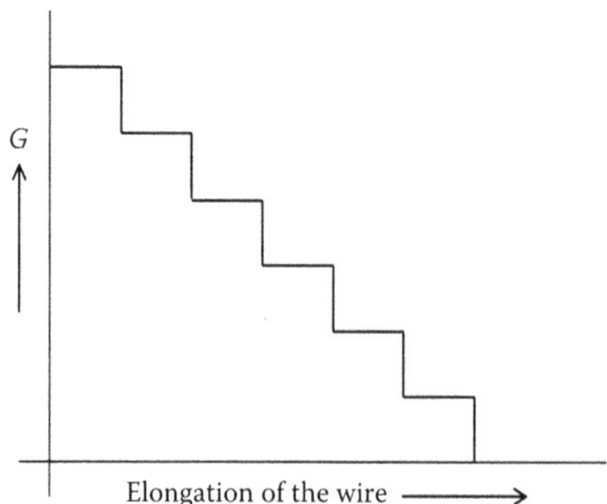

FIGURE 12.18
Schematic plot of the conductance of a nanowire with elongation.

12.4.3 First-Principles Calculations

The jellium model is an approximate model that ignores the ionic structure and therefore cannot be expected to give accurate results. Therefore, first-principles electronic structure calculations play an important role in understanding the structural issues, and magnetic, optical, and transport properties of nanowires. First-principles molecular dynamic simulations have been done on various systems to understand atomic rearrangement and breakage when the nanowire is pulled and stability-related issues.

We shall first discuss metallic nanowires. First-principles calculations have been done on nanowires of several systems such as Na, Al, Cu, Ag, Au, 3*d* and 5*d* transition metals, and so on to understand electronic structure and structural stability. Among these systems, perhaps Au nanowires have drawn maximum attention, experimentally as well as theoretically. The transmission spectroscopy measurements (TEM) showed the Au–Au spacing in Au nanowire to be around 3.0–3.5 Å while the theoretical results suggest the spacing to be around 2.7 Å. However, other measurements found the Au–Au spacing in the range of 2.3–2.9 Å, which is in good agreement with theoretical calculations. Ribeiro and Cohen (2003) did frozen phonon calculations using the *ab initio* pseudopotential method to study the stability of infinite monoatomic chains of Au, Al, Ag, Pd, Rh, and Ru as a function of strain. They found that zero-strain linear chains were not stable but within a window of strains, the Au, Al, Ag, Pd, and Rh linear chains were stable with respect to $q = \pi/a$ deformation. For large strains, all chains were found to dimerize. All these chains were found to exhibit at least one zero-strain zigzag stable equilibrium configuration. More recently, Tavazza et al. (2010) studied the electronic structure and structural changes during the formation of single-atom gold chains under tensile deformation. The electronic structures of nanowires of many 3*d*, 4*d*, and 5*d* transition metals have been studied by several workers (e.g., Nautiyal et al. 2004). One interesting question that was investigated by Delin and Tosatti (2003) was when and how the magnetism appears in nanowires of 5*d* transition metals.

Several first-principles molecular dynamics simulations have been done on nanowires. The main problems that have been investigated are the breaking of a nanowire and the conductance fluctuations. Barnett and Landman (1997) studied the conductance fluctuations and breakage in sodium nanowire. The conductance was calculated by using the Kubo formula. Nakamura et al. (1999) also studied the conductance fluctuations and breaking of an atomic-sized sodium wire using the first-principles molecular dynamics simulation. Anglada et al. (2007) have conducted first-principles simulations of the formation of monoatomic gold chain with different impurities (H, C, O, S). These simulations have shed light on the microscopic mechanism of breakage and conductance fluctuations.

Recently, nanowires of semiconductors, particularly of silicon, have attracted a lot of attention. Silicon nanowires look very promising because they can be used for connecting nano devices. They can be easily integrated with existing Si devices and thus provide an extra advantage. Si nanowires have potential for applications in future mesoscopic electronics and optical devices such as light-emitting diodes (LEDs), field-effect transistors (FETs), inverters, and nanoscale sensors. Silicon nanowires grow along well-defined crystallographic orientations and have a single crystalline core. They can be metal depending on their orientation. They show strong anisotropy in their properties at small diameters. For example, band gaps, the Young's modulus, electrical resistance, and specific heat are different for wires grown along different orientations. The band gap can be direct and tuned by varying the diameter. This could be very important for their use in optoelectronics. We shall not discuss it in detail and refer to the excellent review article of Rurali (2010).

EXERCISES

12.1 For the spherical jellium model of a cluster described by Equation 12.1, find the energy eigenvalues. Show that the energy eigenvalues for this model are in the order of 1*s*, 1*p*, 1*d*, 2*s*, 1*f*, 2*p*, 1*g*, 2*d*, 3*s*, and so on.

12.2 For the jellium model of a nanowire, find α_{mn} and energy bands using Equation 12.16. Plot the result and check if it looks similar to Figure 12.16.

12.3 For the jellium model of a nanowire, calculate the DOS using Equation 12.16. Plot the result and check if it looks similar to Figure 12.17.

12.4 Find the ground-state structure of an Si_4 cluster using simulated annealing and first-principles molecular dynamics. You may use an appropriate code from Appendix A. Compare your results with that of Figure 12.2.

12.5 Find the ground-state structure of C_3 and C_4 clusters using simulated annealing and first-principles molecular dynamics. You may use an appropriate code from Appendix A. Compare your results with that of Raghavachari and Binkley (1987).

12.6 Find the ground-state structure of an SiH_4 cluster and its photoabsorption cross section. Compare your results with Figure 12.9. You may use an appropriate code from Appendix A.

Further Reading

Bowler D. R. 2004. Atomic-scale nanowires: Physical and electronic structure. *J. Phys.: Condens. Matter* 16: R721–R754.

Dresselhaus M. S., Dresselhaus G., and Eklund P. C. 1996. *Science of Fullerenes and Carbon Nanotubes*. New York: Academic Press.

Johnston R. J. 2002. *Atomic and Molecular Clusters*. London: Taylor & Francis.

Khanna S. N. and Castleman A. W. (Eds.) 2003. *Quantum Phenomena in Clusters and Nanostructures*. Berlin: Springer-Verlag.

13

Surfaces, Interfaces, and Superlattices

13.1 Introduction

Until now, we have been considering mostly three-dimensional systems and ignoring surfaces. A surface is a plane of atomic sites where a crystal terminates. When a crystal meets a plane of another crystal, the plane is called an interface. Surfaces and interfaces play an important role in physics, chemistry, and materials science and form a rich field of study that has been growing very fast. Surfaces and interfaces are two-dimensional systems that provide an ideal platform to study physics in two dimensions such as quantum Hall effects, melting transition, etc. Also, they are important from a practical point of view and play a vital role in technological applications such as microelectronics, catalysis, corrosion, lubrication, and photography.

In spite of the importance of surfaces, experimental study of clean crystal surfaces at the atomic level became possible only after the advent of ultrahigh vacuum systems. Only in an ultrahigh vacuum was it possible to create and maintain clean surfaces. This is because the surfaces, in general, are quite reactive. They react with the surrounding atmosphere and in the process get covered with a large amount of foreign atoms. The atomic structure of a surface can be determined by methods such as low-energy electron diffraction (LEED), reflection high energy diffraction (RHEED), x-ray scattering, and STM. Using STM, it is possible to determine the surface structure and look at it in detail at the atomistic level. This technique has revolutionized the field, and for inventing this technique, G. Binnig and H. Rohrer were awarded the Nobel Prize in 1986.

First-principles electronic structure calculations have played an important role in understanding the various phenomena at surfaces such as surface states, surface reconstruction, catalysis, and adsorption. It is not possible to cover all these topics in this chapter; therefore, we shall focus only on surface states and surface relaxation. In Section 13.2, we will briefly discuss the surface geometry and various terminologies used in the field. In Section 13.3, we shall discuss surface states, and in Section 13.4, surface relaxation. In

Section 13.5, we shall discuss interfaces, and in Section 13.6, we shall discuss superlattices.

13.2 Geometry of Surfaces

An ideal surface, which is periodic in two dimensions, can be characterized by two primitive translational vectors $\vec{a}_1$ and $\vec{a}_2$ on the surface plane. These two lattice vectors span the whole two-dimensional lattice and form a surface unit cell. The reciprocal space is spanned by primitive translational vectors $\vec{b}_1, \vec{b}_2$, which are given by

$$\vec{b}_i \cdot \vec{a}_j = 2\pi\delta_{ij} \quad (i, j = 1, 2) \tag{13.1}$$

The vectors $\vec{b}_1$ and $\vec{b}_2$ span the reciprocal lattice whose surface BZ can be obtained by using the standard procedure.

An adsorbate layer on the ideal surface or reconstructed surface can be described in terms of $\vec{a}_1$ and $\vec{a}_2$. Let $\vec{c}_1$ and $\vec{c}_2$ be the lattice vectors corresponding the adsorbate layer. Then, $\vec{c}_1$ and $\vec{c}_2$ can be written as

$$\begin{aligned} \vec{c}_1 &= S_{11}\vec{a}_1 + S_{12}\vec{a}_2 \\ \vec{c}_2 &= S_{21}\vec{a}_1 + S_{22}\vec{a}_2 \end{aligned} \tag{13.2}$$

If the angles between $\vec{c}_1$ and $\vec{c}_2$ are the same as that between $\vec{a}_1$ and $\vec{a}_2$, the shorthand notation of Wood is generally used. In this notation, the lattice spanned by $\vec{c}_1, \vec{c}_2$ can be described as

$$\left(\frac{c_1}{a_1} \times \frac{c_2}{a_2}\right) R\theta \tag{13.3}$$

where θ is the relative rotation of the two lattices. If $\theta = 0$, the angle is omitted. For example, (2×1) structure means c_1 is $2a_1$ and $c_2 = a_2$ and the relative angle is zero. $(\sqrt{3} \times \sqrt{3})R30^\circ$ structure means $c_1 = \sqrt{3}a_1$, $c_2 = \sqrt{3}a_2$, and $\vec{c}_1$ and $\vec{c}_2$ are rotated by 30° with respect to $\vec{a}_1$ and $\vec{a}_2$. A surface is created by terminating a crystal. Therefore, it is identified by the Miller index of the crystal plane to which it belongs. These Miller indices generally refer to the conventional unit cell rather than the primitive unit cell. For example, for fcc and bcc lattices, the Miller indices are obtained by using the conventional cubic unit cell.

13.3 Surface Electronic Structure

The termination of a solid at a surface causes an abrupt change in the potential seen by an electron. This in turn leads to changes in the electronic structure and atomic arrangement near the surface. In this section, we shall focus on the effects concerning the electronic structure. The effects on atomic structure will be considered in the next section.

Let us consider a jellium model in which the ionic charge has been smeared out to a constant value inside the bulk. Let us denote the ionic charge density by $n^+(x)$, where x is perpendicular distance from the surface.

$$n^+(x) = \begin{cases} n^+ = \text{constant} & x \le 0 \\ 0 & x > 0 \end{cases} \tag{13.4}$$

The electronic charge density $n^-(x)$ does not fall to zero abruptly like ionic charge density. It spills out in the vacuum and goes to zero as shown in Figure 13.1. Thus, the region just outside the surface is slightly negatively charged while the region just inside the surface is slightly positively charged. This gives rise to a surface dipole layer also called a double layer, which plays an important role in determining the work function (see below). Note that the electronic charge density does not fall to a constant value inside the solid monotonically but shows oscillations. These oscillations are called Friedel oscillations, which arise because the electrons try to screen the sharp features of the ionic potential at the surface (Lang and Kohn 1970).

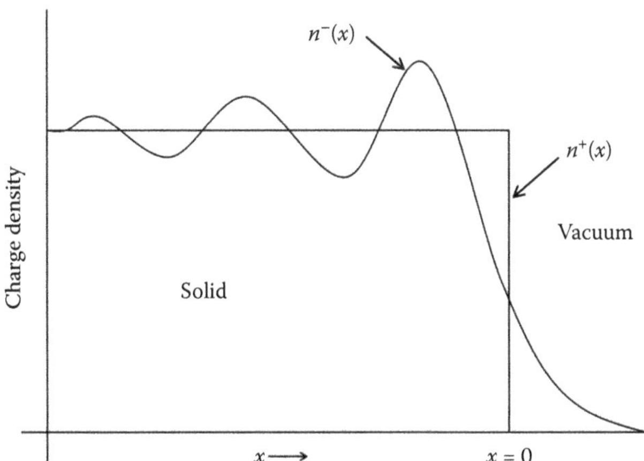

FIGURE 13.1
Schematic showing ionic charge density $n^+(x)$ and electronic charge $n^-(x)$ as a function of x. The surface is placed at $x = 0$.

The work function W is defined as the minimum work that must be done to remove an electron from a solid at 0 K (Ashcroft and Mermin 1976).

$$W = W_s - \varepsilon_F \tag{13.5}$$

where W_s is work done by the field in the double layer and ε_F is the Fermi energy. Note that the work function depends on the surface.

13.3.1 Surface States

Surface states are the electronic states that are localized near the surface of a solid. This happens because the termination of the surface causes an abrupt change in the potential, which can produce such states. To see how this can happen, let us consider a simple one-dimensional model of a solid with lattice spacing a, which is abruptly terminated at $x = 0$.

Let us assume that an electron moving in the solid sees the potential

$$\begin{aligned} V(x) &= V_0 + 2V_1\cos\left(\frac{2\pi x}{a}\right) \qquad & x < 0 \\ &= W & x > 0 \end{aligned} \tag{13.6}$$

where W is the potential barrier at the surface and V_0, V_1 are constants. The potential (Equation 13.6) is schematically shown in Figure 13.2. Assuming that V_1 is small, we can use the nearly free electron model discussed in Chapter 5 to find energy bands and wave function for $x < 0$. The wave function for $x > 0$ can be easily found and will be matched at $x = 0$.

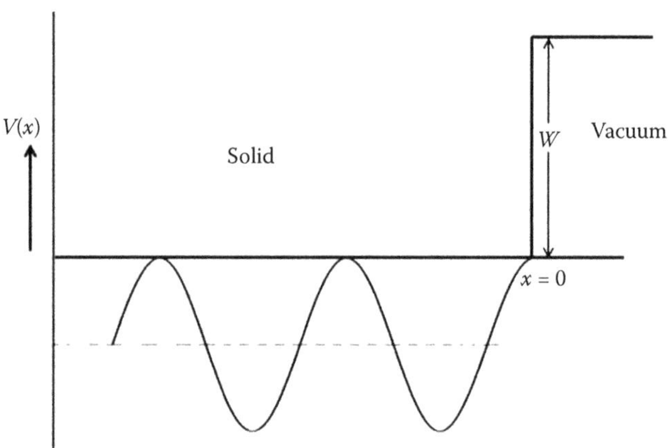

FIGURE 13.2
Potential $V(x)$ as a function of x.

For $x < 0$, the nearly free electron model will predict an energy gap equal to $2V_1$ at BZ boundary $k = \pi/a$. Let us measure k from the BZ boundary and write

$$q = k - \pi/a \tag{13.7}$$

Then, we can show that energy eigenvalue (Exercise 13.1)

$$\varepsilon_q = V_0 + \frac{\hbar^2}{2m_e}\left[q^2 + \frac{\pi^2}{a^2}\right] \pm \sqrt{V_1^2 + \left(\frac{\hbar^2}{2m_e}\, q\pi/a\right)^2} \tag{13.8}$$

and wave function

$$\psi_q = e^{iqx}\,(A\, e^{i\pi x/a} + B\, e^{-i\pi x/a}) \tag{13.9}$$

where A and B are constants. The bands near the zone boundary π/a are shown schematically in Figure 13.3.

For $x > 0$, the wave function for energy $\varepsilon < W$ must decay exponentially and is given by

$$\psi = C'\, e^{-\kappa x} \tag{13.10}$$

where C' and κ (positive) are constants.

Since there is no translational symmetry, the wave function ψ need not satisfy Bloch's theorem. This means q in Equation 13.9 need not be a real

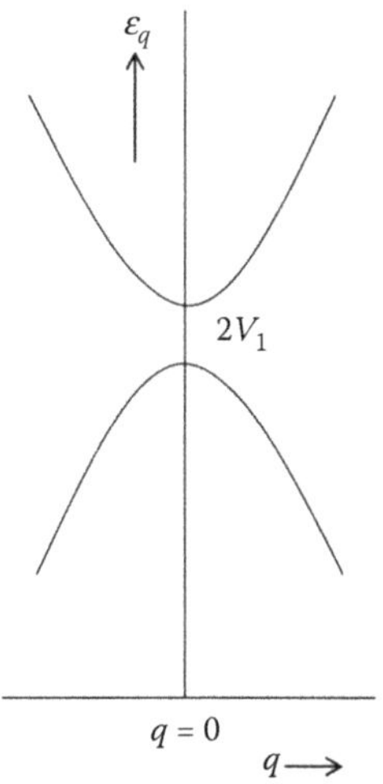

FIGURE 13.3
Schematic plot of energy bands near the zone boundary.

number and can take complex values. For an imaginary value of q, we see from Equation 13.9 that the wave function decays as $e^{-q|x|}$ inside the solid and from Equation 13.10 as $e^{-\kappa x}$ outside the solid. Thus, an electron in this state is localized near the surface. We can see from Equation 13.8 that such an electron with imaginary q has energy in the band gap shown in Figure 13.3. Thus, we have shown that the abrupt change in potential caused by a surface can produce surface states that are localized near the surface. Such localized states can exist only in the bulk band gap. The surface can produce some other states that lie in the band and get mixed with the bulk states. Such states are not localized near the surface and are called surface resonances.

The above result can be generalized to three dimensions (3D). In 3D, there is periodicity in the surface plane but no periodicity in the direction perpendicular to the surface. Thus, the Bloch theorem is valid only in the surface plane implying that the surface states are characterized only by $\vec{k}_{\|}$, the wave vector corresponding to the periodic surface plane. This is in contrast to the bulk states that are characterized by $\vec{k}$, which corresponds to the 3D periodicity. In 3D, the localized surface states can only exist if they lie in a band gap of the bulk band structure. Surface states can be probed by ARPES and STM. The surface states play an important role in catalysis and in topological insulators.

13.3.2 First-Principles Calculations of Surface States

Surface state calculations have been done for many metals, semiconductors, and alloys. As examples, here, we shall discuss surface states on Cu(100) and an ideal Si(100) surface. The most common method is to use slab geometry, which was discussed in Chapter 6. In this method, a large supercell is used that is much longer in the perpendicular direction to the surface. In the supercell, a vacuum is introduced in addition to Cu layers in the perpendicular direction to the surface. It has been found that a supercell consisting of 11 Cu layers plus six vacuum layers is a good choice. We have calculated the (100) surface band structure of copper using the LDA and VASP code (Kresse and Furthmüller 1996a, 1996b), which is shown in Figure 13.4. The surface band structure we calculated is in good agreement with the results of Baldacchini et al. (2003) who used the FP-LMTO method. The surface states/resonances are indicated by dots. We see that there are well-defined surface states at $\bar{M}$ and $\bar{X}$. These surface states have been also seen in ARPES experiments. Surface states have also been observed on Cu(110) and Cu(111) surfaces. Surface states can also exist on the surfaces of disordered alloys (see, e.g., Pessa et al. 1981, Asonen et al. 1982, Prasad et al. 1991). As another example, we show in Figure 13.5 the surface electronic structure of an ideal unconstructed Si(100) surface along symmetry lines of the surface BZ using the GGA and VASP. In the figure, projected bulk bands are also shown in light color while the surface bands are in black. We see that there are surface

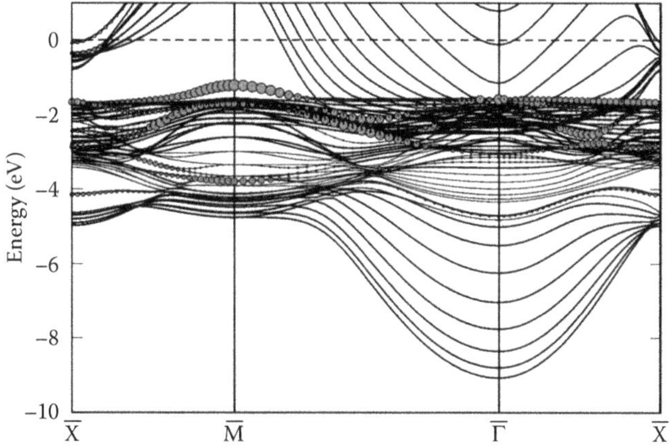

FIGURE 13.4
Surface electronic structure of copper along some symmetry directions of the surface BZ obtained by using the VASP code of Kresse and Furthmüller (1996a, 1996b). Dots indicate surface bands. The Fermi energy is placed at 0 eV. Copper lattice constant was taken to be 3.6309 Å. The surface band structure was calculated using the slab geometry with slab thickness of 25.7961 Å (11 Cu layers) with vacuum layer of 15 Å thickness. The energy cut-off of 400 eV was used with Monkhorst–Pack 18 × 18 × 1 *k*-mesh.

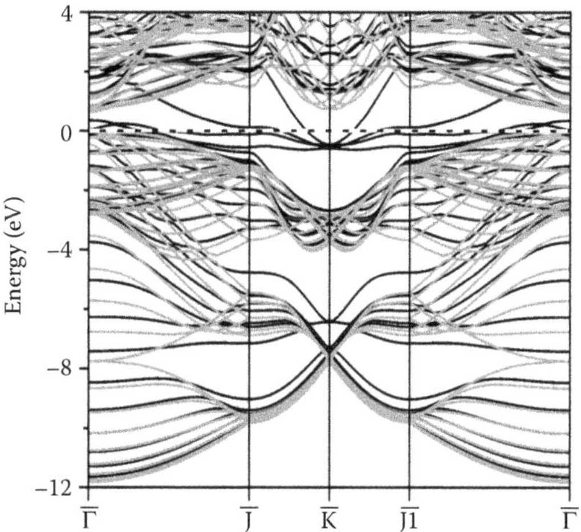

FIGURE 13.5
Surface electronic structure of ideal Si(100) surface using the VASP code of Kresse and Furthmüller (1996a, 1996b). Light-colored bands are bulk bands and the surface bands are shown in black. The zero of energy is placed at the top of the valence band. Silicon lattice constant was taken to be 5.43 Å. The surface band structure was calculated using the slab geometry with slab thickness of 17 Å (13 Si layers) with vacuum layer of 12 Å thickness. The energy cut-off of 250 eV was used with Monkhorst–Pack 12 × 12 × 1 *k*-mesh.

states near $\bar{K}$ point. These surface bands merge with bulk bands away from $\bar{K}$ and become surface resonances.

13.4 Surface Relaxation and Reconstruction

The surface atoms may not have the same atomic arrangement as in the bulk. This is because the presence of the surface causes an abrupt change in the potential. In response to this change in potential, the system would minimize energy by adopting a different atomic arrangement at and near the surface. This causes the phenomena of surface relaxation and surface reconstruction. Surface reconstruction can have a large effect on physical and chemical properties of the surface.

Surface relaxation refers to the phenomena when the surface layer moves closer to the inner layers of the solid, that is, in a perpendicular direction to the surface as shown in Figure 13.6. Thus, the interlayer distance at the surface becomes shorter and the periodicity of atoms parallel to the surface remains unchanged. The surface reconstruction refers to the phenomenon when the surface atoms move parallel to the surface also and may arrange themselves such that their periodic arrangement is different from the bulk atoms. For example, the unit cell of the surface atoms may double in one direction as shown in Figure 13.7b. In the Wood's nomenclature discussed in Section 13.2, this is (2×1) structure.

As an example, we shall discuss the surface reconstruction of (100) surface of silicon, which has been studied extensively. The reason why this has drawn so much interest is that electronic devices grown out of silicon use the Si(100) surface as a substrate. A (2×1) reconstruction of Si(100) was first proposed by Schlier and Farnsworth (1959). In this reconstruction, the surface atoms form

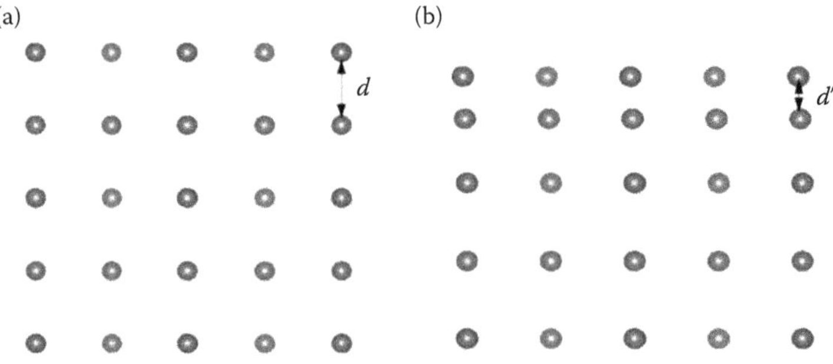

FIGURE 13.6
View normal to the surface. (a) Unrelaxed case, the interlayer spacing d is the same between all layers. (b) Relaxed case, the top layer moves closer to the inner layer so that spacing between the top two layers $d' < d$.

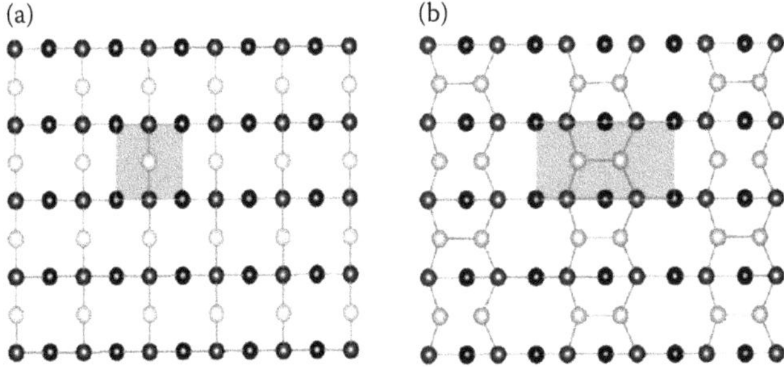

FIGURE 13.7
Top view of Si(100) surface. (a) Ideal unconstructed surface. (b) (2×1) symmetric reconstructed surface. Light-colored atoms form the topmost layer. The shaded cell denotes the unit cell.

pairs of atoms called dimers, and this phenomenon of forming basis is called dimerization. This can be understood in terms of broken covalent bonds. In the bulk, every Si atom is bonded to four Si atoms and thus has four covalent bonds. When the (100) surface is formed as shown in Figure 13.7, the surface atom is bonded to only two Si atoms and thus has two broken covalent bonds or dangling bonds. If two atoms move closer and form a dimer, one of the dangling bonds will be removed and they will be left with only one dangling bond. This results in considerable lowering in energy. It has been shown that this reconstruction lowers the energy of the surface by approximately 2 eV. Dimers have been observed directly by STM experiments.

However, there are problems with the above type of reconstruction in which dimers lay flat on the surface, which is called symmetric (2×1) reconstruction denoted by $p(2 \times 1)s$ (p = primitive). Appelbaum et al. (1976) showed that this model showed a metallic surface although experimentally it is seen to be semiconducting. To resolve this problem, Chadi (1979) proposed an asymmetric dimer model in which the dimers buckle out of the plane of the surface. On the basis of his empirical tight-binding calculation, he showed that such a surface reconstruction indeed gives a semiconducting surface. This kind of surface reconstruction is called asymmetric (2×1) reconstruction and is denoted by $p(2 \times 1)a$. The $p(2 \times 1)$ asymmetric reconstruction has been supported by various experiments. First-principles calculations also show that this reconstruction has lower energy compared to the symmetric one, typically by 0.1 eV per dimer.

The (100) surface of Si also shows other complex reconstructions whose periodicity is larger than $p(2 \times 1)$. Other periodicities that have been reported are $p(2 \times 2)$, $c(4 \times 2)$, $c(8 \times 8)$, and $(2 \times n)$, where n is between 2 and 11. The energy differences between some of these reconstructions come out to be very small and as low as 1 meV per surface atom. Some of these reconstructions may be metastable or stabilized by the presence of impurities. We shall not go into the details of these discussions and refer to the papers by Ramstad et al. (1995) and Roberts and Needs (1990).

13.5 Interfaces

An interface refers to a small number of atomic planes that separates two solids in close contact with each other. At the interface, properties such as interatomic distances differ from those of the bulk materials it separates. The surface of a solid is an interface between a solid and vacuum. Grain boundaries are another example of interfaces that occur between two crystals of the same type. Another example is an interface between two semiconductors, which is called a heterojunction. Heterojunctions have been extensively studied due to their importance in device applications and also because almost perfect interfaces can be grown with atomic precision. The combination of multiple heterojunctions is called a heterostructure although the two terms are sometimes used interchangeably. For developing semiconductor heterostructures used in high-speed electronics and optoelectronics, Kroemer and Alferov were awarded the Nobel Prize in Physics in 2000.

13.5.1 Band Offsets in Heterojunctions

Suppose we make a heterojunction of two semiconductors A and B with energy band gap ε_g^A and ε_g^B. For simplicity, we assume that A and B have the same structure and the same lattice constants. In Figure 13.8, we schematically show a typical band diagram as we go from A to B. We see that the band gap, as well as the valence and conduction bands, shows discontinuous jumps at the interface. In the figure, we have the bands as flat, although, in

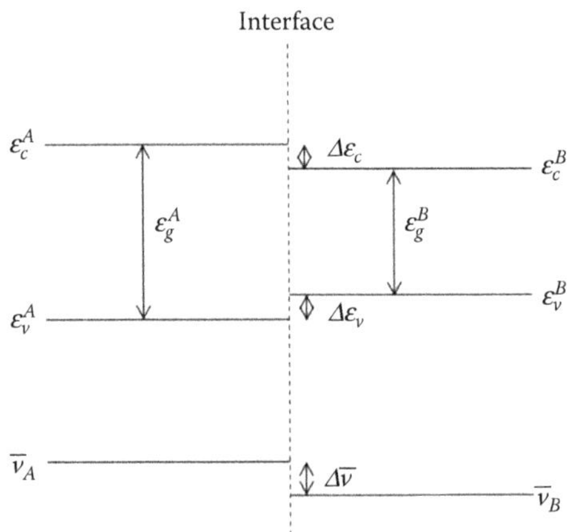

FIGURE 13.8
Schematic band diagram near the interface of A and B semiconductors. ε_v^A and ε_v^B denote the top of the valence bands of A and B. Similarly, ε_c^A and ε_c^B denote the bottom of the conduction bands of A and B.

general, there will be some band bending. This is justified as at the scale of few atomic distances around the interface, bending is negligibly small. The discontinuities of valence and conduction bands at the interface are called band offsets. The band offsets are the most important parameters that characterize a heterojunction as they form a barrier for carrier transport across the interface and therefore are essential for calculating the transport properties.

Suppose we know the band structure of *A* with respect to some reference level $\overline{v}_A$ and similarly of *B* with respect to some reference level $\overline{v}_B$ as shown in Figure 13.8. To find band offsets $\Delta\epsilon_c$ and $\Delta\epsilon_v$, as shown in the figure, we have to line up the two band structures with respect to one another. This can be done only if we can find $\Delta\overline{v} = v_A - v_B$. However, the reference levels $\overline{v}_A$ and $\overline{v}_B$ cannot be obtained by doing two separate band calculations for the two semiconductors *A* and *B*. This is because there is no absolute reference energy level in a band calculation for an infinite solid. Therefore, it is not possible to compare two band calculations of two different solids. This has been attributed to the long-range nature of the Coulomb potential (Kleinman 1981).

From the above discussion, it is clear that we cannot obtain band offsets by only two separate band calculations of *A* and *B* solids as $\Delta\overline{v}$ is unknown. Thus, we have to do a band calculation of a system consisting of *A* and *B* with the given interface. This ensures that the potentials of *A* and *B* are placed with respect to the same reference level and then it is possible to find $\Delta\overline{v}$. Such a calculation can be done by using supercell geometry and has been carried out for many semiconductor heterojunctions (Van de Walle and Martin 1987). It is assumed that the periodicity is maintained perpendicular to the surface. Let $v(x, y, z)$ denote the one-electron potential and $\hat{z}$ the direction perpendicular to the surface. For deriving band line up, one is only interested in the behavior of the potential in the $\hat{z}$ direction. Thus, one defines a quantity $\overline{v}(z)$ in which x and y coordinates are integrated out, that is

$$\overline{v}(z) = \frac{1}{S}\int_S v(x,y,z)\,dx\,dy \tag{13.11}$$

where *S* is the area of the unit cell in the plane of the interface. In Figure 13.9, we see that $\overline{v}(z)$ shows periodic oscillations as a function of z. On both sides of the interface, it quickly recovers the bulk behavior of each of the two materials *A* and *B*. We see that the positions of the averages on each side are shifted with respect to each other. This shift is $\Delta\overline{v}$, which we need to calculate the band offsets. Thus, if we know the bands of *A* and *B* from separate band calculations, the band offsets can be found. The band offsets of many heterojunctions have been found using this procedure and for more details we refer to the papers by Van de Walle and Martin (1987) and Franciosi and Van de Walle (1996).

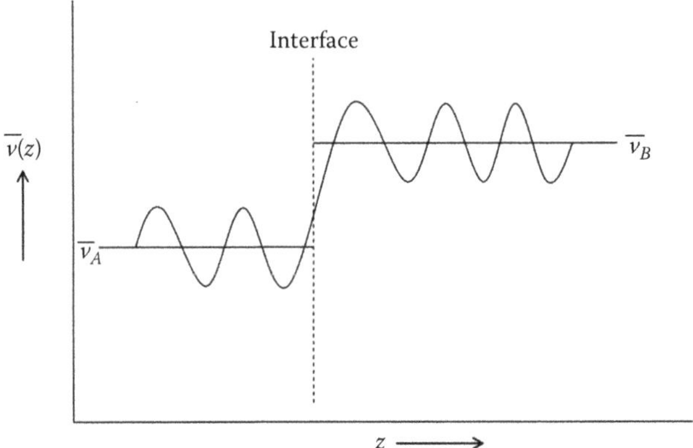

FIGURE 13.9
Schematic plot of $\overline{v}(z)$ near the interface of A and B.

13.6 Superlattices

Superlattices are artificially prepared structures in which m layers of material A and n layers of material B are periodically repeated as shown in Figure 13.10. The thickness of each layer can be of the order of few nanometers. Such a structure can be grown by techniques such as molecular beam epitaxy (MBE). If the growth direction is the z direction, then one has periodicity in the z direction, giving the structure a one-dimensional behavior. However, the electrons can also exhibit two-dimensional behavior as they can freely move in the layers in the x–y plane. Thus, this kind of structure can show new properties not shown by A or B materials. A superlattice of A and B materials will be denoted by A/B.

There has been a surge of activity in the field of semiconductor superlattices following the pioneering work of Esaki and Tsu (1970). Semiconductor superlattices such as $GaAs/GaAs_{1-x}P_x$, $GaAs/Ga_{1-x}Al_xAs$, HgTe/CdTe, and Si/Ge have been extensively studied. These systems show unusual transport

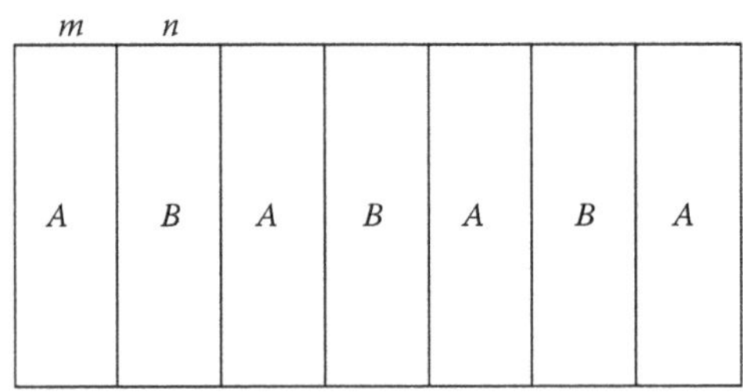

FIGURE 13.10
Schematic of a superlattice of A and B materials.

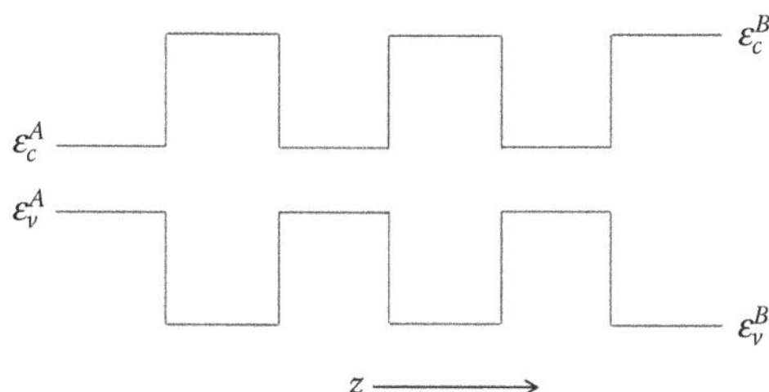

FIGURE 13.11
Potential wells and barriers in a superlattice.

properties and novel behavior such as Wannier–Stark ladders (see, e.g., Mendez and Bastard 1993). In such superlattices, potential wells and barriers are formed as shown in Figure 13.11 because band gaps of *A* and *B* semiconductors are different. Thus, an electron sees a periodic potential in the *z* direction, which is similar to the Kronig–Penney model. Thus, such structures are also called quantum well structure.

Several *ab initio* calculations have been done on superlattices (see, e.g., Smith and Mailhiot 1990). Supercell geometry is most suited for such calculations and, therefore, a majority of the reported *ab initio* calculations use this technique. As an example, we shall discuss the electronic structure of $SrTiO_3/BiFeO_3$ superlattice using the pseudopotential method (Roy et al. 2011a). The calculation was done using the VASP code within the framework of LSDA + U with U = 4.5 eV for Fe 3*d* states only. The calculation was started with experimental lattice constant of cubic $SrTiO_3$ (STO). Using the same lattice constant and ionic positions as STO for bismuth ferrite (BFO), we stacked alternate layers of STO and BFO as shown in Figure 13.12a and b. To consider all possible magnetic structures, a bigger ($\sqrt{2} \times \sqrt{2} \times 2$) supercell was constructed as shown in Figure 13.12c. Four magnetic structures were considered, namely, ferromagnetic, A-, C-, and G-type antiferromagnetic. These structures were relaxed such that the pressure on the cell was nearly zero and the forces on the ions were <0.005 eV/Å. The G-type antiferromagnetic

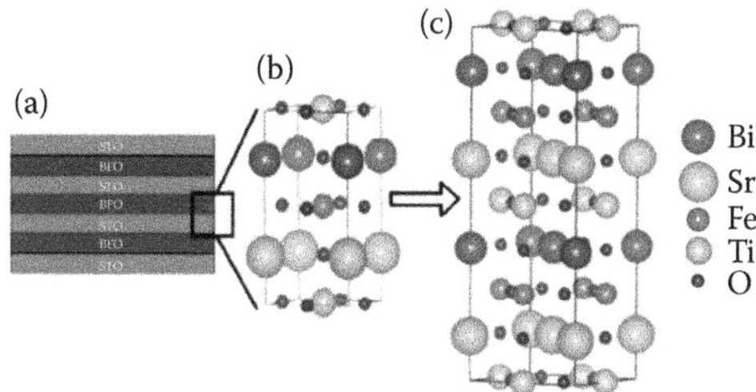

FIGURE 13.12
(a) Schematic showing stacking of STO and BFO layers. (b) Supercell constructed by stacking STO and BFO unit cells. (c) ($\sqrt{2} \times \sqrt{2} \times 2$) supercell constructed from (b).

structure was found to have the lowest energy and therefore is the ground-state structure.

In Figure 13.13a, the band structure of the STO/BFO superlattice is shown along the high symmetry direction. The band structure shows the material to be an insulator with a band gap ~1.7 eV. This is further confirmed by the total density of states (DOS) plot as shown in Figure 13.13b. Near the top of the valence band, it was found that the DOS comprises mainly the Fe 3*d*, Ti 3*d*, and O 2*p* states. This indicates that there is hybridization of the Fe 3*d*, Ti 3*d*, and O 2*p* states, which will impart a covalent character to the Fe–O and Ti–O bonds.

In Figure 13.14, the electronic charge densities in the (001) plane at different layers are shown. Considerable charge sharing is observed between Ti 3*d* and O 2*p* and Fe 3*d* and O 2*p* states at TiO_2 and FeO_2 layers, respectively, on the (001) plane. This indicates the existence of a covalent character of Ti–O and Fe–O bonds as also shown by DOS. Charge densities of Bi and Sr ions are nearly spherically symmetric, showing no significant charge

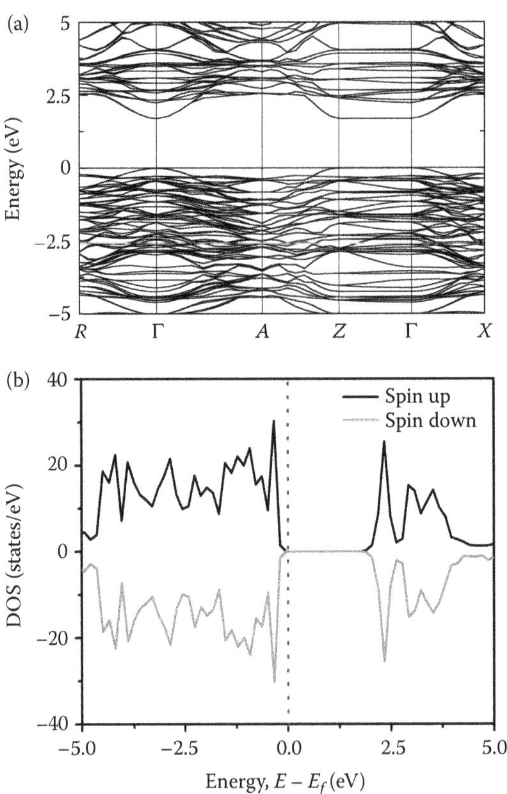

FIGURE 13.13
(a) Band structure and (b) DOS of STO/BFO superlattice. Zero of energy is placed at the top of occupied levels. (From Roy A. et al. 2011a. *Nanotechnology* 2: 112–115.)

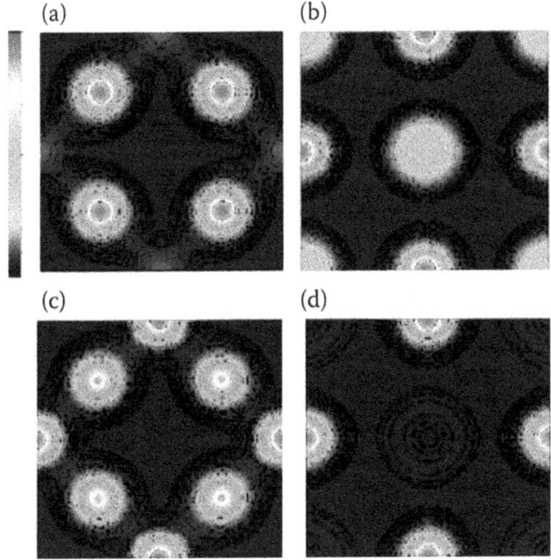

FIGURE 13.14
Charge density in (001) planes of (a) TiO_2, (b) SrO, (c) FeO_2, and (d) BiO layers. (From Roy A. et al. 2011a. *Nanotechnology* 2: 112–115.)

sharing with the surrounding O ions. Thus, Sr and Bi ions do not form covalent bonds with O ions.

EXERCISES

13.1 Show that the energy bands near the zone boundary for the one-dimensional model (Equation 13.6) (without barrier *W*) are given by

$$\varepsilon_q = V_0 + \frac{\hbar^2}{2m_e}\left[q^2 + \frac{\pi^2}{a^2}\right] \pm \sqrt{V_1^2 + \left(\frac{\hbar^2}{2m_e} q\pi / a\right)^2}$$

where $q = k - \pi/a$.

13.2 Show that the wave function of an electron near the zone boundary for the one-dimensional model (Equation 13.6) (without barrier *W*) is given by

$$\psi_q = e^{iqx}\,(A\,e^{i\pi x/a} + B\,e^{-i\pi x/a})$$

where *A* and *B* are constants.

13.3 Calculate the surface electronic structure of Cu(100) surface using slab geometry and identify the surface states. Compare your results with Figure 13.4. You may use one of the codes mentioned in Appendix A.

13.4 Calculate the surface electronic structure of an ideal (unconstructed) Si(100) surface using the slab geometry and identify the surface states. Compare your results with Figure 13.5. You may use one of the codes mentioned in Appendix A.

13.5 Calculate the surface electronic structure of a (2 × 1) reconstructed Si(100) surface using slab geometry and identify the surface states. You may use one of the codes mentioned in Appendix A.

Further Reading

Bastard G. 1998. *Wave Mechanics Applied to Semiconductor Heterostructures*. Les Ulis: Les Editions de Physique.

Kaxiras E. 2003. *Atomic and Electronic Structure of Solids*. Cambridge: Cambridge University Press.

Kittel C. 1986. *Introduction to Solid State Physics* (Sixth Edition). New York: John Wiley & Sons.

Zubko P., Gariglio S., Gabay M., Ghosez P., and Triscone J. M. 2011. Interface physics in complex oxide heterostructures. *Annu. Rev. Condens. Matter Phys.* 2:141–165.

14

Graphene and Nanotubes

14.1 Introduction

In this chapter, we shall discuss the electronic structure of graphene and nanotubes, which are topics of great interest due to interesting physics as well as technological applications. For example, graphene provides the first example of "Dirac material" as its low-energy quasi-particles mimic the behavior of fermions described by the Dirac equation. From the technological point of view, these materials offer great potential for fabrication of new devices at the nanometer scale. We shall discuss the electronic structure of graphene in Section 14.2.

Carbon nanotubes can be considered as rolled-up cylindrical tubes made of graphene sheets and may be capped by fullerenes. These nanotubes have unusual properties that make them very useful for applications in nanotechnology and electronics. We shall discuss the electronic structure of carbon nanotubes in Section 14.3.

14.2 Graphene

14.2.1 Structure and Bands

Graphene is a single layer of graphite and has a honeycomb structure as shown in Figure 14.1. It was isolated from graphite by Andre Geim and Konstantin Novoselov (Figure 14.2a and b) in 2004, for which they were awarded the Nobel Prize in 2010 (Novoselov et al. 2004). They used a very ingenious and simple technique called the "Scotch Tape technique" to extract graphene from bulk graphite and then transferred it to a silicon substrate. This has generated much interest in the physics community due to its novel electronic structure and properties (Castro Neto et al. 2009).

An ideal graphene sheet has honeycomb structure as shown in Figure 14.1. It can be described with a triangular Bravais lattice with a two-atom

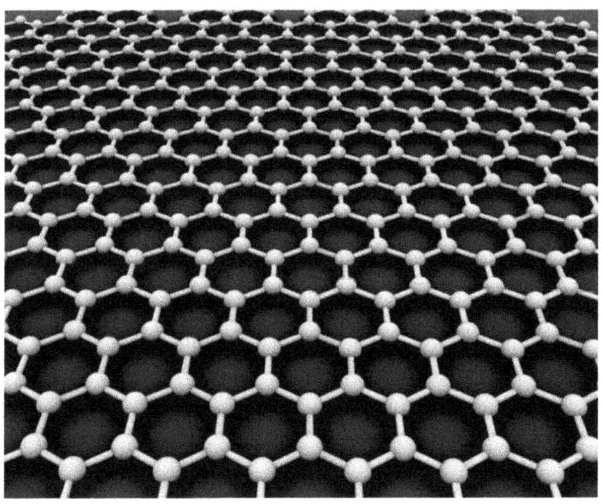

FIGURE 14.1
Honeycomb structure of graphene. (Courtesy of AlexanderAlUS, Wikimedia Commons.)

basis consisting of A and B atoms as shown in Figure 14.3. In other words, there are two interpenetrating triangular sublattices: A sublattice and B sublattice.

The carbon–carbon distance is $a = 1.42$ Å. The lattice vectors $\vec{a}_1$ and $\vec{a}_2$, as shown in the figure, can be written in terms of their components along the x

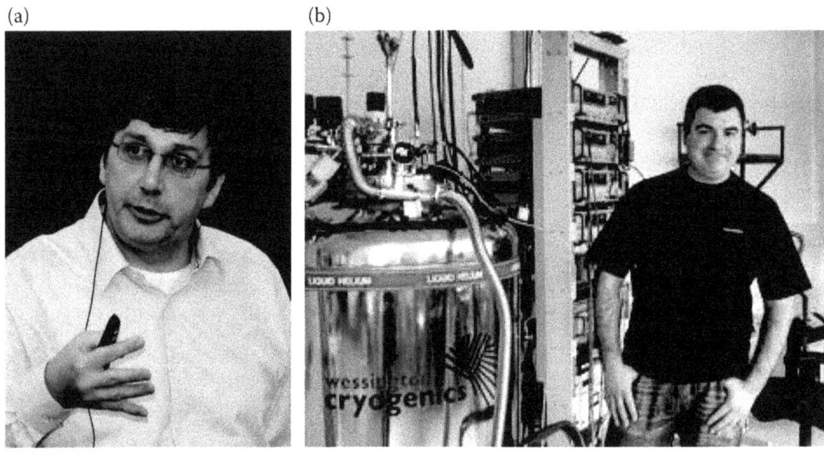

FIGURE 14.2
(a) Andre Geim (1958–) and (b) Konstantin Novoselov (1974–) shared the Nobel Prize in 2010 for their groundbreaking work on graphene. (Geim photo courtesy of Holger Motzkau, 2010, Wikimedia Commons; Novoselov photo courtesy of Zp2010, Wikimedia Commons.)

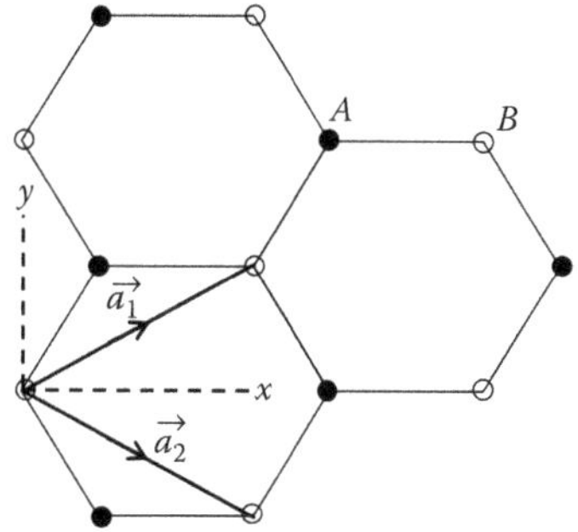

FIGURE 14.3
Triangular lattice with two-atom basis.

and y axes as shown in Figure 14.3 as

$$\vec{a}_1 = \frac{a}{2}(3, \sqrt{3})$$
$$\vec{a}_2 = \frac{a}{2}(3, -\sqrt{3}) \tag{14.1}$$

The three nearest-neighbor vectors are given by $a(1, \sqrt{3})/2$, $a(1, -\sqrt{3})/2$, and $-a\,(1, 0)$ while the six second neighbors are at $\pm\vec{a}_1$, $\pm\vec{a}_2$, and $\pm\,(\vec{a}_2 - \vec{a}_1)$.

The reciprocal lattice vectors are given by

$$\vec{b}_1 = \frac{2\pi}{3a}(1, \sqrt{3})$$
$$\vec{b}_2 = \frac{2\pi}{3a}(1, -\sqrt{3}) \tag{14.2}$$

The unit cell in the reciprocal space and the BZ are shown in Figure 14.4. There are six corners of the hexagonal BZ, out of which only two are

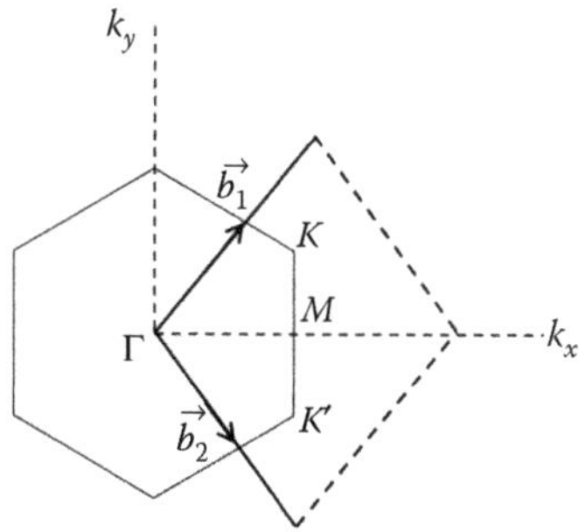

FIGURE 14.4
Brillouin zone of graphene.

independent, which are denoted by K and K', that is, the position vectors of the remaining ones can be expressed in term of $\vec{K}$ and $\vec{K}'$ by adding suitable combinations of $\vec{b}_1$ and $\vec{b}_2$. The position vectors $\vec{K}$ and $\vec{K}'$ are given by

$$\vec{K} = \left(\frac{2\pi}{3a}, \frac{2\pi}{3\sqrt{3}a}\right)$$

$$\vec{K}' = \left(\frac{2\pi}{3a}, -\frac{2\pi}{3\sqrt{3}a}\right)$$

We shall now discuss the electronic structure of graphene using the tight-binding approximation, which was first calculated by Wallace (1947). Let us start with the carbon atom, which has six electrons with the electronic configuration $1s^2$, $2s^2\ 2p^2$. The innermost $1s^2$ electrons, which are quite deep in energy, can be considered as core electrons and do not take part in bonding. Only $2s$ and $2p$ electrons, which are four in number, take part in bonding. If we assume the plane of the graphene sheet as x–y plane, $2s$, $2p_x$, $2p_y$ orbitals can mix and form three hybridized orbitals called sp^2 orbitals (see Appendix G). These sp^2 orbitals overlap and form bonding and antibonding orbitals, which give rise to sp^2 bonding and antibonding bands. The overlap of p_z orbitals gives rise to the valence band (bonding band) and conduction band (antibonding band). The sp^2 bonding bands take away three electrons per carbon atom, and one electron per carbon atom is contributed to the valence band formed by the p_z orbitals. So far as the bonding and structure of graphene is concerned, it mainly comes from sp^2 bonding bands, but the electronic properties mainly arise due to the band formed by the overlap of p_z orbitals. Therefore, in our discussion, we shall only consider p_z orbitals.

We want to write the Bloch wave function $\psi_{\vec{k}}(\vec{r})$ in terms of atomic $2p_z$ orbitals of carbon centered on different sites. Let $\phi(\vec{r})$ denote a $2p_z$ orbital of carbon. Since there are two sublattices A and B, $\psi_{\vec{k}}(\vec{r})$ can be written as

$$\psi_{\vec{k}}(\vec{r}) = C_A\,\psi_{\vec{k}}^{A}(\vec{r}) + C_B\,\psi_{\vec{k}}^{B}(\vec{r}) \tag{14.3}$$

where

$$\psi_{\vec{k}}^{A}(\vec{r}) = \frac{1}{\sqrt{N}}\sum_{A} e^{i\vec{k}\cdot\vec{R}_A}\,\phi(\vec{r}-\vec{R}_A) \tag{14.4}$$

and

$$\psi_{\vec{k}}^{B}(\vec{r}) = \frac{1}{\sqrt{N}}\sum_{B} e^{i\vec{k}\cdot\vec{R}_B}\,\phi(\vec{r}-\vec{R}_B) \tag{14.5}$$

where C_A and C_B are constants, $\vec{R}_A$ and $\vec{R}_B$ denote lattice vectors of A and B sublattices, respectively, and N is the number of unit cells in the graphene sheet. Note that in Equations 14.4 and 14.5, the sum is over lattice vectors of A and B sublattices, respectively. Now, we want to solve the eigenvalue equation

$$H\psi_{\vec{k}}(\vec{r}) = \varepsilon_{\vec{k}}\psi_{\vec{k}}(\vec{r}) \tag{14.6}$$

where H is the Hamiltonian. Substituting Equation 14.3 in Equation 14.6, we get

$$C_A H\psi_{\vec{k}}^A + C_B H\psi_{\vec{k}}^B = \varepsilon_{\vec{k}}(C_A\psi_{\vec{k}}^A + C_B\psi_{\vec{k}}^B) \tag{14.7}$$

By multiplying $\psi_{\vec{k}}^{A^*}$ from the right and integrating over $\vec{r}$, we get

$$C_A H^{AA} + C_B H^{AB} = \varepsilon_{\vec{k}} C_A \tag{14.8}$$

where we have assumed $\psi_{\vec{k}}^A$ and $\psi_{\vec{k}}^B$ to be normalized and approximately orthogonal to each other.

Here

$$H^{AA} = \int \psi_{\vec{k}}^{A^*} H\psi_{\vec{k}}^A d^3r = \varepsilon_0 = H^{BB} \tag{14.9}$$

and

$$H^{AB} = \int \psi_{\vec{k}}^{A^*} H\psi_{\vec{k}}^B d^3r \tag{14.10}$$

Substituting Equations 14.4 and 14.5 for $\psi_{\vec{k}}^A$ and $\psi_{\vec{k}}^B$ in Equation 14.10, we get

$$H^{AB} = \frac{1}{N}\sum_{R_A,R_B} e^{i\vec{k}\cdot(\vec{R}_B-\vec{R}_A)} \int \phi^*(\vec{r}-\vec{R}_A)H\phi(\vec{r}-\vec{R}_B)d^3r \tag{14.11}$$

Now we assume that the integral is finite only between the nearest neighbors, which we denote by $-t$, that is

$$-t = \int \phi^*(\vec{r}-\vec{R}_A)H\phi(\vec{r}-\vec{R}_B)d^3r \tag{14.12}$$

As ϕ represents a $2p_z$ orbital and is real, therefore, t is real. Substituting Equation 14.12 into Equation 14.11, we get

$$H^{AB} = -t\,f(\vec{k}) \tag{14.13}$$

where

$$f(\vec{k}) = \sum_{nn} e^{-i\vec{k}\cdot\vec{R}_A}$$
$$= e^{ik_x a} + 2e^{-ik_x a/2}\cos(\sqrt{3}k_y a/2) \tag{14.14}$$

where nn indicates that the sum is over the nearest neighbors only. Substituting Equations 14.9 and 14.13 into Equation 14.8, we get

$$C_A(\varepsilon_0 - \varepsilon_{\vec{k}}) - C_B\,t\,f(\vec{k}) = 0 \tag{14.15}$$

Similarly, by multiplying Equation 14.7 by $\psi_{\vec{k}}^{B}$ from right and integrating, we get

$$-C_A\;t\,f^*(\vec{k}) + C_B(\varepsilon_0 - \varepsilon_{\vec{k}}) = 0 \tag{14.16}$$

Equations 14.15 and 14.16 are two equations for C_A and C_B. For the nontrivial solution, we must have

$$\begin{vmatrix} \varepsilon_0 - \varepsilon_{\vec{k}} & -t\,f(\vec{k}) \\ -t\,f^*(\vec{k}) & \varepsilon_0 - \varepsilon_{\vec{k}} \end{vmatrix} = 0 \tag{14.17}$$

This gives two values for $\varepsilon_{\vec{k}}$

$$\varepsilon_{\vec{k}}^{\pm} = \varepsilon_0 \pm t|f(\vec{k})| \tag{14.18}$$

where

$$|f(\vec{k})|^2 = 3 + 2\cos\left(\sqrt{3}k_y a\right) + 4\cos\left(\frac{3}{2}k_x a\right)\cos\left(\sqrt{3}k_y a/2\right) \tag{14.19}$$

For simplicity, we choose our zero of energy scale at ϵ_0 so that $\varepsilon_0 = 0$. The value of t is about 2.8 eV. Thus

$$\varepsilon_{\vec{k}}^{\pm} = \pm t|f(\vec{k})| \tag{14.20}$$

Using Equation 14.20, it is quite simple to get the band structure, which is shown in Figure 14.5 along symmetry directions in the Brillouin zone. We see that the bands near points K and K' vary linearly with $\vec{k}$. The band

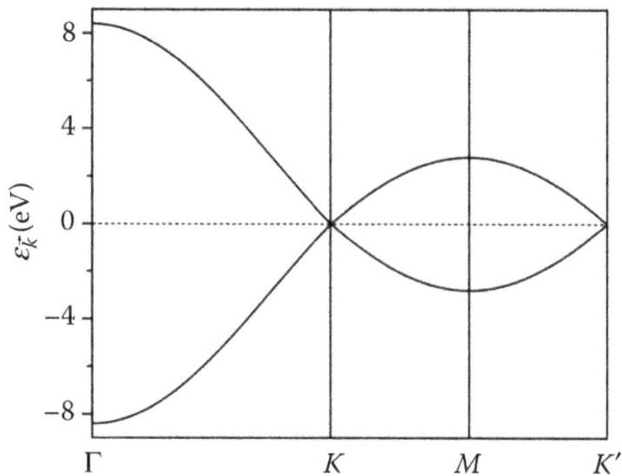

FIGURE 14.5
Band structure of graphene.

structure near K and K' is shown schematically as a function of k_x and k_y in Figure 14.6. Because of the linear behavior, we see the cone-type structure of bands in $k_x - k_y$ plane around K and K'. The linear behavior of the bands near K and K' can be shown analytically by writing $\vec{k} = \vec{K} + \vec{q}$ and expanding Equation 14.19 up to q^2 for $q << K$. Note that $f(\vec{K}) = 0$. Thus

$$|f(\vec{k})|^2 = \frac{9a^2}{4}(q_x^2 + q_y^2) = \frac{9}{4}a^2q^2$$

Therefore

$$\varepsilon_q^{\pm} = \pm v_F|q| \tag{14.21}$$

where $v_F = 3ta/2$ is the Fermi velocity with a value ~1×10^6 m/s, which is much smaller than the light velocity and falls in the nonrelativistic regime. Note that the Fermi velocity is a constant and ε_q is linear in q, very similar

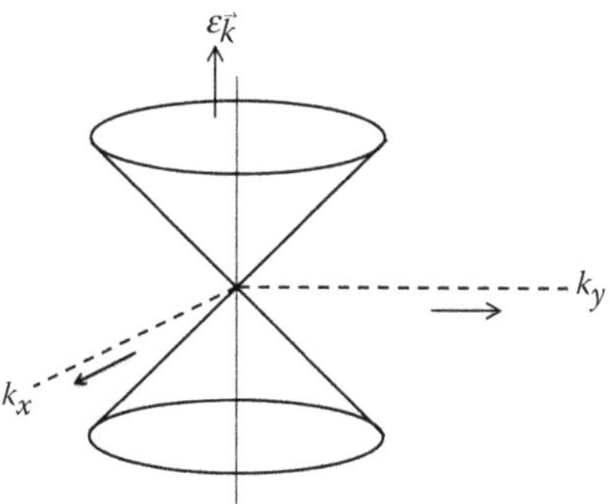

FIGURE 14.6
Schematic plot of the band structure near K and K' as a function of k_x and k_y.

to the relation $E = cp$ for a photon or a massless particle in relativity, v_F playing the role of the light velocity c. This is a very remarkable result and gives graphene a very special place in condensed matter physics, where in general, the electron dispersions are quadratic. In the next section, we shall show that because of this property, it is possible to write the Hamiltonian near K and K' in a form that resembles the Dirac Hamiltonian for massless particles where Fermi velocity plays the role of the light velocity in the Dirac Hamiltonian. Thus, one can do Dirac physics at nonrelativistic velocities!

Owing to the reason stated above, K and K' are known as the Dirac points and the cone-type structure in Figure 14.6 around K and K' points is known as the Dirac cone. We see that the Fermi energy $\varepsilon_F = 0$ and touches valence and conduction bands at K and K' in Figure 14.5. This implies that graphene is a semimetal. The simple tight-binding method gives a good description of the band structure of graphene near K and K' point around $\varepsilon = 0$. However, away from this energy, the tight-binding band structure shows deviations from the *ab initio* band structure (Reich et al. 2002).

14.2.2 Dirac Fermions, Pseudospin, and Chirality

We shall now show that around Dirac points K and K', the effective Hamiltonian can be written in a form that resembles the 2D Dirac Hamiltonian for a massless particle. Let us begin with Equations 14.15 and 14.16 and rewrite them as

$$H\begin{pmatrix} C_A \\ C_B \end{pmatrix} = \varepsilon_{\vec{k}}\begin{pmatrix} C_A \\ C_B \end{pmatrix} \tag{14.22}$$

where

$$H = -t\begin{pmatrix} 0 & f(\vec{k}) \\ f^*(\vec{k}) & 0 \end{pmatrix} \tag{14.23}$$

where we have set $\varepsilon_0 = 0$ as zero of energy. Now, we expand $f(\vec{k})$ around $\vec{k} = \vec{K} = ((2\pi/3a), (2\pi/3a\sqrt{3}))$. Using Taylor's expansion of Equation 14.14

$$\begin{aligned} f(\vec{K} + \vec{q}) &= f(\vec{K}) + \vec{q} \cdot \vec{\nabla}_{\vec{k}} f(\vec{k})\Big|_{\vec{k}=\vec{K}} \\ &= 0 + q_x \frac{\partial f(\vec{k})}{\partial k_x}\Big|_{\vec{k}=\vec{K}} + q_y \frac{\partial f(\vec{k})}{\partial k_y}\Big|_{\vec{k}=\vec{K}} \\ &= q_x \left\{ ia\, e^{ik_x a} - ia\, e^{-ik_x a/2} \cos\left(\sqrt{3}\, \frac{k_y a}{2}\right)\right\}\Big|_{\vec{k}=\vec{K}} \\ &\quad - q_y \left\{ e^{-ik_x a/2} \sin(\sqrt{3}\, k_y a/2)\sqrt{3}\, a\right\}\Big|_{\vec{k}=\vec{K}} \\ &= C(q_x - iq_y) \end{aligned} \tag{14.24}$$

where

$$C = \frac{3}{2} a e^{i\theta} \tag{14.25}$$

where $\theta = -\pi/6$, which can be absorbed by a rotation about the z axis. Thus, we can write by substituting Equation 14.24 in Equation 14.23 as

$$H_K = -\frac{3at}{2}\begin{pmatrix} 0 & q_x - iq_y \\ q_x + iq_y & 0 \end{pmatrix} \tag{14.26}$$

$$= -\frac{3at}{2}(\sigma_x q_x + \sigma_y q_y)$$

$$= -v_F \vec{\sigma} \cdot \vec{q} \tag{14.27}$$

where σ_x and σ_y are Pauli matrices

$$\sigma_x = \begin{pmatrix} 0 & 1 \\ 1 & 0 \end{pmatrix} \text{ and } \sigma_y = \begin{pmatrix} 0 & -i \\ i & 0 \end{pmatrix} \tag{14.28}$$

and v_F is the Fermi velocity. We see that the diagonalization of Equation 14.26 gives the electron dispersion (Equation 14.21) around point K. Similarly, it can be shown that the effective Hamiltonian around K' can be written as

$$H_{K'} = -v_F \vec{\sigma}^* \cdot \vec{q} \tag{14.29}$$

The eigenfunctions of Equation 14.26 can be written as

$$\psi_{\pm K} = \frac{1}{\sqrt{2}}\begin{pmatrix} e^{-i\theta_q/2} \\ \pm e^{i\theta_q/2} \end{pmatrix} \tag{14.30}$$

where

$$\theta_q = \tan^{-1}\frac{q_y}{q_x}$$

where the ± sign corresponds to eigenenergies $\varepsilon_q = \pm v_F q$.

Similarly, the eigenfunction of Equation 14.29 around K' can be written as

$$\psi_{\pm K'} = \frac{1}{\sqrt{2}}\begin{pmatrix} e^{i\theta_q/2} \\ \pm e^{-i\theta_q/2} \end{pmatrix} \tag{14.31}$$

Equation 14.27 implies that the spatial part of the wave function of the quasi-particles around K must vary as $e^{i\vec{q}\cdot\vec{r}}$ as plane waves are the only eigenfunction of momentum. Thus, the full quasi-particle wave functions around K and K' can be written as

$$\psi_{\pm K}(\vec{r}) = C_n e^{i\vec{q}\cdot\vec{r}} \frac{1}{\sqrt{2}} \begin{pmatrix} e^{-i\theta_q/2} \\ \pm e^{i\theta_q/2} \end{pmatrix} \tag{14.32}$$

and

$$\psi_{\pm K'}(\vec{r}) = C_n e^{i\vec{q}\cdot\vec{r}} \frac{1}{\sqrt{2}} \begin{pmatrix} e^{i\theta_q/2} \\ \pm e^{-i\theta_q/2} \end{pmatrix} \tag{14.33}$$

where C_n is a normalization constant. Equations 14.27 and 14.32 imply that in real space, the effective Hamiltonian at K must have a form

$$H_K = iv_F\vec{\sigma}\cdot\vec{\nabla} \tag{14.34}$$

which is similar to the 2D massless Dirac Hamiltonian. Thus, quasi-particles near K and K' move with Fermi velocity as if with zero mass and obey the Dirac Hamiltonian. Note that we have not included relativity in our discussion. This surprising result is basically a consequence of the linear dispersion of graphene around K and K'.

Another surprise is the spinor-like behavior of the wave functions (14.32) and (14.33). Note that we have not included spin in our discussion. The two-component behavior of the wave function is associated with the pseudospin that arises because we have two sublattices A and B. Each component of the pseudospin is the projection of the quasi-particle wave function onto sublattice A (spin up) or sublattice B (spin down). We see from Equation 14.32 that if θ changes by 2π, the wave function changes sign indicating a phase change of π. This phase change of π under rotation is a characteristic of spinor. Thus, it is appropriate to call the two components of the wave function as pseudospin.

An important quantity that is used to characterize the wave function is chirality, which is defined as the projection of the momentum operator along the pseudospin direction. It is associated with the quantum mechanical operator

$$\hat{h} = \frac{1}{2}\vec{\sigma}\cdot\frac{\vec{q}}{|q|} \tag{14.35}$$

From this definition, we see that $\psi_K(\vec{r})$ and $\psi_{K'}(\vec{r})$, which are eigenfunctions of the effective Hamiltonian, are also eigenfunctions of $\hat{h}$. Thus

$$\hat{h}\psi_K(\vec{r}) = \pm\frac{1}{2}\psi_K(\vec{r}) \tag{14.36}$$

and

$$\hat{h}\psi_{K'}(\vec{r}) = \pm\frac{1}{2}\psi_{K'}(\vec{r}) \tag{14.37}$$

Thus, the electrons ($\varepsilon_q > 0$) have positive chirality and the holes ($\varepsilon_q < 0$) have negative chirality. We see that the chirality is a good quantum number near K and K'. The chiral nature of the quasi-particles has been observed in a quantum Hall experiment (Novoselov et al. 2005).

One consequence of the chiral nature of the quasi-particle is that there is no back scattering due to impurities for intravalley transitions. To show this, we calculate the matrix element for this transition from $\vec{q}$ to $-\vec{q}$ if the impurity potential is $V(\vec{r})$.

$$\langle -\vec{q}|\, H_{imp} |\, \vec{q}\rangle \propto \int e^{2i\vec{q}\cdot\vec{r}} \left(e^{i\theta_{-\vec{q}}/2}, e^{-i\theta_{-\vec{q}}/2}\right) \begin{pmatrix} V(\vec{r}) & 0 \\ 0 & V(\vec{r}) \end{pmatrix} \begin{pmatrix} e^{-i\theta_{\vec{q}}/2} \\ e^{i\theta_{\vec{q}}/2} \end{pmatrix} = 0 \tag{14.38}$$

This is because $\theta_{-\vec{q}} = \theta_{\vec{q}} + \pi$. However, intervalley scattering is possible. This is important for understanding transport properties in graphene.

Another important consequence is that if a particle with energy ε, governed by the Dirac equation, impinges upon a potential barrier $V > \varepsilon$, then it does not have tunneling with exponential damping but for a certain angle of incidence, the tunneling probability approaches unity as $V \to \infty$. This is known as the Klein paradox, which does not occur for nonrelativistic particles but has been observed in graphene. There are many interesting results on graphene but we shall not discuss them and refer the reader to the excellent review articles and papers (see e.g., Castro Neto et al. 2009; Gundra and Shukla 2011).

14.3 Carbon Nanotubes

Carbon nanotubes are cylindrical structures made of carbon atoms, which have diameters of the order of nanometers and length of the order of several microns. The length-to-diameter ratio can be as large as 10^4, making them almost a one-dimensional system. Carbon nanotubes can be divided into two categories, single-wall nanotubes (SWNTs) and multiwall nanotubes

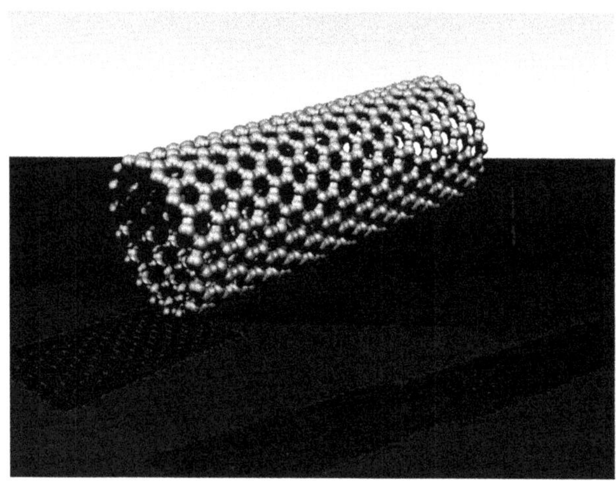

FIGURE 14.7
Zigzag single-wall nanotube (SWNT). (Courtesy of Arnero, Wikimedia Commons.)

(MWNTs). An SWNT consists of a single graphene sheet rolled into a form of hollow cylinder as shown in Figure 14.7. It is only one atomic layer thin and generally has a diameter <2 nm. An MWNT, shown in Figure 14.8, has several concentric layers of carbon sheets. The credit of discovery of the carbon nanotubes is generally given to Iijima (1991) who found them in 1991, but there are also reports of earlier observations. These nanotubes have novel properties that make them very useful for many applications in nanotechnology, electronics, and optics (see, e.g., Saito et al. 1998).

We shall focus on the electronic structure of the SWNT, which is a rolled-up graphene sheet in the form of a hollow cylinder. Owing to this reason, SWNTs are labeled in terms of the graphene lattice vectors. In Figure 14.9, we show an unrolled graphene sheet, which has a honeycomb structure with

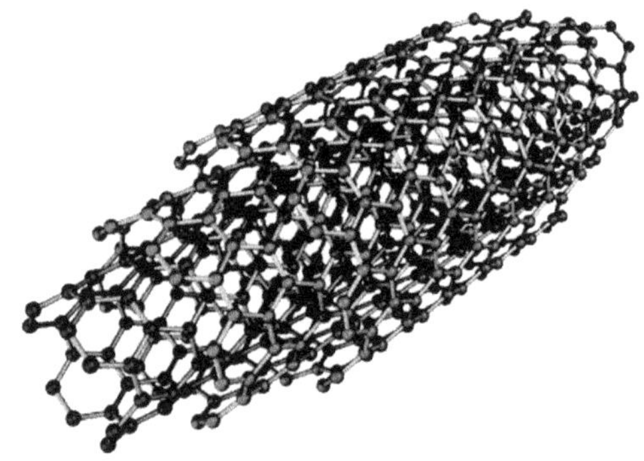

FIGURE 14.8
Multiwall nanotube (MWNT). (Courtesy of Eric Wieser, Wikimedia Commons.)

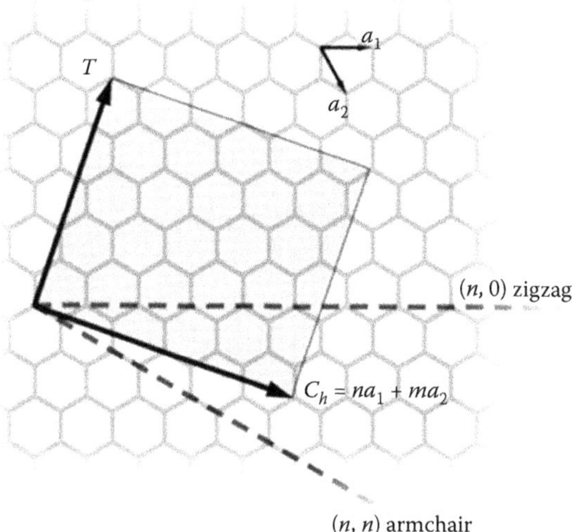

FIGURE 14.9
Unrolled graphene sheet. (Courtesy of Kebes, Wikimedia Commons.)

primitive lattice vectors $\vec{a}_1$ and $\vec{a}_2$. An SWNT is obtained by rolling the graphene sheet along a vector $\vec{C}_h$, called the chiral vector, which uniquely specifies the nanotube. The chiral vector $\vec{C}_h$ defines the circumference of the tube and can be written as

$$\vec{C}_h = n\vec{a}_1 + m\vec{a}_2 \tag{14.39}$$

where n and m are integers and we need to consider only $0 < |m| < n$ due to hexagonal symmetry of the graphene lattice. The pair of integers (n, m) uniquely specifies the geometry of the SWNT and is designated as (n, m) nanotubes. Its diameter is given by

$$d = \frac{|\vec{C}_h|}{\pi} = \frac{a_h}{\pi}\sqrt{n^2 + m^2 + nm} \tag{14.40}$$

where a_h is the lattice constant of the honeycomb network ($a_h = \sqrt{3}a$, where a is the carbon–carbon distance). The angle θ, which $\vec{C}_h$ makes with $\vec{a}_1$, is called the chiral angle and is given by

$$\cos\theta = \frac{\vec{C}_h \cdot \vec{a}_1}{|\vec{C}_h||\vec{a}_1|} = \frac{n a_h^2 + m(a_h^2/2)}{a_h^2(n^2 + m^2 + nm)^{1/2}}$$

$$= \frac{2n + m}{2(n^2 + m^2 + nm)^{1/2}} \tag{14.41}$$

FIGURE 14.10
Armchair SWNT. (Courtesy of Arnero, Wikimedia Commons.)

The angle θ is restricted in the range of $0 \leq \theta \leq 30°$ because of the hexagonal symmetry of the graphene lattice. The nanotubes with $\theta = 0$, that is, of the type $(n, 0)$, are called zigzag tubes because they show a zigzag pattern along the circumference. The nanotubes with $\theta = 30°$ or of the type (n, n) are called armchair tubes because they show an armchair pattern along the circumference. All other tubes with $0 < \theta < 30°$ or (n, m) with $0 < |m| < n$ are called chiral tubes. In Figure 14.7, we show a zigzag tube, in Figure 14.10, an armchair tube, and in Figure 14.11, a chiral tube.

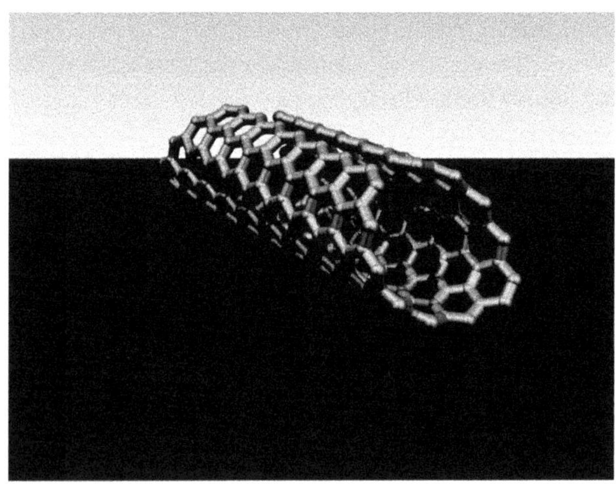

FIGURE 14.11
Chiral SWNT. (Courtesy of Arnero, Wikimedia Commons.)

The smallest graphene lattice vector $\vec{T}$ perpendicular to $\vec{C}_h$ in Figure 14.9 defines the primitive translational vector along the axis of the tube. It can be expressed in terms of $\vec{a}_1$ and $\vec{a}_2$ as

$$\vec{T} = t_1\vec{a}_1 + t_2\vec{a}_2 \tag{14.42}$$

where t_1 and t_2 are integers that can be found by using

$$\vec{C}_h \cdot \vec{T} = 0 \tag{14.43}$$

This gives

$$t_1 = \frac{2m+n}{d_R}$$
$$t_2 = -\frac{2n+m}{d_R} \tag{14.44}$$

where d_R is the greatest common divisor of $(2m + n)$ and $(2n + m)$. The unit cell of the nanotube is the rectangle defined by the vectors $\vec{C}_h$ and $\vec{T}$. The number of graphene unit cells in the rectangle is

$$N = \frac{\left|\vec{C}_h \times \vec{T}\right|}{\left|\vec{a}_1 \times \vec{a}_2\right|} = \frac{2(m^2 + n^2 + nm)}{d_R} \tag{14.45}$$

Let $\vec{K}_h$ and $\vec{K}_T$ denote the primitive reciprocal lattice vectors corresponding to $\vec{C}_h$ and $\vec{T}$, respectively. These can be obtained by using the relations

$$\begin{aligned} \vec{C}_h \cdot \vec{K}_h &= 2\pi \quad \vec{T} \cdot \vec{K}_h = 0 \\ \vec{C}_h \cdot \vec{K}_T &= 0 \quad \vec{T} \cdot \vec{K}_T = 2\pi \end{aligned} \tag{14.46}$$

which gives

$$\vec{K}_h = \frac{1}{N}(-t_2\vec{b}_1 + t_1\vec{b}_2) \quad \vec{K}_T = \frac{1}{N}(m\vec{b}_1 - n\vec{b}_2) \tag{14.47}$$

where $\vec{b}_1$ and $\vec{b}_2$ are the primitive lattice vectors of graphene reciprocal lattice. Here, we have used Equations 14.44 and 14.45 to obtain Equation 14.47.

We shall now discuss the electronic structure of SWNTs using the single-band tight-binding model of graphene, described in the last section. For simplicity, we shall use a zone-folding approach, which ignores the curvature of the system. In this approach, one uses the periodic boundary condition in the circumferential direction. As a result, the wave vector associated with the

$\vec{C}_h$ direction becomes quantized, that is, they can only have a set of discrete values. On the other hand, the wave vectors associated with the tube axis $\vec{T}$ remain quasi-continuous. If these allowed wave vectors of an SWNT are shown on to the BZ of graphene, one gets a series of parallel lines (Charlier et al. 2007). The length, number, and orientation of these lines depend on indices of (n, m) of an SWNT. Thus, the energy bands of an SWNT can be viewed as the superposition of the graphene energy bands along the corresponding allowed $\vec{k}$ wave vectors. If a wave vector falls outside the BZ of the SWNT, it is folded back to the BZ. This folding results in additional bands in the BZ of the SWNT. This is similar to what we had seen in Chapter 5, when plotting empty lattice bands.

We shall illustrate the method by calculating the energy bands of a zigzag nanotube. For a zigzag tube ($\vec{C}_h = (n,0)$), we shall choose the y direction along the $\vec{C}_h$ vector and the x direction along the $\vec{T}$ vector. Using the periodic boundary condition in the y direction, we get

$$n k_y a_h = 2\pi l \quad (l = 1,\ldots,2n) \quad \text{or} \quad k_y = 2\pi(l/n\sqrt{3})a \tag{14.48}$$

where l is an integer that varies up to $2n$ as the unit cell of the SWNT has $2n$ graphene unit cells. Here, a is the C–C bond length. Substituting Equation 14.48 into Equation 14.20, we get

$$\begin{aligned}\varepsilon_{\vec{k}}^{\pm} &= \pm t|f(\vec{k})| \\ &= \pm t\left[3 + 2\cos\left(\frac{2\pi l}{n}\right) \pm 4\cos\left(\frac{l\pi}{n}\right) \times \cos\left(\frac{3}{2}ka\right)\right]^{1/2} \quad (l = 1,\ldots,2n)\end{aligned} \tag{14.49}$$

Here, $k_x = k$ that varies from $-2\pi/|T|$ to $+2\pi/|T|$ or from $-(\pi/3a)$ to $+(\pi/3a)$. The plus–minus sign in the square brackets corresponds to the unfolded and folded energy bands (Saito et al. 1992a).

In Figures 14.12 and 14.13, we show energy bands for (9, 0) and (10, 0) zigzag tubes calculated using Equation 14.49. In these figures, the Fermi level is placed at 0 energy. We see that there is no energy gap at 0 energy for the (9, 0) SWNT but for the (10, 0) SWNT, there is a gap, indicating that the (9, 0) SWNT is metallic while (10, 0) is semiconducting. For a general $(n, 0)$ zigzag tube, if n is a multiple of 3, the energy gap is zero and the tube is metallic, otherwise it is semiconducting. In general, for an (n, m) SWNT, the condition for metallicity is that $(2n + m)$ is a multiple of 3. Thus, armchair SWNT denoted by (n, n) is always metallic. Therefore, about 1/3 tubes are metallic and 2/3 tubes are semiconducting.

In Figure 14.14, we show the density of states (DOS) for (9, 0) and (10, 0) zigzag tubes (Saito et al. 1992b). We see that for the (9, 0) SWNT, the DOS is finite near the Fermi energy, indicating metallic behavior, while for the (10, 0)

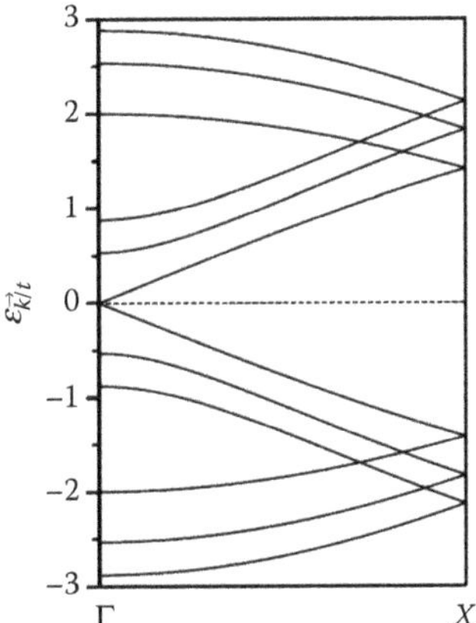

FIGURE 14.12
Energy bands for (9, 0) SWNT. Fermi energy is placed at 0.

SWNT, the DOS is zero, showing the semiconducting nature of the tube. We also notice sharp peaks in the DOS that are due to the van Hove singularities and are characteristic of 1D system as seen in the DOS of a nanowire in Chapter 12.

Another interesting result for the SWNT is that the energy gap ε_g is inversely proportional to its diameter d irrespective of the chiral angle:

$$\varepsilon_g = |t|\frac{a}{d}$$

where a is the carbon–carbon distance. This result has been verified by experiments (see, e.g., Saito et al. 1998).

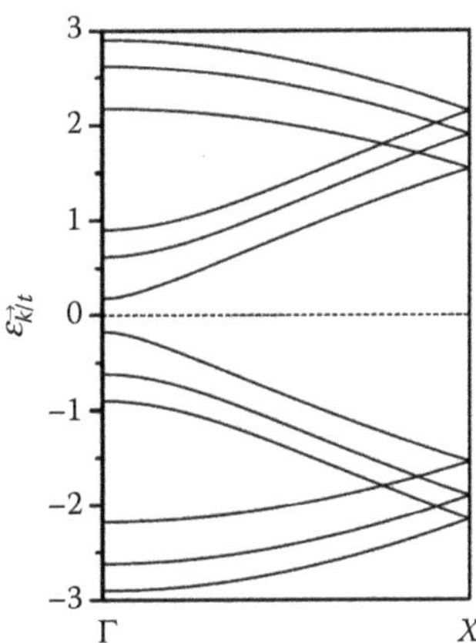

FIGURE 14.13
Energy bands for (10, 0) SWNT. Fermi level is placed at 0.

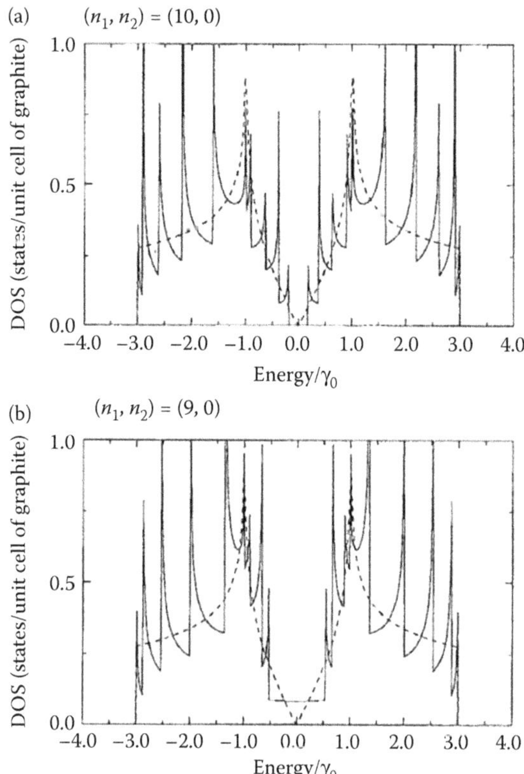

FIGURE 14.14
DOS for (a) (10, 0) zigzag SWNT and (b) (9, 0) zigzag SWNT. The Fermi level is placed at 0 energy. (Reprinted with permission from Saito, R., et al. 1992. Electronic structure of chiral graphene tubules. *Appl. Phys. Lett.* 60:2204–2206. Copyright 1992, American Institute of Physics.)

In the above discussion of the electronic structure of SWNT, we have neglected the curvature of the SWNT, which can become important if the diameter is small. The effect is large for nanotubes of diameter less than that of C_{60} but is not large for nanotubes in the range of $d > 0.7$ nm, which is the range that has been observed experimentally (Saito et al. 1998).

EXERCISES

14.1 Show that the reciprocal lattice vectors of the triangular lattice are given by Equation 14.2. Also find the position vectors of K and K'.

14.2 Show that

$$f(\vec{k}) = e^{ik_x a} + 2e^{-\frac{ik_x a}{2}} \cos\left(\frac{\sqrt{3}k_y a}{2}\right)$$

and

$$|f(\vec{k})|^2 = 3 + 2\cos(\sqrt{3}k_y a) + 4\cos\left(\frac{3}{2}k_x a\right)\cos\left(\frac{\sqrt{3}k_y a}{2}\right)$$

14.3 By writing $\vec{k} = \vec{K} + \vec{q}$ and expanding $|f(\vec{k})|^2$ up to q^2 for $q \ll K$, show that

$$|f(\vec{k})|^2 = \frac{9}{4}a^2q^2$$

14.4 Show that for small q, $f(\vec{K} + \vec{q})$ can be written as

$$f(\vec{K} + \vec{q}) = C(q_x - iq_y)$$

14.5 Show that the eigenfunctions of H_K are given by Equation 14.30.

14.6 Calculate the DOS for a two-dimensional material that has a band with linear and isotropic dispersion with k.

14.7 Show that $\vec{K}_h$ and $\vec{K}_T$ are given by Equation 14.47.

14.8 Show that the energy bands of a zigzag carbon nanotube are given by Equation 14.49.

14.9 Calculate the energy bands and DOS for graphene using supercell geometry and an *ab initio* method. Compare your results with the results of the tight-binding method. You may use one of the codes given in Appendix A.

14.10 Calculate the energy bands and DOS for a zigzag nanotube using supercell geometry and an *ab initio* method. Compare your results with the results of the tight-binding method. You may use one of the codes given in Appendix A.

Further Reading

Geim A. K. and MacDonald A. H. 2007. Graphene: Exploring carbon flatland. *Physics Today* 60:35–41.

Geim A. K. and Novoselov K. S. 2007. The rise of graphene. *Nature Materials* 6:183–191.

Peres N. M. R. 2010. Colloquium: The transport properties of graphene: An introduction. *Rev. Mod. Phys.* 82:2673–2700.

Saito R., Dresselhouse G., and Dresselhouse M. S. 1998. *Physical Properties of Carbon Nanotubes.* London: Imperial College Press.

15

Quantum Hall Effects and Topological Insulators

15.1 Introduction

The discovery of the quantum Hall effect in 1980 by von Klitzing (Figure 15.1) has been one of the most remarkable events that made a great impact on the development of condensed matter physics. He found that the Hall conductance is quantized in integer units of e^2/h; therefore, the effect is known as the integer quantum Hall effect (IQHE). For this work, von Klitzing was awarded the Nobel Prize in Physics in 1985. This discovery was followed by another important one by Tsui et al. in 1982, who found the existence of steps in Hall conductance in units of rational fractions, known as the fractional quantum Hall effect (FQHE). Laughlin was able to explain FQHE based on a wave function proposed by him. Tsui, Stormer, and Laughlin were awarded the Nobel Prize in Physics in 1998 for their work on FQHE.

Another effect known as the quantum spin Hall effect (QSHE) has been proposed by Kane and Mele (2005) and Bernevig et al. (2006), and has been observed by Konig et al. in 2007. This effect is shown by two-dimensional (2D) topological insulators that have recently generated much excitement. We shall briefly discuss these effects beginning with the classical Hall effect in Section 15.2. In Section 15.3, we discuss Landau levels and in Section 15.4, we discuss the IQHE and FQHE. In Section 15.5, we discuss the QSHE and in Section 15.6, we discuss the topological insulators.

15.2 Classical Hall Effect

The Hall effect was discovered by Edwin Hall in 1879. The geometry of the experiment is shown in Figure 15.2. We assume that we have a 2D sample of length L_x and width L_y, which is placed in a constant magnetic field $\vec{B}$ pointing in the z direction and a constant electric field $\vec{E}$ in the x direction,

FIGURE 15.1
Klaus von Klitzing (1943–) discovered the integer quantum Hall effect for which he was awarded the Nobel Prize in 1985. (Courtesy of Ramuman, Wikimedia Commons.)

supplied by a battery. The voltage across its length is V_x and I is the current flowing in the x direction in the steady state. A voltage V_y develops across its width in the y direction, an effect known as the Hall effect. We define resistances

$$R_{xx} = \frac{V_x}{I} \tag{15.1}$$

$$R_{yx} = \frac{V_y}{I} \tag{15.2}$$

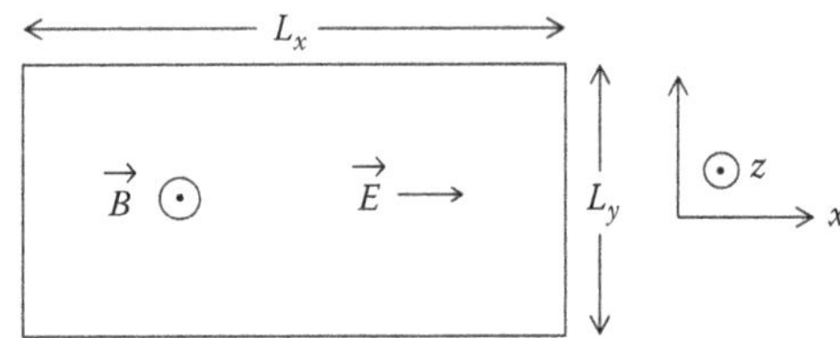

FIGURE 15.2
Geometry of the Hall effect experiment.

We assume that the current carriers are free electrons that obey the classical equation of motion

$$\frac{d\vec{p}}{dt} = \vec{f} - \frac{\vec{p}}{\tau} \tag{15.3}$$

Here, $\vec{p}$ is the electron's momentum, τ is the relaxation time, and $\vec{f}$ is the external force given by

$$\vec{f} = -e\left[\vec{E} + \vec{v} \times \vec{B}\right] \tag{15.4}$$

where $\vec{v}$ is the electron's velocity. Thus, we get

$$\frac{d\vec{p}}{dt} = -e\,[\vec{E} + \vec{v} \times \vec{B}] - \frac{\vec{p}}{\tau} \tag{15.5}$$

In the steady state, $(d\vec{p}/dt) = 0$. This implies

$$\frac{dp_x}{dt} = 0 = -e\,E_x - \frac{p_x}{\tau} \tag{15.6}$$

and

$$\frac{dp_y}{dt} = 0 = -e\,[E_y - v_x B] \tag{15.7}$$

where we have assumed that $v_y = 0$. Thus, $v_x = v$ and $p_x = p$.

From Equation 15.7, we get

$$E_y = vB$$

and

$$V_y = E_y L_y = vB\,L_y \tag{15.8}$$

From Equation 15.6, we get

$$eE_x = -\frac{p}{\tau}$$

or

$$E_x = -\frac{m_e v}{e\tau} \tag{15.9}$$

where m_e is the electron mass. The current density, that is, current per unit width is

$$j = -en_e v$$

where n_e is the number of electrons per unit area. Thus, the current flowing in the x direction is

$$I = j\,L_y = -en_e vL_y$$

Thus

$$\begin{aligned} R_{xx} &= \frac{V_x}{I} = \frac{E_x L_x}{-en_e vL_y} \\ &= +\frac{m_e vL_x}{e^2 \tau n_e vL_y} = \frac{m_e L_x}{e^2 n_e \tau L_y} \\ &= \frac{L_x}{\sigma_0 L_y} \end{aligned} \tag{15.10}$$

where $\sigma_0 = (n_e e^2 \tau/m_e)$ is the dc conductivity of the free electron gas. We see from Equation 15.10 that R_{xx} does not depend on B. Similarly, using Equations 15.2 and 15.8, R_{yx} can be computed as

$$\begin{aligned} R_{yx} &= \frac{V_y}{I} = \frac{vB\,L_y}{j\,L_y} = \frac{vBL_y}{-n_e evL_y} \\ &= -\frac{B}{n_e\,e} \end{aligned} \tag{15.11}$$

which varies linearly with B as shown schematically in Figure 15.3.

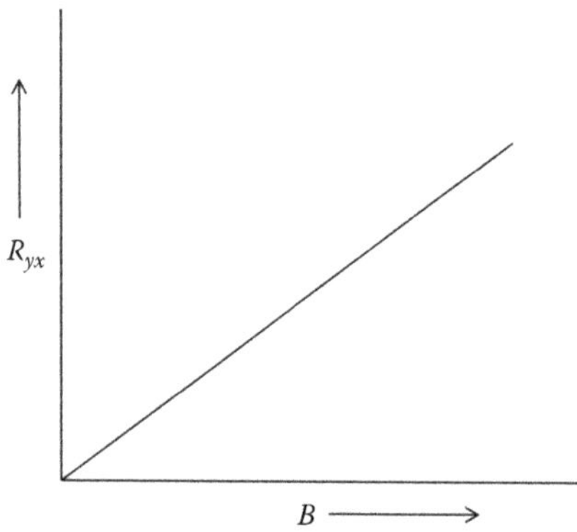

FIGURE 15.3
R_{yx} as a function of B in the classical Hall experiment.

15.3 Landau Levels

Let us solve the problem of 2D electron gas in the magnetic field quantum mechanically. The magnetic field $\vec{B}$ can be expressed in terms of vector potential $\vec{A}$ as

$$\vec{B} = \vec{\nabla} \times \vec{A} \tag{15.12}$$

We assume $\vec{B}$ to be constant and in the z direction. There are various ways of choosing $\vec{A}$ that will give this result for $\vec{B}$. We choose it to be

$$\vec{A} = B(-y,0,0) \tag{15.13}$$

The one-electron Hamiltonian in the presence of magnetic field $\vec{B}$ can be written as

$$H = \frac{1}{2m_e}(\vec{p} + e\vec{A})^2 - \vec{\mu}\cdot\vec{B} \tag{15.14}$$

where $\vec{\mu}$ is the magnetic moment of the electron. Since $\vec{\mu}\cdot\vec{B}$ does not depend on space coordinates, it is just a constant that we can drop for the moment as it amounts to a constant shift of the eigenvalues of H. Thus, Equation 15.14 can be written as

$$H = \frac{1}{2m_e}(p_x - eBy)^2 + \frac{1}{2m_e}p_y^2 \tag{15.15}$$

where we have used Equation 15.13. We note that H commutes with p_x, whose eigenfunctions are $e^{ik_x x}$ with eigenvalues $\hbar k_x$. Therefore, the eigenfunction of H can be chosen of the form

$$\psi(x,y) = \frac{e^{ik_x x}}{\sqrt{L_x}} f(y) \tag{15.16}$$

where L_x is the length of the sample in the x direction and $f(y)$ is a function to be found. We want to solve the equation

$$H\psi = \varepsilon\psi \tag{15.17}$$

Substituting Equation 15.16 in Equation 15.17, we get

$$H\psi = \left[\frac{1}{2m_e}(p_x - eBy)^2 + \frac{p_y^2}{2m_e}\right]\frac{e^{ik_x x}}{\sqrt{L_x}} f(y)$$

$$= \frac{e^{ik_x x}}{\sqrt{L_x}}\left[\frac{1}{2m_e}(\hbar k_x - eBy)^2 + \frac{p_y^2}{2m_e}\right] f(y)$$

$$= \varepsilon \frac{e^{ik_x x}}{\sqrt{L_x}} f(y) \tag{15.18}$$

Now, the cyclotron frequency is given by

$$\omega_c = \frac{eB}{m_e} \tag{15.19}$$

Using Equation 15.19, we can rewrite Equation 15.18 as

$$\left[\frac{p_y^2}{2m_e} + \frac{1}{2} m_e \omega_c^2 (y - y_0)^2\right] f(y) = \varepsilon f(y) \tag{15.20}$$

where

$$y_0 = \frac{\hbar k_x}{eB} \tag{15.21}$$

We see that Equation 15.20 is the equation for the simple harmonic oscillator with eigenvalues

$$\varepsilon_n = \hbar\omega_c\left(n + \frac{1}{2}\right) \tag{15.22}$$

and eigenfunction

$$f(y) = \phi_n(y - y_0)$$

where ϕ_n is a simple harmonic oscillator wave function.

Thus, the eigenvalues of the Hamiltonian in Equation 15.14 can be written as

$$\varepsilon_{n,m_s} = \hbar\omega_c\left(n + \frac{1}{2}\right) + g_s m_s \mu_B B \tag{15.23}$$

where the last term in Equation 15.23 is the Zeeman splitting, g_s is the spin g factor, $m_s = \pm 1/2$, and μ_B is the Bohr magneton. Landau levels are the levels with different values of n. The spacing between the Landau levels is $\hbar\omega_c$, which depends on the magnetic field B. We note that the Landau levels do not depend on k_x, which means that a large number of eigenfunctions with different values of k_x correspond to one Landau level. Thus, there is a high degree of degeneracy. Let us try to find out the degeneracy of a Landau level with quantum number n. We see from Equation 15.20 that y_0 is the center of the electron's orbit that has energy given by Equation 15.23. This center must lie within the sample, that is

$$0 \le y_0 \le L_y \tag{15.24}$$

or

$$0 \le \frac{\hbar k_x}{eB} \le L_y$$

or

$$0 \le k_x \le \frac{eB}{\hbar} L_y$$

or

$$0 \le \frac{2\pi n_x}{L_x} \le \frac{eB}{\hbar} L_y$$

or

$$0 \le n_x \le \frac{eB}{2\pi\hbar} L_x L_y$$

or

$$0 \le n_x \le m_e\omega_c \frac{L_x L_y}{2\pi\hbar} = n_{\max} \tag{15.25}$$

Here, we have used $k_x = (2\pi n_x/L_x)$, where n_x is an integer. Equation 15.25 gives the degeneracy of a Landau level. The degeneracy per unit area of the sample d can be written as

$$d = \frac{m_e\omega_c}{2\pi\hbar} = \frac{m_e eB}{2\pi\hbar m_e} = \frac{eB}{h} \tag{15.26}$$

Now, we can evaluate the density of states per unit area

$$\rho(\varepsilon) = \frac{1}{L_x L_y} \sum_{n,n_x,m_s} \delta(\varepsilon - \varepsilon_{n,m_s})$$

$$= \frac{m_e \omega_c}{L_x L_y} \frac{L_x L_y}{2\pi\hbar} \sum_{n,m_s} \delta(\varepsilon - \varepsilon_{n,m_s})$$

$$= \frac{m_e \omega_c}{2\pi\hbar} \sum_{n,m_s} \delta(\varepsilon - \varepsilon_{n,m_s}) \tag{15.27}$$

where we have used Equation 15.25.

The number of electrons per unit area at finite temperature, n_e, is given by

$$n_e = \int d\varepsilon \, \rho(\varepsilon) f(\varepsilon) \tag{15.28}$$

where $f(\varepsilon)$ is the Fermi–Dirac occupation number. Substituting Equation 15.27 in Equation 15.28, we get

$$n_e = \frac{m_e \omega_c}{2\pi\hbar} \int d\varepsilon \sum_{n,m_s} \delta(\varepsilon - \varepsilon_{n,m_s}) f(\varepsilon)$$

$$= \frac{m_e \omega_c}{2\pi\hbar} \sum_{n,m_s} f(\varepsilon_{n,m_s}) \tag{15.29}$$

$$= \frac{eB}{2\pi\hbar} \nu = d\nu \tag{15.30}$$

where

$$\nu = \sum_{n,m_s} f(\varepsilon_{n,m_s}) \tag{15.31}$$

is called the filling factor and determines the number of occupied Landau levels.

15.4 Integer and Fractional Quantum Hall Effects (IQHE and FQHE)

A quasi-2D gas can be realized by using a semiconductor microstructure such as MOSFET (metal–oxide–semiconductor field-effect transistor) or

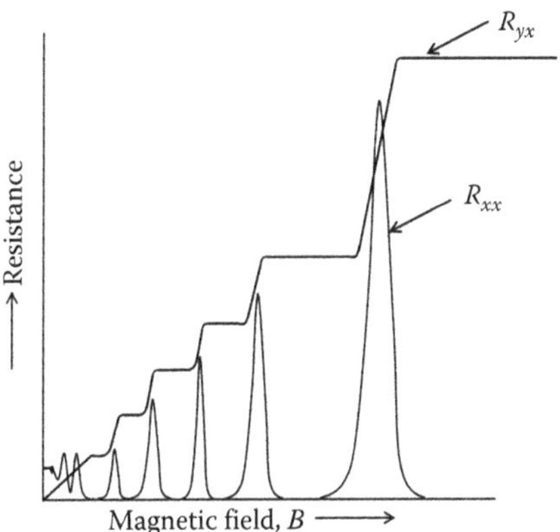

FIGURE 15.4
R_{yx} and R_{xx} as a function of B in the integer quantum Hall effect.

quantum well. The experimental geometry is similar to that employed in the classical Hall effect experiment. However, when the transverse resistance R_{yx} is plotted as a function of magnetic field, a very surprising result was found by von Klitzing et al. (1980); the plot of R_{yx} has a staircase pattern showing a series of flat steps as shown schematically in Figure 15.4. This is in contrast to the linear behavior (Figure 15.3) observed in the classical Hall effect experiment. The longitudinal resistance R_{xx} tends to zero during these steps. The steps in R_{yx} occur at very precise values of resistance, $R_{yx} = R_0/\nu$, where ν is an integer and $R_0 = h/e^2$, that is, the resistance is quantized in units of h/e^2 divided by an integer. This result, which is independent of the material of the sample or its size, is known as the IQHE. Note that the quantum of resistance ($R_0 = h/e^2$) depends only on the fundamental constants. For this remarkable work, von Klitzing was awarded the Nobel Prize in 1985.

IQHE can be understood based on the filling of the Landau levels as the magnetic field is varied and by using Equation 15.11. For simplicity, we shall assume that the Zeeman splitting is quite large compared to the separation of Landau levels, so that the occupied Landau levels have only spin-up electrons; therefore, we do not worry about the spin degrees of freedom. This is reasonable to assume since the effective g factor gets enhanced due to exchange interaction in real systems (Englert et al. 1982), thus, further increasing Zeeman splitting. We start with a very large magnetic field so that the degeneracy d is large compared to n_e (see Equations 15.26 and 15.30). Therefore, all electrons lie in the lowest Landau level, that is, $\nu = (n_e/d) < 1$. Now, let us start decreasing the magnetic field, so that the lowest Landau level is completely filled and $\nu = 1$. Let us call this field as B_1. Again, let us start decreasing the field that will force some electrons in the second Landau level. Ultimately, a magnetic field B_2 is reached when the second Landau level is completely filled and $\nu = 2$. Thus, for integer filling $\nu = j$, there is a special magnetic field B_j such that j Landau levels are completely filled and all other levels are completely empty. Let us evaluate Hall resistance R_{yx} at these special magnetic fields B_j. We know from Equation 15.11 that

$$|R_{yx}| = \frac{B}{n_e e} \tag{15.32}$$

If we substitute n_e from Equation 15.30, we get

$$\begin{aligned} |R_{yx}| &= \frac{h}{e^2 j} \\ &= \frac{R_0}{j} \end{aligned} \tag{15.33}$$

Thus, we have shown that R_{yx} takes quantized values R_0/j at integral values of filling factor ν at some special value of magnetic field B_j. However, for magnetic fields away from B_j, Equation 15.33 will not hold as ν will not be an integer and the curve between R_{yx} and B will still be a straight line as per Equation 15.32. Thus, Equation 15.33 does not explain the flat plateaus occurring in IQHE.

To explain flat plateaus in IQHE, we have to consider the effects of disorder and finite size sample; a real sample always has finite length and width, and it always has impurities that give rise to disorder. The inclusion of these effects in theoretical treatment is very difficult. Therefore, we shall try to understand the IQHE qualitatively based on the picture of Landau levels discussed in Section 15.3. There are two major effects produced by the disorder: (1) Landau levels get broadened and form bands as shown in Figure 15.5, and (2) the disorder introduces localized states in the band tails, as we

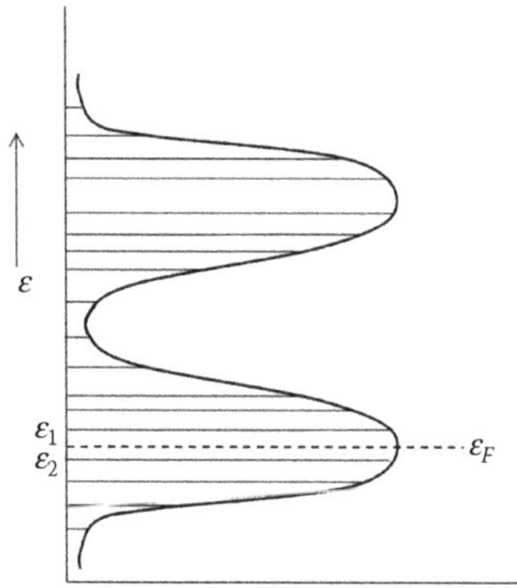

FIGURE 15.5
Schematic showing localized levels in Landau bands.

had seen in Chapter 11, shown by horizontal lines in the figure. The localized states occupy most of the band except in a narrow energy region between the mobility edges denoted by ε_1 and ε_2. Thus, there are very few extended states in the middle of the band in a narrow region between ε_1 and ε_2. Note that the localized states do not contribute to the current and only the extended states contribute to the current and therefore to the resistance. With this picture in mind, let us try to understand how plateaus arise in R_{yx} and why R_{xx} goes to zero during the plateaus. When we change the magnetic field, the degeneracy as well as the spacing between the Landau levels changes. Consequently, these levels pass through the Fermi level as B changes. When the field is such that the Fermi level lies between ε_1 and ε_2 in the extended states, R_{yx} and R_{xx} will change with the magnetic field. But when the Fermi level falls in the region of the localized states, R_{yx} remains at the same value that it had at ε_2 or ε_1, depending on whether B is increasing or decreasing. This is because ν extended states are completely filled and ν does not change when the Fermi level is in the region of localized states. Thus, R_{yx} has the constant value R_0/ν until the Fermi level falls in the extended states region. Since the energy range of the localized states is quite wide and the energy range of extended states is very narrow, we get flat plateaus in R_{yx} with changing B. Thus, the inclusion of disorder is important for explaining the IQHE.

To understand why R_{xx} goes to zero during Hall plateaus, we have to bring in finite size effect into consideration. We know that in a finite sample, there are edge states or surface states that can carry current and can play a very important role. It has been shown (see, e.g., Halperin 1982) that in Hall geometry, the current can be carried only along the edges as shown schematically in Figure 15.6, which shows a rectangular sample placed in a high magnetic field B. This can be more easily understood using a classical argument as explained in Figure 15.7. When the magnetic field is applied to a moving electron, it experiences a force at right angles to its motion. Thus, an electron moves in a circle as shown schematically in Figure 15.7 and does not contribute to the current flowing across the sample. However, an electron near the edges cannot complete the circle. It is bounced off from the boundary and suffers successive scattering from the boundary as shown in Figure 15.7. The electrons deep inside the sample carry no current while the electrons near the edges carry the current. At the top edge, the electrons are moving from left to right, and at the bottom edge, the electrons are moving from

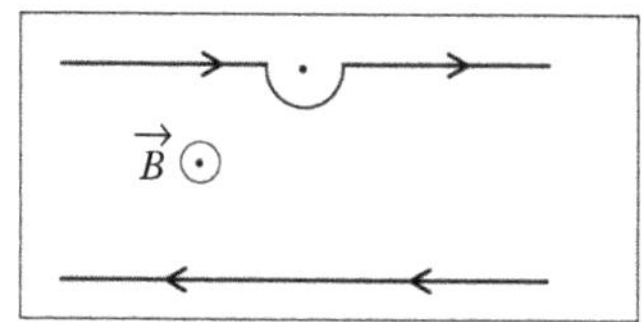

FIGURE 15.6
Schematic showing flow of electrons in a quantum Hall system.

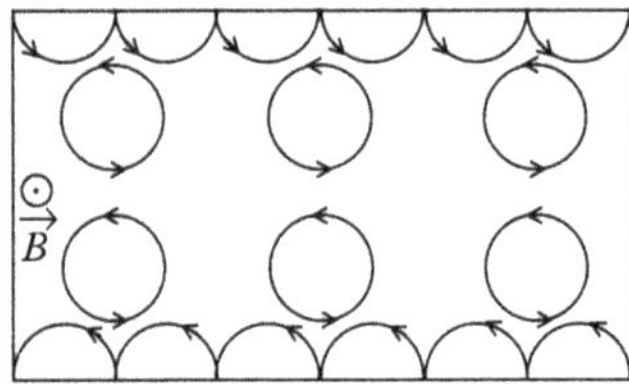

FIGURE 15.7
Classical description of edge currents.

right to left and there is no current inside the sample. It is as if the traffic is separated into two lanes. Thus, the effect of the magnetic field is to separate forward-moving electrons and backward-moving electrons into two channels as shown in Figure 15.6. If the electron encounters a nonmagnetic impurity, it just bypasses it and keeps moving in the same direction as it has no way to turn back. This is because at low temperatures, the accessible empty states into which the electron can get scattered lie at a much higher energy. Note that during the flat Hall plateaus, the Fermi level lies in the region of the localized states and the extended states are at much higher energies. The absence of impurity scattering in a quantum Hall system gives rise to a dissipationless current.

It was seen by Tsui et al. (1982) that the plateaus in R_{yx} occur at R_0/ν where ν is a fraction such as 4/3, 5/3, 2/3, 4/5, and so on. This effect is known as FQHE. This was explained by Laughlin (1983) who proposed a correlated wave function. In 1998, Tsui, Stormer, and Laughlin were awarded the Nobel Prize for the discovery and explanation of the FQHE. FQHE has also been explained by Jain who proposed the composite fermion theory to explain the effect (see, e.g., Jain 2007). The explanation is quite involved; therefore, we refer the reader to the original papers and books by Mahan (2000), Prange and Girvin (1987), and Jain (2007). Note that IQHE as well as FQHE has been observed in graphene (see, e.g., Peres 2010).

15.5 Quantum Spin Hall Effect (QSHE)

We saw in Section 15.4 that the quantum Hall effect occurs when a strong magnetic field is applied to a 2D electron gas in a semiconductor sample placed in the Hall geometry. We also noticed that the current can be carried only along the edges as shown in Figure 15.6 and the flow of electrons was separated into two lanes by the magnetic field. This separation of electron traffic into two lanes and dissipationless current are very interesting features of the quantum Hall effect and could have applications in semiconductor devices. But this occurs only at high magnetic fields, which is not practical.

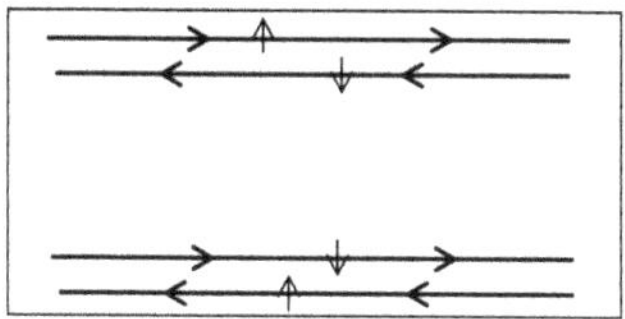

FIGURE 15.8
Flow of electrons in a quantum spin Hall system.

Can this separation of electron traffic be achieved without the application of magnetic field? To answer this question, let us bring spin of the electron into consideration. Note that in the above discussion of quantum Hall effect, spin played no role. So, let us consider a system of forward- and backward-moving electrons with up and down spins as shown in Figure 15.8, which has four channels, two at the top edge and two at the bottom edge. Note that there is no transport of charge but there is a net transport of spin. Such states are known as quantum spin Hall states and the corresponding phenomenon is known as QSHE. This was predicted by Kane and Mele (2005), Bernevig et al. (2006), and Bernevig and Zhang (2006) and has been observed in 2D HgTe quantum wells. In such systems, the spin–orbit interaction is quite large and is responsible for separation of electron traffic into channels (see next section). Thus, the spin–orbit interaction in these systems plays a similar role as magnetic field in quantum Hall systems.

We saw in Chapter 4 that the spin–orbit interaction can be written as

$$H_{SO} = \frac{e}{2m_e^2c^2}\vec{S}\cdot(\vec{E}\times\vec{p}) \tag{15.34}$$

where $\vec{S}$ is the spin angular momentum of the electron, $\vec{p}$, the linear momentum, and $\vec{E}$, the electric field. If the spin–orbit interaction is so strong that the other terms in the Hamiltonian can be neglected, Equation 15.34 implies that $\vec{S}$ is perpendicular to $\vec{p}$. To see this, let us assume that both $\vec{E}$ and $\vec{p}$ lie in the x–y plane. This implies that their cross product lies along the z axis. Thus

$$H_{SO} = C\vec{S}\cdot\hat{z} = CS_z \tag{15.35}$$

where C is a constant. Since the eigenvectors of S_z are $|\uparrow\rangle$ and $|\downarrow\rangle$, Equation 15.35 implies that the spin points in the z direction, which is perpendicular to $\vec{p}$.

Also, we see that H_{SO} has time-reversal symmetry. Thus, if the Hamiltonian of the system has time-reversal symmetry, Kramers' theorem implies that $\varepsilon_{\vec{k}\uparrow} = \varepsilon_{-\vec{k}\downarrow}$. This implies that in an elastic scattering from a nonmagnetic impurity, an electron cannot be backscattered as it involves flipping of the spin, which cannot be done by a nonmagnetic impurity. Thus, similar to the quantum Hall case, an electron just goes around the impurity without getting scattered and the current is dissipationless.

15.6 Topological Insulators

The discovery of topological insulators was inspired by the existence of conducting edge states in quantum Hall systems that were protected against scattering by nonmagnetic impurities. Topological insulators are recently discovered materials that have a bulk energy band gap such as ordinary band insulators but have metallic states on their surfaces or edges. These surface states are protected against scattering from nonmagnetic impurities, which is made possible due to the combination of spin–orbit interaction and time-reversal symmetry as we had seen in Section 15.5. In other words, the interior bulk states carry no current and the current is carried only by the surface states or edge states, which is nearly dissipationless. These materials form a new class of materials and are very different from the ordinary band insulators. These are characterized by a topological invariant, an index similar to genus in topology, which is determined by the bulk band structure. There has been an intense research activity in this field not only due to the possibility of new physics, but also due to potential applications of these materials in spintronics and quantum computing.

The electronic structure plays a very important role in understanding these materials. To understand some important features of the electronic structure, let us consider a simple model of a 2D topological insulator as shown in Figure 15.8. We focus on the conducting states that are localized near the top and bottom edges along the x direction. We assume that the spin–orbit interaction is quite strong and is the dominant term in the Hamiltonian, so that other terms can be neglected. Thus, the Hamiltonian of the system is

$$H = C\vec{S}\cdot(\vec{E}\times\vec{p}) \tag{15.36}$$

where C is a constant, electric field $\vec{E} = -\vec{\nabla}V$, and V is the electric potential experienced by the electrons. Note that H is invariant under time reversal as well as space inversion, which is not the case in quantum Hall effect. Near the edge, the potential V shows a large variation due to the surface barrier; therefore, its gradient will point in the y direction. As $\vec{p}$ lies in the x or $-x$ direction, the cross product $\vec{E}\times\vec{p}$ is in the z direction. Thus, H can be written as

$$H = C\vec{S}\cdot Ep\hat{z} = CES_z p = C'\sigma_z p \tag{15.37}$$

where C' is another constant and σ_z is a Pauli spin matrix whose eigenvalues are +1 for spin up and –1 for spin down. The momentum lies in the x direction so that the spin direction is always perpendicular to $\vec{p}$. Thus, the energy eigenvalues of H are

$$\begin{aligned} \varepsilon_p &= C'p \quad \text{for } p > 0, \text{ spin up and } p < 0, \text{ spin down} \\ &= -C'p \quad \text{for } p > 0, \text{ spin down and } p < 0, \text{ spin up} \end{aligned} \tag{15.38}$$

The energy dispersion for the edge states at the top edge is plotted in Figure 15.9. There will be a similar figure for edge states at the bottom edge (spin reversed). We note that the dispersion curves are linear in p and that there is no band gap. Also, the electrons moving near the top edge in the $+x$ direction ($p > 0$) have spin up, while the electrons moving near the top edge in the $-x$ direction ($p < 0$) have spin down. Similar is the case at the bottom edge, as shown in Figure 15.8. Thus, there is a net transport of spin but no transport of charge. Thus, the system is in the quantum spin Hall state.

The above model can be generalized to a three-dimensional (3D) topological insulator. Let us assume that the top surface lies in the x–y plane. The electric field is due to the gradient of the potential near the surface and will point in the z direction. Only the electrons near the x–y plane will experience this electric field. Thus, the Hamiltonian can be written as

$$\begin{aligned} H &= C\vec{S}\cdot(\vec{E}\times\vec{p}) = C'\vec{E}\cdot(\vec{\sigma}\times\vec{p}) = C'E\hat{z}\cdot(\vec{\sigma}\times\vec{p}) \\ &= C'E(\sigma_x p_y - \sigma_y p_x) \\ &= C'E\begin{pmatrix} 0 & p_x + ip_y \\ p_x - ip_y & 0 \end{pmatrix} \end{aligned} \tag{15.39}$$

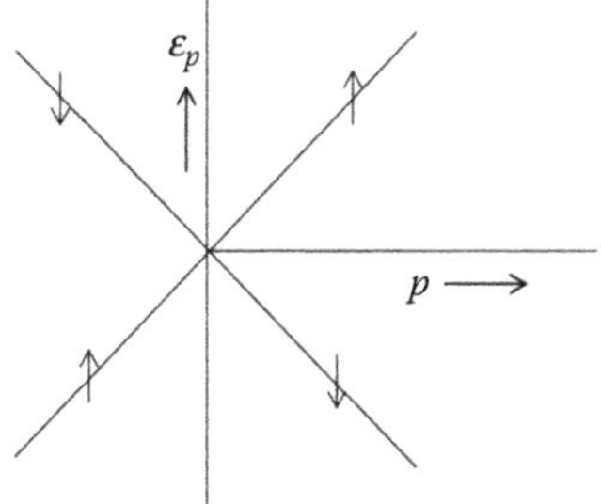

FIGURE 15.9
Schematic energy dispersion curves for edge states (top edge) for 2D topological insulator. The arrow denotes the spin direction.

Note that Equation 15.39 is very similar to the effective Hamiltonian of graphene in Equation 14.26 and represents a Dirac particle in 2D. It can be proved, as in Section 15.5, that the spin direction is always perpendicular to $\vec{p}$. The energy eigenvalues of H can be written as

$$\varepsilon_p = \pm C''\sqrt{(p_x^2 + p_y^2)} = \pm C''p \tag{15.40}$$

Similar to graphene, we get a linear dispersion relation and there is no band gap. Thus, around $p = 0$, the band structure will have a Dirac cone as shown in Figure 15.10. Note that the figure corresponds to the surface states on the top face, and there will be a similar figure (spin reversed) for the bottom face. The existence of a Dirac cone is an essential feature of the band structure of a topological insulator.

As mentioned earlier, a topological insulator is characterized by a topological invariant that depends on the topology of the band structure. This invariant is insensitive to various kinds of perturbations such as disorder. This is analogous to topology in mathematics where closed surfaces can be characterized by their genus, g, which counts the number of holes. For example, for a sphere, $g = 0$ while for a doughnut-shaped surface, $g = 1$. In the case of the topological insulator, the invariant is called the Z_2 invariant denoted by ν. The theory of the Z_2 invariant is quite complicated; therefore, we shall not discuss it here but refer to the review article by Hasan and Kane (2010). In short, the invariant ν can take integer values equal to 0 or 1. For $\nu = 0$, we get an ordinary band insulator and for $\nu = 1$, we get a topological insulator, which is characterized by an odd number of Dirac cones for surface states. This is because there is no intravalley scattering of electrons, but there could be intervalley scattering as we had proved in the case of graphene. The intervalley scattering between a pair of Dirac cones can induce gaps between the surface bands, thus destroying their metallic character. Thus, for $\nu = 0$ or for

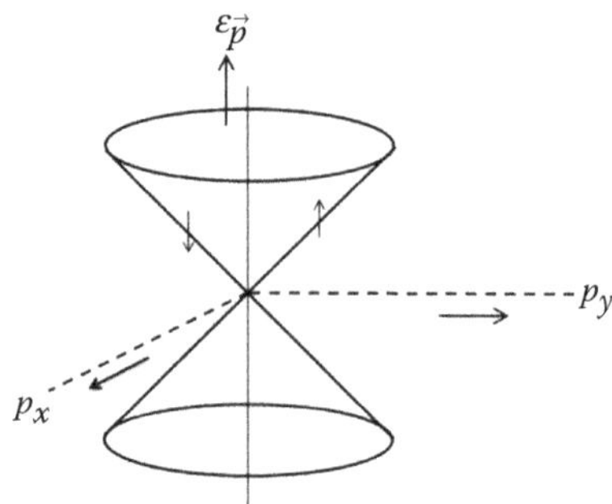

FIGURE 15.10
Schematic energy bands for surface states on top face plotted in the (p_x, p_y) plane for a 3D topological insulator.

an even number of Dirac cones, one does not get scattering-free conducting surface states or a topological insulator. For $\nu = 1$ or an odd number of Dirac cones, the intervalley scattering will cause gaps in pairs of Dirac cones, leaving at least one Dirac cone ungapped. Thus, for $\nu = 1$ or an odd number of Dirac cones, we will always get a topological insulator. The number ν or odd/even number of Dirac cones is not affected by perturbations such as nonmagnetic disorder.

In Figure 15.11, we show a schematic electronic structure of an ideal 3D topological insulator. The bulk energy bands are similar to that of the ordinary band insulator in which the conduction band and the valence band are separated by an energy gap. However, within this gap, there are surface states whose energy bands form a Dirac cone-like structure as in graphene. For simplicity, in Figure 15.11, we assume only one Dirac cone although as mentioned above, surface states of a 3D topological insulator, in general, form an odd number of Dirac cones. These surface states remain metallic, even in the presence of nonmagnetic impurities, due to the spin–orbit interaction and time-reversal symmetry.

The discovery of topological insulators has an interesting history as all the topological insulators known until now were first predicted by theoretical calculations and then verified experimentally. The first prediction was made by Bernevig, Hughes, and Zhang in 2006 that a (Hg,Cd)Te quantum well would behave as a 2D topological insulator and would have topologically protected edge states. This was later verified experimentally by Konig et al. in 2007. Fu and Kane (2007) predicted that $Bi_{1-x}Sb_x$ would be a 3D topological insulator, which was observed experimentally by Hsieh et al. (2008) by using an ARPES experiment. Thus, $Bi_{1-x}Sb_x$ is the first experimentally observed 3D topological insulator. Later, many other systems were predicted

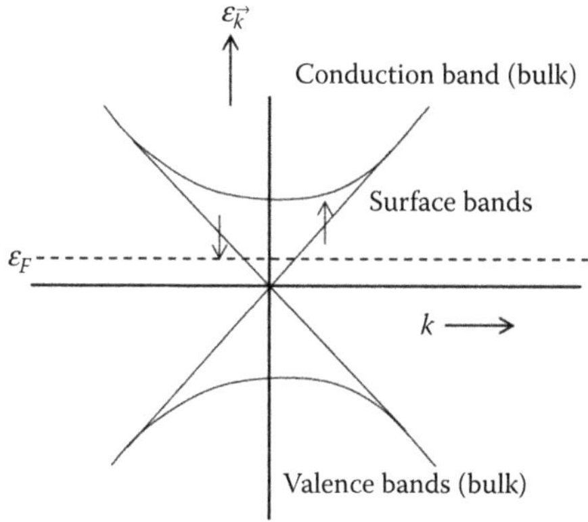

FIGURE 15.11
Schematic band structure of an ideal 3D topological insulator. Surface bands correspond to the top face.

to be topological insulators and were experimentally confirmed. Such systems include Bi_2Se_3, Bi_2Te_3, Sb_2Se_3, and thallium-based ternary chalcogenides. Many more systems have been predicted to be topological insulators but experimental confirmation is yet to be made.

Recently, there has been much research activity on thallium-based ternary chalcogenides as these systems exhibit several phenomena such as topological phase, quantum phase transitions, and Weyl semimetal (Lin et al. 2010, Sato et al. 2010b, Chen et al. 2010, Xu et al. 2011, Singh et al. 2012). These systems have also generated much excitement as they offer the possibility of providing a simple condensed matter system for exploring the Higgs mechanism (Sato et al. 2011). Therefore, we shall briefly discuss the electronic structure of these systems and its relation with these exciting phenomena. These compounds have a rhombohedral crystal structure as shown in Figure 15.12. Singh et al. (2012) calculated the electronic structure of these compounds using a pseudopotential method (VASP) and GGA. They found $TlSbSe_2$, $TlSbTe_2$, $TlBiSe_2$, and $TlBiTe_2$ to be topological insulators, with a single Dirac cone at Γ point, and $TlSbS_2$ and $TlBiS_2$ to be band insulators. For example, Figure 15.13 shows the band structures of $TlBiTe_2$ and $TlBiS_2$ using

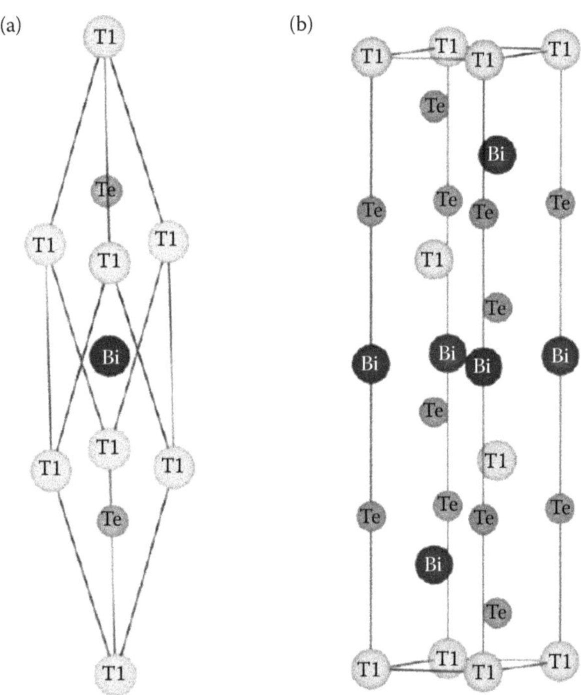

FIGURE 15.12
Unit cell of $TlBiTe_2$. (a) Primitive (rhombohedral) and (b) conventional (hexagonal). (Adapted from Singh B. et al. 2012. *Phys. Rev. B* 86: 115208.)

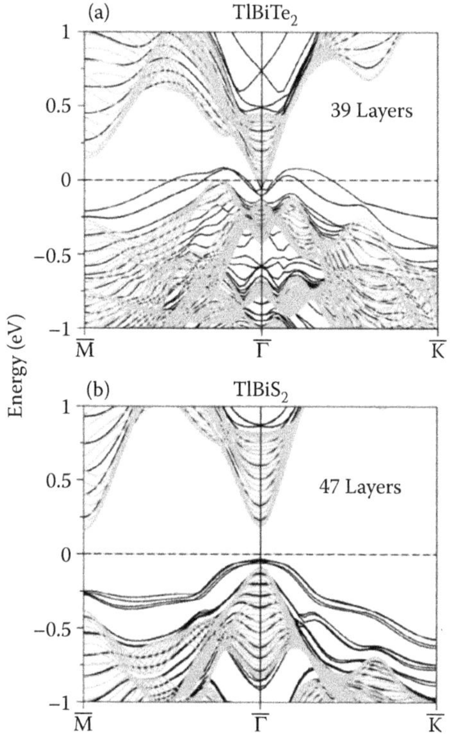

FIGURE 15.13
Electronic structure of $TlBiTe_2$ and $TlBiS_2$ using the slab geometry. The shaded bands are bulk bands. (Adapted from Singh B. et al. 2012. *Phys. Rev. B* 86: 115208.)

slab geometry. The bands in the shaded regions are bulk bands. The $TlBiTe_2$ band structure shows a perfect Dirac-like surface state with Dirac point 0.1 eV below the Fermi energy. This implies that $TlBiTe_2$ is a topological insulator. These results were found to be in good agreement with the available ARPES results. On the other hand, $TlBiS_2$ does not have any metallic surface state in the bulk gap region, which implies that it is a band insulator.

We saw that $TlBiTe_2$ is a topological insulator but $TlBiS_2$ is a band insulator. As these are different quantum phases of the electronic system, it is possible to study quantum phase transition in an alloy $TlBi(S_{1-x}Te_x)_2$ by varying concentration x as the system is a band insulator at $x=0$ and a topological insulator at $x=1$. This transition was found to occur at $x=0.5$. At this critical concentration, Singh et al. (2012) explicitly broke the inversion symmetry with a layer growth in the order of Tl–Te–Bi–S and looked at the bulk band structure. It is very interesting that by changing the c/a ratio they discovered a Weyl semimetal, which is a new topological phase of matter. In this phase, the bulk valence and conduction band touch at certain points, called Weyl points, which are separated in momentum space. The

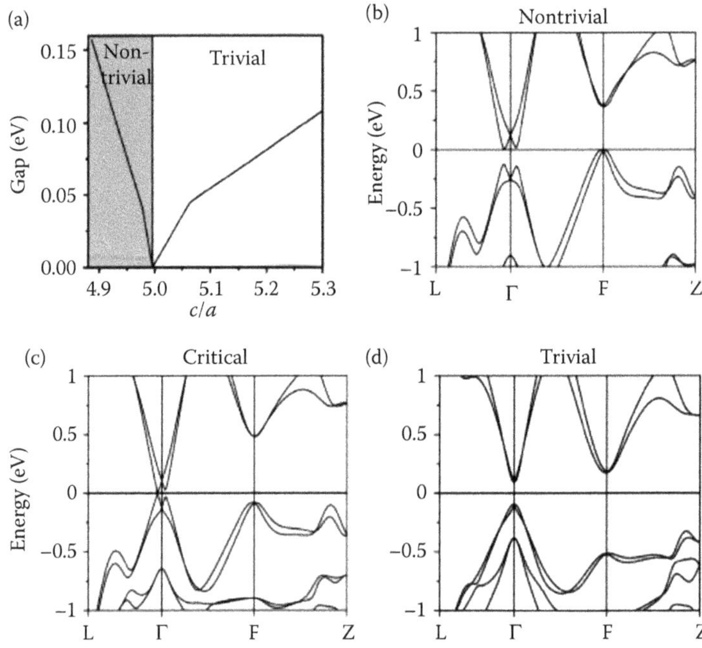

FIGURE 15.14
(a) Phase diagram of topological transition in TlBiTeS as a function of c/a. Panels (b) through (d) show the evolution of the band structure as the c/a ratio is increased. (Adapted from Singh B. et al. 2012. *Phys. Rev. B.* 86: 115208.)

bulk energy bands have linear dispersion relations around these points, just like graphene. Thus a Weyl semimetal could be viewed as a 3D analog of graphene. To see Weyl phase in $TlBi(S_{1-x}Te_x)_2$, they started with a normal insulator with a large value of c/a and varied the lattice parameters to a small value of c/a. This is illustrated in Figure 15.14. We see that for $c/a < 5$, the system is a topological insulator (nontrivial) whereas for $c/a > 5$, it is a band insulator (trivial). The bulk band structure shows band gap on both sides of critical values of $c/a = 5$. But at the critical value of $c/a = 5$, the band gap is zero and the valence and conduction bands touch at a k-point away from Γ at (0:0; 0:0; 0:025) with linear dispersion making it a Weyl semimetal. Recently, Sato et al. (2011) observed that Dirac fermions in the solid solution system $TlBi(S_{1-x}Se_x)_2$ acquire mass as x is varied without explicitly breaking the time-reversal symmetry. They used ARPES to look at the evolution of the electronic state as the system goes through topological to nontopological transition. They found that the Dirac state, which is massless, switches to a massive state before it disappears in the nontopological phase as the system passes through the quantum phase transition. This is very exciting as it is similar to the Higgs mechanism in which the particles acquire mass through spontaneous symmetry breaking.

EXERCISES

15.1 Show that for the Landau level problem in the symmetric gauge

$$\vec{A} = \frac{B}{2}(-y, x, 0)$$

the eigenfunction ψ_m can be written as

$$\psi_m(\rho, \phi) = \rho^{|m|} e^{im\phi} e^{-(\rho^2/4)} v(\rho)$$

where $\rho^2 = x^2 + y^2$, $v(\rho)$ is a function of ρ, and $\tan\phi = y/x$.

15.2 Show that for IQHE, the Hall conductance is quantized in units of e^2/h.

15.3 Prove that $\vec{\sigma}\cdot\vec{p} = 0$ for edge states in a topological insulator.

15.4 In deriving energy eigenvalues (Equation 15.38) for a 2D topological insulator, we assumed that the spin–orbit interaction was the dominating term in the Hamiltonian. Suppose we retain the kinetic energy term in the Hamiltonian. How will the eigenvalues get modified?

15.5 In deriving energy eigenvalues in Equation 15.40 for 3D topological insulators, we assumed that the spin–orbit interaction was the dominating term in the Hamiltonian. Suppose we retain the kinetic energy term in the Hamiltonian. How will the eigenvalues get modified?

15.6 Show that a band gap in the surface state bands of a 3D topological insulator can be opened if you coat the surface with a ferromagnetic material with magnetization perpendicular to the surface. (Hint: see the paper by Rosenberg and Franz (2012).)

15.7 Calculate the surface electronic structure of Bi_2Se_3 using supercell geometry and one of the codes mentioned in Appendix A. Show that the metallic surface states lie in the bulk band gap of the material and that the surface energy bands form a Dirac cone.

16

Ferroelectric and Multiferroic Materials

16.1 Introduction

There has been much excitement during the last decade regarding ferroelectric and multiferroic materials because of their interesting properties that can be controlled by the application of electric or magnetic fields. Ferroelectric materials are insulating materials that can exist in two or more stable polarization states. They have spontaneous polarization that can be reversed by the application of an electric field. Out of 32 point groups, only 10 groups with a unique polar axis can demonstrate spontaneous polarization. Above a certain temperature, known as the ferroelectric Curie temperature, the material loses spontaneous polarization and becomes paraelectric. In general, at Curie temperature, there is a structural change signaling a first-order phase transition, although there are some exceptions. Owing to their interesting properties, ferroelectric materials find applications in various devices such as transducers, actuators, and nonvolatile ferroelectric memories.

Multiferroic materials exhibit at least two out of three properties in the same phase, the properties being ferroelectricity, ferromagnetism or some magnetic order, and ferroelasticity. In this chapter, we shall focus on materials that have magnetic order and ferroelectricity. These materials are exciting because one can control their properties by applying an electric or magnetic field. However, the number of such known materials is very small and now an intense research by several groups is going on to find such materials that may find applications in future devices and technologies.

Spontaneous polarization is an important feature of these materials. Therefore, we begin with how to calculate polarization in insulating materials using a first-principles approach in Section 16.2. In Section 16.3, we define Born effective charge that plays an important role in ferroelectric and multiferroic materials. Then, we shall discuss electronic structure and properties of ferroelectric and multiferroic materials in Sections 16.4 and 16.5.

16.2 Polarization

When a static electric field is applied to an insulator, it polarizes the material, that is, it induces a dipole moment in the system. For a finite system of volume Ω, if the induced dipole moment is $\vec{d}$, polarization $\vec{P}$ is defined as

$$\vec{P} = \frac{\vec{d}}{\Omega} = \frac{1}{\Omega}\int \vec{r}\rho_T(\vec{r})d^3r \tag{16.1}$$

where $\rho_T(\vec{r})$ is the total charge density due to electrons and ions at point $\vec{r}$. One can see that the definition is valid only for a finite sample, because for an infinite sample, Equation 16.1 is ill defined as $\vec{r}$ is unbounded. However, derivatives of $\vec{P}$ or changes in $\vec{P}$ with respect to some reference state can be defined for an infinite bulk system and we shall see that the derivatives of $\vec{P}$ do have a well-defined value in the thermodynamic limit.

The total polarization $\vec{P}$ in Equation 16.1 can be divided into ionic and electronic parts

$$\vec{P} = \vec{P}_{ion} + \vec{P}_{el} \tag{16.2}$$

Let us first focus on the electronic part. We assume the KS Hamiltonian to be

$$H = T + V_{KS}(\lambda) \tag{16.3}$$

where T is the kinetic energy operator and $V_{KS}(\lambda)$ is the KS potential that depends on a parameter λ (Resta 1994, King-Smith and Vanderbilt 1993). Therefore, one-particle KS orbitals $\psi_{\vec{k}n}(\vec{r})$ (n is band index) and electron charge density $\rho(\vec{r})$ also depend on λ. The electronic charge density at $\vec{r}$ is given by

$$\rho(\vec{r}) = e\sum_{\vec{k}m}\left|\psi_{\vec{k}m}(\vec{r})\right|^2 \tag{16.4}$$

where the sum is only over the occupied bands. Thus, using Equation 16.1, $\vec{P}_{el}$ can be written as

$$\vec{P}_{el} = \frac{1}{\Omega}\int \vec{r}\rho(\vec{r})d^3r = \frac{e}{\Omega}\sum_{\vec{k}m}\int \vec{r}\left|\psi_{\vec{k}m}\right|^2 d^3r \tag{16.5}$$

Taking its derivatives with respect to λ, we get

$$\frac{\partial \vec{P}_{el}}{\partial \lambda} = \frac{e}{\Omega} \sum_{\vec{k}m} \int \vec{r}\, (\psi^{*}_{\vec{k}m}\, \psi'_{\vec{k}m} + \psi'^{*}_{\vec{k}m}\, \psi_{\vec{k}m})\, d^3r$$

where the prime denotes the derivative with respect to λ. This can be rewritten using bra and ket notations as

$$\frac{\partial \vec{P}_{el}}{\partial \lambda} = \frac{e}{\Omega} \sum_{\vec{k}m} [\langle \psi_{\vec{k}m} | \vec{r} | \psi'_{\vec{k}m} \rangle + \langle \psi'_{\vec{k}m} | \vec{r} | \psi_{\vec{k}m} \rangle] \tag{16.6}$$

To evaluate $|\psi'_{\vec{k}m}\rangle$, we use the first-order perturbation theory and get

$$|\psi'_{\vec{k}m}\rangle = \sum_{n \neq m} \frac{\langle \psi_{\vec{k}n} | V'_{KS}(\lambda) | \psi_{\vec{k}m} \rangle}{\varepsilon_{\vec{k}m} - \varepsilon_{\vec{k}n}} |\psi_{\vec{k}n}\rangle$$

where $\varepsilon_{\vec{k}m}$ are the one-electron energies that also depend on λ. In the above equation, the sum is over unoccupied states n, as m denotes an occupied state. Thus, Equation 16.6 can be written as

$$\frac{\partial \vec{P}_{el}}{\partial \lambda} = \frac{e}{\Omega} \sum_{\vec{k}m} \sum_{n \neq m} \frac{\langle \psi_{\vec{k}m} | \vec{r} | \psi_{\vec{k}n} \rangle \langle \psi_{\vec{k}n} | V'_{KS}(\lambda) | \psi_{\vec{k}m} \rangle}{\varepsilon_{\vec{k}m} - \varepsilon_{\vec{k}n}} + cc \tag{16.7}$$

where cc stands for the complex conjugate. Now, using the commutator of $\vec{r}$ with H, we can write (Exercise 16.1)

$$\langle \psi_{\vec{k}m} | \vec{r} | \psi_{\vec{k}n} \rangle = -\frac{i\hbar}{m_e} \frac{\langle \psi_{\vec{k}m} | \vec{p} | \psi_{\vec{k}n} \rangle}{\varepsilon_{\vec{k}m} - \varepsilon_{\vec{k}n}} \tag{16.8}$$

where m_e is the electron mass. Substituting Equation 16.8 in Equation 16.7, we get

$$\frac{\partial \vec{P}_{el}}{\partial \lambda} = -\frac{ie\hbar}{\Omega m_e} \sum_{\vec{k}m} \sum_{n \neq m} \frac{\langle \psi_{\vec{k}m} | \vec{p} | \psi_{\vec{k}n} \rangle \langle \psi_{\vec{k}n} | V'_{KS} | \psi_{\vec{k}m} \rangle}{(\varepsilon_{\vec{k}m} - \varepsilon_{\vec{k}n})^2} + cc \tag{16.9}$$

Equation 16.9 can be simplified by making a transformation to the Bloch functions $u_{\vec{k}n}$ that are related to $\psi_{\vec{k}n}$ by

$$\psi_{\vec{k}n}(\vec{r}) = e^{i\vec{k}\cdot\vec{r}} u_{\vec{k}n}(\vec{r}) \tag{16.10}$$

and satisfy the equation (see Exercise 4.6)

$$H_{\vec{k}}\, u_{\vec{k}n} = \varepsilon_{\vec{k}n} u_{\vec{k}n} \tag{16.11}$$

where

$$H_{\vec{k}} = \frac{1}{2m_e}\left(-i\hbar\vec{\nabla} + \hbar\vec{k}\right)^2 + V_{\mathrm{KS}} \tag{16.12}$$

For simplifying Equation 16.9, we use the relation (Exercise 16.2)

$$\langle \psi_{\vec{k}m} | \vec{p} | \psi_{\vec{k}n} \rangle = \frac{m_e}{\hbar} \langle u_{\vec{k}m} | [\vec{\nabla}_{\vec{k}}, H_{\vec{k}}] | u_{\vec{k}n} \rangle \tag{16.13}$$

$$= \frac{m_e}{\hbar} \langle u_{\vec{k}m} | [\vec{\nabla}_{\vec{k}} H_{\vec{k}} - H_{\vec{k}} \vec{\nabla}_{\vec{k}}] | u_{\vec{k}n} \rangle$$

$$= -\frac{m_e}{\hbar} (\varepsilon_{\vec{k}m} - \varepsilon_{\vec{k}n}) \langle u_{\vec{k}m} | \vec{\nabla}_{\vec{k}} | u_{\vec{k}n} \rangle \tag{16.14}$$

where we have used Equation 16.11. Also, we use the relation (Exercise 16.3)

$$\langle \psi_{\vec{k}n} | V'_{\mathrm{KS}} | \psi_{\vec{k}m} \rangle = \left\langle u_{\vec{k}n} \left| \left[\frac{\partial}{\partial \lambda}, H_{\vec{k}} \right] \right| u_{\vec{k}m} \right\rangle \tag{16.15}$$

$$= (\varepsilon_{\vec{k}m} - \varepsilon_{\vec{k}n}) \left\langle u_{\vec{k}n} \left| \frac{\partial}{\partial \lambda} \right| u_{\vec{k}m} \right\rangle \tag{16.16}$$

Substituting Equations 16.14 and 16.16 in Equation 16.9, we get

$$\frac{\partial \vec{P}_{el}}{\partial \lambda} = \frac{ie\hbar}{\Omega m_e} \frac{m_e}{\hbar} \sum_{\vec{k}m} \sum_{n \neq m} \langle u_{\vec{k}m} | \vec{\nabla}_{\vec{k}} | u_{\vec{k}n} \rangle \left\langle u_{\vec{k}n} \left| \frac{\partial}{\partial \lambda} \right| u_{\vec{k}m} \right\rangle + cc$$

$$= \frac{ie}{\Omega} \sum_{\vec{k}m} \left\langle u_{\vec{k}m} \left| \vec{\nabla}_{\vec{k}} \frac{\partial}{\partial \lambda} \right| u_{\vec{k}m} \right\rangle + cc$$

$$= \frac{ie}{8\pi^3\hbar} \sum_m \int \left\langle u_{\vec{k}m} | \vec{\nabla}_{\vec{k}} | \frac{\partial}{\partial \lambda} u_{\vec{k}m} \right\rangle d^3k + cc$$

$$= \frac{ie}{8\pi^3\hbar} \sum_m \int \frac{\partial}{\partial \lambda} \left\langle u_{\vec{k}m} | \vec{\nabla}_k | u_{\vec{k}m} \right\rangle d^3k \tag{16.17}$$

Note that in Equation 16.17, the sum is over only occupied states m as the sum over conduction states n has disappeared. Also, the integral in Equation 16.17 is over the BZ and is finite, in contrast to Equation 16.5, where the integral is over the whole volume of the solid. Thus, we see from this equation that $\partial\vec{P}_{el}/\partial\lambda$ has a finite value in the thermodynamic limit in contrast to $\vec{P}_{el}$ in Equation 16.5. From Equation 16.17, we can write change in polarization, $\Delta\vec{P}_{el}$, in going from $\lambda = 0$ to $\lambda = 1$ as

$$\Delta\vec{P}_{el} = \int_0^1 \frac{\partial\vec{P}_{el}}{\partial\lambda} d\lambda \tag{16.18}$$

$$= \frac{ie}{8\pi^3}\sum_m \int \langle u_{\vec{k}m} | \vec{\nabla}_{\vec{k}} | u_{\vec{k}m}\rangle d^3k \Big|_0^1$$

$$= \vec{P}_{el}(1) - \vec{P}_{el}(0) \tag{16.19}$$

where

$$\vec{P}_{el}(\lambda) = \frac{ei}{8\pi^3}\sum_m \int \langle u_{\vec{k}m} | \vec{\nabla}_{\vec{k}} | u_{\vec{k}m}\rangle d^3k \tag{16.20}$$

Equation 16.19 is the central result of this theory and has a very elegant form. It can be connected to the Berry phase theory; $i\langle u_{\vec{k}m} | \vec{\nabla}_{\vec{k}} | u_{\vec{k}m}\rangle$ can be recognized as Berry's connection whose integral over a closed manifold (here BZ) gives Berry phase (Pancharatnam 1956, Berry 1984).

Let us now discuss a special case when the Hamiltonians at $\lambda = 0$ and 1 are identical (King-Smith and Vanderbilt 1993). This implies that the Bloch function $u^{(0)}_{km}$ and $u^{(1)}_{km}$ corresponding to $\lambda = 0$ and 1, respectively, can at most differ by a phase factor so that

$$u^{(1)}_{\vec{k}m}(\vec{r}) = e^{i\theta_{\vec{k}m}} u^{(0)}_{\vec{k}m}(\vec{r}) \tag{16.21}$$

In this special case, Equation 16.19 gives

$$\Delta\vec{P}_{el} = \vec{P}_{el}(1) - \vec{P}_{el}(0)$$

$$= \frac{e}{8\pi^3} i \sum_m \int [\langle u^{(1)}_{\vec{k}m} | \vec{\nabla}_{\vec{k}} | u^{(1)}_{\vec{k}m}\rangle - \langle u^{(0)}_{\vec{k}m} | \vec{\nabla}_{\vec{k}} | u^{(0)}_{\vec{k}m}\rangle] d^3k$$

$$= \frac{e}{8\pi^3} i \sum_m \int [\langle e^{i\theta_{\vec{k}m}} u^{(0)}_{\vec{k}m} | (i\vec{\nabla}_{\vec{k}}\theta_{\vec{k}m}) e^{i\theta_{\vec{k}m}} u^{(0)}_{\vec{k}m} + e^{i\theta_{\vec{k}m}} \vec{\nabla}_{\vec{k}} u^{(0)}_{\vec{k}m}\rangle$$

$$- \langle u^{(0)}_{\vec{k}m} | \vec{\nabla}_{\vec{k}} | u^{(0)}_{\vec{k}m}\rangle] d^3k$$

$$= -\frac{e}{8\pi^3}\sum_m \int \vec{\nabla}_{\vec{k}}\, \theta_{\vec{k}m} d^3k \tag{16.22}$$

where we have assumed that the wave functions are normalized. Since wave functions are periodic in $\vec{k}$ space, that is

$$\psi_{\vec{k}+\vec{G},m}(\vec{r}) = \psi_{\vec{k},m}(\vec{r}),$$

it implies that

$$u_{\vec{k}m}(\vec{r}) = e^{i\vec{G}\cdot\vec{r}}\, u_{\vec{k}+\vec{G},m}(\vec{r}) \tag{16.23}$$

Equations 16.21 and 16.23 imply that $\theta_{\vec{k}m}$ must be periodic in $\vec{k}$ space (Exercise 16.4). In the most general term, this can be expressed as

$$\theta_{\vec{k}m} = \beta_{\vec{k}m} + \vec{k}\cdot\vec{R}_m \tag{16.24}$$

where $\beta_{\vec{k}m}$ is periodic in $\vec{k}$ and $\vec{R}_m$ is a direct lattice vector. Substituting Equation 16.24 in Equation 16.22, we get

$$\begin{aligned}\Delta\vec{P}_{el} &= -\frac{e}{8\pi^3}\sum_m \int d^3k\, \vec{R}_m \\ &= -\frac{e}{\Omega_{cell}}\sum_m \vec{R}_m \end{aligned} \tag{16.25}$$

where Ω_{cell} is the volume of a unit cell. This means that the change in polarization for paths where the Hamiltonian returns to itself is quantized in units of $(e/\Omega_{cell})\vec{R}$. This is a very important result. Also, this means that the total polarization is uncertain within a factor of $(e/\Omega_{cell})\vec{R}$.

The total change in polarization can be written as

$$\Delta\vec{P} = \Delta\vec{P}_{ion} + \Delta\vec{P}_{el} \tag{16.26}$$

where $\Delta\vec{P}_{el}$ is given by Equation 16.19. $\Delta\vec{P}_{ion}$ can be evaluated classically using the relation

$$\Delta\vec{P}_{ion} = \frac{e}{\Omega_{cell}}\sum_s Z_s^{ion}\,\vec{r}_s \tag{16.27}$$

where Z_s^{ion} is the net positive charge of the sth ion and $\vec{r}_s$ is the displacement of ions with respect to the $\lambda = 0$ structure. Ferroelectric materials have spontaneous polarization, that is, polarization in the absence of any applied electric field. This can be calculated from Equation 16.26, assuming that $\lambda = 0$

state corresponds to the centrosymmetric (paraelectric) structure and $\lambda = 1$ state corresponds to ferroelectric low-symmetry structure. In such calculation, it is understood that the material remains insulating in going from $\lambda = 0$ to $\lambda = 1$ state.

16.3 Born Effective Charge

The Born effective charge (BEC) tensor is a measure of the coupling between polarization and sublattice displacement in zero electric field. Let α, β define two directions (such as x, y) and i denotes an atom; then, the BEC tensor $Z^*_{i,\alpha\beta}$ of the ith atom is defined as

$$Z^*_{i,\alpha\beta} = \frac{\Omega_{cell}}{e}\frac{\partial P_\alpha}{\partial u_{i\beta}} \tag{16.28}$$

where P_α is the component of the polarization in the αth direction, $u_{i\beta}$ is the periodic displacement of the ith atom in the β direction, and Ω_{cell} is the unit cell volume. The definition of the BEC given in Equation 16.28 links the two vector quantities, namely, polarization and displacement. Thus, BEC is a second-rank tensor whose elements are restricted by the symmetry of the ionic site. For instance, cubic symmetry allows isotropic values of the elements of the BEC tensor with off-diagonal terms to be zero. Upon reduction in the site symmetry, off-diagonality appears and the principal elements are different from each other.

BEC can be easily calculated by using Equations 16.28 and 16.26 from first-principles. Alternatively, the polarization with respect to the paraelectric phase can be calculated if the BECs are known. Using Equation 16.28, it can be shown that (Exercise 16.5)

$$P_\alpha = \frac{e}{\Omega_{cell}}\sum_{i,\beta} Z^*_{i,\alpha\beta}\, u_{i\beta} \tag{16.29}$$

The BEC tensor plays an important role in ferroelectrics and multiferroics and is an important ingredient in the lattice dynamics of ferroelectric materials. Recent first-principles calculations show anomalous values of the elements of BEC tensors of the constituent ions of ferroelectric materials. Such anomaly is often related to the increasing covalent character of cation–anion bonds in these materials. For most perovskite ABO_3 structures, in contrast to the static ionic charges of 1 or 2 for A ions, 4 or 5 for B ions and –2 for O ions, recent first-principles calculations demonstrate almost two times an increase/decrease in BEC values that are explained in terms of strong hybridization between O 2*p* and transition metal 3*d*/4*d* states. BEC is also considered a fundamental quantity in lattice dynamics because it governs the amplitude of long-range Coulomb interaction between nuclei and the splitting between longitudinal and transverse optic phonon modes.

16.4 Ferroelectric Materials

A large number of compounds exhibit ferroelectric behavior. This includes $BaTiO_3$, $PbTiO_3$, $SrTiO_3$, $KNbO_3$, $LiNbO_3$, $Bi_4Ti_3O_{12}$, $PbZr_{1-x}Ti_xO_3$ (PZT), and so on (Rabe et al. 2007). Many of these materials have been studied by the first-principles electronic structure methods, which have greatly contributed to the understanding of ferroelectricity in these materials. Here, as an example, we shall briefly discuss some results pertaining to $BaTiO_3$ and $Bi_4Ti_3O_{12}$.

$BaTiO_3$ is a prototypical example of the perovskite compounds and has been studied extensively by several groups (Rabe et al. 2007). It has ideal perovskite structure in the paraelectric phase at high temperatures (above 393 K) as shown in Figure 16.1a, which is a cubic structure with a five atom basis with space group $Pm\bar{3}m$. Ba atoms are at the corners of the cube, Ti atoms are at the body-center, and O atoms are at the face centers of the cube. As the temperature is lowered, it transforms into a tetragonal structure (P4*mm*) as shown in Figure 16.1b and becomes ferroelectric with polarization in the [001] direction. As the temperature is lowered further, this phase remains stable until 278 K at which it transforms into a ferroelectric orthorhombic phase (A*mm*2). With further lowering of temperature, there is another phase transition at 183 K, when it goes into a rhombohedral phase (R3*m*).

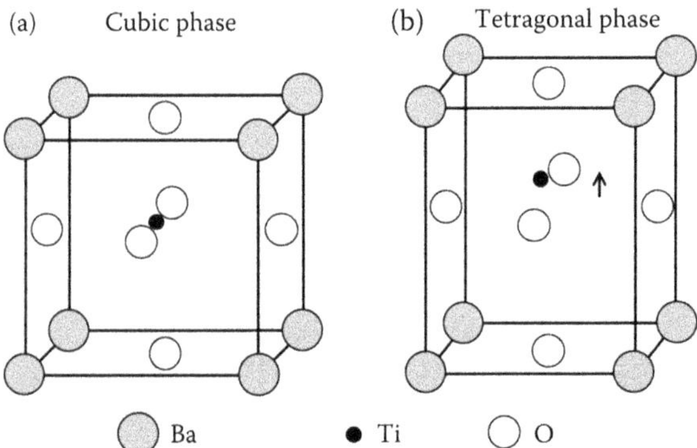

FIGURE 16.1
Schematic view of the crystal structure of perovskite $BaTiO_3$. (a) High temperature paraelectric cubic structure. (b) Ferroelectric tetragonal structure.

Several first-principles calculations have been reported for $BaTiO_3$ using various first-principles methods (e.g., Zhong et al. 1994, Ghosez et al. 1998, Filippetti and Spaldin 2003). All these calculations give the lattice constant of the cubic phase within 1% of the experimental value (4.0 Å). This is indeed a very good agreement. BECs for $BaTiO_3$ in cubic structure are given in Table 16.1 by various workers using different approaches and are compared with experimental values. Again, we see a good agreement between theory and experiment. We see from the table that the BECs for Ti and O are much larger compared to their nominal charges. This is because of the strong hybridization between O 2*p* and Ti 3*d* orbitals, leading to a strong covalent character of the bonding between the two atoms. This strong hybridization is what gives rise to ferroelectricity in this material. The spontaneous polarizations for tetragonal, orthorhombic, and rhombohedral phases of $BaTiO_3$ as reported by Zhong et al. (1994), using the first-principles pseudopotential method, are 0.30, 0.26, and 0.44 C/m^2, respectively, which are in good agreement with the corresponding experimental values of 0.27, 30, and 33 C/m^2. Thus, the first-principles calculations not only explain the experimental data on this material but also shed light on the origin of ferroelectricity.

Now, we shall discuss bismuth titanate, $Bi_4Ti_3O_{12}$ (BiT), which is a layered compound and belongs to the Aurivillius family of phases represented by a general formula $Bi_2A_{x-1}B_xO_{3x+3}$ or $(Bi_2O_2)^{2+}$ $(A_{x-1}B_xO_{3x+1})^{2-}$. Here, A is a mono-, di-, or trivalent cation such as Na^+, Sr^{2+}, Bi^{3+}, and so on, B is a transition metal ion such as Ti^{4+}, Ta^{5+}, Nb^{5+}, and so on, and x varies between 2 and 4. Its unit cell consists of alternate stacking of $(Bi_2O_2)^{2+}$ and a perovskite-like $(A_{x-1}B_xO_{3x+1})^{2-}$ layers arranged along the *c*-axis as shown in Figure 16.2. BiT undergoes a displacive type of ferroelectric to paraelectric transition across the Curie temperature at $T_c \sim 948$ K. The high-temperature paraelectric phase has a tetragonal crystal structure with space group I4/*mmm*. However, until recently, there has been considerable ambiguity regarding the crystal structure of the low-temperature ferroelectric phase since experimental studies reported both orthorhombic and monoclinic structures. But now, several first-principles calculations have shown that the low-temperature ferroelectric phase is B1*a*1 monoclinic structure (Shrinagar et al. 2008, Roy et al. 2010, Singh et al. 2010). The optimized values of lattice parameters for both

TABLE 16.1

BECs, Z^*, for $BaTiO_3$ Using Various Methods

Z^*_{Ba}	Z^*_{Ti}	$Z^*_{O\parallel}$	$Z^*_{O\perp}$	Z^*_{Ba}/Z_{Ba}	Z^*_{Ti}/Z_{Ti}	Method
2.77	7.25	−5.71	−2.15	1.39	1.81	DFT (LDA) (Ghosez et al. 1998)
2.75	7.16	−5.69	−2.11	1.38	1.79	DFT (LDA) (Zhong et al. 1994)
2.61	5.88	−4.43	−2.03	1.31	1.47	Pseudo-SIC (Filippetti and Spaldin 2003)
2.9	6.7	−4.8	−2.4	1.45	1.68	Experiment (Axe 1967)

Z denotes the nominal charge.

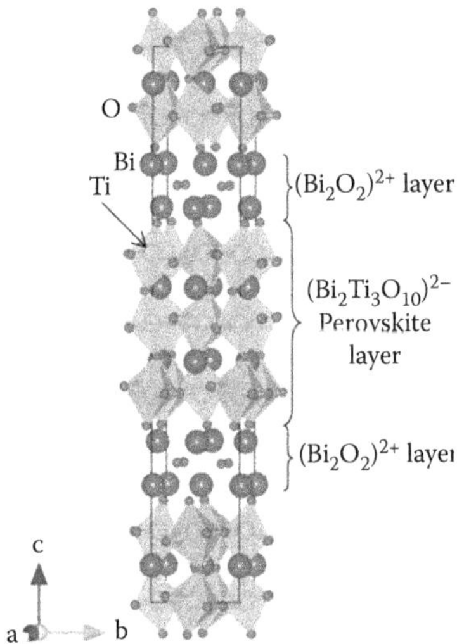

FIGURE 16.2
Unit cell of bismuth titanate.

structures are shown in Table 16.2 as obtained by Roy et al. (2010) using the VASP code and PAW potentials and are compared with experimental values. We see that the LDA underestimates the lattice parameters but GGA values are quite close to the experimental ones (Rae et al. 1990).

Figures 16.3 and 16.4 show the band structures of BiT in paraelectric and ferroelectric phases, respectively (Roy et al. 2010). These figures show that BiT remains insulator in both ferroelectric and paraelectric phases. It is seen that the material in its paraelectric phase possesses an indirect band gap of 1.53 eV along $M \to \Gamma$ and a direct band gap of 2.22 eV at Γ. The indirect and direct band gaps in the ferroelectric (B1*a*1) phase are 2.17 and 2.30 eV compared to the experimental values, 3.27 and 3.60 eV, respectively. We see that the calculated gaps are much smaller than the experimental gaps. This

TABLE 16.2
Structural Parameters of B1*a*1 and I4/*mmm* Phases

Lattice Parameter	GGA		LDA		Experiment (Rae et al. 1990)
	B1*a*1	**I4/*mmm***	**B1*a*1**	**I4/*mmm***	**B1*a*1**
a (Å)	5.4289	3.8082	5.3149	3.7356	5.450
b (Å)	5.4059	3.8082	5.2924	3.7356	5.4059
c (Å)	32.8762	32.8163	32.1860	32.1906	32.832

Source: Roy A. et al. 2010. *J. Phys.: Condens. Matter* 22: 165902.

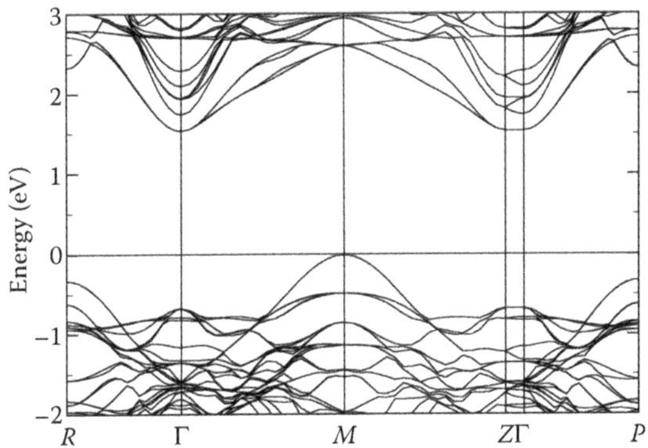

FIGURE 16.3
Band structure of paraelectric BiT. (From Roy A. et al. 2010. *J. Phys.: Condens. Matter* 22: 165902.)

is because of the use of LDA and GGA, which are known to underestimate the band gaps.

The calculation of PDOS showed that there is a strong hybridization between Ti 3*d* and O 2*p* states, which may be responsible for ferroelectricity in this material. Figure 16.5 shows the charge density contours in the (001) plane containing Ti and O atoms, indicating strong hybridization of Ti and O states, giving a strong covalent character to the Ti–O bond. One also sees that for the paraelectric phase, the charge density contours are quite symmetric, while for the ferroelectric phase, they become unsymmetrical and distorted, which is due to displacements of Ti and O ions. This asymmetry in charge density gives rise to spontaneous polarization in the ferroelectric phase.

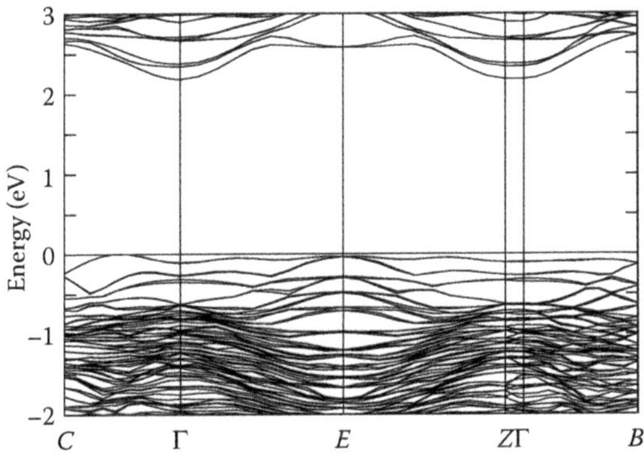

FIGURE 16.4
Band structure of ferroelectric BiT. (From Roy A. et al. 2010. *J. Phys.: Condens. Matter* 22: 165902.)

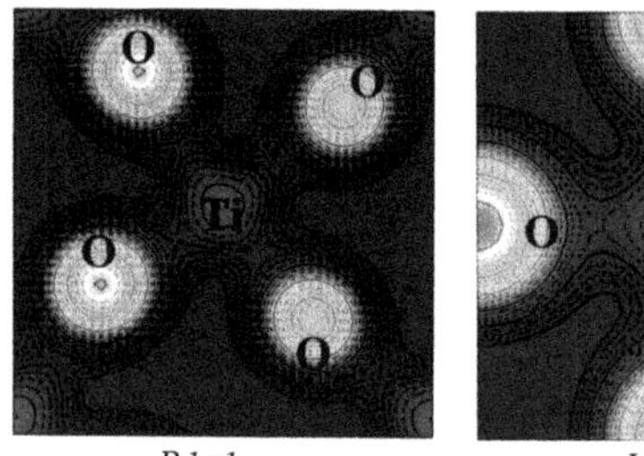

FIGURE 16.5
Charge density contours in (001) plane of ferroelectric (B1*a*1) and paraelectric BiT.

The BECs for various ions in ferroelectric phase using GGA are given in Table 16.3. For simplicity, we are showing only the diagonal part of the BEC tensor, although Roy et al. (2010) reported all elements of the tensor. We see that Ti and some O ions have very large values compared to their nominal charges of +3 and –2, respectively, which indicates strong covalent character of the bonding. Using the BECs, Roy et al. obtained the average value of spontaneous polarization as $55 \pm 13\ \mu$ C/cm^2, which is in good agreement with the experimental value of $50 \pm 10\ \mu$C/cm^2. Roy et al. found that most of the contribution to spontaneous polarization comes from the Ti and O ions and very little from the Bi ion. This is further clarified by looking at the displacements of the ions relative to the paraelectric structure. It was found that the Ti and O ions showed large displacements compared to the Bi ion and thus gave a large contribution to the spontaneous polarization.

TABLE 16.3
BECs for Ferroelectric BiT

	GGA		
Ion (Nominal Charge)	Z_{xx}	Z_{yy}	Z_{zz}
Bi1 (+3)	5.10	5.35	5.15
Bi2 (+3)	4.83	5.12	4.81
Ti1 (+4)	6.60	7.23	5.81
Ti2 (+4)	5.84	6.39	6.32
O1 (–2)	–3.77	–6.15	–2.22
O2 (–2)	–3.13	–3.14	–2.70
O3 (–2)	–2.93	–3.12	–4.58
O4 (–2)	–2.09	–2.33	–4.65
O5 (–2)	–3.47	–3.02	–2.37
O6 (–2)	–3.62	–4.30	–2.34

Source: Roy A. et al. 2010. *J. Phys.: Condens. Matter* 22: 165902.

16.5 Multiferroic Materials

As we mentioned in the beginning, we shall discuss only those multiferroic materials that are magnetic and ferroelectric. This field is a very active area of research and to date several such materials have been discovered, for example, $BiFeO_3$, $GaFeO_3$, $BiMnO_3$, $YMnO_3$, $HoMnO_3$, $TbMnO_3$, $Tb_2Mn_2O_5$, $LuFe_2O_4$, and so on. Most of these compounds are oxides containing a transition metal ion. This is because the covalent nature of transition metal 3*d* and O 2*p* bonds provides strong polarizability and an unfilled *d* shell of transition metal ion is needed for magnetism. The requirements of having covalent bond and an unfilled *d* shell are somewhat contradictory and that is the reason why there are relatively small number of multiferroic materials. Owing to the presence of an unfilled *d* shell on the transition metal ion, strong correlations play an important role in these materials, unlike simple ferroelectrics.

There have been many first-principles calculations on these materials, which have greatly contributed to their understanding. First-principles calculations on these materials are more difficult because of the presence of magnetism. Not only does one have to choose a large supercell, one also has to find a way of handling electron correlations that are quite important due to the presence of a transition metal ion.

First, we shall briefly discuss $BiFeO_3$ (BFO), which is the most studied multiferroic material to date (see, e.g., Neaton et al. 2005 and references therein, Ederer and Spaldin 2005). BFO at room temperature is simultaneously ferroelectric ($T_c \sim 1100$ K) and antiferromagnetic ($T_N \sim 650$ K). In the ferroelectric phase, it has a rhombohedrally distorted perovskite structure with space group *R*3*c* as shown in Figure 16.6. The primitive unit cell contains two formula units and the two O octahedra are rotated with respect to each other along the [111] directions. Bi, Fe, and O are displaced relative to each other along this axis. The largest displacements are for Bi relative to O. Above the Curie temperature, the structure changes to a high-symmetry cubic phase.

Initial measurements on bulk BFO showed a very small value of the spontaneous polarization (~6 $\mu C/cm^2$). However, more recent measurements on thin films have shown rather large polarization (~50–90 $\mu C/cm^2$). This has been explained by the first-principles calculations done by Neaton et al. (2005) who did LSDA and LSDA+U calculations using VASP code. Both LSDA and LSDA+U methods gave insulating ground state. The lattice parameters that they obtain for various values of U are very close and are insensitive to the choice of U and are in good agreement with the experimental values. However, the band gap depends on the choice of U. The BECs for Bi, Fe, and O ions in cubic phase are, respectively, 6.32, 4.55, and –3.06 and in *R*3*c* 4.37, 3.49, and –2.61 compared to their nominal charges 4, 3, and –2. We see that Bi has a large value of BEC compared to its nominal charge. If we compare it to the BEC of Ba in $BaTiO_3$ in Table 16.2, we see that Ba has a rather small value but Ti has a large value. Also, Bi in the ferroelectric phase has a larger

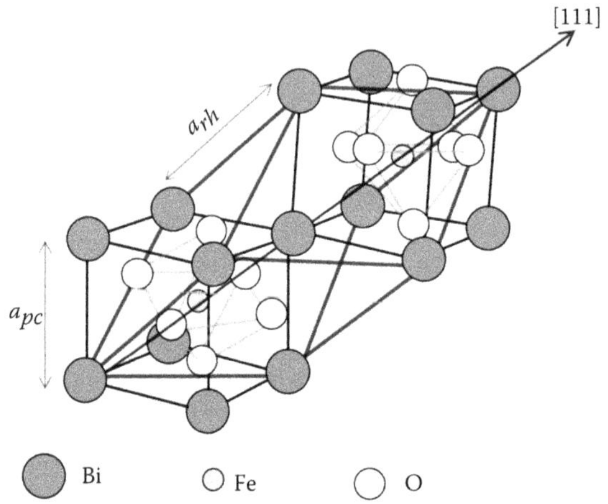

FIGURE 16.6
Schematic view of *R*3*c* structure built from two perovskite unit cells. Oxygen octahedra rotate with alternating sense around the [111] direction.

displacement from their positions in the paraelectric phase (cubic) compared to Fe. This indicates that in $BiFeO_3$, it is Bi that gives larger contribution to the spontaneous polarization, in contrast to Ba in $BaTiO_3$ or Bi in BiT. In BFO, Neaton et al. (2005) obtained a spontaneous polarization in the [111] direction of 123.1 $\mu C/cm^2$ when they used BECs of the cubic phase. However, they obtained a polarization of 87.3 $\mu C/cm^2$ when they used BECs of the *R*3*c* phase. These results are in good agreement with recent experiments. The coupling between ferroelectricity and magnetic order was explored by Ederer and Spaldin (2005) using the LSDA and LSDA+U approaches. They found that the weak ferromagnetism of Dzyaloshinskii–Moriya type exists in this material.

Gallium ferrite ($GaFeO_3$ or GFO) is another multiferroic material that has received considerable attention recently (Roy et al. 2011b, 2012a,b). It is piezoelectric and ferromagnetic with its magnetic transition temperature T_c close to room temperature (RT). Its T_c can be tuned close to or above RT depending upon the Ga:Fe ratio and processing conditions. GFO has an orthorhombic structure that is retained over a wide temperature range of 4–700 K. First-principles studies have shown that GFO is an A-type antiferromagnet at 0 K. Ga and Fe have almost similar ionic size, which makes it possible for Ga and Fe to interchange their sites resulting in substantial cation site disorder. This leads to an uneven distribution of magnetic moments and is believed to be the cause of ferrimagnetism in GFO with substantial magnetic moment below T_c.

Roy et al. (2011b, 2012a,b) have done extensive study of GFO using the first-principles PAW method. They used VASP code and the LSDA+U method with U = 5 eV and J = 1 eV for Fe d states. A plane wave energy cutoff of

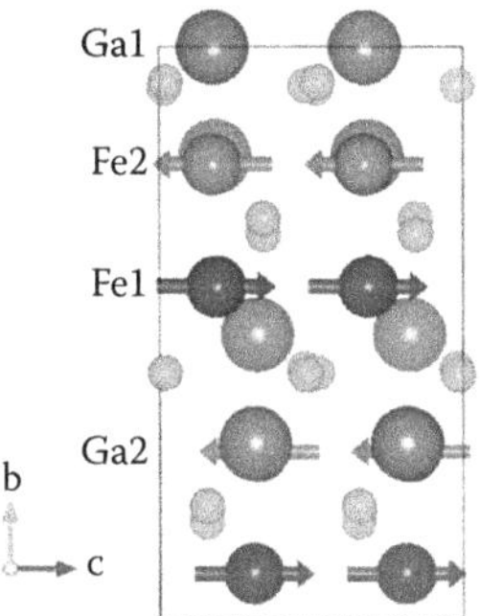

FIGURE 16.7
Schematic showing antiferromagnetic gallium ferrite. Arrows show the magnetic moment directions.

550 eV was used with Monkhorst–Pack 7 × 7 × 12 *k*-point mesh. Ground-state crystal structure was found to be orthorhombic $Pc2_1n$ symmetry with A-type antiferromagnetic configuration as shown in Figure 16.7. These calculations assumed stoichiometric GFO, that is, no partial occupancies of the constituent ions. Calculated ground-state lattice parameters were found to be $a = 8.6717$ Å, $b = 9.3027$ Å, and $c = 5.0403$ Å, which agree well with experiments. Calculated ionic positions, cation–oxygen, and cation–cation bond lengths also agreed well with experiments.

The electronic band structure of GFO using LSDA+U is shown in Figure 16.8 along the high-symmetry directions. Also shown in the figure is the total

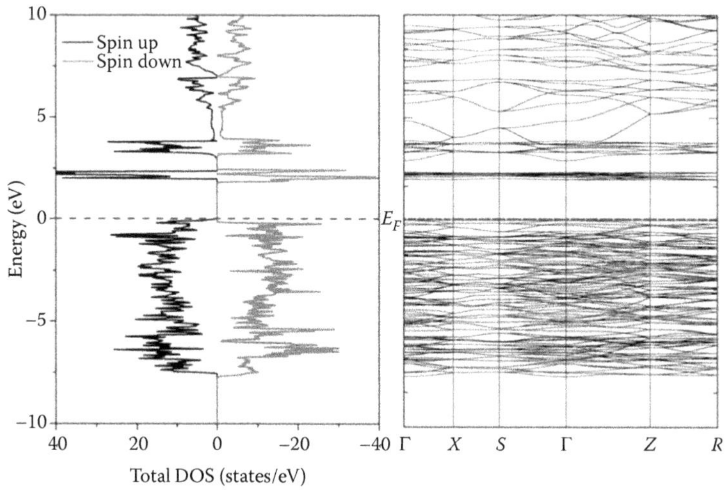

FIGURE 16.8
Density of states and band structure of gallium ferrite. E_F denotes the top of the valence band. (From Roy A. et al. 2011b. *J. Phys.: Condens. Matter*: 325902.)

density of states. Note that the highest occupied state is at zero energy. Thus, the figure clearly shows that GFO is an insulator with a direct band gap (E_g) of ~2.0 eV. However, experimental studies based on the optical absorption spectra of GFO reported a band gap of 2.7–3.0 eV. The difference between the calculated band gap and the experimental band gap is expected as the LSDA is known to underestimate the band gap.

Roy et al. (2011b) also calculated charge density and concluded that the bonds in GFO were largely ionic. The BECs for Fe, Ga, and O were not large compared to their nominal charges. This is because these bonds have mainly ionic character and very small covalent character. However, the spontaneous polarization was found to be ~58.63 μC cm^{-2} along the b direction, which was mainly attributed to the structural distortion. Roy et al. (2012a) explained the origin of ferromagnetism in GFO and attributed it to the cation site disorder. They further predicted the presence of magneto-structural coupling, which was observed experimentally. They also calculated optical properties and found that inclusion of site disorder improves the agreement between theory and experiment, thus confirming the presence of site disorder in the sample used in the experiment (Roy et al. 2012b).

EXERCISES

16.1 Prove the identity

$$\langle \psi_{\vec{k}m} | \vec{r} | \psi_{\vec{k}n} \rangle = -\frac{i\hbar}{m_e} \frac{\langle \psi_{\vec{k}m} | \vec{p} | \psi_{\vec{k}n} \rangle}{\varepsilon_{\vec{k}m} - \varepsilon_{\vec{k}n}}$$

16.2 Prove that

$$\langle \psi_{\vec{k}m} | \vec{p} | \psi_{\vec{k}n} \rangle = \frac{m_e}{\hbar} \langle u_{\vec{k}m} | [\vec{\nabla}_k, H_{\vec{k}}] | u_{\vec{k}n} \rangle$$

where $u_{\vec{k}n}$ is a Bloch function defined in Equation 16.10.

16.3 Show that

$$\langle \psi_{\vec{k}n} | V'_{KS} | \psi_{\vec{k}m} \rangle = \left\langle u_{\vec{k}n} \left| \left[\frac{\partial}{\partial \lambda}, H_{\vec{k}} \right] \right| u_{\vec{k}m} \right\rangle$$

16.4 Show that $\theta_{\vec{k}m}$ given by

$$\theta_{\vec{k}m} = \beta_{\vec{k}m} + \vec{k} \cdot \vec{R}_m$$

is periodic in $\vec{k}$-space, where $\beta_{\vec{k}m}$ is periodic in $\vec{k}$-space and $\vec{R}_m$ is a direct lattice vector.

16.5 Assuming polarization in the paraelectric phase to be zero, show that change in the αth component of the polarization P_α with respect to the paraelectric phase in the linear approximation can be expressed in terms of BEC as

$$P_\alpha = \frac{1}{\Omega_{cell}} \sum_{i,\beta} Z^*_{i,\alpha\beta}\, u_{i\beta}$$

16.6 Using one of the first-principles codes mentioned in Appendix A, calculate the electronic structure, BECs, and spontaneous polarization in the ferroelectric phase of $BaTiO_3$. Compare your results with those of Table 16.1.

16.7 Using one of the first-principles codes mentioned in Appendix A, calculate the electronic structure, ground-state structure, BECs, and spontaneous polarization in the multiferroic $BiFeO_3$. Compare your results with those of Neaton et al. (2005).

Further Reading

Martin R. M. 2004. *Electronic Structure, Basic Theory and Practical Methods*. Cambridge: Cambridge University Press.

Spaldin N. A., Cheong S. W., and Ramesh R. 2010. Multiferroics: Past, present and future. *Physics Today* 63(10): 38–43.

17

High-Temperature Superconductors

17.1 Introduction

Superconductivity was discovered by Kamerlingh Onnes in 1911 when he found that mercury loses its resistance below a certain temperature, known as transition temperature, T_c, which is 4.2 K for mercury. This generated a lot of excitement and there was an intense search for superconducting materials. Subsequently, many other superconductors were found such as Pb, Nb, NbC, Nb_3Ge, and so on. The technological applications of these materials were quickly recognized as these materials have no power dissipation as the electrical resistance is zero. However, T_c of these materials was very low and for useful applications in various devices, one would like T_c to be around room temperature. Later, many more superconducting materials were found but their T_c was rather low. Until 1980, the maximum T_c was found for Nb_3Ge, which was about 23 K. However, this changed in 1986, when Bednorz and Müller (1986) (Figure 17.1a and b) found superconductivity in the La–Ba–Cu–O system with T_c around 35 K. Soon after this discovery, many other systems such as YaBaCuO were discovered with much higher T_c. All these systems have layered structures with Cu–O planes and are called cuprates. At present, the highest T_c obtained in cuprates is about 135 K in mercury barium calcium copper oxide at ambient pressure, which reaches about 160 K at high pressure (Gao et al. 1994).

Recently, superconductivity has been observed in another class of materials known as pnictides, which have generated much interest. These are iron-based superconductors and have layers of Fe and a pnictogen such as arsenic or phosphorus. In this chapter, we shall focus on the electronic structure of some materials representing cuprates and pnictides only, although some materials such as MgB_2 are sometimes also included under high-T_c materials. Although many models for explaining the mechanism of superconductivity in these materials have been proposed, the correct mechanism of superconductivity is still elusive. Even the normal state of these materials is not well understood. In these materials, electronic correlations play an important role and therefore it is difficult to get a good description of the normal state using the one-electron approximation such as LDA or GGA. Nonetheless, the LDA

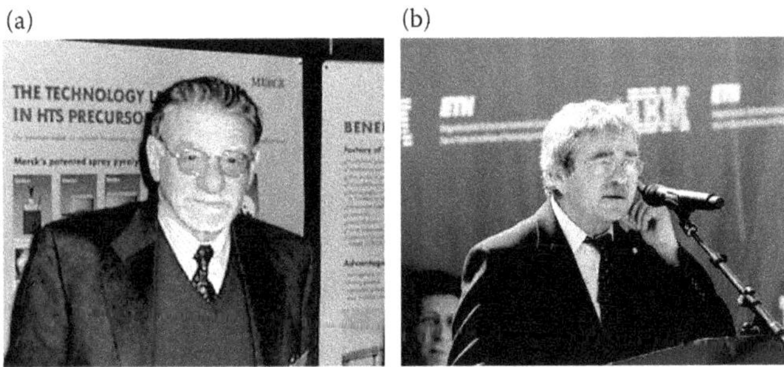

FIGURE 17.1
(a) Karl Müller (1927–) and (b) J. G. Bednorz (1950–) discovered high temperature superconductivity in La–Ba–Cu–O system, for which they were jointly awarded the Nobel Prize in 1987. (Muller photo courtesy of Armin Kübelbeck, Wikimedia Commons. Bednorz photo courtesy of Michael Lowry, Wikimedia Commons.)

electronic structure still provides a convenient starting point for understanding the normal state. Hopefully, if the normal state is well understood, it can pave the way for understanding the superconducting state in these materials.

17.2 Cuprates

Here, we shall discuss only La–Cu–O- and YbBaCuO-based systems (for a review, see, e.g., Pickett 1989). At low temperatures, pristine La_2CuO_4 is insulating and antiferromagnetic and has orthorhombic structure below 500 K. Only upon doping with elements such as Sr, the system becomes superconducting. At high temperatures, the system has tetragonal structure (bct).

The unit cell of pristine La_2CuO_4 is shown in Figure 17.2 for high-temperature tetragonal structure (*I4/mmm*). The lattice parameters are $a = 3.79$ Å and $c = 13.12$ Å. We see that the system has layered structure with Cu–O planes separated by La and O. The corresponding BZ is shown in Figure 17.3.

The band structure of paramagnetic tetragonal La_2CuO_4 has been calculated by many workers (see, e.g., Mattheiss 1987). We have calculated the band structure of paramagnetic tetragonal La_2CuO_4 using the LAPW method as implemented in WIEN2k code (Schwarz and Blaha 2003). The band structure is shown in Figure 17.4 along various symmetry lines in the BZs. The band structure that we calculated is in good agreement with that of Mattheiss (1987). Although the band structure in Figure 17.4 shows many bands, the band that plays the important role is the one that intersects the Fermi energy ε_F. This band arises from strong interaction between Cu 3*d* orbitals of $3d_{x^2-y^2}$ symmetry and O 2*p* orbitals. It is interesting to compare this band to a band arising from a 2D single-band cubic tight-binding model as discussed in Chapter 5. In this model, this band can be obtained by using

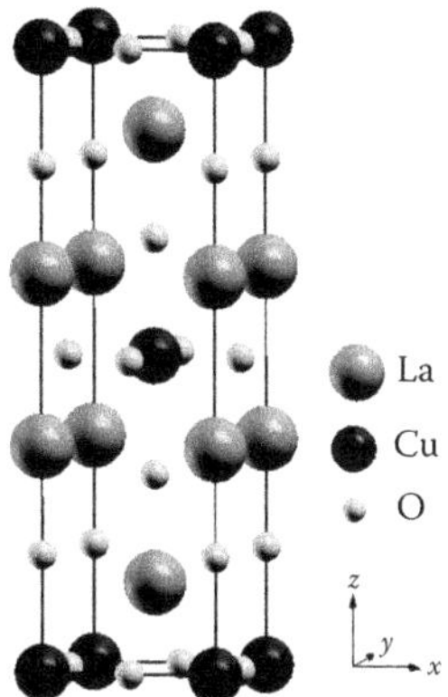

FIGURE 17.2
Unit cell of La_2CuO_4.

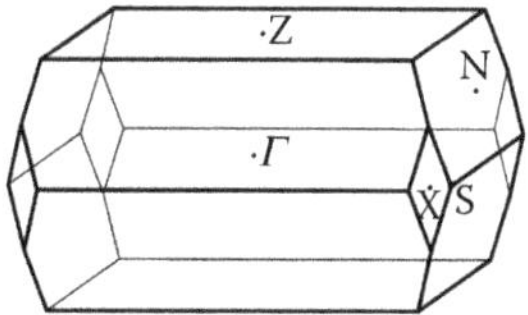

FIGURE 17.3
Brillouin zone of bct La_2CuO_4.

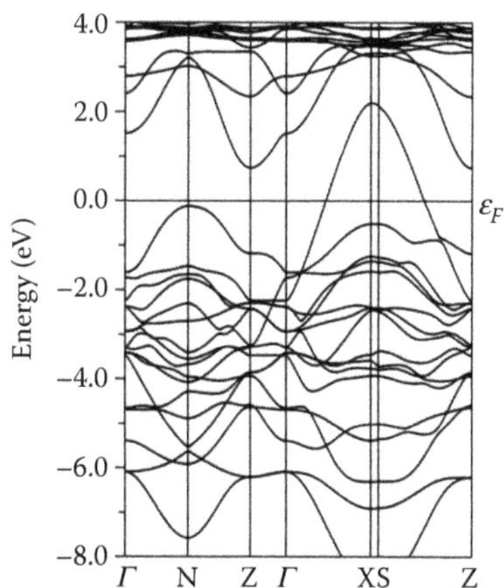

FIGURE 17.4
Band structure of La_2CuO_4 along the symmetry lines of the BZ obtained by using the WIEN2k code. (Schwarz K. and Blaha P. 2003. *Comput. Mater. Sci.* 28: 259–273.)

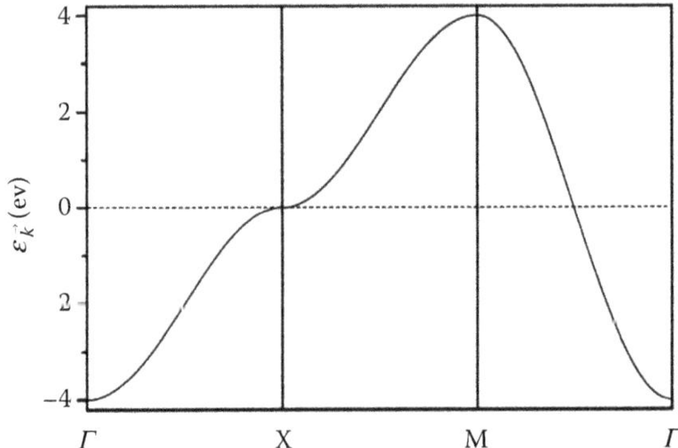

FIGURE 17.5
Tight-binding band structure for a square lattice along the symmetry lines of the BZ using Equation 17.1 assuming $t = 1$ eV.

$$\varepsilon_{\vec{k}} = -2t\,(\cos k_x a + \cos k_y a) \tag{17.1}$$

where t is the effective overlap matrix element between two neighboring Cu $3d_{x^2-y^2}$ orbitals (O $2p$ orbitals can be eliminated, see below). This band is plotted in Figure 17.5, which looks similar to the band intersecting ε_F in Figure 17.4.

For a half-filled band, which is the case, Equation 17.1 implies a square Fermi surface as shown in Figure 17.6. This bears some resemblance to the LAPW Fermi surface of La_2CuO_4 as shown in Figure 17.7. Thus, a 2D tight-binding model approximately describes the band intersecting ε_F in Figure 17.4, the differences mainly coming from second and third nearest-neighbor overlap integrals.

The LDA band structure shows that the system is metallic although La_2CuO_4 is insulating. This is because in these materials, electron correlations are very important. This can be easily seen by using a two-band Hubbard model (Zhang and Rice 1988). We have seen that the band intersecting ε_F in Figure 17.4 arises due to the interaction of Cu $3d$ electrons of $3d_{x^2-y^2}$ symmetry and O $2p$ electrons. Thus, we focus on these electrons, which are responsible for low-energy physics. In Figure 17.8 we show $d_{x^2-y^2}$ and $2p_x$, $2p_y$ orbitals,

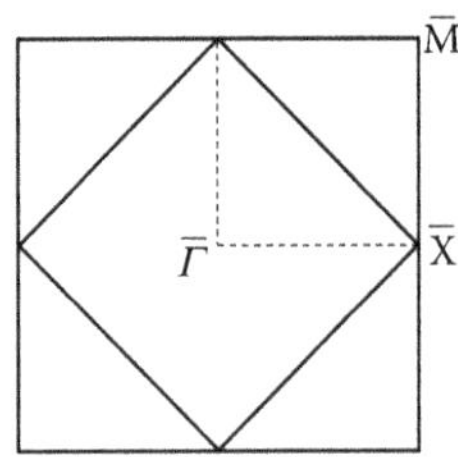

FIGURE 17.6
Fermi surface of half-filled tight-binding model of square lattice.

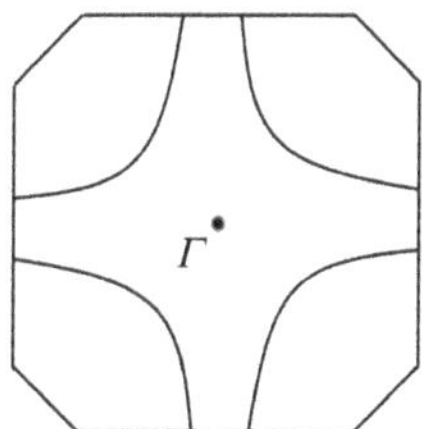

FIGURE 17.7
Fermi surface of La_2CuO_4 in the central plane of the BZ.

which are centered on a square lattice made of Cu and O atoms. We see that the lobes of $d_{x^2-y^2}$ orbitals are directed in the x or y direction and overlap with O p_x or p_y orbitals, respectively. Thus, in x and y directions, the hopping element between $d_{x^2-y^2}$ and p orbitals is the same. Let ε_d and ε_p denote the atomic energies of these orbitals and U the onsite Coulomb interaction between the d electrons. The Hamiltonian for the system can be written as

$$H = \sum_{i,\sigma} \varepsilon_d d_{i\sigma}^{\dagger} d_{i\sigma} + \sum_{l,\sigma} \varepsilon_p p_{l\sigma}^{\dagger} p_{l\sigma} + \mathrm{U} \sum_{i,\sigma} d_{i\uparrow}^{\dagger} d_{i\uparrow} d_{i\downarrow}^{\dagger} d_{i\downarrow} + \sum_{i,l} (v_{il} d_{i\sigma}^{\dagger} p_{l\sigma} + h.c.) \quad (17.2)$$

where v_{il} denotes hybridization between Cu 3d electrons at site i and O 2p electrons on site l and $h.c.$ indicate the Hermitian conjugate. Only the nearest-neighbor interaction is assumed. Thus, l sum in the last term of Equation 17.2 is restricted to the nearest O neighbors of Cu at site i. In Equation 17.2, vacuum is defined as filled Cu $3d^{10}$ and filled O p^6 orbitals. Thus, $d_{i\sigma}^{\dagger}$, $d_{i\sigma}$ are the creation and annihilation operators for d holes at site i and similarly, $p_{l\sigma}^{\dagger}$, $p_{l\sigma}$ are the creation and annihilation operators for p holes at site l. For simplicity, we assume $\varepsilon_d = 0$ and $\varepsilon_p > 0$. It can be shown that for large U and small hybridization, the

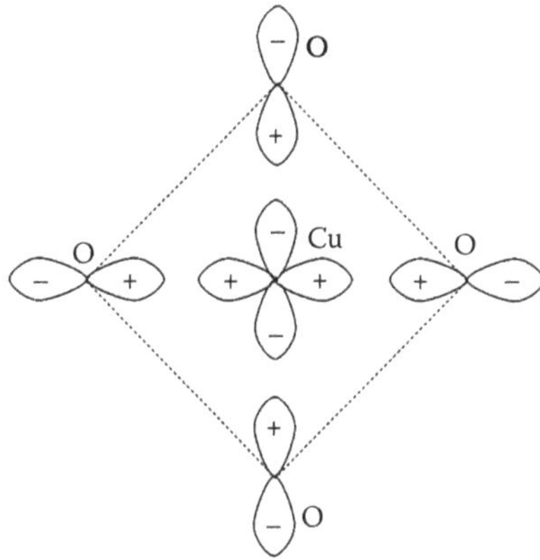

FIGURE 17.8
Schematic showing $d_{x^2-y^2}$ and $2p_x$, $2p_y$ orbitals centered on Cu and O atoms.

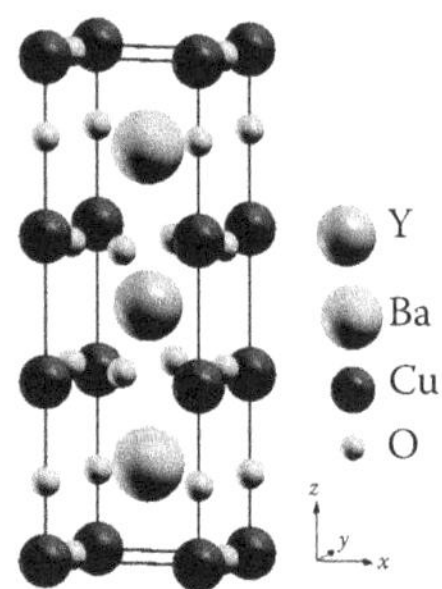

FIGURE 17.9
Unit cell of orthorhombic $YBa_2Cu_3O_7$.

Hamiltonian (17.2) gives insulating and antiferromagnetic ground state for pristine La_2CuO_4. This is because the virtual hopping between Cu 3*d* and O 2*p* levels creates antiferromagnetic superexchange between holes at neighboring Cu sites (Zhang and Rice 1988). Zhang and Rice further showed that O 2*p* orbitals can be eliminated from the picture and the model (Equation 17.2) can be reduced to an effective single-band model for describing low-energy physics. This provides the justification for describing the band intersecting ε_F in Figure 17.4, by a single tight-binding band (Equation 17.1). The work of Zhang and Rice clearly shows that electron correlations play a very important role in cuprates. The electron correlations in band structure calculations can be taken into account approximately by schemes such as LSDA+U. Indeed, LSDA+U calculations show that the system comes out to be insulating and antiferromagnetic (see, e.g., Wei and Qi 1994).

Now, we shall briefly discuss the electronic structure of $YBa_2Cu_3O_7$ (YBCO), which has T_c about 90 K and was the first material to achieve superconductivity above the boiling point of nitrogen. The superconducting phase has orthorhombic structure *Pmmm* as shown in Figure 17.9. The primitive cell consists of one formula unit, that is, it has 1 Y, 2 Ba, 3 Cu, and 7 O. As shown in the figure, Y and Ba ions are along the *z* axis. A very interesting feature of the structure is that Cu and O order either on a plane or along a chain as shown. The BZ of the structure is orthorhombic and is shown in Figure 17.10 with the symmetry points.

The band structure of YBCO has been calculated by several groups. In Figure 17.11, we show the band structure calculated by Bansil et al. (1988),

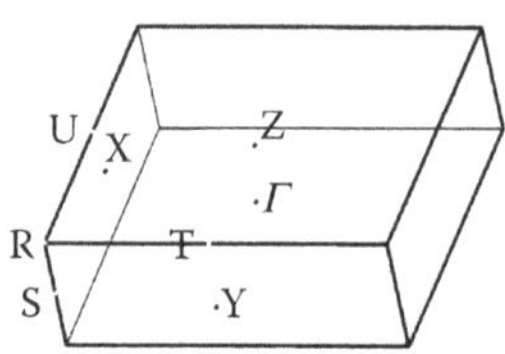

FIGURE 17.10
Brillouin zone of orthorhombic $YBa_2Cu_3O_7$.

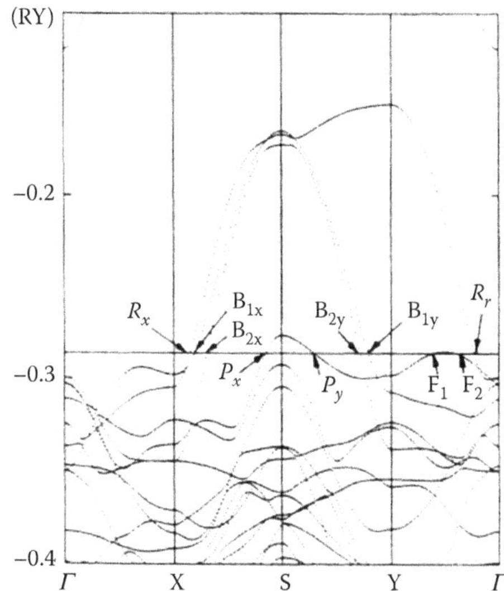

FIGURE 17.11
Band structure of $YBa_2Cu_3O_7$ from Bansil et al. (1988). The Fermi energy is shown by the horizontal straight line.

which is in good agreement with LAPW calculations (Yu et al. 1987, Krakauer et al. 1988). The Fermi energy intersects four bands at various points, which are marked in the figure.

The band marked by R_x and $R_Γ$ is due to Cu–O chains while the nearly degenerate bands marked by B_{1x}, B_{2x}, B_{1y}, B_{2y} arise due to Cu–O planes.

The Fermi surface of YBCO has been calculated by several groups (e.g., Yu et al. 1987, Krakauer et al. 1988). In Figure 17.12, we show schematically the Fermi surface cross sections with $k_z = 0.0$ plane. The Fermi surface consists of four sheets: (i) a sheet shown in the top left panel associated with the Cu–O chain band, which is completely occupied along $Γ$X but is almost empty along $Γ$Y. This gives an electron "ridge" running along $Γ$X (shown by R_x and $R_Γ$ in Figure 17.11; (ii) a pair of closely placed S-centered hole sheets shown in the bottom left and top right panels arising primarily from the states in the Cu–O planes, which has the shape of inner and outer surfaces of a barrel (marked B_{1x}, B_{2x}, B_{1y}, B_{2y} in Figure 17.11); (iii) a hole sheet shown in the bottom right panel related to the band, which straddles the Fermi energy and gives a hole "pillbox" around S (marked P_x, P_y in Figure 17.11); and (iv) a hole "butterfly" along $Γ$y (marked F_1, F_2 in Figure 17.11).

Fermi surface of YBCO has been observed in various 2D ACAR experimental studies using positron. Here, we discuss the results of Bansil et al. (1988) who calculated electron–positron momentum density $N_{2r}(\vec{p})$, which is measured in the experiment. Its derivative with momentum dN_{2r}/dp is expected to show a jump at the Fermi surface crossing. In Figure 17.13,

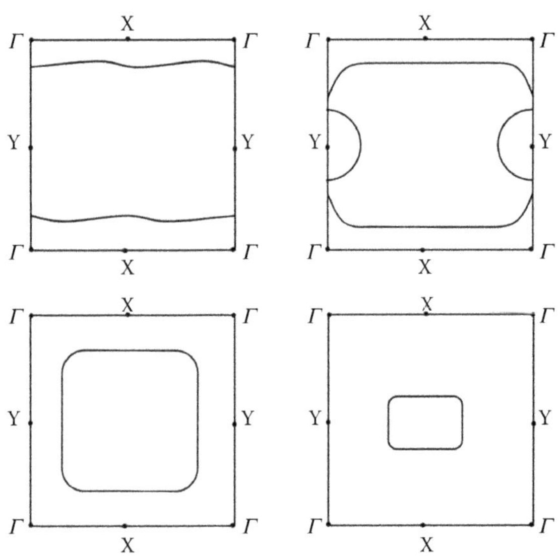

FIGURE 17.12
Schematic Fermi surface of $YBa_2Cu_3O_7$ in $k_z = 0$ plane of the BZ.

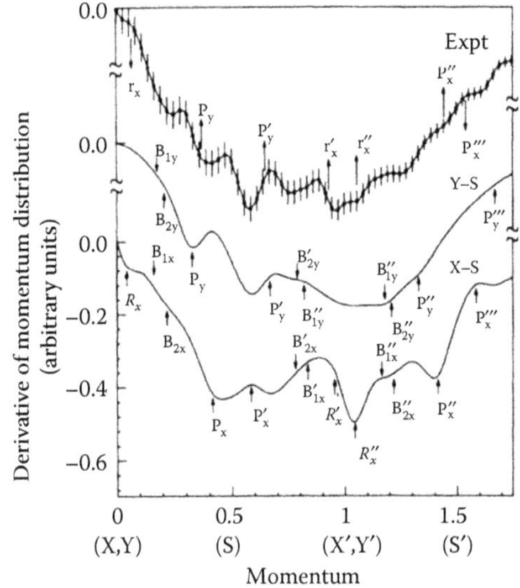

FIGURE 17.13
Derivative of the electron–positron momentum distribution, dN_{2r}/dp, from Bansil et al. (1988). The top curve is the experimental curve while the bottom two curves are theoretical curves (see text).

we show the comparison of experimental and theoretical results for dN_{2r}/dp. Since the experiment was done on twinned crystals, the experimental result along XS would correspond to the theoretical results along XS as well as YS directions. Therefore, there are two theoretical curves corresponding to one experimental curve in Figure 17.13. We see that there is a good agreement between theory and experiment. Many features in the curve have been marked such as P_x, P_y, and so on. We see that some features are clearly seen in the experiment, particularly the features coming from the chains.

17.3 Iron-Based Superconductors

Superconductivity was first reported in an iron-based compound LaOFeP in 2006 with $T_c = 3.5$ K by Kamihara et al. (2006). By replacing P with As and some oxygen with fluorine, T_c could be raised up to 26 K. By substituting La by other rare earth compounds, much higher T_c was obtained and in $SmO_{1-x}F_xFeAs$, it reached 55 K. These compounds show some remarkable similarities with cuprates. Similar to cuprates, in pristine form ($x = 0$), these compounds are antiferromagnetic insulators and only on doping do they show superconductivity. They also have layered structure as shown in Figure 17.14. In these systems, FeAs layers are separated by LaO layers, similar to cuprates where CuO_2 layers are separated by La or Y–Ba layers.

We shall briefly discuss the electronic structure of LaOFeAs, which was first calculated by Singh and Du (2008) by using the LAPW method, which we have reproduced using the WIEN2k code (Schwarz and Blaha 2003). LaOFeAs has a tetragonal structure as shown in Figure 17.14. The lattice parameters used were experimental lattice parameters $a = 4.0355$ Å and $c = 8.7393$ Å. The internal coordinates of La and As atoms were determined by energy minimization

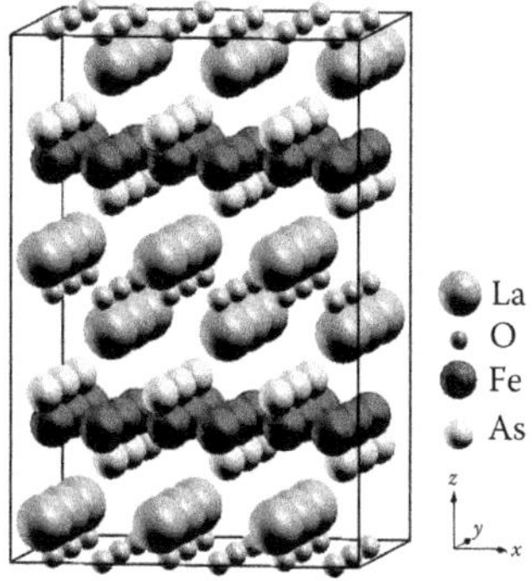

FIGURE 17.14
Crystal structure of LaOFeAs.

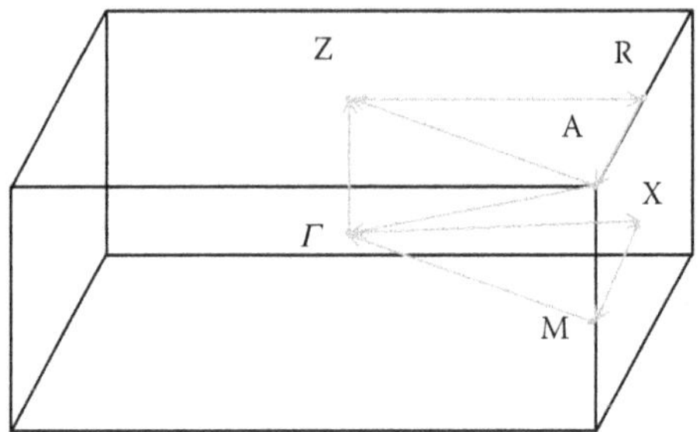

FIGURE 17.15
Brillouin zone of LaOFeAs.

yielding $z_{La} = 0.1418$ and $z_{As} = 0.6926$. The BZ of LaOFeAs is tetragonal and is shown in Figure 17.15 along with various symmetry points. In Figures 17.16 and 17.17, we show the density of states (DOS) and band structure of LaOFeAs. The contribution to the DOS between −5.5 and −2.1 eV mainly comes from O *p* and As *p* states. The As *p* contribution is mainly concentrated above −3.2 eV. The Fe *d* states-derived bands contribute to the DOS between −2.2 and 2.2 eV with La-derived bands contributing at higher energies.

The Fermi surface of LaOFeAs is shown in Figure 17.18, which comprises of five sheets: (i) two electron cylinders around the zone edge around M; (ii) two-hole cylinder around the zone center Γ; and (iii) a hole pocket centered at Z. The hole pocket centered at Z is derived from Fe $d_z{}^2$ states, which hybridize with As *p* and La orbitals. All sheets are of two-dimensional character except the hole pocket centered at Z, which is three dimensional. The electron cylinders are derived from in-plane Fe *d* orbitals.

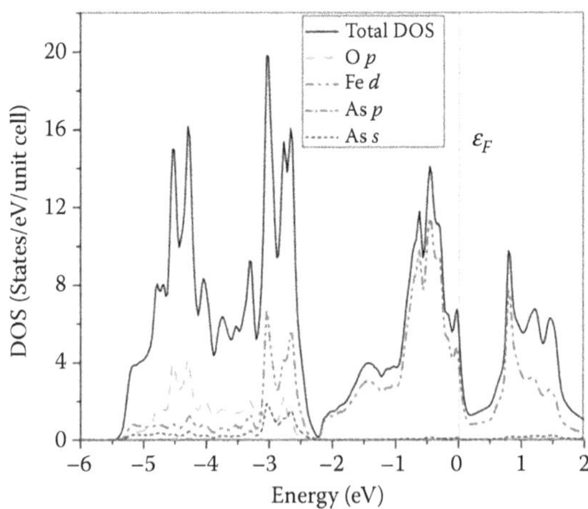

FIGURE 17.16
DOS of LaOFeAs obtained by using WIEN2k code. (From Schwarz K. and Blaha P. 2003. *Comput. Mater. Sci.* 28: 259–273.)

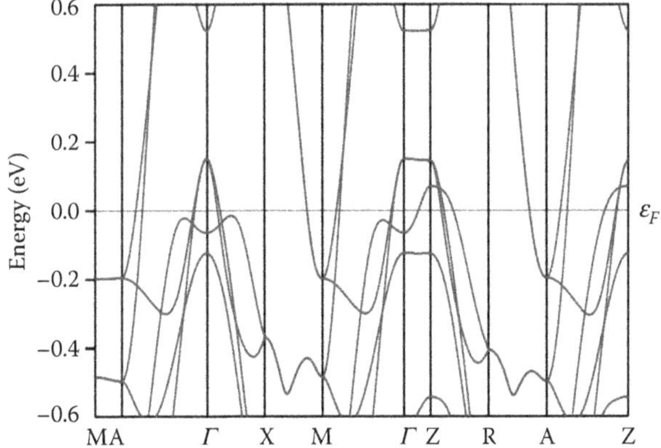

FIGURE 17.17
Band structure of LaOFeAs obtained by using WIEN2k code. (From Schwarz K. and Blaha P. 2003. *Comput. Mater. Sci.* 28: 259–273.)

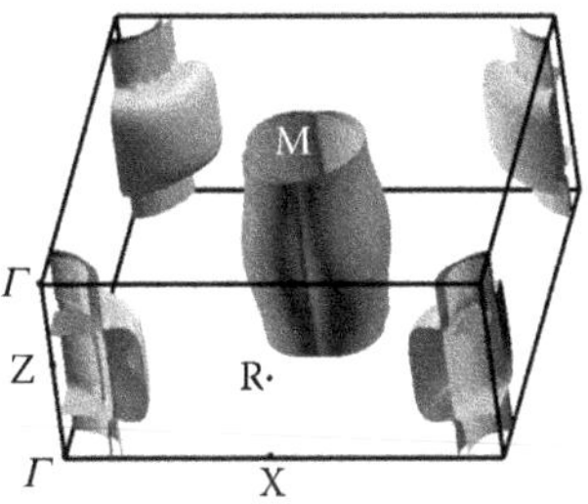

FIGURE 17.18
Fermi surface of LaOFeAs.

EXERCISES

17.1 Calculate the band structure of the tight-binding model of a square lattice given by Equation 17.1 along the symmetry lines of the BZ. Compare your results with Figure 17.5.

17.2 Show that the tight-binding band structure given by Equation 17.1 gives a square Fermi surface for half-filling of the band.

17.3 Using one of the codes mentioned in Appendix A, calculate the band structure of La_2CuO_4 and compare it with Figure 17.4. Also calculate the Fermi surface and compare it with Figure 17.7.

17.4 Calculate the band structure of La_2CuO_4 using LSDA+U method and show that it gives insulating and antiferromagnetic ground state.

17.5 Using one of the codes mentioned in Appendix A, calculate the band structure of $YBa_2Cu_3O_7$ and compare it with Figure 17.11.

Also calculate the Fermi surface and show that it is similar to Figure 17.12.

17.6 Using one of the codes mentioned in Appendix A, calculate the band structure of LaOFeAs and compare it with Figure 17.17. Also calculate the Fermi surface and compare it with Figure 17.18.

17.7 Use the two-band tight-binding model of Raghu et al. (2008) to fit the Fermi surface of LaOFeAs obtained in Exercise 17.6. Calculate the band structure using the fitted parameters and compare it with Figure 17.17.

Further Reading

Izyumov Y. and Kurmaev K. 2010. *High-Tc Superconductors Based on FeAs Compounds*. Berlin: Springer.

Poole C. P. Jr., Farach H. A., Creswick R. J., and Prozorov R. 2007. *Superconductivity*. Amsterdam: Elsevier.

18

Spintronic Materials

18.1 Introduction

The word "spintronics" is made from two words, spin and electronics and refers to the emerging field of spin-polarized transport in solids. It is based on the control and manipulation of an electron's spin leading to novel spintronic devices such as spin-field transistors, spin storage/memory devices, or spin quantum computers. Since the electron's spin can be switched from one state to another much faster than charge can move around a circuit, spintronic devices are expected to operate much faster. Furthermore, such devices are expected to produce less heat than conventional electronic devices. Also spintronic devices are expected to be of much smaller size compared to conventional microelectronic devices leading to further miniaturization.

The field of spintronics essentially began in 1988 when Albert Fert and Peter Grünberg independently discovered giant magnetoresistance (GMR) in Fe/Cr magnetic multilayers (Baibich et al. 1988, Binasch et al. 1989). For this discovery, Fert and Grünberg were awarded the Nobel Prize in Physics in 2007. They found that the electrical resistance of multilayers consisting of thin ferromagnetic (FM) layers separated by thin nonmagnetic layers can be altered significantly by applying a magnetic field. The GMR effect is already being used in the present magnetic read heads in computer hard discs. There is another similar effect known as tunnel magneto-resistance (TMR), which arises when two FM layers are separated by a thin layer of insulating material. In TMR, the spin-polarized electrons tunnel quantum mechanically through the barrier layer.

Most current electronic devices are based on semiconductors, in particular, on silicon and gallium arsenide. Thus, the challenge is to develop spintronic devices around these materials. For this to happen, one needs to find a way to inject, manipulate, and detect the spin of the electron in semiconductors. FM metals could be used for spin injection but are often incompatible with existing semiconductor technology. For this purpose, dilute magnetic semiconductors (DMS) have received a lot of attention. These are dilute alloys where a small fraction of constituent semiconducting atoms has been replaced by magnetic transition metal atoms such as Mn. These alloys are semiconducting but can be magnetic and may be used to inject electrons with well-defined spin in adjacent nonmagnetic semiconductors.

In a spintronic device, one also needs a source of spin-polarized electrons in addition to a DMS material that forms an interface between magnetic and nonmagnetic semiconductors. An FM electrode can serve as a source of spin-polarized electrons. For this purpose, one needs a material with a high value of spin polarization. Therefore, half-metallic ferromagnets (HMF) that have a spin polarization of 100% have attracted a lot of attention. Examples of HMFs are CrO_2, $La_{0.7}Sr_{0.3}MnO_3$, Sr_2FeReO_6, Fe_3O_4, and some Heusler alloys.

The electronic structure calculations have played an important role in search of suitable spintronic materials. In this chapter, we shall discuss some electronic structure calculations of some such materials. In Section 18.2, we discuss the electronic structure of magnetic multilayers, Section 18.3 discusses half-metallic ferromagnets, and Section 18.4 explains dilute magnetic semiconductors.

18.2 Magnetic Multilayers

Magnetic multilayers are artificially designed materials that have several thin layers (of the order of few nanometers) of magnetic material like Fe separated by thin layers of nonmagnetic material in a periodic arrangement as shown in Figure 18.1. These materials show important phenomena like oscillatory interlayer exchange coupling and GMR and have found important applications in magnetic sensors, read heads, magnetic storage media such as computer discs and random access memory.

Two FM layers separated by a nonmagnetic layer can interact magnetically, and this interaction is called interlayers exchange coupling (IEC) and will be denoted by J (Grünberg et al. 1986). It was found by Parkin et al. (1990) that J shows damped oscillatory behavior with thickness d of the nonmagnetic layers as shown schematically in Figure 18.2. This means that two FM layers separated by a nonmagnetic layer can align ferromagnetically or

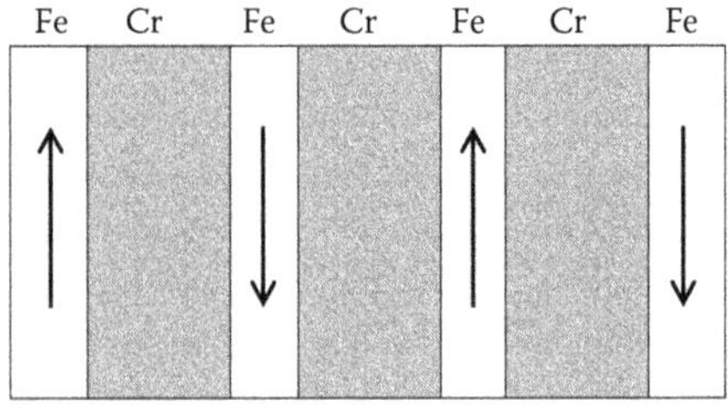

FIGURE 18.1
Schematic diagram of Fe/Cr magnetic multilayer.

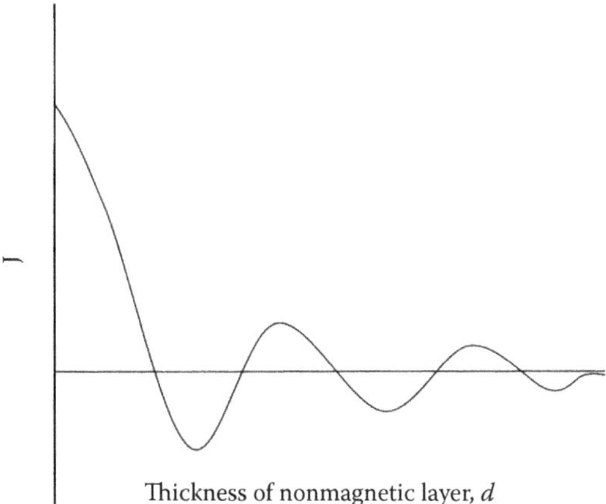

FIGURE 18.2
Schematic plot of J as a function of *d*.

antiferromagnetically depending on the thickness *d* of the nonmagnetic layers. This effect has been observed for many multilayer structures such as Fe/Cr, Co/Ru, Co/Cr, and so on. This is an important result that generated much interest in these materials.

Another remarkable phenomenon in magnetic multilayers is the large change in the electrical resistance due to the application of magnetic field (Baibich et al. 1988, Binasch et al. 1989) and is known as giant magnetoresistance (GMR). This is shown schematically in Figure 18.3, which shows a big drop in resistance of Fe/Cr multilayers due to the application of magnetic field. In the absence of magnetic field, the magnetic layers are aligned antiferromagnetically. When sufficient magnetic field is applied, the magnetic layers get ferromagnetically aligned resulting in a large drop in resistance. GMR was observed in 1988 by Grünberg and independently by Fert in Fe/Cr

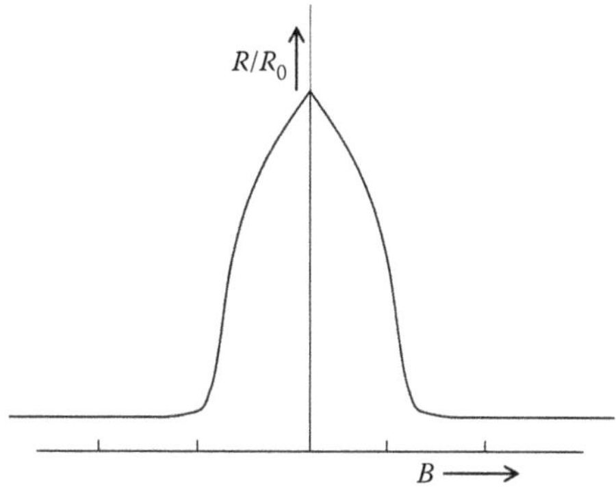

FIGURE 18.3
Schematic plot of resistance *R* as a function of magnetic field *B*.

multilayers for which they were awarded the Nobel Prize in 2007. The application potential of the GMR was immediately recognized and as a result an intense search for materials showing GMR or IEC was started.

The GMR is defined as

$$GMR = (R_{AP} - R_P)/R_P \tag{18.1}$$

where R_{AP} and R_P are resistances in antiparallel and parallel configurations of the magnetic layers. Thus for GMR one needs a large difference in R_{AP} and R_P, which, for a multilayer, can be understood based on spin-dependent scattering. According to Mott, the total conductivity in the presence of the magnetic field can be expressed as $\sigma = \sigma_\uparrow + \sigma_\downarrow$ where $\sigma_\uparrow$ and $\sigma_\downarrow$ represent the conductivity of spin $\uparrow$ and spin $\downarrow$ electrons (see, e.g., Mott 1964). We know that electrical resistivity $(1/\sigma)$ is proportional to electron scattering rates. In magnetic multilayers, the scattering rates for $\uparrow$ and $\downarrow$ spin electrons could be quite different as can be seen from Figure 18.4. In the left panel, we show two magnetic layers separated by a nonmagnetic layer when both of them have magnetization in the same direction. In the figure, M denotes the magnetic layers and NM denotes the nonmagnetic layer. For simplicity, we assume that only those electrons whose spins are antiparallel to the local magnetization at the interface are scattered. Thus, a spin $\uparrow$ electron, whose spin is parallel to the magnetization M, will flow rather freely but spin $\downarrow$ electrons will suffer scattering at the interfaces. Thus in parallel configuration the resistance for the spin $\uparrow$ electrons will be much less compared to spin $\downarrow$ electrons. In the antiparallel configuration as shown in the right panel of Figure 18.4, both spin $\uparrow$ and spin $\downarrow$ electrons will suffer scattering at the interfaces. In the parallel configuration, the spin $\uparrow$ electrons essentially short circuit the current, therefore, the resistance in the antiparallel configuration, R_{AP}, will be much larger compared to that of the parallel configuration, R_P, leading to GMR.

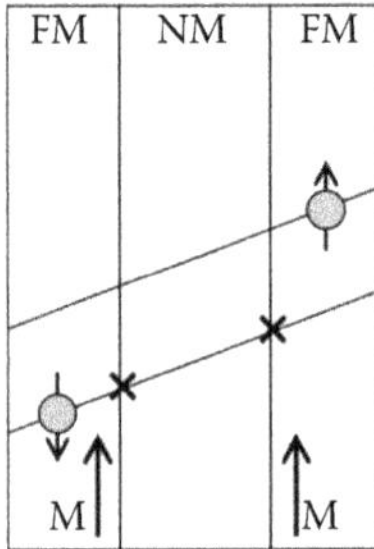

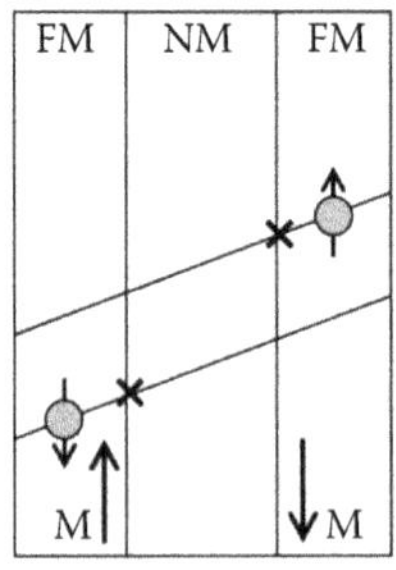

FIGURE 18.4
Schematic showing spin-dependent scattering. The left panel corresponds to the case when the magnetization in the magnetic layers is in the same direction and the right panel, when the magnetization is in the opposite directions. A dot with an arrow denotes an electron and the arrow denotes its spin. A cross denotes that the electron suffers scattering at the interface.

The interlayer exchange coupling at absolute zero of temperature can be defined as (see, e.g., Shukla and Prasad 2004)

$$J = E_{AP} - E_P \tag{18.2}$$

where E_{AP} and E_P are total energies per unit cell of antiparallel and parallel configurations of the magnetic layers. For understanding IEC, various models have been proposed such as the Ruderman–Kittel–Kasuya–Yosida (RKKY) model and the quantum well (QW) model (see, e.g., van Schilfgaarde and Harrison, 1993). In the RKKY model, IEC is similar to the indirect exchange interaction between two magnetic impurities placed in a nonmagnetic metal (see, e.g., Yafet 1987). The first impurity polarizes the surrounding electron gas, which then interacts with the other impurity producing the RKKY interaction. This interaction decays as cos $k_F r/r^3$ where r is the separation between the two impurities and k_F is the Fermi wave vector and thus shows a damped oscillatory nature as a function of r. A similar coupling mechanism is possible between two magnetic layers separated by a nonmagnetic layer of thickness d and has been shown to show damped oscillations as a function of d. In another model, known as the quantum well (QW) model, the coupling arises due to the spin-dependent confinement of electrons inside the nonmagnetic layer (see, e.g., Edwards et al. 1991, van Schilfgaarde and Harrison, 1993). A QW can be formed in the nonmagnetic layer because of potential jumps at the interfaces. These QWs will have different depths for up and down spin electrons as the band structure of the FM layer is spin dependent. The confinement of electrons inside the nonmagnetic layer leads to discrete energy levels, which depends on the thickness d of the nonmagnetic layer. By increasing d, these levels can move below the Fermi energy and get populated. Thus, the total energy of the system depends on d. It is possible that for a certain d the energy of antiparallel configuration is lower compared to the parallel configuration. From Equation 18.2, this will lead to antiferromagnetic (AFM) coupling between the two FM layers. Similarly, for some other d, the energy of the parallel configuration may be lower, giving rise to the FM coupling between the layers. Thus, the magnetic alignment of the magnetic layers depends on d, as a result, IEC oscillates with thickness d of the nonmagnetic layers.

First-principles calculations could be very useful for searching magnetic multilayers which show GMR and IEC. Also such calculations could be used to explore fundamental issues such as mechanism of GMR and IEC. Here, we shall briefly discuss the results obtained by first-principles calculations on Fe/Cr and Fe/Nb multilayers. The Fe/Cr system has been studied by a number of researchers (e.g., Levy et al. 1990, Herman et al. 1991, Xu and Freeman 1993, van Schilfgaarde and Harrison 1993). Levy et al. (1990) studied the electronic structure, IEC and GMR of Fe_m/Cr_n ($m = 3, 4$ and $n = 3, 4, 5$) multilayers stacked in the (001) direction. They used supercell geometry and calculated the total energy difference between parallel configuration (i.e., FM) and antiparallel configuration (AFM) as a function of thickness of a nonmagnetic layer. From these energy differences, they found that the iron layers had AFM coupling except for $m = 3, n = 3$. Using the LMTO method, Herman et al. (1991) found an alternating sign change in the total energy difference between FM and AFM configuration as a function of thickness of Cr layers. Using the self-consistent LMTO method with combined correction term, Xu and Freeman (1993) studied the electronic structure of Fe_m/Cr_n layers in the (001) stacking direction. They found that for systems containing three or more layers, FM ordering dominates over AFM ordering, but for systems containing a single Fe layer the AFM interaction may slightly exceed the FM interactions.

Now, we shall discuss one such example, where first-principles calculation has been used to explore the mechanism of IEC in a Fe/Nb magnetic multilayer (Shukla and Prasad 2004, Shukla et al. 2007). Fe/Nb is an interesting system because Fe is a strong ferromagnet while the spacer layer can be a superconductor at low temperatures, offering the possibility of studying the interplay of superconductivity and magnetism in this system. To explore the mechanism of IEC in Fe/Nb system, the total energy calculations were done using the LMTO method and the GGA. Various tetragonal cells with different numbers of magnetic layers and different thicknesses of nonmagnetic layers were constructed as shown in Figure 18.5. The IEC corresponds to the

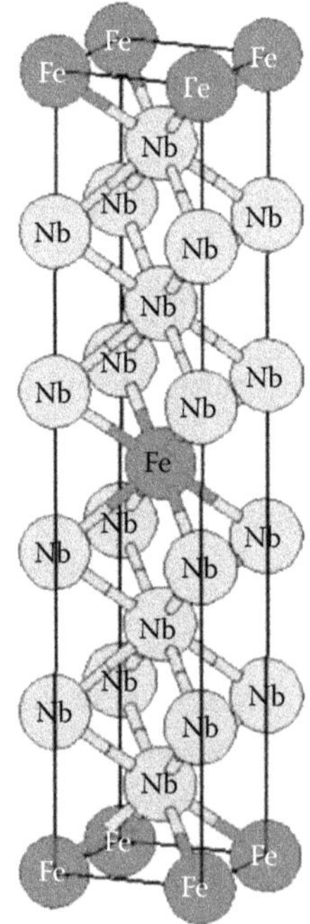

FIGURE 18.5
Supercell of Fe/Nb magnetic multilayer with one layer of Fe and four layers of Nb. (From Shukla N. N. and Prasad R. 2004. *Phys. Rev.* B 70: 014420.)

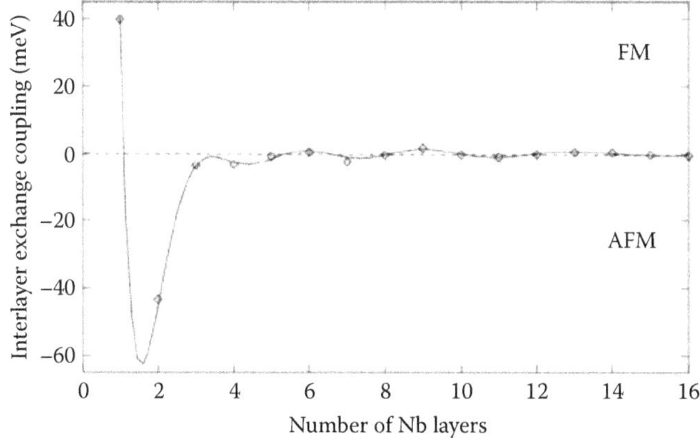

FIGURE 18.6
IEC for Fe/Nb magnetic multilayers as a function of Nb thickness. (From Shukla N. N., Sen A. and Prasad R. 2007. *Phys. Rev.* B 76, 174417.)

energy difference between FM and AFM configurations per unit cell and is shown as a function of Nb thickness in Figure 18.6. Rapid oscillations are seen in up to nine monolayers of Nb thickness after which the IEC becomes appreciably weak. The IEC changes from FM to AFM at about 2, 7, and 10 Nb monolayer thickness. These results were analyzed to elucidate the RKKY or QW nature of the coupling. The results indicate that the QW model gives a better description of the IEC in an Fe/Nb system.

18.3 Half-Metallic Ferromagnets

Half-metallic ferromagnets (HMFs) are materials which show the properties of metals as well as insulators at the same time in the same material depending on the spin direction. This happens because electrons responsible for metallic behavior have the same spin while the opposite spin electrons show insulating behavior. This unusual behavior can be understood by drawing a schematic density of states (DOS) diagram as shown in Figure 18.7. Figure 18.7a shows schematically the DOS for a metal that has a partially filled band. Figure 18.7b shows DOS for a semiconductor that has a completely filled band and the Fermi energy ε_F lies in the band gap. Figure 18.7c shows the DOS for spin $\uparrow$ and spin $\downarrow$ electrons in a hypothetical ferromagnet. Due to the exchange interactions, the DOS for $\uparrow$ and $\downarrow$ electrons are different. Now suppose the Fermi energy ε_F lies in the band gap of $\downarrow$ spin electrons but intersects the conduction band for $\uparrow$ spin electrons as shown in Figure 18.7c. We see that spin $\uparrow$ electron band is partially filled while spin $\downarrow$ electron band

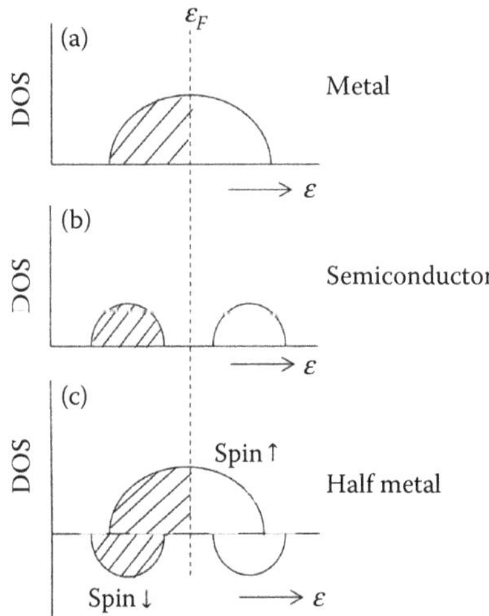

FIGURE 18.7
Schematic showing DOS for a metal, semiconductor, and half-metal. Note that in (c) the DOS for spin ↑ electrons is shown in the upper half-plane while for spin ↓ electrons, it is in the lower half-plane.

is completely filled. Therefore, the system is metallic but the current carriers are only spin ↑ electrons because for ↓ spin electrons the ε_F falls in the band gap. Thus, the material is metallic for ↑ spin electrons but an insulator for ↓ spin electrons. Such a material is called a HMF or simply a half-metal.

If $N_\uparrow$ and $N_\downarrow$ are the numbers of spin ↑ and spin ↓ electrons/unit cell in the conduction band respectively, spin polarization P_s is defined as

$$P_s = \frac{N_\uparrow - N_\downarrow}{N_\uparrow + N_\downarrow} \tag{18.3}$$

For a paramagnet $N_\uparrow = N_\downarrow$ and one gets $P_s = 0$. For an ideal HMF as shown in Figure 18.7, $N_\downarrow = 0$, therefore, $P_s = 1$. That is, the spin polarization is 100%. Such materials will serve as an ideal source of spin-polarized electrons for spintronic devices. It should be noted that 100% spin polarization is an idealization that can be approached only at low temperatures and by neglecting spin–orbit interaction.

One interesting consequence of the half-metallicity is that the magnetic moment is quantized in units of Bohr magneton. This is because the total number of electrons/unit cell, $N = N_\uparrow + N_\downarrow$ is an integer. Therefore, $N_\uparrow = N - N_\downarrow$ is also an integer for an ideal half-metal because $N_\downarrow$ is either zero (as in Figure 18.7) or an integer. Therefore, magnetic moment per unit cell $M = \mu_B (N_\uparrow - N_\downarrow)$ is also an integer times Bohr magneton. For an ordinary ferromagnet like Fe or Ni, M is not an integer times Bohr magneton.

Many experimental techniques have been used to probe half-metallicity. The most direct measurement is spin-resolved positron annihilation, but this is an expensive and tedious technique. Another promising technique is spin-resolved photoemission spectroscopy in which electrons are photo-emitted from the surface of the HMF, but this method is very sensitive to surface properties. Another way of measuring spin polarization is by using spin-polarized tunneling between a ferromagnet and a superconductor. Andreev reflection experimentation has also been used to probe half-metallicity.

In 1983, de Groot et al. (1983) found by using *ab initio* calculations that a half-Heusler alloy NiMnSb is half-metallic. Since then, several materials have been found half-metallic, such as full-Heusler alloys (e.g., Co_2MnSi), CrO_2, Fe_3O_4, $La_{0.7}Sr_{0.3}MnO_3$, Sr_2FeReO_6 and so on. DMS such as (Ga, Mn)As form another important class of half-metals and will be discussed in the next section. All of these systems contain a magnetic atom that imparts magnetic character to the system.

First-principles electronic structure calculations have played an important role in the discovery and understanding the behavior of these materials. We shall focus on some Heusler alloys because these are among the most promising materials for spintronic applications, such as spin–injector devices, spin filters, tunnel junctions, and GMR devices. There are two main advantages of Heusler alloys over other systems such as CrO_2 and Fe_3O_4. First, they have relatively high Curie temperatures (T_c), for example, for NiMnSb it is about 730 K and for Co_2MnSi about 980 K while T_c for other compounds is near room temperature. This means that these materials will have high polarization at well above room temperature. Second, they are structurally similar to zinc blend structure, which is adopted by binary semiconductors such as GaAs used in the electronics industry. This makes them easier to integrate with binary semiconductors.

Heusler alloys are named after Friedrich Heusler, who studied such alloys in 1903. These alloys can be classified into two categories, full Heusler and half-Heusler (or semi-Heusler) alloys. Full Heusler alloys are of the form X_2YZ, where X is a high valence transition metal or noble metal, Y a low valence transition metal, and Z an *sp* element. An example of a full Heusler alloy is Co_2MnSi. They crystallize in $L2_1$ structure which consists of four fcc sublattices as shown in Figure 18.8. The Heusler alloys of the second kind, that is, half-Heusler, are alloys of the form XYZ and crystallize in $C1_b$ structure which consists of three fcc sublattices as shown in Figure 18.9. For example, NiMnSb is a half-Heusler alloy. The electronic structure of half-Heusler alloy NiMnSb was first studied by de Groot et al. (1983) and subsequently many calculations have appeared (e.g., Öğüt and Rabe 1995; Galanakis et al. 2002; Nanda and Dasgupta 2003; Yamasaki et al. 2006; Galanakis and Mavropoulos 2007). Figure 18.10 shows the band structure of NiMnSbs, which we calculated by using the FP-LAPW method with WIEN2k code (Schwarz and Blaha 2003). We see that the band structure clearly shows the half-metallic character. The Fermi energy ε_F intersects the spin ↑ bands while it lies in the band

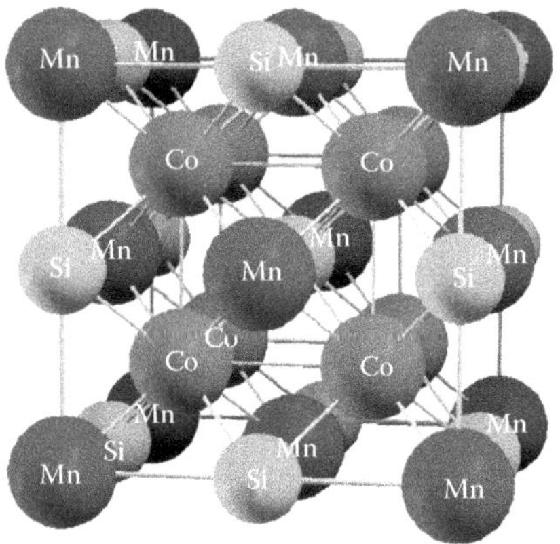

FIGURE 18.8
Crystal structure of full-Heusler alloy Co_2MnSi with $L2_1$ structure having *fcc* Bravais lattice. The Co atoms are located at (3/4, 3/4, 3/4) and (1/4, 1/4, 1/4) whereas Mn and Si atoms are at (0,0,0) and (1/2, 1/2, 1/2), respectively.

gap of spin ↓ bands. This is further confirmed by the DOS calculation as shown in Figure 18.11. The total magnetic moment per unit cell comes out to be 4 μ_β.

The electronic structure of the full Heusler alloy Co_2MnSi has been studied by Galanakis et al. (2006). Figure 18.12 shows the DOS for spin ↑ and ↓ electrons that we calculated by using the WIEN2k code (Schwarz and Blaha 2003). We notice that the Fermi energy ε_F lies in the band gap for spin ↓ electrons, while for spin ↑ electrons, it intersects a band. This clearly shows that it is a

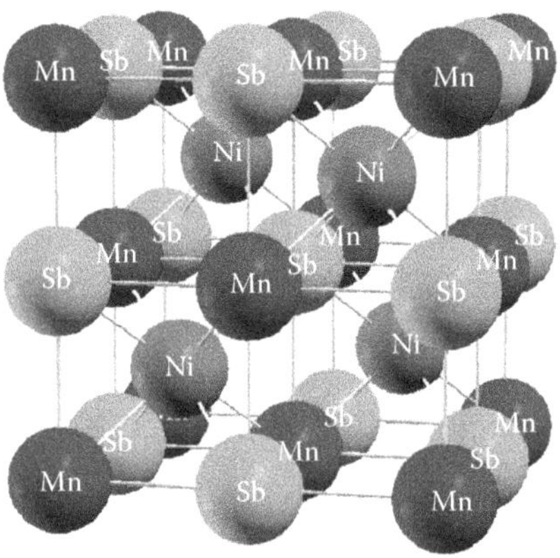

FIGURE 18.9
Crystal structure of half-Heusler alloy NiMnSb with $C1_b$ structure having *fcc* Bravais lattice. Ni atom is located at (3/4, 3/4, 3/4) whereas Mn and Sb atoms are at (0,0,0) and (1/2, 1/2, 1/2), respectively.

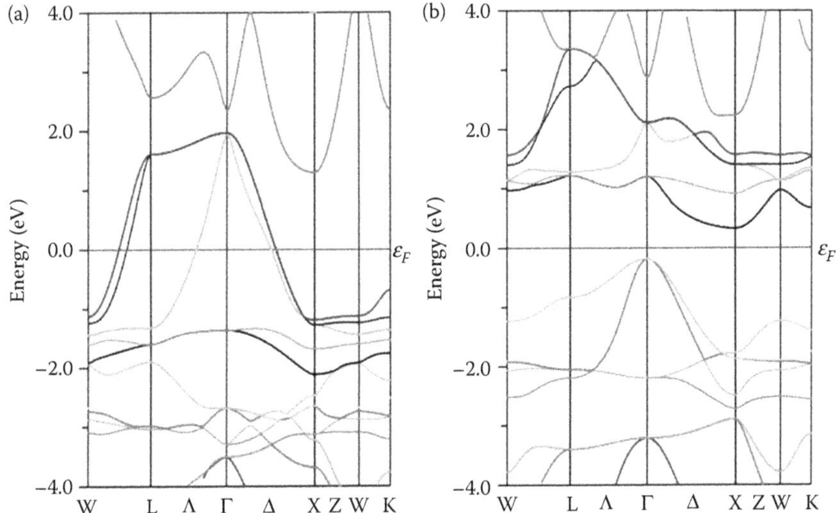

FIGURE 18.10
Spin polarized band structure of NiMnSb for majority spin (a) and minority spin (b) using the FP-LAPW method and GGA exchange correlation potential. WIEN2k code of Schwarz and Blaha (2003) was used with 12 × 12 × 12 k mesh. The experimental lattice parameter of 5.903 Å was used.

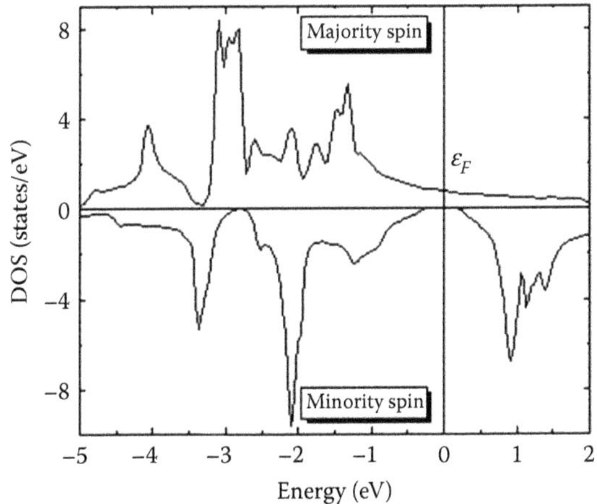

FIGURE 18.11
DOS of half-Heusler alloy NiMnSb using the FP-LAPW method and GGA. WIEN2k code of Schwarz and Blaha (2003) was used.

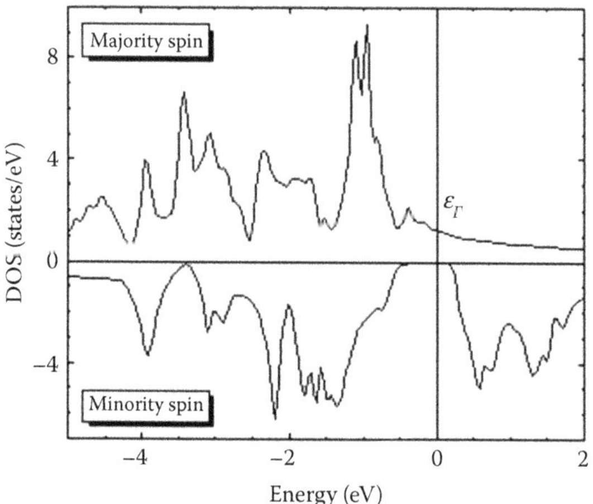

FIGURE 18.12
DOS of full Heusler alloy Co_2MnSi using the FP-LAPW method and GGA. WIEN2k code of Schwarz and Blaha (2003) was used. The experimental lattice parameter 5.656 Å was used for Co_2MnSi.

half-metal. The magnetic moment per unit cell is about 4.94 μ_β. For further details we refer to the review articles (e.g., Katsnelson et al. 2008; Das and Dasgupta 2008).

18.4 Dilute Magnetic Semiconductors

A dilute magnetic semiconductor (DMS) is a dilute alloy of a semiconductor and a magnetic transition metal and shows semiconducting as well as magnetic properties. Typical examples are Mn-doped GaAs denoted as (Mn,Ga)As, (Mn,Ga)N, (Cd,Mn)Te, Mn-doped ZnO, Co-doped TiO_2, and so on. As mentioned in the introduction, DMSs are considered to be potential candidates for injecting spin-polarized electrons in spintronic devices and will form building blocks of novel technologies. Thus over the last decade, the field of DMS has attracted much attention and grown very rapidly into an important branch of materials science.

Spintronic devices are expected to operate at room temperature, therefore one of the goals of DMS research has been to search for FM semiconductors with Curie temperature well above room temperature. Curie temperature (T_c) for [Mn,Ga]As, which is one of the most studied systems, is reported to be 190 K. Several DMS materials have been reported to have ordering temperature above room temperature, including Mn-doped GaN, Co-doped TiO_2, and Mn-doped ZnO. However, such observations have not been confirmed by other studies, which at times report much lower ordering temperatures. The large differences in ordering temperatures reported by various groups may be attributed to sample preparation, clustering of magnetic impurities, defects, and so on.

To find DMS material with T_c above room temperature, it is important to understand the mechanism of ferromagnetism in these materials. Dietl et al. (2000) explained the ferromagnetism of these materials using Zener's theory. For example in (Mn,Ga)As, where Mn is doped substitutionally at the Ga site, Mn provides localized spins and also creates itinerant holes. The localized spins and holes interact via a *p–d* exchange interaction that in turn produces coupling between the localized spins. In semiconductors, this is similar to the RKKY approach that brings about interaction between two localized spins in a metal. Dietl et al. (2000) used this theory and the mean-field theory to estimate T_c of several DMS materials. The theory predicted several materials with large T_c such as Mn-doped GaN, Mn-doped ZnO, and so on. It turns out that this theory overestimates T_c because it neglects thermal fluctuations and disorder effects that can drastically reduce T_c.

First-principles calculations using DFT have played an important role in understanding and searching new DMS materials. Using the first-principles methods, one can calculate exchange coupling constants and by assuming Heisenberg model and Monte Carlo method, one can calculate T_c. It turns out that such calculations give much more realistic values of T_c and the values one gets are much lower compared with the predictions of Dietl et al. (2000). Based on the *ab initio* calculations of T_c, it seems that prospects for finding DMS materials with T_c well above room temperature are rather low.

As an example, here we shall briefly discuss the electronic structure of Mn-doped GaN, which is a prototypical wide-band-gap DMS. Dietl et al. (2000) predicted high T_c ~400 K for this material with 5% Mn doping, based on Zener's model and the mean-field theory. This result triggered several experimental studies. However, experimental results on this system have been controversial. Some experiments claim to have observed a high T_c value whereas some claim the material to be nonmagnetic. (For a review, see, Bonanni 2007, Sato et al. 2010a.)

Several *ab initio* electronic structure calculations have been reported for Mn-doped GaN (e.g., Sato and Katayama-Yoshida 2002, Sanyal et al. 2003). Here, we shall briefly discuss the *ab initio* electronic structure calculations by Sanyal et al. (2003). Although they reported LSDA as well as LSDA + U calculations, here we shall discuss only the LSDA calculation. For GaN, a wurtzite structure was used. The LSDA calculation used plane wave code (VASP) and Vanderbilt ultra-soft pseudo potentials. Experimental lattice parameters, $a = 3.189$ Å, $c = 5.185$ Å with $c/a = 1.626$ were used. The DMS $Mn_xGa_{1-x}N$ was considered as an ordered alloy with $x = 0.06255$, 0.125, 0.25, and 0.5. Mn was assumed to be at Ga site as has been supported by experiments. The calculations were done using supercell geometry.

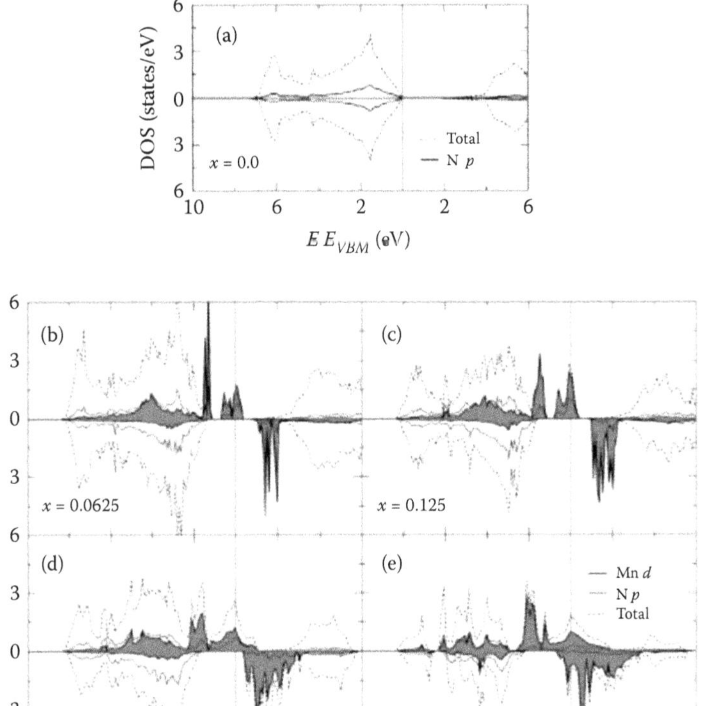

FIGURE 18.13
Spin-resolved DOS for Mn-doped GaN. (a) Pure GaN, (b) 6.25% Mn, (c) 12.5% Mn, (d) 25% Mn, and (e) 50% Mn. (Reprinted with permission from Sanyal B., Bengone O., and Mirbt S. 2003. Electronic structure and magnetism of Mn-doped GaN. *Phys. Rev.* B 68: 205210. Copyright (2003) by the American Physical Society.)

In Figure 18.13a, the DOS for GaN is shown. It shows a band gap of about 1.9 eV, which is much smaller than the experimental value of 3.4 eV. The underestimation can be attributed to the use of LSDA. In Figure 18.13b–e, the DOS for $Mn_xGa_{1-x}N$ is shown for various values of Mn concentration x. For small values of $x = 0.0625$ and $x = 0.125$, an impurity band due to Mn d states appears in the band gap. Up to $x = 0.25$, the Fermi level cuts only the DOS for up spin and the system is a half-metal. As x increases, the impurity band gets wider and at $x = 0.5$ it covers the whole band gap region, and the material becomes an FM metal. For further details, we refer to the original paper by Sanyal et al. (2003). We also refer to an excellent review by Sato et al. (2010), which covers electronic structure of a very large number of DMS materials and related issues.

EXERCISES

18.1 Calculate the DOS and total energy for Fe_1Cr_3(001) multilayer for FM as well as AFM configuration. Use supercell geometry and one of the codes mentioned in Appendix A. Compare the DOS and the total energies of the two configurations.

18.2 Calculate the DOS and total energy for Fe_1Nb_3(001) multilayer for FM as well as AFM configuration. Use supercell geometry and one of the codes mentioned in Appendix A. Compare the DOS and the total energies of the two configurations.

18.3 Calculate the band structure and the DOS of half-Heusler alloy NiMnSb using one of the codes mentioned in Appendix A. Compare your results with Figures 18.10 and 18.11.

18.4 Calculate the band structure and the DOS of full Heusler alloy Co_2MnSi using one of the codes mentioned in Appendix A. Compare your DOS with Figure 18.12

18.5 Calculate the DOS for 25% Mn doped GaN using supercell geometry and one of the codes mentioned in Appendix A. Compare your results with Figure 18.13.

Further Reading

Bandyopadhyay S. and Cahay M. 2008. *Introduction to Spintronics*. Boca Raton, FL: CRC Press.

Dietl T. 2010. A ten-year perspective on dilute magnetic semiconductors and oxides. *Nat. Mater.* 9: 965–974.

Grünberg P. 2008. Nobel Lecture: From spin waves to giant magnetoresistance and beyond. *Rev. Mod. Phys.* 80: 1531–1540.

Pickett W. E. and Moodera J. S. 2001. Half metallic magnets. *Physics Today* 54(5): 39–44.

Sato. K. et al. 2010. First-principles theory of dilute magnetic semiconductors. *Rev. Mod. Phys.* 82: 1633–1690.

19

Battery Materials

19.1 Introduction

With the widespread use of laptop computers and mobile phones, the global demand for batteries is increasing day by day. With the shortage of petroleum and greater concern for the environment, there is a likely shift from oil to batteries for powering cars and the demand is likely to increase further. Owing to these reasons, there is a great interest in the search for developing cheaper, safer, and environment-friendly batteries.

Batteries convert chemical energy into electrical energy. A battery may consist of several electrochemical cells joined serially or in parallel or a single electrochemical cell. Here, we shall assume that it is a single cell. The batteries may further be classified as primary or secondary batteries. A primary battery is nonrechargeable as the electrochemical reaction in the battery is irreversible. A Daniel cell and alkaline batteries are examples of primary batteries. A secondary battery is rechargeable as the electrochemical reaction in the battery is reversible. Ni–Cd batteries or lithium-ion batteries are examples of secondary batteries.

Among rechargeable batteries, the Li-ion batteries are the batteries of choice at present because they have high-energy density and are lightweight, which makes them very suitable for portable devices such as laptops and mobiles. As its name suggests, in Li-ion batteries, the charge flow inside the batteries occurs via Li^+ ions. It consists of two electrodes, anode and cathode, that are separated by an electrolyte as shown schematically in Figure 19.1. In early batteries, the anode was LiC_6 and the cathode consisted of $LiCoO_2$. The electrolyte consisted of $LiPF_6$ dissolved in a mixture of propylene carbonate and diethyl carbonate (Palacín 2009). This electrolyte has the property that it is a good Li conductor but a poor electron conductor.

The principle of the Li-ion battery is quite simple. The anode and cathode are chosen such that the anode is at a higher chemical potential than the cathode. During discharge, Li^+ ions move from the anode to the cathode. At

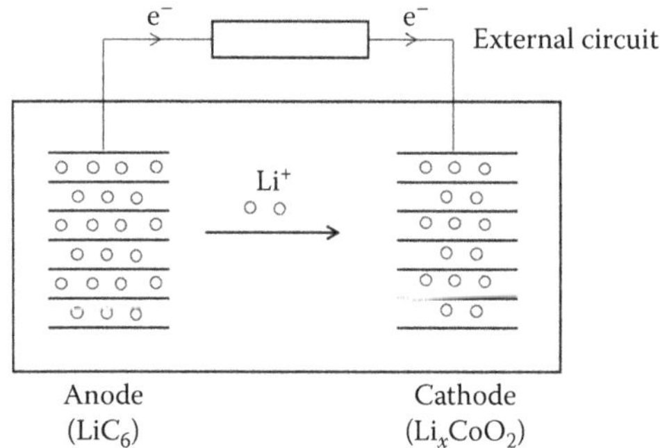

FIGURE 19.1
Schematic diagram of a lithium-ion battery.

the same time, the electron travels through the external circuit and performs the work. The following chemical reactions take place in the electrodes:

$$Li_xC_6 \rightleftarrows x\,Li^+ + 6\,C + x\,e^- \tag{19.1}$$

$$x\,Li^+ + Li_{1-x}CoO_2 + x\,e^- \rightleftarrows LiCoO_2 \tag{19.2}$$

The above reactions are reversible. During the charging process, an external potential is applied to the electrode in such a way that Li^+ moves from the cathode to the anode.

Both $LiCoO_2$ and LiC_6, which are used as cathode and anode, are intercalation compounds and have layered structures as indicated in Figure 19.1. As we can see, these compounds store Li^+ in their structures. The layered structure makes it easy to insert Li^+ ions into their structure and extract from it. This is called the Li insertion/extraction process that takes place during charging/discharging of the battery.

There are several material-related issues in Li-ion batteries on which intense research is going on. These issues are related to the cost and life of a battery, phase stability of the materials involved, safety, and environmental issues. For example, $LiCoO_2$ is expensive, which makes the cost of the battery high. Therefore, there is an intense search for a cheaper substitute to cut down the cost of the battery. Also, Co is toxic; therefore, a material that does not have Co would be preferable. Further, Li is very reactive and explodes when it comes in contact with water. Thus, one has to take all these factors into account when selecting the appropriate material for designing a battery. In solving many material-related issues, computational methods, which we have discussed in the earlier chapters, can be very useful and have been used by several workers. In particular, electronic structure calculations can give accurate information about total energies, phase stability, alloy formation energies, vacancy formation energies, average cell voltage, diffusion constants, and so on. All this information can help in choosing the right materials and guide the design of a battery. As an example, we will discuss in Sections 19.2 and 19.3 the use of computational methods to understand the phase stability problems in layered $LiMnO_2$ and $LiMn_2O_4$, which are important materials for cathodes in Li-ion batteries.

19.2 $LiMnO_2$

There is an intense search for an alternative cathode material in Li-ion batteries because $LiCoO_2$, which is mostly used as a cathode, is an expensive and toxic material. There are many materials such as $LiMnO_2$, $LiFePO_4$, $LiNiO_2$, $LiMn_2O_4$, etc. that could be used as cathode materials. In this section, we shall focus on $LiMnO_2$. $LiMnO_2$ is chemically similar to $LiCoO_2$ and has a great cost advantage over $LiCoO_2$ as it costs about 1% of the cost of $LiCoO_2$. However, it poses several problems if one wants to use it as a cathode material. The stable form of $LiMnO_2$ is orthorhombic, which is less favorable for Li insertion/extraction cycling than layered $LiCoO_2$. However, a metastable phase of layered $LiMnO_2$ has been synthesized (Armstrong and Bruce 1996), but the problem is that this phase is not stable during battery operation, in which Li insertion/extraction process occurs in the cathode. Thus, the problem of phase stability has to be solved before this material could be used as a cathode material. The first-principles calculations can help in understanding this problem and thus can guide in the design of a suitable cathode material.

The phase stability problem in layered $LiMnO_2$ is caused due to the Jahn–Teller (JT) distortion. Before proceeding further, we will take a pause and understand what JT distortion is. According to the JT theorem, if the ground state of an ion in an ion complex is orbitally degenerate, it is energetically favorable for the ion complex to distort in such a way so as to remove the orbital degeneracy. To understand this, let us assume that we have a Mn^{3+} ion with configuration $3d^4$ at the center of an octahedral ion complex as shown in Figure 19.2. Let us assume that at the vertices of the octahedron, we have O^{2-} ions. Mn d levels, which are fivefold degenerate in the atom, get split into t_{2g} (threefold degenerate) and e_g (twofold degenerate) levels in the cubic crystal field as shown in Figure 19.3. If we distort the ion complex such that it is elongated in the z direction so that the symmetry is now tetragonal, these t_{2g} and e_g levels further split as shown in Figure 19.3.

Since the highest level in the tetragonal environment is unoccupied, there would be a gain in electronic energy if the ion complex gets distorted from the cubic symmetry. If the gain in electronic energy overcomes the elastic energy, the system would get distorted. On the other hand, if, instead of a Mn^{3+} ion, we have a Mn^{4+} ion at the center of an ion complex, there would not be any gain in energy due to tetragonal distortion, as the highest level will also be occupied. Thus, the Mn^{3+} ion is a JT-active ion whereas Mn^{4+} is not.

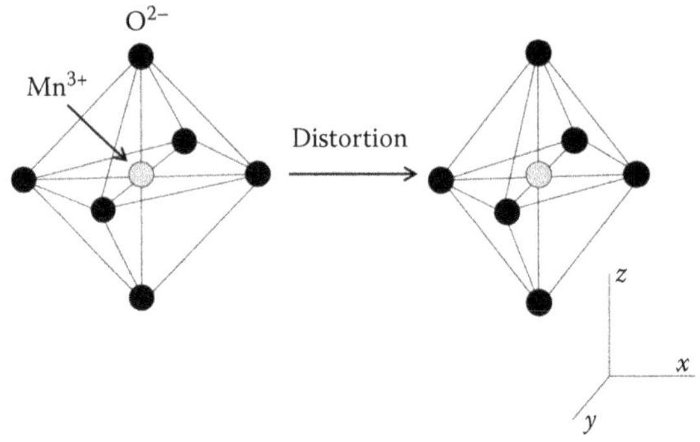

FIGURE 19.2
Schematic showing JT distortion.

Let us return to the discussion of the phase stability of rhombohedral $LiMnO_2$. The oxidation state of Mn in $LiMnO_2$ is 3+ and, therefore, Mn^{3+} is a JT ion in the rhombohedral environment as discussed above. This means that it would lead to JT distortion, which is absent in $LiCoO_2$ because Co^{3+} is not a JT ion. If the oxidation state of Mn could somehow be pushed up to 4+, such a distortion could be suppressed as Mn^{4+} is not a JT ion. This could be achieved by doping Co or Mg at the manganese site. To understand this, Prasad et al. (2003, 2005) performed first-principles calculations of total energy for pure and doped $LiMnO_2$ in rhombohedral and monoclinic phases for several dopants. We shall briefly discuss some results of their calculations.

Monoclinic and rhombohedral structures of $LiMnO_2$ are shown in Figures 19.4 and 19.5, which have $C2/m$ and $R\bar{3}m$ symmetry, respectively. We first consider the monoclinic structure for which the electronic structure calculations were first reported by Singh (1997) using LSDA and LAPW method. He also suggested the magnetic structure of the monoclinic phase by considering

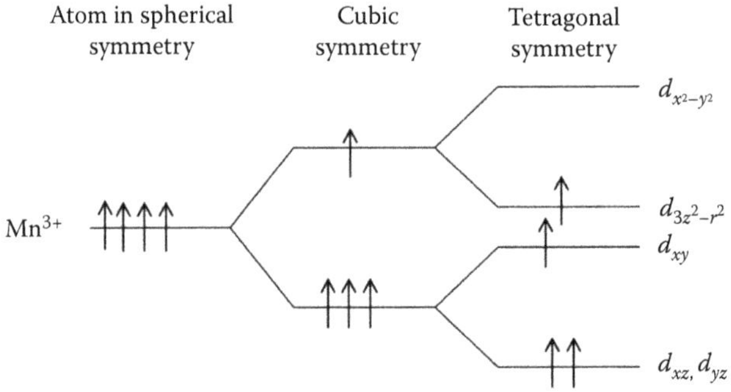

FIGURE 19.3
Schematic diagram showing the splitting of d levels of Mn^{3+} due to JT distortion shown in Figure 19.2.

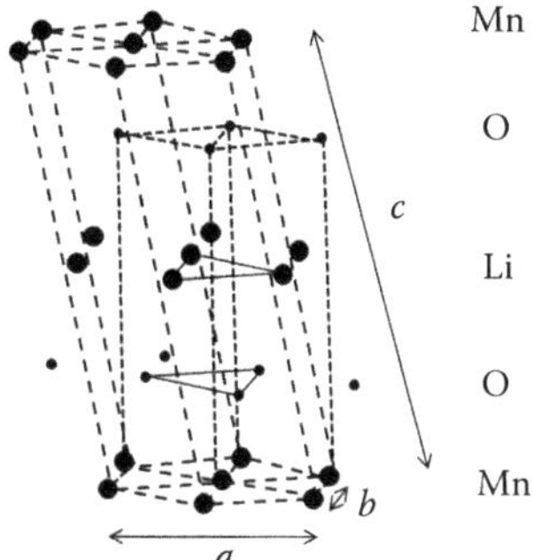

FIGURE 19.4
Atomic structure of monoclinic $LiMnO_2$. (From Prasad R., Benedek R., and Thackeray M. M. 2005. *Phys. Rev.* B 71: 134111.)

various magnetic arrangements such as non-spin polarized, ferromagnetic (FM), and antiferromagnetic (AF) arrangements. Note that Mn atoms form an almost hexagonal lattice in the plane of Mn layers in Figure 19.4. Thus, it is not possible to arrange all spins so that all nearest neighbors are antiferromagnetically aligned. Therefore, Singh (1997) considered three magnetic configurations shown in Figure 19.6 that are AF but all neighbors are not antiferromagnetically aligned. In Figure 19.6a, there is an FM alignment of spins within the layer but the neighboring layers are antiferromagnetically stacked. Let us call this spin arrangement as AF1. In Figure 19.6b, we show another spin arrangement in which spins are ferromagnetically aligned along the chains running in the [010] direction, but the chains are antiferromagnetically

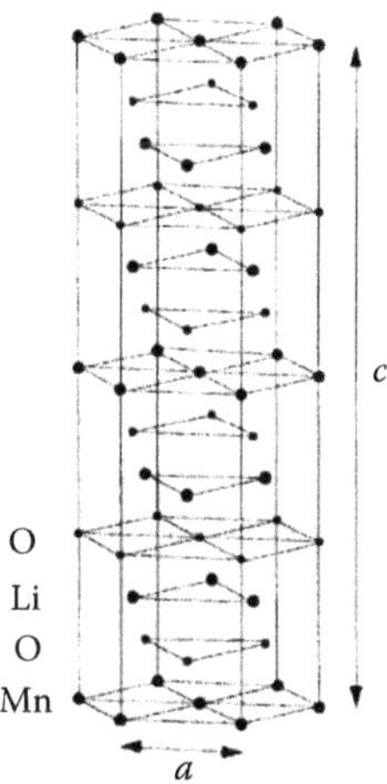

FIGURE 19.5
Atomic structure of rhombohedral $LiMnO_2$. (From Prasad R., Benedek R., and Thackeray M. M. 2005. *Phys. Rev.* B 71: 134111.)

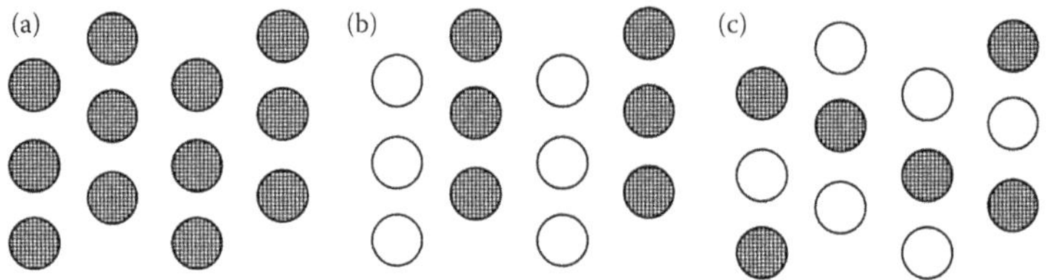

FIGURE 19.6
Magnetic configurations of Mn atoms in Mn layer. The filled circles show spin up atoms and the empty circles show spin down atoms. (a) AF1, (b) AF2, and (c) AF3 configuration.

aligned. We call this configuration AF2. In Figure 19.6c, we show the AF3 configuration in which the spins are ferromagnetically aligned along the chains running in [110] directions, but the chains are antiferromagnetically aligned. Singh showed that out of all the configurations considered, the AF3 configuration (Figure 19.6c) showed the lowest energy. Therefore, we shall assume AF3 magnetic structure in our subsequent discussion.

The nonmagnetic primitive unit cell of monoclinic $LiMnO_2$ has two formula units. Let us call its basis vectors $\vec{a}$, $\vec{b}$, and $\vec{c}$. The primitive unit cell for the magnetic structure can be constructed with basis vectors

$$\vec{A}_1 = \vec{a} + \vec{b}, \quad \vec{A}_2 = \vec{a} - \vec{b}, \quad \vec{A}_3 = \vec{c} \tag{19.3}$$

Note that this primitive cell contains four formula units and is illustrated in Figure 19.7. In the AF3 configuration, spins are ordered ferromagnetically along the $\vec{A}_2$ axis and are antiferromagnetically aligned along the $\vec{A}_1$ axis as shown in the figure. The unit cell has four Mn atoms in the cell; thus, one

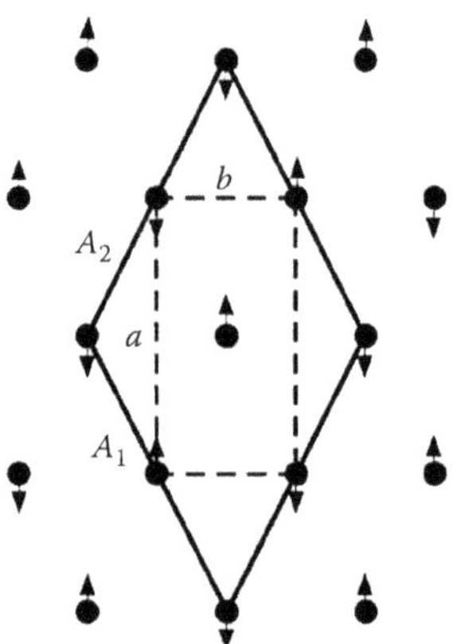

FIGURE 19.7
Magnetic unit cell of monoclinic $LiMnO_2$ in Mn layer is shown by solid lines and the nonmagnetic cell is shown by dashed lines. (From Prasad R., Benedek R., and Thackeray M. M. 2005. *Phys. Rev.* B 71: 134111.)

could study a doped system such as a Co-doped system with 25% Co doping. For a Co-doped system, the unit cell contains three Mn atoms and one Co atom in the transition metal layer. The Mn atom next to Co in the unit cell will be called "a" and the one farthest in the unit cell will be called "b."

As mentioned earlier, pristine $LiMnO_2$ can exist in various phases such as rhombohedral layered, monoclinic layered, and orthorhombic structures. To study the relative phase stability of various phases, Mishra and Ceder (1999) calculated the formation energy of Li_xMnO_2 defined by

$$\Delta E_f = E_{\mathrm{Li}_x\mathrm{MnO}_2} - xE_{\mathrm{LiMnO}_2} - (1-x)E_{\mathrm{MnO}_2} \tag{19.4}$$

where $E_{\mathrm{Li}_x\mathrm{MnO}_2}$ is the total energy of Li_xMnO_2 in the given phase and E_{LiMnO_2}, E_{MnO_2} are the total energies of $LiMnO_2$ and MnO_2 in rhombohedral layered phases. Note that the choice of the rhombohedral layered phase as a reference is arbitrary and does not change any conclusion. A negative value of the formation energy at $x = 1$ (or $x = 0$) implies that the compound is more stable than the rhombohedral layered $LiMnO_2$ (or MnO_2). In Table 19.1, we give the formation energy of various phases of $LiMnO_2$ with respect to the rhombohedral layered FM phase obtained by using GGA and spin-polarized calculations (Mishra and Ceder 1999). We see that the orthorhombic AF phase is the most stable phase. Among layered phases, the monoclinic layered AF phase is the most stable.

The average cell voltage can be obtained from the total energy calculation. The average open-cell voltage is given by (Mishra and Ceder 1999)

$$V = -\frac{\Delta G_r}{F} \tag{19.5}$$

where F is the Faraday constant defined as $F = eN_A$, N_A being the Avogadro number. ΔG_r is the Gibbs free energy for the reaction

$$\mathrm{MnO_2\ (cathode) + Li\ (anode) \rightarrow LiMnO_2\ (cathode)} \tag{19.6}$$

TABLE 19.1

Formation Energies of Various Phases of $LiMnO_2$ with Respect to the Rhombohedral (FM) Phase

System	Formation Energy (meV)
Rhombohedral layered (FM)	0
Rhombohedral layered (AF3)	–36
Monoclinic layered (FM)	–248
Monoclinic layered (AF3)	–375
Orthorhombic (FM)	–241
Orthorhombic (AF)	–407

Source: Data from Mishra S. K. and Ceder G. 1999. *Phys. Rev. B* 59: 6120–6130.

and can be written as

$$\Delta G_r = \Delta E_r + P\Delta V_r - T\Delta S_r \tag{19.7}$$

Here, E_r, ΔV_r, and ΔS_r, respectively, denote change in the total energy, volume, and entropy during the reaction, and P and T, respectively, denote pressure and temperature. Since changes in volume and entropy are very small, at $T = 0$ K, ΔG_r can be approximated by change in the total energy ΔE_r. Using spin-polarized calculations and GGA, Mishra and Ceder (1999) obtained the average cell voltage 2.95 V for the monoclinic-layered antiferromagnetic phase and 3.32 V for the orthorhombic-layered AF phase. In transition metal oxides, electron correlations play an important role. To investigate the effect of electron correlations, Shukla and Prasad (2006) did LSDA and LSDA+U studies of various phases of $LiMnO_2$ using the full-potential LAPW method. The values of U and J in LSDA+U were chosen to be 8.0 and 0.82 eV, respectively. In Figure 19.8a, we show the density of states (DOS) and partial DOS

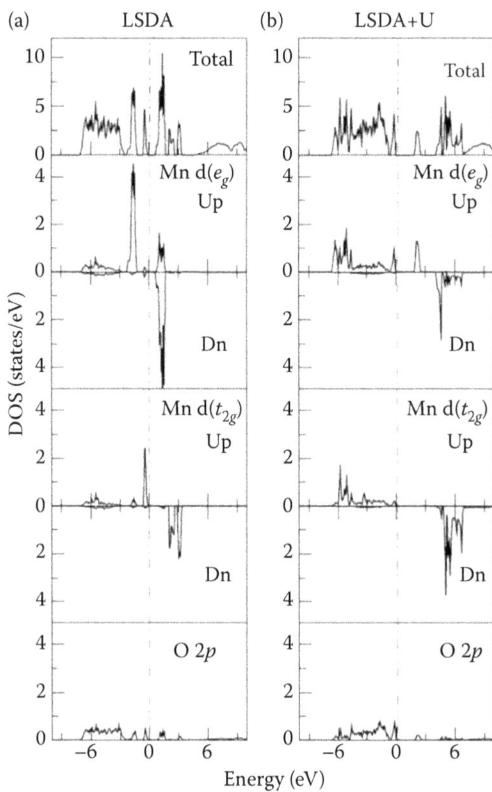

FIGURE 19.8
DOS and partial DOS for the AF3 monoclinic $LiMnO_2$ using (a) LSDA and (b) LSDA+U. (From Shukla N. N. and Prasad R. 2006. *J. Phys. Chem. Solids* 67:1731–1740.)

for the AF3 ordering for monoclinic $LiMnO_2$ using LSDA. We see that the Mn *d* band splits into the e_g and t_{2g} subbands, but there is further splitting of these subbands due to the JT distortion. The band gap at Fermi energy ε_F, shown by the dashed vertical line, arises due to the JT distortion, which is about 0.6 eV. Note that without the JT splitting, the system would be a metal, as is the case for the rhombohedral phase (see Figure 19.9a). In Figure 19.8b, we show the DOS and partial DOS for LSDA+U case. We see that the occupied *d* bands have shifted to lower energy while the unoccupied *d* bands have shifted to higher energy, thus increasing the band gap significantly. The band gap at the Fermi energy has increased to 1.8 eV compared to 0.6 eV for the LSDA.

In Figure 19.9, we show the DOS for the AF3 and FM phase of rhombohedral $LiMnO_2$ using LSDA as well as LSDA+U. Note that the rhombohedral phase of $LiMnO_2$ is not the stable phase but can be stabilized by doping of Co or Mg as we shall see. We see from the figure that LSDA does not produce any gap at Fermi energy for both AF3 and FM phase and thus predicts the system to be a metal. In contrast, LSDA+U produces a gap at the Fermi energy for the AF3 as well as FM phases predicting it to be an insulator. Thus, the inclusion of electron correlations has a profound effect on the electronic structure of this system.

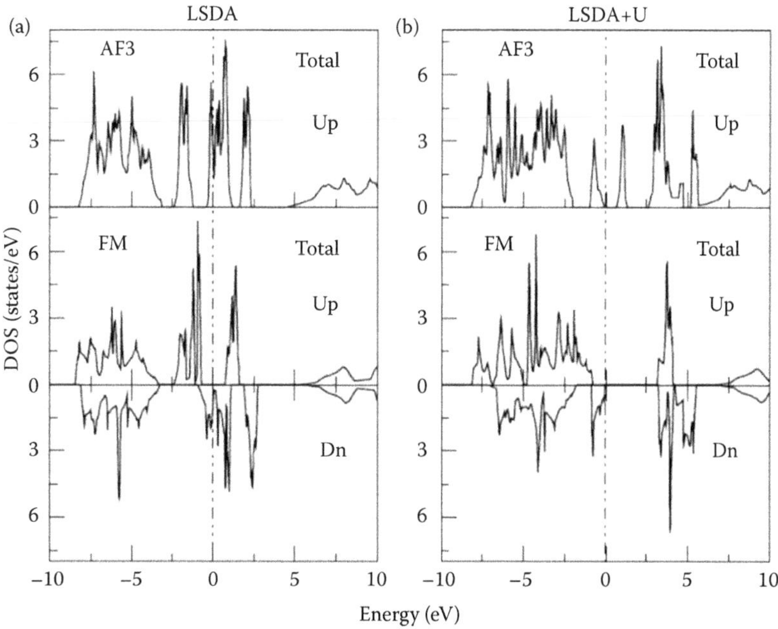

FIGURE 19.9
DOS for the AF3 rhombohedral $LiMnO_2$ using (a) LSDA and (b) LSDA+U. (From Shukla N. N. and Prasad R. 2006. *J. Phys. Chem. Solids* 67:1731–1740.)

We now consider the effect of Co doping on the electronic structure of $LiMnO_2$ and oxidation state of Mn (Prasad et al. 2003, 2005). The calculations were performed with VASP code based on the local spin density GGA. In Figure 19.10, we show the partial *d* DOS for Mn ions in monoclinic $LiMn_{0.75}Co_{0.25}O_2$ and for comparison, Mn partial *d* DOS in monoclinic $LiMnO_2$. The solid lines represent the up spin DOS and the broken lines represent the down spin DOS. We see that the up spin DOS for Mn in the undoped system and for *b*-type Mn in the doped system have t_{2g} and e_g bands with similar weights. However, the *a*-type Mn atom has significantly smaller e_g peak. This behavior indicates charge transfer between the *a*-type Mn atom and Co. The *a*-type Mn atoms are oxidized to 4+ and the neighboring Co atoms are reduced to 2+, while the *b*-type Mn atoms remain in the 3+ state. This has been further confirmed by Co *K*-edge x-ray absorption spectroscopy (XAS) results shown in Figure 19.11. The figure shows results for CoO in which the Co oxidation state is 2+ and $LiCoO_2$, in which the Co oxidation state is 3+. The figure also shows results for $LiMn_{0.9}Co_{0.1}O_2$. In this experiment, the Co *K* edge (near 7.72 keV) is sensitive to the Co oxidation state and thus gives information about the oxidation state of Co in a given compound. We see the Co edge in $LiMn_{0.9}Co_{0.1}O_2$ (near 7.72 keV) lies closer to that of CoO in which the Co oxidation state is the 2+. This implies that Co in $LiMn_{0.9}Co_{0.1}O_2$ is in 2+ oxidation state, in agreement with the theoretical prediction.

Thus, from the above calculations, it is clear that Co doping forces the neighboring Mn atom to a 4+ state. With increased Co doping, the number of Mn^{3+} ions decreases while the number of Mn^{4+} ion increases. Since Mn^{4+} and Co are not JT ions, at certain concentration of Co, the JT distortion is arrested. Once this is understood, one can try other dopants and find out

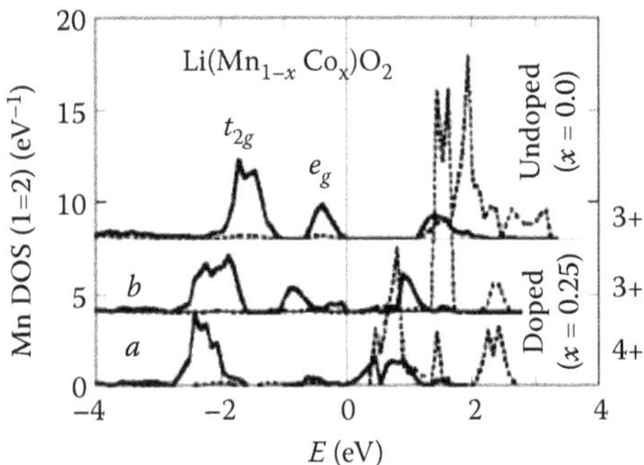

FIGURE 19.10
Partial *d* DOS for Mn ion for monoclinic $Li(Mn_{1-x}Co_x)O_2$ for $x = 0$ (top panel) and $x = 0.25$. The solid lines correspond to the up spins and the broken lines correspond to the down spins. The oxidation states of Mn are indicated by 3+ and 4+ on the right. (From Prasad R. et al. 2003. *Phys. Rev. B* 68: 012101.)

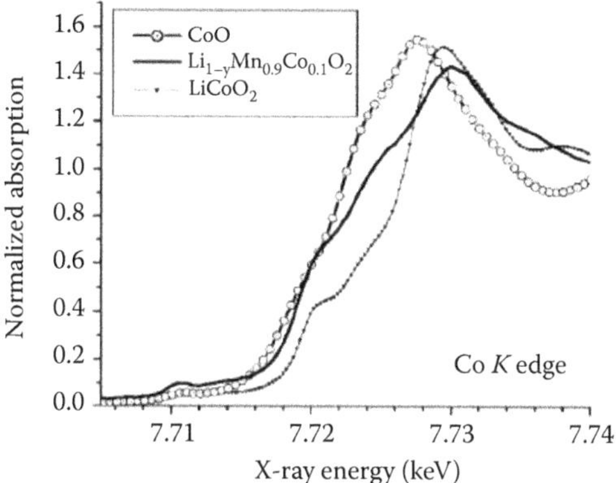

FIGURE 19.11
XAS *K*-edge spectra of Co in $LiMn_{0.9}Co_{0.1}O_2$ and reference systems CoO and $LiMnO_2$. (From Prasad R. et al. 2003. *Phys. Rev. B* 68: 012101.)

which will be the best dopant. This was done by Prasad et al. (2005) who performed first-principles calculations of total energy for pure and doped $LiMnO_2$ for many dopants both in rhombohedral and in monoclinic phases. These calculations show that the divalent dopants such as Mg and Zn are more effective in suppressing the JT distortion. Since electron correlations are important in these materials, Shukla et al. (2008) performed GGA+U calculations for various doped materials using VASP. They found that the divalent dopants are more effective in suppressing the JT distortion than the trivalent dopants as was the case in GGA. However, they found some significant differences such as Mn showing high spin state in rhombohedral $LiMnO_2$ in contrast to GGA, in which it shows low spin state. Similarly, Co and Ni show high spin state in GGA+U than in GGA. In spite of such uncertainties, GGA and GGA+U clearly indicate that the divalent dopants such as Mg are most effective in suppressing the JT distortion and thus stabilizing the rhombohedral structure. Thus, such calculations could be very helpful in guiding the design of suitable materials for cathodes in lithium-ion batteries.

19.3 $LiMn_2O_4$

Spinel $LiMn_2O_4$ is also a cathode material for Li-ion batteries. It has a cubic spinel structure at room temperature. As one lowers the temperature, this structure becomes orthorhombic below 280 K. As this temperature is close to room temperature, the material faces a phase stability problem in a battery, that is, continuous recycling in a battery creates structural changes in the material, causing poor performance. Thus, it is important to stabilize the cubic spinel structure. This can be achieved by appropriate doping of divalent or trivalent atoms as we had seen in the case of $LiMnO_2$. Among various dopants, chromium and magnesium have been studied experimentally and have been found to stabilize the cubic spinel phase. Shi et al. (2003) have studied Cr-doped $LiMn_2O_4$ using LDA. However, spin-polarized calculations using GGA were performed for the end compounds $LiMn_2O_4$ and $LiCrMnO_4$. They have calculated the average cell voltage that is about 3.8 V for $LiMn_2O_4$, and this increases with Cr doping up to 4.5 V for $LiCrMnO_4$. Since electron correlations are important in these materials, Singh et al. (2009) performed GGA and GGA+U calculations to understand the suppression of JT distortion by Cr and Mg doping. We shall briefly discuss their work.

The conventional unit cell of spinel $LiMn_2O_4$ has 56 atoms. But to speed up the calculation, Singh et al. (2009) chose a smaller cell with two formula units containing 14 atoms. For structural optimization and electronic structure calculations, VASP code was used with PAW potentials. For GGA+U calculations, U was chosen to be 4.5 eV for Cr and 5.0 eV for Mn and J was taken to be 1.0 eV. A plane wave cut-off of 550 eV was used and an $(8 \times 8 \times 8)$ Monkhorst–Pack (1976) scheme was used for k point generation.

We shall first discuss the optimized structure and electronic structure of pristine $LiMn_2O_4$. Singh et al. (2009) obtained the optimized relaxed structures by repeated sequential relaxation of volume, ion positions, and shape of the primitive unit cell. They found that the optimized structures for GGA and GGA+U are different. While the final optimized structure for GGA is body-centered tetragonal, for GGA+U, it is face-centered orthorhombic. In Figure 19.12, we show the DOS for cubic, orthorhombic, and tetragonal structures of $LiMn_2O_4$. We see that GGA predicts the ground state to be metallic (panel i(c) tetragonal), while GGA+U predicts the ground state to be insulating (panel ii(b) orthorhombic). The GGA+U prediction is found to be in agreement with experiments.

We shall now discuss the doped systems $LiM_{0.5}Mn_{1.5}O_4$ and $LiMMnO_4$, where M is Cr or Mg and replaces a Mn in pristine $LiMn_2O_4$. It is interesting that both GGA and GGA+U predict the same ground-state structures for these systems. For $LiM_{0.5}Mn_{1.5}O_4$, the ground-state structure is base-centered monoclinic, whereas for $LiMMnO_4$, it is body-centered orthorhombic. For the ground-state structures of these systems, we show the DOS in Figure 19.13 for GGA and GGA+U. GGA predicts that Cr-doped systems remain metallic while GGA+U predicts the system to be metallic for $x = 0.5$ but insulating for $x = 1$. For Mg doping, GGA and GGA+U predict that Mg doping drives the system toward the metallic state.

Singh et al. (2009) further calculated JT amplitudes to understand the suppression of JT distortion due to Cr and Mg doping. They found two types of Mn–O bond lengths in pristine $LiMn_2O_4$ in agreement with the experimental results. This implies the presence of two kinds of Mn ions, Mn^{3+} and Mn^{4+},

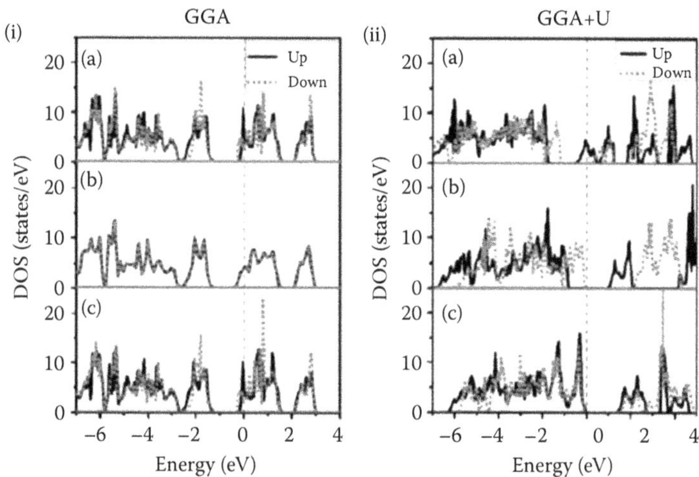

FIGURE 19.12
DOS for (a) cubic, (b) orthorhombic, and (c) tetragonal structures of $LiMn_2O_4$ using (i) GGA and (ii) GGA+U. (From Singh G. et al. 2009. *J. Phys. Chem. Solids* 70:1200–1206.)

with different radii. The JT distortion occurs because Mn^{3+} is a JT-active ion, which will cause JT distortion around it, but Mn^{4+} is not a JT-active ion and does not cause any distortion. Since Cr and Mg are not JT-active ions, their doping is expected to suppress the JT distortion. Further, Singh et al. (2009) found that there is a significant charge transfer from Mn to dopant atoms; therefore, upon doping, some of the Mn^{3+} ions get promoted to Mn^{4+} ions.

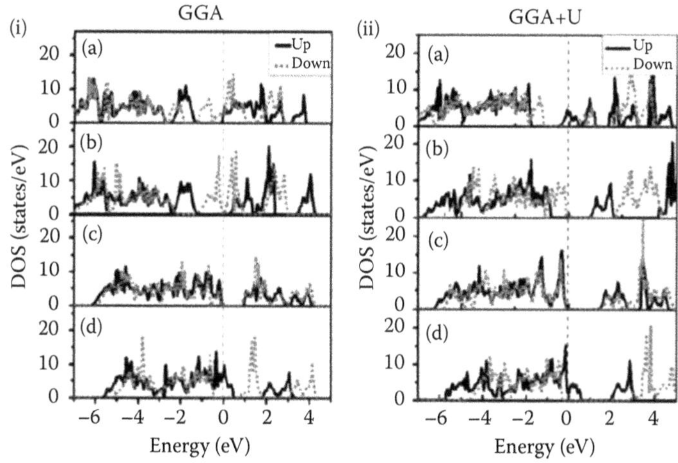

FIGURE 19.13
DOS of (a) $LiCr_{0.5}Mn_{1.5}O_4$, (b) $LiCrMnO_4$, (c) $LiMg_{0.5}Mn_{1.5}O_4$, and (d) $LiMgMnO_4$ using (i) GGA and (ii) GGA+U. (From Singh G. et al. 2009. *J. Phys. Chem. Solids* 70:1200–1206.)

This further suppresses the JT distortion. Thus, doping of $LiMn_2O_4$ with Cr or Mg increases structural stability.

EXERCISES

19.1 Calculate the DOS and the total energy of Li in bcc structure using GGA and one of the codes mentioned in Appendix A.

19.2 Calculate the DOS and the total energy of monoclinic layered $LiMnO_2$ in AF3 configuration using GGA and one of the codes mentioned in Appendix A. Compare your results with those of Mishra and Ceder (1999).

19.3 Calculate the DOS and the total energy of rhombohedral layered $LiMnO_2$ in AF3 configuration using GGA and one of the codes mentioned in Appendix A. Compare your results with those of Mishra and Ceder (1999).

19.4 Calculate the DOS and the total energy of rhombohedral layered MnO_2 in AF3 configuration using GGA and one of the codes mentioned in Appendix A. Compare your results with those of Mishra and Ceder (1999).

19.5 Using the results of Exercises 19.1 through 19.4, calculate the average cell voltage for monoclinic layered $LiMnO_2$ and rhombohedral layered $LiMnO_2$. Compare your results with those of Mishra and Ceder (1999).

19.6 Calculate the formation energies of monoclinic layered $LiMnO_2$(AF3) and rhombohedral layered $LiMnO_2$(AF3) with respect to the FM rhombohedral layered $LiMnO_2$ and compare your results with those given in Table 19.1.

19.7 Calculate the DOS and the total energy for pristine body-centered tetragonal $LiMn_2O_4$ and calculate its average cell voltage.

20

Materials in Extreme Environments

20.1 Introduction

Materials in extreme environments such as high pressure or high temperature can show very interesting and exotic behavior, which may be absent in normal conditions. For example, under high pressure, a material may exhibit insulator to metal transition, metal to superconducting transition, phase transformations, and so on. Materials at high pressures and temperatures are of great interest to understand the earth's core, which presumably has high pressure (~350 GPa) and high temperature (~5700 K). Also, such studies may be of interest to understand various phenomena occurring in planets and stars. There is a great interest in these studies because of their potential applications in various industries related to energy production and space. For example, for fission and future fusion reactors, one needs materials that can operate at very high pressures and temperatures and can withstand a high radiation flux. For spaceships, one needs a material that is light, strong, and can withstand very high temperatures and stress.

Although experimental research has made great advances during the last two decades, such experiments cannot be done in extreme conditions. To study materials in extreme environments, therefore, first-principles calculations are ideally suited. Such calculations do not need any parameters from experiments, have predictive power, do not need an expensive laboratory, and are fast and reliable. In particular, the development of first-principles molecular dynamics has greatly improved the predictive power and opened the way for reliable simulation of phase transformations and chemical processes. The conditions of high pressures and temperatures prevailing in the earth's core, planets, and stars can be easily generated in the computer. In the computer, one can change the pressure and temperature to any value without any fear of explosion. It acts like a virtual laboratory. In this chapter, we shall discuss some examples of first-principles calculations performed in extreme environments. In Section 20.2, we shall discuss materials at high pressures, and in Section 20.3, materials at high temperatures.

20.2 Materials at High Pressures

During the last two decades, the study of materials under high pressure has attracted a lot of attention due to intense scientific and technological interests such as understanding of phase transformations under pressure, earth's and planetary cores, nuclear explosions, discovery of super hard materials, and so on. Also, a rapid progress in both theoretical and experimental methods has given further impetus to the field. Here, we shall focus only on the role of first-principles electronic structure calculations in advancing this field giving few specific examples.

The basic quantity that governs the phase stability of a system is the Gibbs free energy G given by

$$G = E + P\Omega - TS \tag{20.1}$$

where E is the energy, P the pressure, T the temperature, Ω the volume, and S the entropy. In practical application, P and T are applied externally and E, Ω, and S are free to adjust so as to minimize G. Equation 20.1 implies that when pressure is applied, the structures with lower volume can have lower G even if they have higher energy, and thus can exist at higher pressures. Thus, increasing pressure can lead to several phase transformations.

The phase transformation between two phases occurs when the Gibbs energy (Equation 20.1) becomes equal between the two phases. In the present case, when $T = 0$, this implies that the pressure-induced transition will occur along the common tangent line between $E(\Omega)$ curves of the two phases (e.g., see Figure 20.1). The transition pressure P_T is the negative of the slope of the common tangent line (see Equation 20.2).

Pressure can induce many more changes in addition to structural transformations. As the structure of a material changes because of the applied pressure, there will be major changes in the electronic structure. This may lead to insulator–metal transition, superconductivity, and so on. Pressure-induced insulator–metal transitions have been observed in a number of materials. For example, silicon transforms at low temperature from diamond-type insulating phase to metallic β-tin structure at a pressure of ≈11 GPa. Pressure can also induce chemical reactions that do not occur in normal conditions.

We have seen that first-principles calculations can give the total energy per unit cell of a solid accurately as a function of its cell volume Ω_{cell} at temperature $T = 0$. The pressure P and the bulk modulus B can be calculated as (see Exercise 20.1)

$$P = -\frac{dE}{d\Omega_{\text{cell}}} \tag{20.2}$$

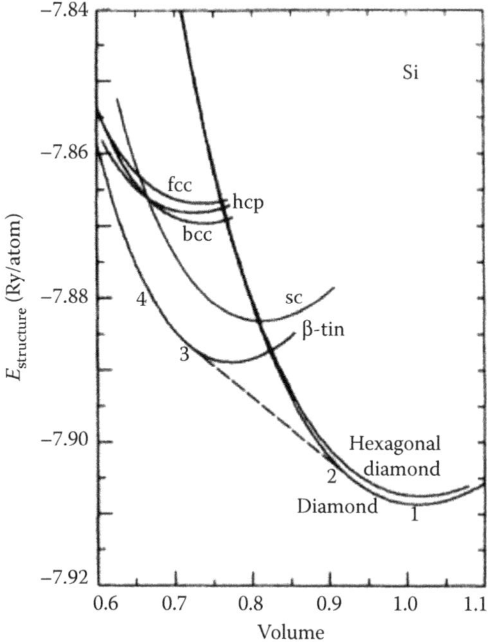

FIGURE 20.1
Total energy curves of the seven phases of Si as a function of the atomic volume. Dashed line is the common tangent of the energy curves for the diamond phase and the β-tin phase. (Reprinted with permission from Yin M. T. and Cohen M. L. Theory of static structural properties, crystal stability, and phase transformations: Application to Si and Ge. *Phys. Rev.* B26: 5668–5687. Copyright 1982, by the American Physical Society.)

$$B = -\Omega_{\text{cell}} \frac{dP}{d\Omega_{\text{cell}}} = \Omega_{\text{cell}} \frac{d^2E}{d\Omega_{\text{cell}}^2} \tag{20.3}$$

The following procedure is generally used to obtain equilibrium cell volume, cohesive energy, bulk modulus, etc. The total energy E is evaluated at many values of Ω_{cell} and the equilibrium volume, cohesive energy, and the bulk modulus are obtained by least-square fitting to Murnaghan's equation of state (Murnaghan 1944, Yin and Cohen 1982)

$$E(\Omega_{\text{cell}}) = \frac{B_0 \Omega_{\text{cell}}}{B_0'} \left[\frac{(\Omega_0/\Omega_{\text{cell}})^{B_0'}}{B_0' - 1} + 1 \right] + \text{constant} \tag{20.4}$$

where Ω_0 is the equilibrium cell volume, B_0 and B_0' are the bulk modulus and its pressure derivative at equilibrium volume Ω_0 (see Exercise 20.2). The minimum energy $E_{\min}$, the equilibrium volume, and the bulk modulus are then deduced from the fitted parameters of Equation 20.4. Pressure P as a function of Ω can also be obtained using Equation 20.2.

We shall now discuss pressure-induced phase transformations in silicon based on the first-principles electronic structure calculations. Silicon is one of the most well-studied systems and many such calculations exist for this material (see, e.g., review by Mujica et al. 2003). The stable form of silicon in normal conditions has a diamond structure. Yin and Cohen (1982) used the first-principles pseudopotential method to obtain total energy and considered seven structures of Si, namely, fcc, bcc, hcp, sc, cubic diamond (CD), hexagonal diamond (HD), and β-tin. For each phase, they calculated the total energies at several lattice constants, which were then least-square fitted to Equation 20.4. The fitted total energy curves for the seven phases of silicon are shown in Figure 20.1. It can be seen that the diamond phase of silicon has the lowest energy and is the most stable of the seven phases of Si, in agreement with the experimental observations. The total energy of the HD phase is slightly more than that of the diamond phase. The remaining five phases have much higher energy and are metallic. Thus, the pressure induces insulator to metal transition, which is in agreement with the experimental results. Other calculations have confirmed these general results and also have considered many other structures also (for a review, see Mujica et al. 2003).

In Figure 20.1, the common tangent line between the diamond and β-tin phases is shown, which corresponds to transition pressure ≈8 GPa compared to the experimental value of ≈11 GPa. The use of the improved exchange-correlation functional increases the transition pressure by a moderate amount. Note that although the hexagonal diamond phase of Si is closest to the diamond phase in energy, the common tangent between the two energy curves either does not exist or has a slope much larger than that of between β-tin and diamond energy curves. Thus, on applying pressure, the diamond phase will transform to the metallic β-tin phase and not to the hexagonal diamond phase.

It is interesting to note that tetrahedral semiconductors like Si and GaAs also show insulator to metal transition on expansion. In Figure 20.2, the density of states (DOS) for GaAs at various lattice constants is shown (Shukla et al. 2004). Figure 20.2a, which corresponds to the equilibrium lattice constant, shows a band gap above the valence band. This band gap decreases as the lattice constant is increased and becomes zero at a = 12.24 a.u., signaling an insulator to metal transition.

Carbon, which is another important material, has been studied extensively using first-principles electronic structure methods (Fahy et al. 1986, Galli et al. 1990, Mailhiot and McMahan 1991, Furthmuller et al. 1994, Scandolo et al. 1995). The stable form of carbon in normal conditions is hexagonal graphite, while diamond is thermodynamically stable above the pressure of 1.7 GPa at 0 K. At normal temperature and pressure, diamond can persist indefinitely in metastable form. It has been found that there is a very small energy difference between the graphite and diamond phases. The graphite to diamond transition has been the subject of many studies (Fahy et al. 1986,

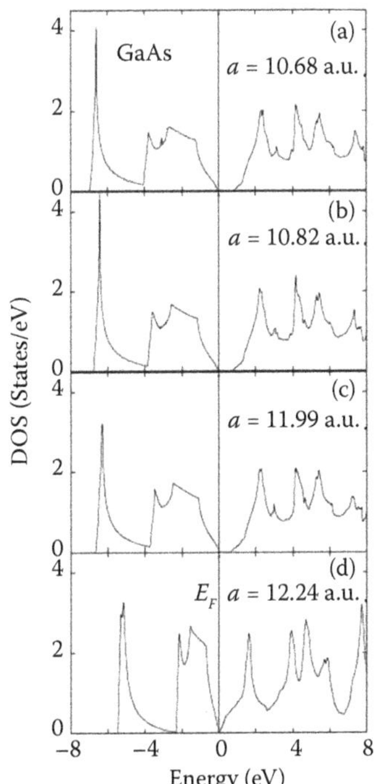

FIGURE 20.2
DOS of GaAs at various lattice constants (a) 10.68 a.u., (b) 10.82 a.u., (c) 11.99 a.u., and (d) 12.44 a.u.

Scandolo et al. 1995). Once in diamond form, no further transformation to other phases has been observed on applying further pressure experimentally. However, first-principles calculations have predicted transformation to bc8 phase at pressure above 1.1–1.2 TPa. Above even higher pressure ~2.7 TPa, there is another transformation to simple cubic structure. We shall not go into details, and the readers are advised to refer the original papers and excellent review article by Mujica et al. (2003).

20.3 Materials at High Temperatures

Understanding materials behavior at high temperatures is needed not only to understand the earth's core and planets, but also for technological applications in the energy industry. Experiments at temperatures prevailing in the core of the earth or planets are not possible. So, the only hope to understand materials at such temperatures is through theoretical computations. The static first-principles calculations are of not much use at finite temperatures.

Thus, most finite temperature calculations are based on molecular dynamics (MD) or Monte Carlo methods. Classical MD results are dependent on the choice of empirical potential, and therefore, are less reliable. Therefore, we shall discuss the results of a few first-principles MD or quantum Monte Carlo calculations on systems such as hydrogen and carbon, which are materials present in many planets and stars.

We shall first discuss some important results for hydrogen, which is the simplest system containing one proton and one electron. However, the behavior of this system as a function of pressure and temperature is far from simple, and has been challenging to theoreticians. This is because the proton has a small mass that gives rise to a large zero point motion. Thus, unlike other systems where zero point motion can be neglected and the nuclei could be treated classically, in hydrogen, protons have to be treated quantum mechanically at the same theoretical footing as the electrons.

In 1935, Wigner and Huntington (1935) predicted that at pressures at about 25 GPa, molecular hydrogen would dissociate into monatomic metal, at temperature $T = 0$ K. In 1968, Ashcroft predicted that this hypothetical monatomic metallic hydrogen may show very unusual property such as high-temperature superconductivity (Ashcroft 1968, Hemley and Ashcroft 1998). However, hydrogen still remains an insulating molecular solid at $T = 0$ K up to the highest pressures achievable experimentally, ~350 GPa (Silvera 2010). Thus, understanding the metallization of hydrogen has been one of the great challenges in condensed matter physics.

Another way to achieve metallization of hydrogen is through a different route by using high pressure at high temperature in the liquid phase. To understand this, we have drawn a schematic phase diagram in Figure 20.3, in which solid H_2, liquid H_2, and metallic liquid H phases are clearly

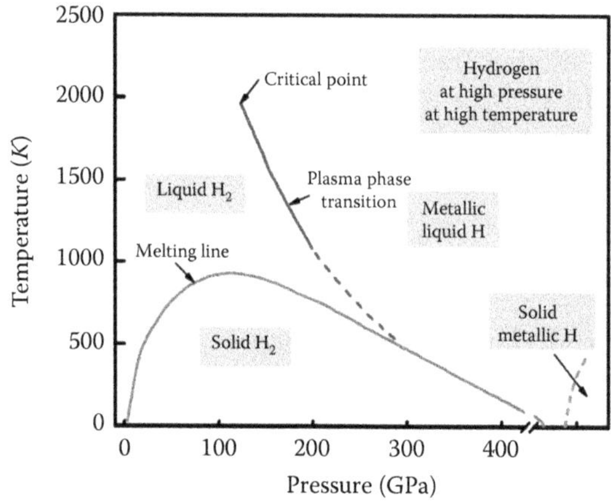

FIGURE 20.3
Schematic phase diagram of hydrogen.

indicated. The melting line that demarcates solid and liquid phases is also shown. Scandolo (2003) and Bonev et al. (2004) used first-principles MD to predict a peak in the melting line of hydrogen that has been observed in experiments. We see in Figure 20.3 that with increasing pressure, beyond the peak in the melting line, the melting temperature decreases. They also predicted that above this melting line, there is a line of dissociation between the liquid H_2 and metallic H phases. This transition from molecular liquid to atomic metallic liquid is called plasma phase transition (PPT). Morales et al. (2010) have found clear evidence of a first-order transition between low conductivity molecular liquid and a high conductivity atomic state using quantum simulation techniques based on their quantum Monte Carlo simulations. Recently, Eremets and Troyan (2011) have claimed to have observed metallic hydrogen at high pressures ~270 GPA in a diamond anvil cell.

We shall now discuss some results of first-principles calculations for carbon and calculation of its phase diagram. Significant quantities of carbon are supposed to exist in giant planets such as Uranus and Neptune. Thus, understanding the behavior of carbon at high pressure and high temperature is essential for predicting the evolution and structure of such planets. Since experiments cannot be done at pressures and temperatures prevailing in these planets, first-principles simulations have played an extremely important role in providing the information under such conditions. Several first-principles simulations have been done on carbon at high temperatures and high pressures (see, e.g., Galli et al. 1990, Grumbach and Martin 1996, Wang et al. 2005, Correa et al 2006, Eggert et al. 2010).

To discuss some of their main results, we draw a phase diagram of carbon in Figure 20.4 based on the calculation of Wang et al. (2005) (for simplicity, ignore dashed and dash-dot curves, which are empirical melting curves from two different groups). The phase diagram has the following four features:

1. The boundary between graphite and diamond regions.
2. Melting line of graphite, which separates the graphite phase from the liquid phase.
3. Melting line of diamond, which separates diamond and liquid phases. Note the change of slope in the upper region.
4. Triple point where all these three lines meet, that is, where diamond, graphite, and liquid phase coexist.

So far, only the graphite/diamond boundary and the melting line of graphite have been determined experimentally with reasonable accuracy. Also, the triple point has been found experimentally in the range of 4500–5000 K and pressure ~12 GPa.

Let us now discuss the melting line of diamond, which has been the subject of debate for a long time. Before 1980, it was thought that the melting

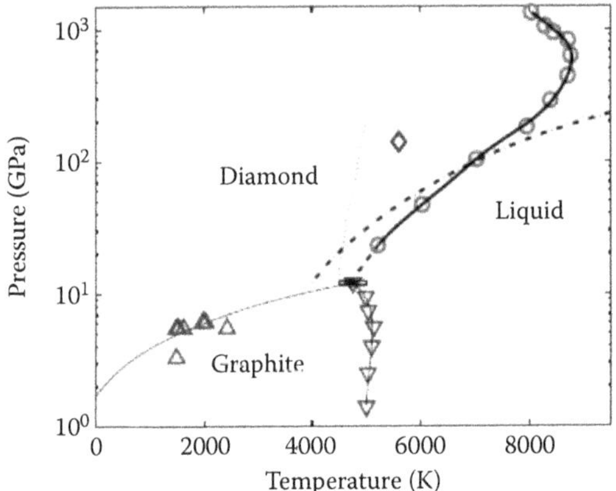

FIGURE 20.4
Proposed phase diagram of carbon. (Reprinted with permission from Wang X.-F. et al. Carbon phase diagram from *ab initio* molecular dynamics. *Phys. Rev. Lett.* 95: 185701. Copyright 2005, by the American Physical Society.)

line would have a negative slope. This was based on the analogy with the Si and Ge phase diagram in which melting lines have negative slopes due to the higher density of the liquid compared to the solid. This picture was brought into question by the experiment of Shaner et al. (1984) and had to be abandoned. Later, Galli et al. (1990) found the positive slope for the melting line based on their first-principles MD simulations. Grumbach and Martin (1996) did an extensive study of carbon over a wide range of pressure (400–4000 GPa) and temperature (2000–36,000 K) using first-principles MD. They predicted a change in the slope of the diamond melting line at high pressures and a maximum in the melting temperature. This was based on their observation that there is a change from fourfold to sixfold coordination in liquid when the pressure is varied from 400 to 1000 GPa. Wang et al. (2005) have also performed first-principles MD on carbon and confirmed that there is indeed a change in slope at high pressure, and there is a maximum in the melting curve at $T_m = 8755$ K and $P_m = 630$ GPa. They, however, do not find a discontinuous change in slope around this point. Thus, they rule out the occurrence of a first-order liquid–liquid phase transition. Based on their calculations, they propose a phase diagram that is shown in Figure 20.4.

Correa et al. (2006) explored the diamond and bc8 phase melting line curves and their phase boundary in the solid phase using first-principles MD simulations. They found maxima in both melting curves with a triple point at ~850 GPa and 7400 K. They also found hot compressed diamond as a semiconductor, which becomes metal on melting. For more details, we refer to the original papers.

EXERCISES

20.1 Using thermodynamical relations, derive Equations 20.2 and 20.3.

20.2 Derive Murnaghan's equation of state. [Hint: write $B = B_0 + B_0'P$ and use Equations 20.2 and 20.3.]

20.3 Calculate the total energy of silicon in diamond and β-tin phases as a function of atomic volume and plot them as in Figure 20.1. Draw the common tangent and find the transition pressure.

20.4 Calculate the DOS of GaAs as a function of change in its lattice constant as shown in Figure 20.2. Show that the insulator to metal transition occurs with lattice compression as well as expansion.

20.5 Repeat Exercise 20.4 for carbon. Does it also show insulator to metal transition for lattice compression as well as lattice expansion?

Further Reading

Hemley R. J., Crabtree G. W., and Buchanan M. V. 2009. Materials in extreme environments. *Physics Today* 62(11): 32–37.

Appendix A: Electronic Structure Codes

Some commonly used first-principles electronic structure codes are listed below with the link which gives detailed information about the code:

1. VASP (Viena *Ab-initio* Simulation package): http://www.vasp.at/

   A first-principles plane wave code that can use norm-conserving, ultrasoft pseudopotentials or the PAW method. It can also do BOMD. It has a database containing pseudopotentials for all elements. It has a license fee.

2. QUANTUM ESSPRESSO: http://www.quantum-espresso.org/

   Another first-principles plane wave code that can use norm-conserving or ultra-soft pseudopotentials and that is distributed free of charge under GNU public license.

3. ABINIT: http://www.abinit.org/

   Another first-principles plane wave code that uses pseudopotentials and that is distributed free of charge under GNU public license.

4. CPMD: http://www.cpmd.org/

   A first-principles pseudopotential plane wave code particularly designed for CPMD that is distributed free of charge to nonprofit organization under CPMD free license.

5. CASTEP: http://www.castep.org/CASTEP/CASTEP

   A first-principles pseudopotential plane wave code. It can simulate a wide range of material properties such as energetics, structure, vibrational properties, and so on. It is free for UK academic groups but has a license fee for users from other countries.

6. SPRKKR code: http://olymp.cup.uni-muenchen.de/index.php?option=com_content&view=article&id=8&catid=4&Itemid=7

   A KKR code and can also do disordered alloy calculations using the KKR-CPA method. It is free of charge under SPRKKR license.

7. WIEN2k: http://www.wien2k.at/

   A full potential LAPW code and has a license fee.

8. FPLO: http://www.fplo.de/

   FPLO is a full potential local orbital minimum basis electronic structure code. It can also handle disordered alloy using CPA. It has a license fee.

9. LMTO: http://www.fkf.mpg.de/andersen/

TB-LMTO code and FP-LMTO code are available free of charge from this site on request.

10. DFTB: http://www.dftb.org/

    The DFTB code is based on the density functional-based tight-binding method, which is very fast and efficient. It is free of charge for non-commercial use.

11. Octopus: http://www.tddft.org/programs/octopus/wiki/index.php/Main_Page

    Octopus is a pseudopotential-based code that can do TD-DFT calculations. It is a free software under GPL license.

12. SIESTA http://www.icmab.es/dmmis/leem/siesta/

    A linear-scaling code with local orbitals and is very efficient for large systems. It is free for academic use.

13. Pseudopotential generation codes: Several sites give free pseudopotential generation codes such as

    i. Vanderbilt Ultra-soft potential site: http://www.physics.rutgers.edu/~dhv/uspp/

    ii. Octopus pseudopotential generator: Norm-conserving Troullier Martins potentials can be generated by the code: http://www.tddft.org/programs/octopus/pseudo.php

    iii. Opium pseudopotential generator: http://opium.sourceforge.net/index.html

Appendix B: List of Projects

1. Electronic structure of Mn-doped GaN dilute magnetic semiconductor.
2. Electronic structure of full Heusler alloy CO_2MnSi.
3. Electronic structure of half-Heusler alloy NiMnSb.
4. Electronic structure of FeNb magnetic multilayers.
5. Electronic structure of high T_c Fe-pnictides and a minimal two-band model.
6. Electronic structure of high T_c cuprate La_2CuO_4.
7. *Ab initio* study of battery material spinel $LiMn_2O_4$.
8. Electronic structure of battery material layered $LiMnO_2$.
9. First-principles calculations of $InGaZnO_4$ crystalline system.
10. Effect of crown ether on dynamics of metal ion hydration from *ab initio* molecular dynamics.
11. Electronic structure of hydrogenated silicon nanowires.
12. Study of Si (100) surface reconstruction.
13. Electronic structure of Cu (111) surface.
14. Electronic structure of Bi_2Se_3 topological insulator.
15. Electronic structure of graphene using the tight-binding method.
16. Electronic structure of carbon nanotubes using the tight-binding method.
17. Band structure of silicon using the tight-binding method.
18. Electronic structure of $BaTiO_3$ in paraelectric and ferroelectric phases.
19. Electronic structure of multiferroic $BiFeO_3$.
20. Structure of Si_7 cluster using simulated annealing and *ab initio* molecular dynamics.
21. Total energy of silicon in diamond and β-tin phases and determination of transition pressure.

Appendix C: Atomic Units

Since we will be working at the atomic scale, the CGS or SI units are too big for our purpose. For example, the interatomic distances are of the order of a few angstroms, which is very small compared to CGS or SI unit of length. Electronic structure calculations for atoms, molecules, and solids are most conveniently done in atomic units (a.u.). There are two kinds of a.u. in use, namely Hartree atomic units and Rydberg atomic units. In this book, we follow Hartree atomic units, which are more commonly used now. In a.u., various equations such as Schrödinger equation take a much simpler form. For a good introduction using scaling arguments, see notes by J. Peatross at the link www.physics.byu.edu/faculty/peatross/Handouts/AtomicUnits.pdf.

In a.u., m_e (electron mass) = 1, $\hbar = 1$, and e (electronic charge) = 1. Also the Coulomb constant $1/(4\pi\varepsilon_0)$ is taken as unity. This implies units of length, energy, time, and so on as follows:

Length: The Bohr radius $a_0 = 4\pi\varepsilon_0(\hbar^2/m_e e^2) = 1$. Thus, length is measured in units of a_0 (0.529177×10^{-10} m).

Mass: Mass is measured in units of electron mass m_e (9.109382×10^{-31} kg).

Velocity: The fine structure constant $\alpha = e^2/4\pi\varepsilon_0\hbar c = 1/137$, which is dimensionless.
Therefore, 1 a.u. of velocity = $c/137 = \alpha c$

Time: 1 a.u. of time = 1 a.u. of length/1 a.u. of velocity = $a_0/\alpha c$ = 2.418884×10^{-17} s

Energy: The units of energy is Hartree = E_h = 2 Ry = 27.2 eV. This is derived as follows:

$$1\,\text{Ry} = \frac{m_e e^4}{(4\pi\varepsilon_0)^2 2\hbar^2} = \frac{1}{2}\text{a.u.} = 13.6\ \text{eV}$$

$$1\ \text{a.u.}\ (E_h = 27.2\ \text{eV}) = 2\ \text{Ry}$$

Charge: Charge is measured in units of electron charge $e = 1.602176 \times 10^{-19}$ C

The values of electronic charge and mass are quoted from *The NIST Reference on Constants, Units and Uncertainty*. National Institute of Standard and Technology. http://physics.nist.gov/cuu/index.html

Appendix D: Functional, Functional Derivative, and Functional Minimization

A function $f(x)$ is a rule that tells us how to go from a variable x to a number $f(x)$. Following are examples of a function:

$$f(x) = x^2, \quad f(x) = \sin(x), \quad f(x) = e^x, \quad \text{etc.}$$

A functional $F[f]$ is a rule that tells us how to go from a function $f(x)$ to a number $F[f]$. A functional of $f(x)$ is generally denoted as $F[f]$ with a square bracket. A functional involves an integral over the function. For example,

$$F[f] = \int \sin(x) f(x) dx \tag{D.1}$$

is a functional of a function $f(x)$. Another example is expectation value of H (Equation 2.18), which is functional of $\psi(x)$.

We define a functional derivative through a differential of a function, δF as

$$\begin{aligned} \delta F[f] &= F[f + \delta f] - F[f] \\ &= \int \frac{\delta F}{\delta f(x)} \delta f(x) dx \end{aligned} \tag{D.2}$$

where $\delta F / \delta f(x)$ is a functional derivative of F with respect to f at point x. For example, functional derivative of Equation D.1 will be

$$\frac{\delta F}{\delta f(x)} = \sin(x)$$

Note that the integral has disappeared after taking the functional derivative. For more details, refer to Parr and Yang (1989).

Another point that we would like to explain regards the minimization of energy functional $E[\psi]$. Since ψ is a complex, minimizing $E[\psi]$ with respect to ψ means minimizing it with respect to the real and imaginary parts of ψ, which will give us two equations. But, since $E[\psi]$, is real, this is equivalent to

$$\frac{\delta E}{\delta \psi^*} = 0 \tag{D.3}$$

as we can see from the following simple example.

Suppose $f(z)$ is a function of complex variable $z = x + iy$ but $f(z)$ is real. Thus, minimizing $f(z)$ with respect to z will mean two equations

$$\frac{\partial f(z)}{\partial x} = 0 \quad \text{and} \quad \frac{\partial f(z)}{\partial y} = 0 \tag{D.4}$$

Now writing

$$x = \frac{1}{2}(z + z^*)$$

and

$$y = \frac{z - z^*}{2i}$$

we have

$$\frac{\partial x}{\partial z^*} = \frac{1}{2} \quad \text{and} \quad \frac{\partial y}{\partial z^*} = \frac{1}{2}i$$

Thus,

$$\begin{aligned}\frac{\partial f(z)}{\partial z^*} &= \frac{\partial f(z)}{\partial x}\frac{\partial x}{\partial z^*} + \frac{\partial f(z)}{\partial y}\frac{\partial y}{\partial z^*} \\ &= \frac{1}{2}\left[\frac{\partial f(z)}{\partial x} + i\frac{\partial f(z)}{\partial y}\right]\end{aligned} \tag{D.5}$$

Since f is real, setting

$$\frac{\partial f(z)}{\partial z^*} = 0 \tag{D.6}$$

gives both equations as shown in Equation D.4. Thus, a single equation (D.6) is equivalent to the two equations given in Equation D.4 and therefore, we just have to minimize function f with respect to z^*.

Appendix E: Orthonormalization of Orbitals in the Car–Parrinello Method

If we could integrate the equation of motion for the orbitals given in Equation 9.31 exactly, the orbitals at any time t would automatically come out to be orthogonal provided that they were initially orthogonal. This is because the orthogonality constraint has already been put in the Lagrangian. However, the equations of motion are integrated numerically on a discretized grid of points that introduces error at each step. The error accumulates as the simulation progresses and as a result orthogonality deteriorates very rapidly. Because of this problem, it is necessary to orthogonalize the orbitals at each time step. The orthogonalization procedure should be such that the total energy given by Equation 9.39 is conserved and is consistent with the expression for the Lagrange multipliers given by Equation 9.38. This can be achieved by using an algorithm such as SHAKE as proposed by Ryckaert et al. (1977) for geometrically constrained classical MD. We shall now briefly discuss how it can be implemented in this case (Car and Parrinello 1989).

We start with numerically integrating equation of motion for orbitals, Equation 9.31, using the Verlet algorithm:

$$\psi_i(\vec{r},t+\Delta t) = 2\psi_i(\vec{r},t) - \psi_i(\vec{r},t-\Delta t) + \frac{\Delta t^2}{\mu}\left(-H_{KS}\,\psi_i(\vec{r},t) + \sum_j \Lambda_{ij}\,\psi_j(\vec{r},t)\right) \tag{E.1}$$

We assume that $\psi_i(\vec{r},t)$ and $\psi_i(\vec{r},t-\Delta t)$ are orthonormal. We define $\bar{\psi}_i(\vec{r},t+\Delta t)$ as the wave function at time $t+\Delta t$, as given by Equation E.1 without the constraint, that is

$$\bar{\psi}_i(\vec{r},t+\Delta t) = 2\psi_i(\vec{r},t) - \psi_i(\vec{r},t-\Delta t) + \frac{\Delta t^2}{\mu}(-H_{KS}\psi_i(\vec{r},t)) \tag{E.2}$$

Thus, from Equations E.1 and E.2, we can write

$$\begin{aligned}\psi_i(\vec{r},t+\Delta t) &= \bar{\psi}_i(\vec{r},t+\Delta t) + \frac{\Delta t^2}{\mu}\sum_j \Lambda_{ij}\,\psi_j(\vec{r},t)\\ &= \bar{\psi}_i(\vec{r},t+\Delta t) + \sum_j X^*_{ij}\,\psi_j(\vec{r},t)\end{aligned} \tag{E.3}$$

where

$$X = \frac{\Delta t^2}{\mu}\Lambda^* \tag{E.4}$$

The matrix X or Λ is obtained by demanding that $\psi_i(\vec{r},t+\Delta t)$ are orthogonal, that is

$$\int \psi_i^*(\vec{r},t+\Delta t)\psi_j(\vec{r},t+\Delta t)d^3r = \delta_{ij} \tag{E.5}$$

Substitution of Equation E.3 into Equation E.5 gives

$$A_{ij} + \sum_\ell X_{j\ell}^* B_{\ell i}^* + \sum_\ell X_{i\ell} B_{\ell j} + \sum_\ell X_{i\ell} X_{j\ell}^* = \delta_{ij} \tag{E.6}$$

where

$$A_{ij} = \int \bar{\psi}_i^*(\vec{r},t+\Delta t)\bar{\psi}_j(\vec{r},t+\Delta t)\, d^3r \tag{E.7}$$

$$= \int \left[\psi_i^*(\vec{r},t+\Delta t) - \frac{\Delta t^2}{\mu}\sum_\ell \Lambda_{i\ell}^* \psi_\ell^*(\vec{r},t)\right]$$

$$\times \left[\psi_j(\vec{r},t+\Delta t) - \frac{\Delta t^2}{\mu}\sum_j \Lambda_{jk} \psi_k^*(\vec{r},t)\right] d^3r$$

$$= \delta_{ij} - O(\Delta t^2) \tag{E.8}$$

and

$$B_{ij} = \int \psi_i^*(\vec{r},t)\bar{\psi}_j(\vec{r},t+\Delta t)\, d^3r \tag{E.9}$$

$$= \int \psi_i^*(\vec{r},t)\left[\psi_j(\vec{r},t) + \Delta t\, \dot{\psi}_j(\vec{r},t)\ldots\right]d^3r$$

$$= \delta_{ij} + O(\Delta t) \tag{E.10}$$

Thus to the lowest order in Δt,

$$A = I + O(\Delta t^2)$$

$$B = I + O(\Delta \mathrm{t}) \tag{E.11}$$

$$X = O(\Delta t^2)$$

Writing Equation E.6 in matrix form, we get

$$A + B^{\dagger} X^{\dagger} + XB + XX^{\dagger} = I \tag{E.12}$$

Note that in Equation E.12, A and B are known and we solve for X. A conventional way to solve Equation E.12 is by iteration. To the lowest order in Δt, Equation E.12 becomes

$$A + (I + O(\Delta t))X^{\dagger} + X(I + O(\Delta t)) = I$$

or

$$A + X^{\dagger} + X = I$$

or

$$X^{(O)} = \frac{1}{2}(I - A) \quad \text{as } X^{\dagger} = X \tag{E.13}$$

For iteration purpose, Equation E.12 can be rewritten as

$$I - A + X(I - B) - X + (I - B^{\dagger})X - X - X^2 = 0$$

or

$$X = \frac{1}{2}\left[I - A + X(I - B) + (I - B^{\dagger})X - X^2\right] \tag{E.14}$$

For numerical iteration, Equation E.14 can be used as

$$X^{(n)} = \frac{1}{2}\left[I - A + X^{(n-1)}(I - B) + (I - B^{\dagger})X^{(n-1)} - X^{(n-1)^2}\right] \tag{E.15}$$

where $X^{(n)}$ is the value of X in the nth iteration. This iteration process generally converges in <10 steps. Once X is known, orthonormal orbital $\psi_i(\vec{r}, t + \Delta t)$ is calculated using Equation E.3.

Appendix F: Sigma (σ) and Pi (π) Bonds

Let us first consider angular dependence of atomic orbitals. An *s* orbital is spherically symmetric and represented by drawing a circle as shown in Figure F.1. There are three *p* orbitals, which can be chosen as

$$|p_x\rangle = f(r)x/r$$

$$|p_y\rangle = f(r)y/r$$

$$|p_z\rangle = f(r)z/r$$

where $f(r)$ is a function of radial distance r from the center of the atom. The angular dependence of the *p* orbitals is governed by x, y, or z. These orbitals are schematically shown in Figure F.2. This is essentially angle dependence of $\cos\theta$, where θ is the angle measured from x-, y-, or z-axis.

Now, let us consider a diatomic molecule. When two atoms such as H atoms are brought closer, they will form a covalent bond between two *s*-orbitals of

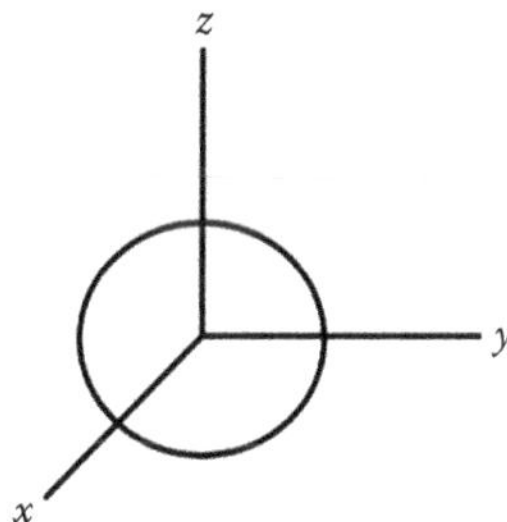

FIGURE F.1
Schematic representation of an *s* orbital.

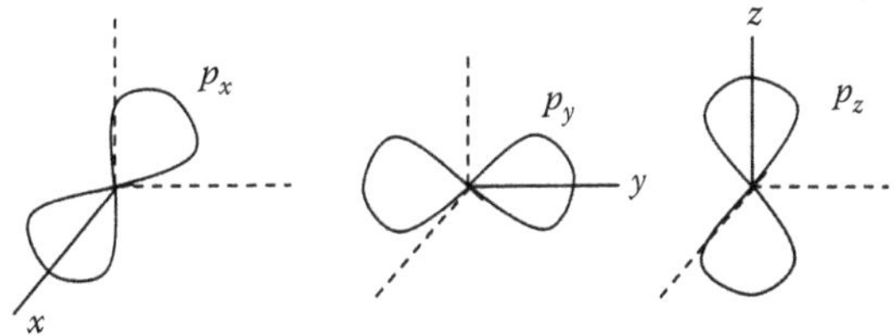

FIGURE F.2
Schematic representation of *p* orbitals.

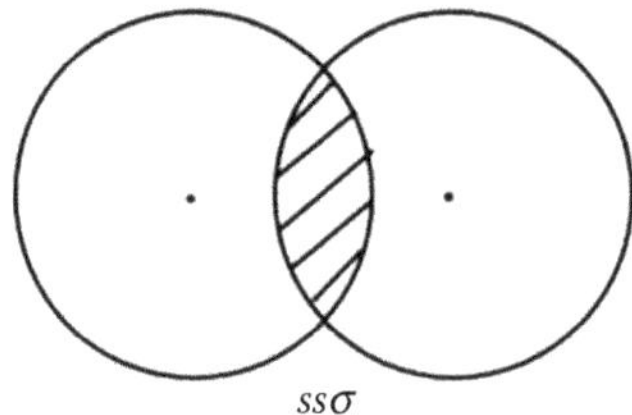

FIGURE F.3
Schematic showing formation of $ss\sigma$ bond.

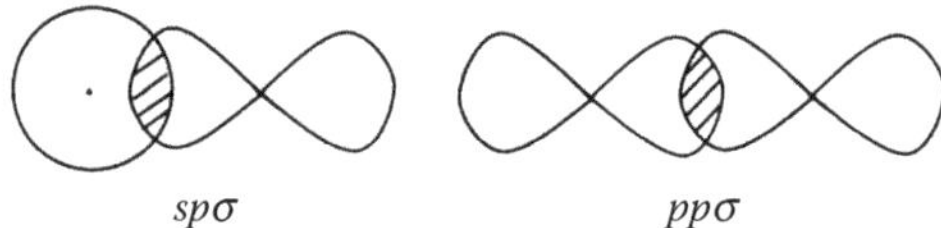

FIGURE F.4
Formation of $sp\sigma$ and $pp\sigma$ bonds.

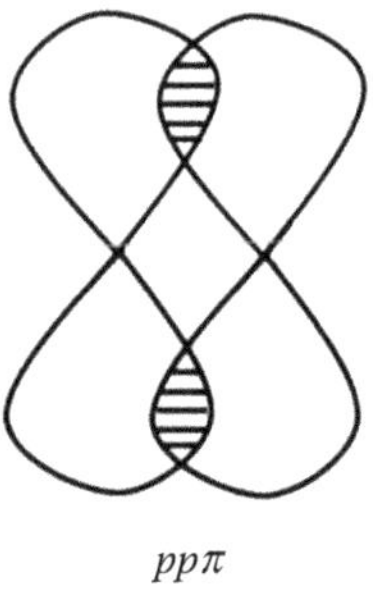

FIGURE F.5
Formation of $pp\pi$ bond.

the atoms as shown in Figure F.3. This type of bond in which the overlap regions lie in between the line joining the two nuclei is known as a σ bond. A σ bond can occur between two s orbitals, or one s and p, or two p orbitals. These bonds are then denoted as $ss\sigma$, $sp\sigma$, and $pp\sigma$, respectively, and their formation is schematically shown in Figures F.3 and F.4. A pi (π) bond occurs between two orbitals, which are placed parallel to each other as shown in Figure F.5. One can see that, in this case, the overlap region is away from the line joining the two nuclei.

Appendix G: sp, sp², *and* sp³ *Hybrids*

In a tight-binding or LCAO calculation, it is sometimes more convenient to use new orbitals known as hybrids, which are formed from a linear combination of atomic orbitals of the same atom. Here, we shall briefly discuss hybrids that are constructed from *s* and *p* atomic orbitals on the same atom. For more details, we refer to Sutton (1993).

1. *sp Hybrids*: A hybrid constructed from one *s* and one *p* orbital say, p_x, without changing other two *p* orbitals is called an *sp* hybrid. In this way, we can construct two linearly independent *sp* hybrids that can be expressed mathematically as

$$h_1 = \frac{1}{\sqrt{2}}(|s\rangle + |p_x\rangle)$$

$$h_2 = \frac{1}{\sqrt{2}}(|s\rangle - |p_x\rangle)$$

These hybrids lie along a line and are schematically shown in Figure G.1. We can see that they have a 180° angle between them and thus are best suited for a linear geometry.

2. *sp² Hybrids*: A hybrid constructed from one *s* orbital and two *p* orbitals, say p_x and p_y, without changing the remaining *p* orbital, is called an *sp*² hybrid. In this way, we can construct three linearly independent *sp*² hybrids that can be written mathematically as

$$h_1 = \frac{1}{\sqrt{3}}(|s\rangle + \sqrt{2}\,|p_x\rangle)$$

$$h_2 = \frac{1}{\sqrt{3}}\left(|s\rangle - \frac{1}{\sqrt{2}}\,|p_x\rangle + \sqrt{\frac{3}{2}}\,|p_y\rangle\right)$$

$$h_3 = \frac{1}{\sqrt{3}}\left(|s\rangle - \frac{1}{\sqrt{2}}\,|p_x\rangle - \sqrt{\frac{3}{2}}\,|p_y\rangle\right)$$

These hybrids are schematically shown in Figure G.2. These hybrids lie in a plane and the angle between them is 120°. These are best suited for a planar geometry.

3. *sp³ Hybrids*: A hybrid constructed from one *s* orbital and three *p* orbitals is called an *sp*³ hybrid. In this way, we can construct four linearly independent *sp*³ hybrids that can be written as

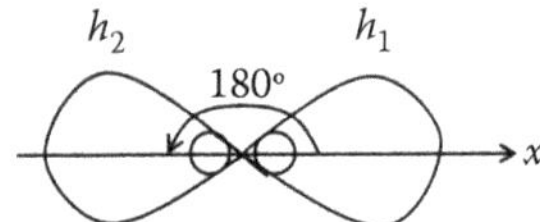

FIGURE G.1
Schematic showing *sp* hybrids.

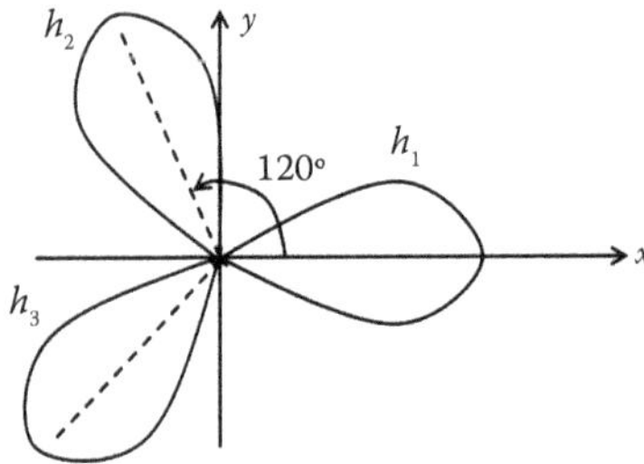

FIGURE G.2
Schematic showing sp^2 hybrids.

$$h_1 = \frac{1}{2}\left(|s\rangle + |p_x\rangle + |p_y\rangle + |p_z\rangle\right)$$

$$h_2 = \frac{1}{2}\left(|s\rangle + |p_x\rangle - |p_y\rangle - |p_z\rangle\right)$$

$$h_3 = \frac{1}{\sqrt{2}}\left(|s\rangle - |p_x\rangle + |p_y\rangle - |p_z\rangle\right)$$

$$h_4 = \frac{1}{\sqrt{2}}\left(|s\rangle - |p_x\rangle - |p_y\rangle + |p_z\rangle\right)$$

These hybrids are schematically shown in Figure G.3 and lie along the lines joining the center of a regular tetrahedron to its vertices. Thus, the angle between these hybrids is 109.5°. These hybrids are best suited for a tetrahedral geometry.

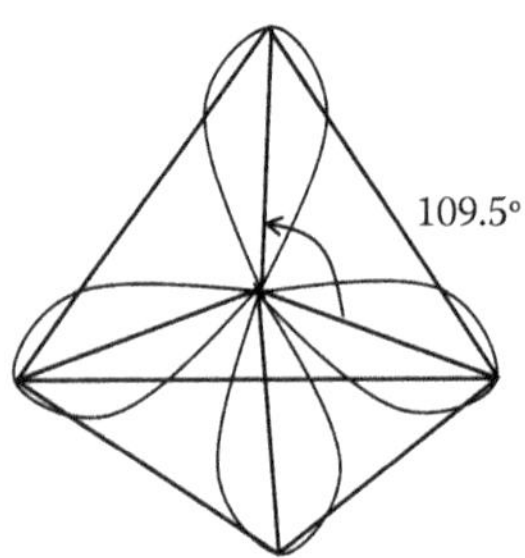

FIGURE G.3
Schematic showing sp^3 hybrids.

References

Abruña H. D., Kiya Y., and Henderson J. C. 2008. Batteries and electrochemical capacitors. *Physics Today* 61(12): 43–47.

Adjaoud O., Steinle-Neumann G., Burton B. P., and van de Walle A. 2009. First-principles phase diagram calculations for the HfC–TiC, ZrC–TiC, and HfC–ZrC solid solutions. *Phys. Rev. B* 80: 134112.

Agarwal S. C. 1996. Heterogeneities in hydrogenated amorphous silicon. *Ind. J. Pure Appl. Phys.* 34: 597–602.

Alder B. J. and Wainwright T. E. 1959. Studies in molecular dynamics. I. General method. *J. Chem. Phys.* 31: 459–466.

Alfe D., Gillan M. J., and Price G. D. 2002. Composition and temperature of the earth's core constrained by combining *ab initio* calculations and seismic data. *Earth Planet. Sci. Lett.* 195: 91–98.

Allen M. P. and Tildesley D. J. 1989. *Computer Simulation of Liquids*. New York, NY: Oxford University Press.

Andersen O. K. 1975. Linear methods in band theory. *Phys. Rev. B* 12: 3060–3083.

Andersen O. K. and Saha-Dasgupta T. 2000. Muffin-tin orbitals of arbitrary order. *Phys. Rev. B* 62: R16219–R16222.

Anderson P. W. 1958. Absence of diffusion in certain random lattices. *Phys. Rev.* 109: 1492–1505.

Andreoni W. 1991. Computer simulations of small semiconductor. *Z. Phys. D* 19: 31–36.

Anglada E. et al. 2007. Formation of gold nanowires with impurities: A first-principles molecular dynamics simulation. *Phys. Rev. Lett.* 98: 096102.

Animosov V. I., Aryasetiawan F., and Lichtenstein A. A. 1997. First principles calculations of the electronic structure and spectra of strongly correlated systems: The LDA+U method. *J. Phys.: Condens. Matter* 9: 767–808.

Antoncik E. 1959. Approximate formulation of the orthogonalized plane-wave method. *J. Phys. Chem. Solids* 10: 314–320.

Appelbaum J. A., Baraff G. A., and Hamann D. R. 1976. The Si (100) surface. III. Surface reconstruction. *Phys. Rev. B* 14: 588–601.

Armstrong A. R. and Bruce P. G. 1996. Synthesis of layered $LiMnO_2$ as an electrode for rechargeable lithium batteries. *Nature* 381: 499–500.

Aryasetiawan F. and Gunnarsson O. 1998. The GW method. *Rep. Prog. Phys.* 61: 237–312.

Asato M., Settels A., Hoshino T., Asada T., Blügel S., Zeller R., and Dederichs P. H. 1999. Full-potential KKR calculations for metals and semiconductors. *Phys. Rev. B* 60: 5202–5210.

Ashcroft N. W. 1968. Metallic hydrogen: A high temperature superconductor? *Phys. Rev. Lett.* 21: 1748–1749.

Ashcroft N. and Mermin N. 1976. *Solid State Physics*. New York, NY: W. B. Saunders Company.

Asonen H., Lindroos M., Pessa M., Prasad R., Rao R. S., and Bansil A. 1982. Angle-resolved photoemission study of (100), (110), and (111) surfaces of Cu0.9Al0.1: Bulk and surface electronic structure of the alloy. *Phys. Rev. B* 25: 7075–7085.

Avron J. E., Osadchy D., and Seiler R. 2003. A topological look at the quantum Hall effect. *Physics Today* 56(8): 38–42.

Axe J. D. 1967. Apparent ionic charges and vibrational eigenmodes of $BaTiO_3$ and other perovskites. *Phys. Rev.* 157: 429–437.

Bachelet G. B., Hamann D. R., and Schlüter M. 1982. Pseudopotentials that work: From H to Pu. *Phys. Rev. B* 26: 4199–4228.

Bagno P., Donà Dalle Rose L. F., and Toigo F. 1991. Electrostatic energy of a complex and infinitely extended electronic system. *Adv. Phys.* 40: 685–718.

Baibich M. N., Broto J. M., Fert A., Nguyen Van Dau F., Petroff F., Eitenne P., Creuzet G., Friederich A., and Chazelas J. 1988. Giant magnetoresistance of (001)Fe/(001)Cr magnetic superlattices. *Phys. Rev. Lett.* 61: 2472–2475.

Balamurugan D. 2004. Ground and excited state properties of small hydrogenated silicon clusters: A first principles electronic structure study. PhD thesis (IIT Kanpur).

Balamurugan D. and Prasad R. 2001. Effect of hydrogen on ground state structures of small hydrogenated silicon clusters. *Phys. Rev. B* 64: 205406.

Balamurugan D. and Prasad R. 2006. Charge induced effects on structures and properties of silane and disilane derivatives. *Phys. Rev. B* 73: 235415.

Balamurugan D., Harbola M. K., and Prasad R. 2004. Symmetry breaking and structural distortion in charged XH4 (X = C, Si, Ge, Pb) molecules. *Phys. Rev. A* 69: 033201.

Baldacchini C. et al. 2003. Cu,100... surface: High-resolution experimental and theoretical band mapping. *Phys. Rev. B* 68: 195109.

Baldereschi A. 1973. Mean value point in the Brillouin zone. *Phys. Rev. B* 7: 5212–5215.

Bandyopadhyay S. and Cahay M. 2008. *Introduction to Spintronics*. Boca Raton, FL: CRC Press.

Bansil A. 1975. Special directions in the Brillouin zone. *Solid State Commun.* 16: 885–889.

Bansil A. 1978. Application of coherent-potential approximation to disordered muffin-tin alloys. *Phys. Rev. Lett.* 41: 1670–1673.

Bansil A. 1979a. Coherent-potential and average *t*-matrix approximations for disordered muffin-tin alloys. I. Formalism. *Phys. Rev. B* 20: 4025–4034.

Bansil A. 1979b. Coherent-potential and average *t*-matrix approximations for disordered muffin-tin alloys. II. Application to realistic systems. *Phys. Rev. B* 20: 4035–4043.

Bansil A. 1987. Modern band theory of disordered alloys. *Lect. Notes Phys.* 283: 273–298. Berlin: Springer.

Bansil A. and Pessa M. 1983. Surface and bulk electronic structure of disordered metallic alloys. *Phys. Scr.* T4: 52–60.

Bansil A., Rao R. S., Prasad R., Asonen H., and Pessa M. 1984. An angle-resolved photoemission study of Cu_{96} Ge_4, and $Cu_{89}Ge_{11}$ single crystals. *J. Phys. F: Met. Phys.* 14: 273–279.

Bansil A., Pankaluoto R., Rao R. S., Mijnarends P. E., Dlugosz W., Prasad R., and Smedskjaer L. C. 1988. Fermi surface, ground state electronic structure and positron experiments in $YBa_2Cu_3O_7$. *Phys. Rev. Lett.* 61: 2480–2483.

Barnett R. N. and Landman U. 1997. Cluster-derived structures and conductance fluctuations in nanowires. *Nature* 387: 788–791.

Bastard G. 1998. *Wave Mechanics Applied to Semiconductor Heterostructures*. Les Ulis: Les Editions de Physique.

Bednorz J. G. and Müller K. A. 1986. Possible high TC superconductivity in the Ba–La–Cu–O system. *Z. Für. Phys. B* 64: 189–193.

Bellisent R., Menelle A., Howells W. S., Wright A. C., Brunier T. M., Sinclair R. N., and Jansen F. 1989. The structure of amorphous Si:H using steady state and pulsed neutron sources. *Physica B* 156 & 157: 217–219.

Bendt P. and Zunger A. 1983. Simultaneous relaxation of nuclear geometries and electric charge densities in electronic structure theories. *Phys. Rev. Lett.* 50: 1684–1688.

Bernevig, B. A. and Zhang S. C. 2006. Quantum spin Hall effect. *Phys. Rev. Lett.* 96: 106802.

Bernevig B. A., Hughes T. A., and Zhang S. C. 2006. Quantum spin Hall effect and topological phase transition in HgTe quantum wells. *Science* 314: 1757–1761.

Berry M. V. 1984. Quantal phase factors accompanying adiabatic changes. *Proc. R. Soc. Lond. A* 392: 45–57.

Binasch G., Grünberg P., Saurenbach F., and Zinn W. 1989. Enhanced magnetoresistance in layered magnetic structures with antiferromagnetic interlayer exchange. *Phys. Rev. B* 39: 4828–4830.

Bishop A. R. and Mookerjee A. 1974. A new approach to the CPA and its generalizations. *J. Phys. C: Solid State Phys.* 7: 2165–2179.

Bloch F. 1928. Über die quantenmechanik der elektronen in kristallgittern. *Z. Physik.* 52: 555–600.

Blöchl P. E. 1994. Projector augmented-wave method. *Phys. Rev. B* 50: 17953–17978.

Blochl P. E., Jepsen O., and Andersen O. K. 1994. Improved tetrahedron method for Brillouin zone integrations. *Phys. Rev. B* 49: 16223–16233.

Bonanni A. 2007. Ferromagnetic nitride-based semiconductors doped with transition metals and rare earths. *Semicond. Sci. Technol.* 22: R41–R56.

Bonev S. A., Schwegler E., Ogitsu T., and Galli G. 2004. A quantum fluid of metallic hydrogen suggested by first-principles calculations. *Nature* 431: 669–672.

Born M. and Oppenheimer J. R. 1927. Zur quantentheorie der molekeln. *Ann. Physik.* 84: 457–484.

Bowler D. R. 2004. Atomic-scale nanowires: Physical and electronic structure. *J. Phys.: Condens. Matter* 16: R721–R754.

Butler W. H. 1973. Self-consistent cluster theory of disordered alloys. *Phys. Rev. B* 8: 4499–4510.

Car R. and Parrinello M. 1985. Unified approach for molecular dynamics and density functional theory. *Phys. Rev. Lett.* 55: 2471–2474.

Car R. and Parrinello M. 1989. The unified approach for molecular dynamics and density functional theory. *Simple Molecular Systems at Very High Density*, eds. Polian A., Loubeyre P., and Boccara N. pp. 455–476. New York, NY: Plenum.

Castro Neto A. H., Guinea F., Peres N. M. R., Novoselov K. S., and Geim A. K. 2009. The electronic properties of graphene. *Rev. Mod. Phys.* 81: 109–162.

Ceder G. 2010. Opportunities and challenges for first-principles materials design and applications to Li battery materials. *MRS Bull.* 35: 693–701.

Ceperley D. M. and Alder B. J. 1980. Ground state of the electron gas by a stochastic method. *Phys. Rev. Lett.* 45: 566–569.

Chacko S., Joshi K., Kanhere D. G., and Blundell S. A. 2004. Why do gallium clusters have a higher melting point than the bulk? *Phys. Rev. Lett.* 92: 135506.

Chadi D. J. 1979. Atomic and electronic structures of reconstructed Si(100) surfaces. *Phys. Rev. Lett.* 43: 43–47.

Chadi D. J. and Cohen M. L. 1973. Special points in the Brillouin zone. *Phys. Rev. B* 8: 5747–5753.

Charlier J. C., Blasé X., and Roche S. 2007. Electronic and transport properties of nanotubes. *Rev. Mod. Phys.* 79: 677–732.

Chelikowsky J. R., Kronik L., and Vasiliev I. 2003. Time-dependent density-functional calculations for the optical spectra of molecules, clusters, and nanocrystals. *J. Phys.: Condens. Matter* 15: R1517–R1547.

Chen Y. T. et al. 2010. Single Dirac cone topological surface state and unusual thermoelectric property of compounds from a new topological insulator family. *Phys. Rev. Lett.* 105: 266401.

Correa A. A., Bonev S. A., and Galli G. 2006. Carbon under extreme conditions: Phase boundaries and electronic properties from first-principles theory. *Proc. Natl. Acad. Sci. USA* 103: 1204–1208.

Courths R. and Hufner S. 1984. Photoemission experiments on copper. *Phys. Rep.* 112: 53–171.

Das G. P. and Dasgupta I. 2008. Spintronics: A revolution in materials science and semiconductor devices. *Phys. News* 38: 46–53.

Dasgupta I., Saha T., and Mookerjee A. 1995. Augmented-space recursive technique for the analysis of alloy phase stability in random binary alloys. *Phys. Rev. B* 51: 3413–3421.

de Groot R. A. et al. 1983. New class of materials: Half-metallic ferromagnets. *Phys. Rev. Lett.* 50: 2024–2027.

Delin A. and Tosatti E. 2003. Magnetic phenomena in 5D transition metal nanowires. *Phys. Rev. B* 68: 144434.

Dickey J. M. and Paskin A. 1969. Computer simulation of lattice dynamics of solids. *Phys. Rev.* 188: 1407–1418.

Dietl T. 2010. A ten-year perspective on dilute magnetic semiconductors and oxides. *Nat. Mater.* 9: 965–974.

Dietl T. et al. 2000. Zener model description of ferromagnetism in zinc-blende magnetic semiconductors. *Science* 287: 1019–1022.

Dirac P. A. M. 1930. Note on exchange phenomenon in the Thomas–Fermi atom. *Proc. Camb. Philos. R. Soc.* 26: 376–385.

Dreizler R. M. and Gross E. K. U. 1990. *Density Functional Theory: An Approach to the Quantum Many-Body Problem*. Berlin: Springer.

Dresselhaus M. S., Dresselhaus G., and Eklund P. C. 1996. *Science of Fullerenes and Carbon Nanotubes*. New York, NY: Academic Press.

Drude P. 1900. Zur elektronentheorie der metalle. *Annalen der Physik.* 306: 566–613.

Economou E. N. 1979. *Green's Function in Quantum Physics*. Berlin: Springer.

Ederer C. and Spaldin N. A. 2005. Weak ferromagnetism and magnetoelectric coupling in $BiFeO_3$. *Phys. Rev. B* 71: 040401.

Edwards D. M. et al. 1991. Oscillations of the exchange in magnetic multilayers as an analog of de Haas–van Alphen effect. *Phys. Rev. Lett.* 67: 493–496.

Eggert J. H., Hicks D. G., Celliers P. M., Bradley D. K., McWilliams R. S., Jeanloz R., Miller J. E., Boehly T. R., and Collins G. W. 2010. Melting temperature of diamond at ultrahigh pressure. *Nat. Phys.* 6: 40–43.

Ehrenreich H. and Schwartz L. M. 1976. Electronic structure of alloys, *Solid State Physics*, eds. Ehrenreich H., Seitz F., and Turbull D., vol 31, pp. 149–286. New York, NY: Academic Press.

Eisberg R. and Resnick R. 1974. *Quantum Physics*. New York, NY: John Wiley & Sons.

Elliot S. R. 1984. *Physics of Amorphous Materials*. London and New York: Longman.

Elstner M. et al. 1998. Self-consistent-charge density-functional tight-binding method for simulations of complex materials properties. *Phys. Rev. B* 58: 7260–7268.

Englert T., Tsui D. C., Gossard A. C., and Uihlein C. 1982. *g*-Factor enhancement in the 2d electron gas in GaAs/AlGaAs heterojunctions. *Surf. Sci.* 113: 295–300.

Eremets M. I. and Troyan I. A. 2011. Conductive dense hydrogen. *Nat. Mater.* 10: 927–931.

Esaki L. and Tsu R. 1970. Superlattice and negative differential conductivity in semiconductors. *IBM J. R. D.* 14: 61–65.

Fahy, S., Louie S. G., and Cohen M. L. 1986. Pseudopotential total-energy study of the transition from rhombohedral graphite to diamond. *Phys. Rev. B* 34: 1191–1199.

Falicov L. M. 1966. *Group Theory and Its Physical Applications.* Chicago: The University of Chicago Press.

Faulkner J. S. 1982. Modern theory of alloys. *Prog. Mater. Sci.* 27: 1–187.

Faulkner J. S. and Stocks G. M. 1980. Calculating properties with the coherent-potential approximation. *Phys. Rev. B* 21: 3222–3244.

Fermi E. 1927. Un metodo statistico per la determinazione di alcune priopreità dell'atom. *Rend. Accad. Naz. Lincei* 6: 602–607.

Feuston B. P. et al. 1991. Electronic and vibrational properties of C_{60} at finite temperature from *ab initio* molecular dynamics. *Phys. Rev. B* 44: 4056–4059.

Feynman R. P. 1959. There's plenty of room at the bottom. Talk given at the APS meeting and reprinted in 1992. *J. Microelectromech. Syst.* 1: 60–66.

Filippetti A. and Spaldin N. A. 2003. Strong correlation effects in Born effective charges. *Phys. Rev. B* 68: 045111.

Fock V. 1930. A method for the solution of many-body problems in quantum mechanics. *Zeitschrift für Physik* 61: 126–148.

Forro L. and Mihaly L. 2001. Electronic properties of doped fullerenes. *Rep. Prog. Phys.* 64: 649–699.

Franciosi A. and Van de Walle C. G. 1996. Heterojunction band offset engineering. *Surf. Sci. Rep.* 25: 1–140.

Fu L. and Kane C. L. 2007. Topological insulators with inversion symmetry. *Phys. Rev. B* 76: 045302.

Furthmuller J., Hafner J., and Kresse G. 1994. *Ab initio* calculation of the structural and electronic properties of carbon and boron nitride using ultrasoft pseudopotentials. *Phys. Rev. B* 50: 15606–15622.

Galanakis I. and Mavropoulos P. 2007. Spin-polarization and electronic properties of half-metallic Heusler alloys calculated from first principles. *J. Phys.: Condens. Matter* 19: 315213.

Galanakis I., Dederichs P. H., and Papanikolaou N. P. 2002. Origin and properties of the gap in the half-ferromagnetic Heusler alloys. *Phys. Rev. B* 66: 134428.

Galanakis, I., Mavropoulos P., and Dederichs P. H. 2006. Electronic structure and Slater–Pauling behaviour in half-metallic Heusler alloys calculated from first principles. *J. Phys. D* 39: 765–775.

Galli G., Martin R. M., Car R., and Parrinello. M. 1990. Melting of diamond at high pressure. *Science* 250: 1547–1549.

Gao L. et al. 1994. Superconductivity up to 164 K in $HgBa_2Ca_{m-1}Cu_mO_{2m+2+\delta}$ ($m = 1, 2,$ and 3) under quasihydrostatic pressures. *Phys. Rev. B* 50: 4260–4263.

Garoufalis C. S., Zdetsis A. D., and Grimme S. 2001. High level *ab initio* calculations of the optical gap of small silicon quantum dots. *Phys. Rev. Lett.* 87: 276402.

Geim A. K. and MacDonald A. H. 2007. Graphene: Exploring carbon flatland. *Physics Today* 60: 35–41.

Geim A. K. and Novoselov K. S. 2007. The rise of graphene. *Nat. Mater.* 6: 183–191.

Gell-Mann M. and Brueckner K. A. 1957. Correlation energy of an electron gas at high density. *Phys. Rev.* 106: 364–368.

Ghosez P., Michenaud J.-P., and Gonze X. 1998. Dynamical effective charges: The case of ABO_3 compounds. *Phys. Rev. B* 58: 6224–6240.

Goedecker S. 1999. Linear scaling electronic structure methods. *Rev. Mod. Phys.* 71: 1085–1122.

Gross E. K. U. and Dreizler R. M. eds. 1995. *Density Functional Theory*. New York, NY: Plenum Press.

Grossman J. C., Mitas L., and Raghavachari K. 1995. Structure and stability of molecular carbon: Importance of electron correlation. *Phys. Rev. Lett.* 75: 3870–3873.

Grumbach M. P and Martin R. 1996. Phase diagram of carbon at high pressures and temperatures. *Phys. Rev. B* 54: 15730–15741.

Grünberg P. 2008. Nobel lecture: From spin waves to giant magnetoresistance and beyond. *Rev. Mod. Phys.* 80: 1531–1540.

Grünberg P., Schreiber R., Pang Y., Brodsky M. B., and Sowers H. 1986. Layered magnetic structures: Evidence for antiferromagnetic coupling of Fe layers across Cr interlayers. *Phys. Rev. Lett.* 57: 2442–2445.

Gundra K. and Shukla A. 2011. Band structure and optical absorption in multilayer armchair graphene nanoribbons: A Pariser–Parr–Pople model study. *Phys. Rev. B* 84: 075442.

Gunnarsson O. and Lundqvist B. I. 1976. Exchange and correlation in atoms, molecules and solids by the spin-density functional formalism. *Phys. Rev. B* 13: 4274–4298.

Gyorffy B. L. 1972. Coherent-potential approximation for a nonoverlapping-muffin-tin-potential model of random substitutional alloys. *Phys. Rev. B* 5: 2382–2384.

Haddon R. C. et al. 1986. Electronic structure and bonding in icosahedral C_{60}. *Chem. Phys. Lett.* 125: 459–464.

Handschuh H. et al. 1995. Stable configurations of carbon clusters: Chains, rings, and fullerenes. *Phys. Rev. Lett.* 74: 1095–1098.

Hafner J. 2008. Materials simulations using VASP—A quantum perspective to materials science. *Comput. Phys. Commun.* 177: 6–13.

Hafner J., Wolverton C., and Ceder G. 2007. Toward computational materials design: The impact of density functional theory on materials research. *MRS Bull.* 31: 659–665.

Hahn T. H. 2002. *International Tables for Crystallography, Volume A: Space Group Symmetry* (5th ed.). Berlin: Springer.

Haile J. M. 1997. *Molecular Dynamics Simulation: Elementary Methods*. New York, NY: John Wiley & Sons.

Halperin B. I. 1982. Quantized Hall conductance, current-carrying edge states, and the existence of extended states in a two-dimensional disordered potential. *Phys. Rev. B* 25: 2185–2190.

Hamann D. R., Schlüter M., and Chiang C. 1979. Norm-conserving pseudopotentials. *Phys. Rev. Lett.* 43: 1494–1497.

Harbola M. K. 1992. Magic numbers for metallic clusters and the principle of maximum hardness. *Proc. Natl. Acad. Sci. USA* 89: 1036–1039.

Harrison W. 1970. *Solid State Theory*. New York, NY: McGraw-Hill Book Company.

Harrison W. 1980. *Electronic Structure and Properties of Solids*. San Francisco: W H Freeman and Company.

Hartree D. R. 1928. The wave mechanics of an atom with a noncoulomb central field. Part I. Theory and methods. Part II. Some results and discussions. Part III. Term values and intensities in series in optical spectra. *Proc. Camb. Philos. Soc.* 24: 89–110, 111–132, 426–437.

Hasan M. Z. and Kane C. L. 2010. Colloquium: Topological insulators. *Rev. Mod. Phys.* 82: 3045–3067.

Hedin L. and Lundqvist S. 1969. *Solid State Physics,* vol. 23, eds. Ehrenreich H., Seitz F., and Turnbull D. pp. 1–181. New York, NY: Academic Press.

Heine V. 1960. *Group Theory in Quantum Mechanics.* London: Pergamon Press.

Hellmann H. 1935. A new approximation method in the problem of many electrons. *J. Chem. Phys.* 3: 61.

Hemley R. J. and Ashcroft N. W. 1998. The revealing role of pressure in the condensed matter sciences. *Physics Today* 51(8): 26–32.

Hemley R. J., Crabtree G. W., and Buchanan M. V. 2009. Materials in extreme environments. *Physics Today* 62(11): 32–37.

Herman F. 1958. Theoretical investigation of the electronic energy band structure of solids. *Rev. Mod. Phys.* 30: 102–121.

Herman F. 1984. Elephants and mahouts—Early days in semiconductor physics. *Physics Today* 37(6): 56–63.

Herman F. and Callaway J. 1953. Electronic structure of the germanium crystal. *Phys. Rev.* 89: 518–519.

Herman F., Van Dyke J. P., and Ortenburger I. B. 1969. Improved statistical exchange approximation for inhomogeneous many-electron systems. *Phys. Rev. Lett.* 22: 807–811.

Herman F. et al. 1991. Spin-polarized band structure of magnetically coupled multilayers. *J. Appl. Phys.* 69: 4783–4785.

Herring C. 1940. A new method for calculating wave functions in crystals. *Phys. Rev.* 57: 1169–1177.

Ho K. et al. 1998. Structures of medium-sized silicon clusters. *Nature* 392: 582–585.

Hoddeson L., Braun E., Teichmann J., and Weart S. 1992. *Out of the Crystal Maze, Chapters from the History of Solid State Physics.* New York, NY: Oxford University Press.

Hohenberg P. and Kohn W. 1964. Inhomogeneous electron gas. *Phys. Rev.* 136: B864–B871.

Hsieh D. et al. 2008. A topological Dirac insulator in a quantum spin Hall phase. *Nature* 452: 970–974.

Hubbard J. 1958. The description of collective motions in terms of many-body perturbation theory. II. The correlation energy of a free-electron gas. *Proc. R. Soc. Lond. Ser. A.* 243: 336–352.

Hubbard J. 1963. Electron correlations in narrow energy bands. *Proc. R. Soc. A* 276: 238–257.

Iijima S. 1991. Helical microtubules of graphitic carbon. *Nature* 354: 56–58.

Itoh U., Toyoshima Y., Onuki H., Washida N., and Ibuki T. 1986. Vacuum ultraviolet absorption cross sections of SiH_4, GeH_4, Si_2H_6, and Si_3H_8. *J. Chem. Phys.* 85: 4867–4872.

Izyumov Y. and Kurmaev K. 2010. *High-Tc Superconductors Based on FeAs Compounds.* Berlin: Springer.

Jain J. K. 2007. *Composite Fermions.* New York, NY: Cambridge University Press.

Jelitto R. J. 1969. The density of states of some simple excitations in solids. *Phys. Chem. Solids* 30: 609–626.

Jepsen O. and Andersen O. K. 1971. The electronic structure of h.c.p. ytterbium. *Solid State Commun.* 9: 1763–1767.

Johnston R. J. 2002. *Atomic and Molecular Clusters*. London and New York, NY: Taylor & Francis.

Jones R. 1999. Density functional study of carbon clusters C2n, $2 < n < 16$…. I. Structure and bonding in the neutral clusters. *J. Chem. Phys.*110: 5189–5200.

Kamihara Y. et al. 2006. Iron-based layered superconductor: LaOFeP. *J. Am. Chem. Soc.* 128: 10012–10013.

Kane C. L. and Mele E. J. 2005. Quantum spin Hall effect in graphene. *Phys. Rev. Lett.* 95: 226801.

Kaprzyk S. and Bansil A. 1990. Green's function and a generalized Lloyd formula for the density of states in disordered muffin-tin alloys. *Phys. Rev. B* 42: 7358–7362.

Kaprzyk S. and Mijnarends P. E. 1986. A simple linear analytic method for Brillouin zone integration of spectral functions in the complex energy plane. *J. Phys. C: Solid State Phys.* 19: 1283–1292.

Katsnelson M. I. et al. 2008. Half-metallic ferromagnets: From band structure to many-body effects. *Rev. Mod. Phys.* 80: 315–378.

Kaxiras E. 2003. *Atomic and Electronic Structure of Solids*. Cambridge: Cambridge University Press.

Kaxiras E. and Jackson K. 1993. Shape of small silicon clusters. *Phys. Rev. Lett.* 71: 727–730.

Khanna S. N. and Castleman A. W. eds. 2003. *Quantum Phenomena in Clusters and Nanostructures*. Berlin: Springer-Verlag.

King-Smith R. D. and Vanderbilt D. 1993. Theory of polarization of crystalline solids. *Phys. Rev. B* 47: 1651–1654.

Kirkpatrick S., Gelatt C. D., and Vecchi M. P. 1983. Optimization by simulated annealing. *Science* 220: 671–680.

Kittel C. 1986. *Introduction to Solid State Physics* (6th edition). New York, NY: John Wiley & Sons.

Kittel C. 1987. *Quantum Theory of Solids* (2nd edition). New York, NY: John Wiley & Sons.

Kleinman L. 1981. Comment on the average potential of a Wigner solid. *Phys. Rev. B* 24: 7412–7414.

Kleinman L. and Bylander D. M. 1982. Efficacious form for model pseudopotentials. *Phys. Rev. Lett.* 48: 1425–1428.

Knight W. D., Clemenger K., de Heer W. A., Saunders W. A., Chou M. Y., and Cohen M. L. 1984. Electronic shell structure and abundances of sodium clusters. *Phys. Rev. Lett.* 52: 2141–2143.

Kohanoff J. 2006. *Electronic Structure Methods for Solids and Molecules*. Cambridge: Cambridge University Press.

Kohn W. 1996. Density functional and density matrix method scaling linearly with the number of atoms. *Phys. Rev. Lett.* 76: 3168–3171.

Kohn W. 1999. Nobel lecture: Electronic structure of matter wave functions and density functionals. *Rev. Mod. Phys.* 71: 1253–1266.

Kohn W. and Rostoker N. 1954. Solution of the Schrödinger equation in periodic lattices with an application to metallic lithium. *Phys. Rev.* 94: 1111–1120.

Kohn W. and Sham L. J. 1965. Self-consistent equations including exchange and correlation effects. *Phys. Rev.* 140: A1133–A1138.

Kondo Y. and Takayanagi K. 1997. Gold nanobridge stabilized by surface structure. *Phys. Rev. Lett.* 79: 3455–3458.

Konig M., Wiedmann S., Brune C., Roth A., Buhmann H., Molenkamp L. H., Qi X. L., and Zhang S. C. 2007. Quantum spin Hall insulator state in HgTe quantum wells. *Science* 318: 766–770.

Korringa J. 1947. On the calculation of the energy of a Bloch wave in a metal. *Physica* 13: 392–400.

Kosimov D. P., Dzhurakhalov A. A., and Peeters F. M. 2010. Carbon clusters: From ring structures to nanographene. *Phys. Rev. B* 81: 195414.

Krakauer H., Pickett W. E., and Cohen R. E. 1988. Analysis of electronic structure and charge density of high temperature superconductor $YBa_2Cu_3O_7$. *J. Supercond.* 1: 111–141.

Kresse G. and Furthmüller J. 1996a. Efficiency of *ab-initio* total energy calculations for metals and semiconductors using a plane-wave basis set. *Comput. Mat. Sci.* 6: 15–50.

Kresse G. and Furthmüller J. 1996b. Efficient iterative schemes for *ab initio* total-energy calculations using a plane-wave basis set. *Phys. Rev. B* 54: 11169–11186.

Kroto et al. 1985. C60: Buckminsterfullerene. *Nature* 318: 162–163.

Kumar V. and Kawazoe Y. 2001. Metal-encapsulated fullerene like and cubic caged clusters of silicon. *Phys. Rev. Lett.* 87: 045503.

Landau L. D. 1957. The theory of a Fermi liquid. *Sov. Phys. JETP* 3: 920–925.

Landau L. D. 1959. On the theory of a Fermi liquid. *Sov. Phys. JETP* 8: 70–74.

Landau L. D. and Lifshitz E. M. 1977. *Quantum Mechanics.* Oxford: Pergamon Press.

Lang N. D. and Kohn W. 1970. Theory of metal surfaces: Charge density and surface energy. *Phys. Rev. B* 1: 4555–4568.

Laughlin R. B. 1983. Anomalous quantum Hall effect: An incompressible quantum fluid with fractionally charged excitations. *Phys. Rev. Lett.* 50: 1395–1398.

Lee P. A. and Ramakrishnan T. V. 1985. Disordered electronic systems. *Rev. Mod. Phys.* 57: 287–337.

Levy P. M. et al. 1990. Theory of magnetic superlattices: Interlayer exchange coupling and magnetoresistance of transition metal structures. *J. Appl. Phys.* 67: 5914–5919.

Lin H., Markiewicz R. S., Wray L. A., Fu L., Hasan M. Z., and Bansil A. 2010. Single-Dirac-cone topological surface states in the $TlBiSe_2$ class of topological semiconductors. *Phys. Rev. Lett.* 105: 036404.

Lorentz H. A. 1909. *The Theory of Electrons and Its Applications to the Phenomena of Light and Radiant Heat.* Leipzig: Teubner.

Loucks T. 1967. *The Augmented Plane Wave Method.* New York, NY: Benjamin.

Lu G. and Kaxiras E. 2005. Overview of multiscale simulations of materials. *Handbook of Theoretical and Computational Nanotechnology*, eds. Rieth M. and Schommers W., vol. 10, pp. 1–33. Valencia: American Scientific Publishers.

Lundqvist S. and March N. H. eds. 1983. *Theory of the Inhomogeneous Electron Gas.* New York, NY: Plenum Press.

Lyo I. W. and Plummer E. W. 1988. Quasiparticle band structure of Na and simple metals. *Phys. Rev. Lett.* 60: 158–1561.

Mahan G. D. 2000. *Many-Particle Physics.* New York, NY: Kluwer Academic/Plenum Publishers.

Mailhiot C. and McMahan A. K. 1991. Atmospheric-pressure stability of energetic phases of carbon. *Phys. Rev. B* 44: 11 578–11591.

Mardar M. 2010. *Condensed Matter Physics*. New York, NY: John Wiley and Sons.

Marques M. A. L., Castro A., and Rubio A. 2001. Assessment of exchange-correlation functionals for the calculation of dynamical properties of small clusters in time-dependent density functional theory. *J. Chem. Phys.* 115: 3006–3014.

Martin R. M. 2004. *Electronic Structure, Basic Theory and Practical Methods*. Cambridge: Cambridge University Press.

Marx D. and Hutter J. 2009. *Ab Initio Molecular Dynamics: Basic Theory and Advanced Methods*. Cambridge: Cambridge University Press.

Matsumoto I., Kawata H., and Shiotani N. 2001. Fermi-surface geometry of the Cu—27.5 at.% Pd disordered alloy and short-range order. *Phys. Rev. B* 64: 195132.

Mattheiss L. F. 1987. Electronic band properties and superconductivity in $La_{2-y}X_yCuO_4$. *Phys. Rev. Lett.* 58: 1028–1030.

Mendez E. E. and Bastard G. R. 1993. Wannier–Stark ladders and Bloch oscillations in superlattices. *Physics Today* 46: 34–42.

Mermin N. D. 1965. Thermal properties of the inhomogeneous electron gas. *Phys. Rev.* 137: A1441–A1443.

Merzbacher E. 1998. *Quantum Mechanics*. New York, NY: John Wiley & Sons.

Methfessel M. and Paxton A. T. 1989. High precision sampling for Brillouin-zone integration in metals. *Phys. Rev. B* 40: 3616–3621.

Mihaly L. and Martin M. C. 1996. *Solid State Physics, Problems and Solutions*. New York, NY: John Wiley & Sons.

Mills R. and Ratanavararaksa P. 1978. Analytic approximation for substitutional alloys. *Phys. Rev. B* 18: 5291–5308.

Mishra S. K. and Ceder G. 1999. Structural stability of lithium manganese oxides. *Phys. Rev. B* 59: 6120–6130.

Monkhorst H. J. and Pack J. D. 1976. Special points for Brillouin-zone integrations. *Phys. Rev. B* 13: 5188–5192.

Mookerjee A. 1973. A new formalism for the study of configuration-averaged properties of disordered systems. *J. Phys. C: Solid State Phys.* 6: L205–L208.

Mookerjee A. and Prasad R. 1993. Generalized augmented space theorem and cluster coherent potential approximation. *Phys. Rev. B* 48: 17724–17731.

Morales M. A., Pierleoni C., Schwegler E., and Ceperley D. M. 2010. Evidence for a first order liquid–liquid transition in high pressure hydrogen from *ab-initio* simulations. *Proc. Natl. Acad. Sci.* 107: 12799–12803.

Moruzzi V. L., Janak J. F., and Williams A. R. 1978. *Calculated Electronic Properties of Metals*. New York, NY: Pergamon Press.

Mott N. F. 1949. The basis of the electron theory of metals, with special reference to the transition metals. *Proc. Phys. Soc. A* 62: 416–422.

Mott N. F. 1964. Electrons in transition metals. *Adv. Phys.* 13: 325–422.

Mujica A., Rubio A., Munoz A., and Needs R. J. 2003. High-pressure phases of group-IV, III–V, and II–VI compounds. *Rev. Mod. Phys.* 75: 863–912.

Murnaghan F. D. 1944. The compressibility of media under extreme pressures. *Proc. Natl. Acad. Sci.* 30: 244–247.

Nakamura A. et al. 1999. Density functional simulation of a breaking nanowire. *Phys. Rev. Lett.* 82: 1538–1541.

Nanda B. R. K. and Dasgupta I. 2003. Electronic structure and magnetism in half-Heusler compounds. *J. Phys.: Condens. Matter* 15: 7307–7323.

Nautiyal T., Rho T. H., and Kim K. S. 2004. Nanowires for spintronics: A study of transition-metal elements of groups 8–10. *Phys. Rev. B* 69: 193404.

Neaton J. B. et al. 2005. First-principles study of spontaneous polarization in $BiFeO_3$. *Phys. Rev. B* 71: 014113.

Nickel B. G. and Butler W. H. 1973. Problems in strong-scattering binary alloys. *Phys. Rev. Lett.* 30: 373–377.

Nieminen R. M. 2002. From atomistic simulation towards multiscale modelling of materials. *J. Phys.: Condens. Matter* 14: 2859–2876.

Norris C. and Williams G. P. 1978. Photoelectron measurements of α-phase alloys CuZn, CuGa, and CuGe. *Phys. Status Solidi (B)* 85: 325–330.

Novoselov K. S. et al. 2004. Electric field effect in atomically thin carbon films. *Science* 306: 666–669.

Novoselov K. S. et al. 2005. Two-dimensional gas of massless Dirac fermions in graphene. *Nature* 438: 197–200.

Oganov A. R. and Ono S. 2004. Theoretical and experimental evidence for a post-perovskite phase of $MgSiO_3$ in earth's *D* layer. *Nature* 430: 445–448.

Öğüt S. and Rabe K. M. 1995. Band gap and stability in the ternary intermetallic compounds NiSnM (M=Ti, Zr, Hf): A first-principles study. *Phys. Rev. B* 51: 10443–10453.

Ohnishi H., Kondo Y., and Takayanagi K. 1998. Quantized conductance through individual rows of suspended gold atoms. *Nature* 395: 780–783.

Palacin M. R. 2009. Recent advances in rechargeable battery materials: A chemist's perspective. *Chem. Soc. Rev.* 38: 2565–2575.

Pancharatnam S. 1956. Generalized theory of interference, and its applications. Part I. Coherent pencils. *Proc. Indian Acad. Sci. A* 44: 247–262.

Pant M. M. and Rajagopal A. 1972. Theory of inhomogeneous magnetic electron gas. *Solid State Commun.* 10: 1157–1160.

Parkin S. S. P., More N., and Roche K. P. 1990. Oscillations in exchange coupling and magnetoresistance in metallic superlattice structures: Co/Ru, Co/Cr, and Fe/Cr. *Phys. Rev. Lett.* 64: 2304–2307.

Parr R. G. and Yang W. 1989. *Density Functional Theory of Atoms and Molecules.* New York, NY: Oxford University Press.

Payne M. C., Teter M. P., Allan D. C., Arias T. A., and Joannopoulos J. D. 1992. Iterative minimization techniques for *ab initio* total-energy calculations: Molecular dynamics and conjugate gradients. *Rev. Mod. Phys.* 64: 1045–1097.

Perdew J. P. 1991. Unified theory of exchange and correlation beyond the local density approximation. *Electronic Structure of Solids '91*, eds. Ziesche P. and Eschrig H. pp. 11–20. Berlin: Akademie Verlag.

Perdew J. P. and Zunger A. 1981. Self-interaction correction to density-functional approximations for many-electron systems. *Phys. Rev. B* 23: 5048–5079.

Perdew J. P., Burke K., and Ernzerhof M. 1996. Generalized gradient approximation made simple. *Phys. Rev. Lett.* 77: 3865–3868.

Peres N. M. R. 2010. Colloquium: The transport properties of graphene: An introduction. *Rev. Mod. Phys.* 82: 2673–2700.

Pessa M., Asonen H., Rao R. S., Prasad R., and Bansil A. 1981. Observation of a Tamm-type state on Cu0.9Al0.1(100) surface: Disorder effects and bulk electronic structure of an alloy. *Phys. Rev. Lett.* 47: 1223–1226.

Pettifor D. G. 2003. Electron theory in materials modeling. *Acta Mater.* 51: 5649–5673.

Phillips J. C. and Kleinman L. 1959. New method for calculating wave functions in crystals and molecules. *Phys. Rev.* 116: 287–294.

Pickett W. E. 1989. Electronic structure of the high temperature superconductors. *Rev. Mod. Phys.* 61: 433–512.

Pickett W. E. and Moodera J. S. 2001. Half metallic magnets. *Physics Today* 54(5): 39–44.

Poole C. P. Jr., Farach H. A., Creswick R. J., and Prozorov R. 2007. *Superconductivity*. Amsterdam: Elsevier.

Prange R. E. and Girvin S. M. eds. 1987. *The Quantum Hall Effect*. New York, NY: Springer.

Prasad R. 1994. KKR approach to random alloys. *Methods of Electronic Structure Calculations*, eds. Andersen O. K., Kumar V., and Mookerjee A. pp. 211–230. Singapore: World Scientific.

Prasad R. 2005. Hydrogenated silicon clusters. *Proc. Indian Natl. Sci. Acad.* 71A: 371–376.

Prasad R. and Bansil A. 1980. Special directions for Brillouin-zone integration: Application to density of states calculations. *Phys. Rev. B* 21: 496–503.

Prasad R. and Bansil A. 1982. Non-linear composition dependence of the Fermi surface dimensions in alpha-phase copper–germanium alloys. *Phys. Rev. Lett.* 48: 113–116.

Prasad R. and Shenoy S. R. 1996. Staebler–Wronski effect in hydrogenated amorphous silicon. *Phys. Lett. A* 218: 85–90.

Prasad R., Auluck S., and Joshi S. K. 1975. Giant Kohn anomaly in a quasi-one dimensional conductor. *J. Phys.* C8: L139–L141.

Prasad R., Benedek R., and Thackeray M. M. 2005. Dopant induced stabilization of rhombohedral $LiMnO_2$ against Jahn–Teller distortion. *Phys. Rev. B* 71: 134111.

Prasad R., Joshi S. K., and Auluck S. 1977. Fermi surface and band structure of ferromagnetic nickel. *Phys. Rev. B* 16: 1765–1767.

Prasad R., Papadopoulos S. C., and Bansil A. 1981. Fermi surface properties of disordered a-phase alloys of copper with zinc. *Phys. Rev. B* 23: 2607–2613.

Prasad R., Serageldin A. Y., and Bansil A. 1991. Shockley-type surface states on low-index faces of Cu-based alloys with polyvalent solutes. *J. Phys: Condens. Matter* 3: 801–812.

Prasad R. et al. 2003. Divalent dopant criterion for the suppression of Jahn–Teller distortion in Mn oxides. *Phys. Rev. B* 68: 012101.

Pulay P. 1969. *Ab initio* calculation of force constants and equilibrium geometries in polyatomic molecules. *Mol. Phys.* 17: 197–204.

Qi X. L. and Zhang S. C. 2010. The quantum spin Hall effect and topological insulators. *Physics Today* 63(1): 33–38.

Rabe K. M., Ahn C. H., and Triscone J. M. eds. 2007. *Physics of Ferroelectrics*. Berlin: Springer-Verlag.

Rae A. D. et al. 1990. Structure refinement of commensurately modulated bismuth titanate, $Bi_4Ti_3O_{12}$. *Acta Cryst. B* 46: 474–487.

Raghavachari K. and Binkley J. S. 1987. Structure, stability, and fragmentation of small carbon clusters. *J. Chem. Phys.* 87: 2191–2197.

Raghavachari K. and Logovinsky V. 1985. Structure and bonding in small silicon clusters. *Phys. Rev. Lett.* 55: 2853–2856.

Raghu S. et al. 2008. Minimal two-band model of the superconducting iron oxypnictides. *Phys. Rev. B* 77: 220503(R).

Rahaman M. and Mookerjee A. 2009. Augmented-space cluster coherent potential approximation for binary random and short-range. *Phys. Rev. B* 79: 054201.

Rahaman M., Ganguli S., Samal P., Harbola M. K., Saha-Dasgupta T., and Mookerjee A. 2009. A local-density approximation for the exchange energy functional for excited states in bulk semiconductors: The band gap problem. *Physica B* 404: 1137.

Rahman A.1964. Correlations in the motion of atoms in liquid argon. *Phys. Rev.* 136: A405–A411.

Rahman A., Mandell M. J., and McTague J. P. 1976. Molecular dynamics study of amorphous Lennard–Jones solid at low temperatures. *J. Chem. Phys.* 64: 1564–1568.

Raimes S. 1961. *The Wave Mechanics of Electrons in Metals*. Amsterdam: North-Holland Publishing Company.

Rajagopal A. K. and Callaway J. 1973. Inhomogeneous electron gas. *Phys. Rev. B* 7: 1912–1919.

Rajput S. S., Prasad R., Singru R. M., Kaprzyk S., and Bansil A.1996. Electronic structure of disordered Nb–Mo alloys using charge–selfconsistent KKR–CPA method. *J. Phys: Condens. Matter* 8: 2929–2944.

Rajput S. S., Prasad R., Singru R. M., Triftshauser W., Eckert A., Kogel G., Kaprzyk S., and Bansil A. 1993. A study of the Fermi surfaces of lithium and disordered lithium–magnesium alloys: Theory and experiment. *J. Phys.: Condens. Matter* 5: 6419–6432.

Ramstad A., Brocks G., and Kelly P. J. 1995. Theoretical study of the Si(100) surface reconstruction. *Phys. Rev. B* 51: 14504–14523.

Rao R. S., Prasad R., and Bansil A. 1983. Composition dependence of optical gaps in the Cu-based Hume–Rothery alloys. *Phys. Rev. B* 28: 5762–5765.

Rapaport D. C. 2004. *The Art of Molecular Dynamics Simulation*. Cambridge: Cambridge University Press.

Razee S. S. A. and Prasad R. 1992. Configuration averaged Green's function within Korringa–Kohn–Rostoker coherent-potential approximation. *Phys. Rev. B* 45: 3265–3270.

Razee S. S. A. and Prasad R. 1993a. Disordered alloys with short-range order: Korringa–Kohn–Rostoker formulation within cluster coherent-potential approximation. *Phys. Rev. B* 48: 1349–1355.

Razee S. S. A. and Prasad R. 1993b. Cluster coherent-potential approximation in the tight-binding linear-muffin-tin-orbital formalism. *Phys. Rev. B* 48: 1361–1367.

Razee S. S. A., Mookerjee A., and Prasad R. 1991. On the augmented-space cluster coherent-potential approximation and its analytic properties. *J. Phys: Condens. Matter* 3: 3301–3310.

Razee S. S. A., Rajput S. S., Prasad R., and Mookerjee A. 1990. Electronic structure of disordered alloys: Korringa–Kohn–Rostoker cluster coherent potential approximation. *Phys. Rev. B* 42: 9391–9402.

Reich, S., Maultzsch J., Thomsen C., and Ordejón P. 2002. Tight-binding description of graphene. *Phys. Rev. B* 66: 035412.

Remler D. K. and Madden P. A. 1990. Molecular dynamics without effective potentials via the Car–Parrinello approach. *Mol. Phys.* 70: 921–966.

Renker B. et al. 1973. Observation of giant Kohn anomaly in the one-dimensional conductor $K_2Pt(CN)_4Br_{0.3}\cdot 3H_2O$. *Phys. Rev. Lett.* 30: 1144–1147.

Resta R. 1994. Macroscopic polarization in crystalline dielectrics: The geometric phase approach. *Rev. Mod. Phys.* 66: 899–915.

Ribeiro F. J. and Cohen M. L. 2003. *Ab initio* pseudopotential calculations of infinite monatomic chains of Au, Al, Ag, Pd, Rh, and Ru. *Phys. Rev. B* 68: 035423.

Roberts N. and Needs R. J. 1990. Total energy calculations of dimer reconstructions on the silicon (001) surface. *Surf. Sci.* 236: 112–121.
Roman P. 1965. *Advanced Quantum Theory.* Reading, MA: Addison-Wesley.
Rosenberg G. and Franz M. 2012. Surface magnetic ordering in topological insulators with bulk magnetic dopants. *Phys. Rev. B* 85: 195119.
Rosseinsky M. J. et al. 1991. Superconductivity at 28 K in Rb_xC_{60}. *Phys. Rev. Lett.* 66: 2830–2832.
Rowlands D. A. 2009. Short-range correlations in disordered systems: Non-local coherent potential approximation. *Rep. Prog. Phys.* 72: 086501.
Roy A., Garg A., Prasad R., and Auluck S. 2011a. Multiferroic $SrTiO_3/BiFeO_3$ superlattice: A first-principles study. *Nanotechnology* 2: 112–115.
Roy A., Mukherjee S., Gupta R., Auluck S., Prasad R., and Garg A. 2011b. Electronic structure, Born effective charges and spontaneous polarization in magnetoelectric gallium ferrite. *J. Phys.: Condens. Matter* 23: 325902.
Roy A., Mukherjee S., Sarkar S., Auluck S., Prasad R., Gupta R., and Garg A. 2012b. Effect of site-disorder, off-stoichiometry and epitaxial strain on the optical properties of magnetoelectric gallium ferrite. *J. Phys.: Condens. Matter* 24: 435501.
Roy A., Prasad R., Auluck A., and Garg A. 2010. First-principles calculations of Born effective charges and spontaneous polarization of ferroelectric bismuth titanate. *J. Phys.: Condens. Matter* 22: 165902.
Roy A., Prasad R., Auluck A., and Garg A. 2012a. Effect of site-disorder on magnetism and magneto-structural coupling in gallium ferrite: A first-principles study. *J. Appl. Phys.* 111: 043915.
Ruban A. V. and Abrikosov I. A. 2008. Configurational thermodynamics of alloys from first-principles: Effective cluster interactions. *Rep. Prog. Phys.* 71: 046501.
Runge E. and Gross E. K. U. 1984. Density-functional theory for time-dependent systems. *Phys. Rev. Lett.* 52: 997–1000.
Rurali R. 2010. Colloquium: Structural, electronic, and transport properties of silicon nanowires. *Rev. Mod. Phys.* 82: 427–448.
Ryckaert J. P., Ciccotti G., and Berendsen H. J. C. 1977. Numerical integration of the Cartesian equations of motion of a system with constraints: Molecular dynamics of *n*-alkanes. *J. Comput. Phys.* 23: 327–341.
Saha T., Dasgupta I., and Mookerjee A. 1994. Augmented-space recursive method for the study of short-ranged ordering effects in binary alloys. *Phys. Rev. B* 50: 13267–13275.
Saito R., Dresselhouse G., and Dresselhouse M. S. 1998. *Physical Properties of Carbon Nanotubes.* London: Imperial College Press.
Saito R., Fujita M., Dresselhaus G., and Dresselhaus M. S. 1992a. Electronic structure of graphene tubules based on C_{60}. *Phys. Rev. B* 46: 1804–1811.
Saito R., Fujita M., Dresselhaus G., and Dresselhaus M. S. 1992b. Electronic structure of chiral graphene tubules. *Appl. Phys. Lett.* 60: 2204–2206.
Saito S. and Oshiyama A. 1991. Cohesive mechanism and energy bands of solid C_{60}. *Phys. Rev. Lett.* 66: 2637–2640.
Sakurai J. J. 1999. *Modern Quantum Mechanics.* Reading, MA: Addison-Wesley.
Samal P. and Harbola M. K. 2006. Exploring foundations of time-independent excited-state density-functional theory. *J. Phys. B: At. Mol. Opt. Phys.* 39: 4065.
Sanyal B., Bengone O., and Mirbt S. 2003. Electronic structure and magnetism of Mn-doped GaN. *Phys. Rev. B* 68: 205210.
Sato K. and Katayama-Yoshida H. 2002. First principles materials design for semiconductor spintronics. *Semicond. Sci. Technol.* 17: 367–376.

Sato. K. et al. 2010a. First-principles theory of dilute magnetic semiconductors. *Rev. Mod. Phys.* 82: 1633–1690.

Sato T. et al. 2010b. Direct evidence for the Dirac-cone topological surface states in the ternary chalcogenide $TlBiSe_2$. *Phys. Rev. Lett.* 105: 136802.

Sato T. et al. 2011. Unexpected mass acquisition of Dirac fermions at the quantum phase transition of a topological insulator. *Nat. Phys.* 7: 840–844.

Savrasov S. Y., Kotliar G., and Abrahams E. 2000. Correlated electrons in—Plutonium within a dynamical mean-field picture. *Nature* 410: 793–795.

Scandolo S. 2003. Liquid–liquid phase transition in compressed hydrogen from first-principles simulations. *Proc. Natl. Acad. Sci. USA* 100: 3051–3053.

Scandolo, S., Bernasconi M., Chiarotti G. L., Focher P., and Tosatti E. 1995. Pressure-induced transformation path of graphite to diamond. *Phys. Rev. Lett.* 74: 4015–4018.

Schlier R. E. and Farnsworth H. E. 1959. Structure and adsorption characteristics of clean surfaces of germanium and silicon. *J. Chem. Phys.* 30: 917–926.

Schwarz K. and Blaha P. 2003. Solid state calculations using WIEN2k. *Comput. Mater. Sci.* 28: 259–273.

Scuseria G. E. 1991. *Ab initio* theoretical predictions of the equilibrium geometries of C_{60}, $C_{60}H_{60}$ and $C_{60}F_{60}$. *Chem. Phys. Lett.* 176: 423–427.

Seitz F. 1940. *The Modern Theory of Solids*. New York, NY: McGraw-Hill Book Company.

Shaner F. W., Brown J. M., Swenson C. A., and McQueen R. G. 1984. Sound velocity of carbon at high pressures. *J. Phys. (Paris), Colloq.* 45: C8–235–237.

Shi S. et al. 2003. First-principles studies of cation-doped spinel $LiMn_2O_4$ for lithium ion batteries. *Phys. Rev. B* 67: 115130.

Shrinagar A., Garg A., Prasad R., and Auluck S. 2008. Phase stability in ferroelectric bismuth titanate: A first principles study. *Acta Crystallogr. A* 64: 368–375.

Shukla N. N. and Prasad R. 2004. Energy functional dependence of exchange coupling and magnetic properties of Fe/Nb multilayers. *Phys. Rev. B* 70: 014420.

Shukla N. N. and Prasad R. 2006. Electronic structure of $LiMnO_2$: Comparative study of the LSDA and LSDA+*U* methods. *J. Phys. Chem. Solids* 67: 1731–1740.

Shukla N. N., Sen A., and Prasad R. 2007. Quantum well states in Fe/Nb(001) multilayers: First principles study. *Phys. Rev. B* 76: 174417.

Shukla N. N., Shukla S., Prasad R., and Benedek R. 2008. Phase stability of cation-doped $LiMnO_2$ within GGA+*U* approximation. *Model. Simul. Mater. Sci. Eng.* 16: 055008.

Shukla S., Kumar D., Shukla N. N., and Prasad R. 2004. Peculiar insulator-to-metal transition in tetrahedral semiconductors on lattice expansion. *Int. J. Mod. Phys. B* 18: 975–988.

Silvera I. 2010. The insulator–metal transition in hydrogen. *Proc. Natl. Acad. Sci* 107: 12743–12744.

Singh B., Sharma A., Lin H., Hasan M. Z., Prasad R., and Bansil A. 2012. Topological electronic structure and Weyl semimetal in the $TlBiSe_2$ class of semiconductors. *Phys. Rev B* 86: 115208.

Singh D. J. 1997. Magnetic and electronic properties of $LiMnO_2$. *Phys. Rev. B* 55: 309–312.

Singh D. J. and Du M. J. 2008. $LaFeAsO_{1-x}F_x$: A low carrier density superconductor near itinerant magnetism. *Phys. Rev. Lett.* 100: 237003.

Singh D. J. and Nordstrom L. 2006. *Planewaves, Pseudopotentials, and the LAPW Method* (2nd edition). New York, NY: Springer.

Singh D. J., Seo S. S. A., and Lee H. N. 2010. Optical properties of ferroelectric $Bi_4Ti_3O_{12}$. *Phys. Rev. B* 82: 180103(R).

Singh G., Gupta S. L., Prasad R., Auluck S., Gupta R., and Sil A. 2009. Suppression of Jahn–Teller distortion by chromium and magnesium doping in spinel $LiMn_2O_4$: A first-principles study using GGA and GGA + U. *J. Phys. Chem. Solids* 70: 1200–1206.

Singh R., Prakash S., Shukla N. N., and Prasad R. 2004. Sample dependence of structural, vibrational and electronic properties of *a*–Si:H: A density functional based tight-binding study. *Phys. Rev. B* 70: 115213.

Singleton J. 2001. *Band Theory and Electronic Properties of Solids.* New York, NY: Oxford University Press.

Singwi K. S., Tosi M. P., Land R. H., and Sjolander A. 1968. Electron correlations at metallic densities. *Phys. Rev.* 176: 589–599.

Slater J. C. 1929. The theory of complex spectra. *Phys. Rev.* 34: 1293–1322.

Slater J. C. 1934. Electronic energy bands in metals. *Phys. Rev.* 45: 794–801.

Slater J. C. 1937. Wave functions in a periodic potential. *Phys. Rev.* 51: 846–851.

Smith D. L. and Mailhiot C. 1990. Theory of semiconductor superlattice electronic structure. *Rev. Mod. Phys.* 62: 173–234.

Sommerfeld A. 1928. Zur elektronentheorie der metalle auf grund der Fermischen statistik. *Zeitschrift für Physik* 47: 1–32.

Soven P. 1967. Coherent potential model for substitutional disordered alloys. *Phys. Rev.* 156: 809–813.

Spaldin N. A., Cheong S. W., and Ramesh R. 2010. Multiferroics: Past, present and future. *Physics Today* 63(10): 38–43.

Springborg M. 2000. *Methods of Electronic Structure Calculations.* Chichester: John Wiley & Sons Ltd.

Srivastava G. P. and Weaire D. 1987. The theory of the cohesive energies of solids. *Adv. Phys.* 36: 463–517.

Staebler D. L. and Wronski C. R. 1977. Reversible conductivity changes in discharge-produced amorphous Si. *Appl. Phys. Lett.* 31: 292–294.

Stafford C. A., Baeriswyl D., and Bürki J. 1997. Jellium model of metallic nanocohesion. *Phys. Rev. Lett.* 79: 2863–2866.

Stillinger F. H. and Weber T. A. 1985. Computer simulation of local order in condensed phases of silicon. *Phys. Rev. B* 31: 5262–5271.

Stocks G. M., Temmerman W. M., and Gyorffy B. L. 1978. Complete solution of the Korringa–Kohn–Rostoker coherent-potential-approximation equations: Cu–Ni alloys. *Phys. Rev. Lett.* 41: 339–343.

Street R. A. 1991. *Hydrogenated Amorphous Silicon.* Cambridge: Cambridge University Press.

Sugihara M., Buss V., Entel P., and Hafner J. 2004. The nature of the complex counterion of the chromophore in rhodopsin. *J. Phys. Chem. B* 108: 3673–3680.

Sutton A. P. 1993. *Electronic Structure of Materials.* Oxford: Clarendon Press.

Tarascon J. M. and Armand M. 2001. Issues and challenges facing rechargeable lithium batteries. *Nature* 414: 359–367.

Tavazza F., Levine L. E., and Chaka A. M. 2010. Structural changes during the formation of gold single-atom chains: Stability criteria and electronic structure. *Phys. Rev. B* 81: 235424.

Thingna J., Prasad R., and Auluck S. 2011. Photo-absorption spectra of small hydrogenated silicon clusters using the time-dependent density functional theory. *J. Phys. Chem. Solids* 72: 1096–1100.

Thomas L. H. 1927. The calculation of atomic fields. *Proc. Camb. Philos. Soc.* 23: 542–548.

Tinkham M. 1964. *Group Theory and Quantum Mechanics*. New York, NY: McGraw-Hill Book Company.

Tomanek D. and Schülter M. A. 1987. Structure and bonding of small semiconductor clusters. *Phys. Rev. B* 36: 1208–1217.

Tomanek D. and Schlüter M. A. 1991. Growth regimes of carbon clusters. *Phys. Rev. Lett.* 67: 2331–2334.

Tran F. and Blaha P. 2009. Accurate band gaps of semiconductors and insulators with a semilocal exchange-correlation potential. *Phys. Rev. Lett.* 102: 226401.

Troullier N. and Martins J. L. 1991. Efficient pseudopotentials for plane-wave calculations. *Phys. Rev. B* 43: 1993–2006.

Tsui D. C., Stormer H. L., and Gossard A. C. 1982. Two-dimensional magnetotransport in the extreme quantum limit. *Phys. Rev. Lett.* 48: 1559–1562.

Vanderbilt D. 1990. Soft self-consistent pseudopotentials in a generalized eigenvalue formalism. *Phys. Rev. B* 41: 7892–7895.

Van de Walle C. G. and Martin R. M. 1987. Theoretical study of band offsets at semiconductor interfaces. *Phys. Rev. B* 35: 8154–8165.

Van Orden A. and Saykally R. J. 1998. Small carbon clusters: Spectroscopy, structure, and energetics. *Chem. Rev.* 98: 2313–2357.

van Schilfgaarde M. and Harrison W. A. 1993. *Phys. Rev. Lett.* 71: 3870–3873.

Vashishta P. and Singwi K. S. 1972. Electron correlations at metallic densities. *Phys. Rev. B* 6: 875–887.

Vasiliev I., Öğüt S., and Chelikowsky J. R. 1997. *Ab initio* calculations for the polarizabilities of small semiconductor clusters. *Phys. Rev. Lett.* 78: 4805–4808.

Velický S., Kirkpatrick S., and Ehrenreich H. 1968. Single-site approximations in the electronic theory of simple binary alloys. *Phys. Rev.* 175: 747–766.

Verlet L. 1967. Computer "experiments" on classical fluids. I. Thermodynamical properties of Lennard–Jones molecules. *Phys. Rev.* 159: 98–103.

Vitos L., Korzhavyi P. A., and Johansson B. 2002. Elastic property maps of austenitic stainless steels. *Phys. Rev. Lett.* 88: 155501-1–155501-4.

von Barth U. and Hedin L. 1972. A local exchange-correlation potential for the spin polarized case. *J. Phys. C: Solid State Phys.* 5: 1629–1642.

von Klitzing K., Dorda G., and Pepper M. 1980. New method for high-accuracy determination of the fine-structure constant based on quantized Hall resistance. *Phys. Rev. Lett.* 45: 494–497.

Vosko S., Wilk L., and Nusair M. 1980. Accurate spin-dependent electron liquid correlation energies for local spin density calculations: A critical analysis. *Can. J. Phys.* 58: 1200–1211.

Wallace P. R. 1947. The band theory of graphite. *Phys. Rev.* 71: 622–634.

Wang X.-F., Scandolo, S., and Car R. 2005. Carbon phase diagram from *ab initio* molecular dynamics. *Phys. Rev. Lett.* 95: 185701.

Wannier G. H. 1937. The structure of electronic excitation levels in insulating crystals. *Phys. Rev.* 52: 191–197.

Wannier G. H. 1962. Dynamics of band electrons in electric and magnetic fields. *Rev. Mod. Phys.* 34: 645–655.

Wei P. and Qi Z. Q. 1994. Electronic structure of La_2CuO_4 and $YBa_2Cu_3O_6$: A local-spin-density approximation with on-site coulomb-U correlation calculation. *Phys. Rev. B* 49: 12159–12164.

Weizsäcker C. F. V. 1935. Zur theorie der Kernmassen. *Zeitschrift für Physik* 96: 431–458.

Wigner E. P. 1934. On the interaction of electrons in metals. *Phys. Rev.* 46: 1002–1011.

Wigner E. P. and Huntington H. B. 1935. On the possibility of a metallic modification of hydrogen. *J. Chem. Phys.* 3: 764–770.

Wigner E. P. and Seitz F. 1934. On the constitution of metallic sodium. II. *Phys. Rev.* 46: 509–524.

Wigner E. P. and Seitz F. 1955. Qualitative analysis in the cohesion of metals. *Solid State Phys.: Adv. Res. Appl.* 1: 97–126.

Wilson A. H. 1931a. The theory of electronic semi-conductors. *Proc. R. Soc.* A133: 458–491.

Wilson A. H. 1931b. The theory of electronic semi-conductors—II. *Proc. R. Soc.* A134: 277–287.

Wimmer E. et al. 2008. Temperature-dependent diffusion coefficients from *ab initio* computations: Hydrogen, deuterium, and tritium in nickel. *Phys. Rev. B* 77: 134305.

Wooten F., Winer K., and Weaire D. 1985. Computer generation of structural models of Si and Ge. *Phys. Rev. Lett.* 54: 1392–1395.

Xu J. H. and Freeman A. J. 1993. Interlayer-coupling magnetism and electronic structure of Fe/Cr(001) superlattices. *Phys. Rev. B* 47: 165–173.

Xu S. Y. et al. 2011. Topological phase transition and texture inversion in a tunable topological insulator. *Science* 332: 560–564.

Yabana K. and Bertsch G. F. 1996. Time-dependent local-density approximation in real time. *Phys. Rev. B* 54: 4484–4487.

Yabana K. and Bertsch G. F. 1999. Time-dependent local-density approximation in real time: Application to conjugated molecules. *Int. J. Quant. Chem.* 75: 55–66.

Yafet Y. 1987. Ruderman–Kittel–Kasuya–Yosida range function of a one-dimensional free-electron gas. *Phys. Rev. B* 36: 3948–3949.

Yamasaki A., Chioncel L., Lichtenstein A. I., and Andersen O. K. 2006. Model Hamiltonian parameters for half-metallic ferromagnets NiMnSb and CrO_2. *Phys. Rev. B* 74: 024419.

Yang S. et al. 1988. UPS of 2–30-atom carbon clusters: Chains and rings. *Chem. Phys. Lett.* 144: 431–436.

Yin M. T. and Cohen M. L. 1982. Theory of static structural properties, crystal stability, and phase transformations: Application to Si and Ge. *Phys. Rev. B* 26: 5668–5687.

Yu J. et al. 1987. Bonds, bands, charge transfer excitations and superconductivity of $YBa_2Cu_3O_7$–δ. *Phys. Lett. A* 122: 203–208.

Zallen R. 1998. *The Physics of Amorphous Solids*. New York, NY: John Wiley and Sons.

Zeller R., Deutz J., and Dederichs P. H. 1982. Application of complex energy integration to selfconsistent electronic structure calculations. *Solid State Commun.* 44: 993–997.

Zhang F. C. and Rice T. M. 1988. Effective Hamiltonian for the superconducting Cu oxides. *Phys. Rev. B* 37: 3759–3761.

Zhang Q. M., Yi J. Y., and Bernholc J. 1991. Structure and dynamics of solid C_{60}. *Phys. Rev. Lett.* 66: 2633–2636.

Zhong W. et al. 1994. Giant LO–TO splitting in perovskite ferroelectrics. *Phys. Rev. Lett.* 72: 3618–3621.

Ziman J. M. 1969. *Principles of the Theory of Solids*. Cambridge: Cambridge University Press.

Ziman J. M. 1982. *Models of Disorder.* Cambridge: Cambridge University Press.

Zubko P., Gariglio S., Gabay M., Ghosez P., and Triscone J. M. 2011. Interface physics in complex oxide heterostructures. *Annu. Rev. Condens. Matter Phys.* 2: 141–165.

Index

A

Ab initio. See First-principles
Amorphous materials, 179, 241
Amorphous silicon, 244, 248
Anderson localization, 245
Anderson, P.W., 241, 245, 246
Anderson transition, 248
Armchair tube, 310
Atomic clusters, 255
Atomic pseudopotential, 119
  construction, 133
  first-principles, 127, 132, 133
Atomic units, 413
Augmented plane wave (APW) method, 172
Average *t*-matrix approximation (ATA), 185

B

Band insulator, 86, 335
Band offsets, 290
Band structure, 84, 174
  copper, 85
  Cu–Ge alloys, 204
  graphene, 303
  La_2CuO_4, 358, 359
  LaOFeAs, 366, 367
  NiMnSb, 377, 379
  nanowire, 275
  silicon, 87
  STO/BFO, 294
  $TlBiS_2$, 335
  $TlBiTe_2$, 334
  YBCO, 362, 363
  zigzag tube, 312
Band theory, 67
Battery
  primary, 385
  secondary, 385
Battery materials, 385
Bednorz J. G., 357, 358
Berry phase, 343
Beyond CPA, 206
Beyond LDA, 61
Beyond the Hartree–Fock theory, 37
Bloch, F., 2, 69
Bloch's theorem, 69
BOMD. *See* Born–Oppenheimer MD
BOPES, 18
Born, M., 12
Born effective charge, 345
Born–Oppenheimer
  approximation, 12
  MD, 209, 214
  potential energy surface. *See* BOPES
Brillouin zone (BZ), 73
Brillouin zone integration, 92
Bulk modulus, 400, 401

C

C_{60}, 269
Calculation of forces, 225
Calculation of physical properties, 212
Calculation of total energy, 142
Cancellation theorem, 117
Carbon clusters, 268
Carbon nanotubes, 307
Car–Parrinello MD, 6, 209, 215, 417
Cell voltage, 391
Charge density, 90
  copper, 91
  silicon, 91
  STO/BFO, 294, 295
Chiral tube, 310
Chirality, 304, 307
Classical Hall effect, 317
Classical MD, 209, 210
Coherent potential approximation (CPA), 185
Cohesive energy, 401
Common tangent line, 400
Complex energy bands, 203
  Cu–Al alloy, 204
  Cu–Ge alloy, 204

Configuration interaction (CI) method, 30
Continuous random network (CRN), 244
Correlation energy, 37
Correlations, 29
Coulomb correlations, 29
CPMD. *See* Car–Parrinello MD
Crystal potential, 68
Cuprates, 357, 358

D

Dense random packing (DRP), 243
Density functional theory, 41
Density of states, 33, 87
  amorphous Si, 251, 252
  copper, 89
  Co_2MnSi, 378, 380
  cubic single band tight binding model, 108, 109
  Cu–Ge alloy, 202
  GaN, 382
  LaOFeAs, 366
  $LiMn_2O_4$, 396, 397
  monoclinic $LiMnO_2$, 392
  nanowire, 275
  NiMnSb, 378, 379
  rhombohedral $LiMnO_2$, 393
  silicon, 90
  STO/BFO, 294
  van Hove singularities, 108
  zigzag tube, 312, 314
DFT. *See* Density functional theory
Diffusion constant, 213
Dilute magnetic semiconductor (DMS), 369, 380
Dirac cone, 304, 332
Dirac fermions, 304
Dirac points, 304
Disordered alloys, 179, 205
  Fermi surface, 205
  general theory, 184
  single band tight binding model, 194
  substitutional, 179
DOS. *See* Density of states
Drude, P., 1, 2
Dyson's equation, 160

E

Electron correlations, 22
Empty lattice approximation, 95
Empty lattice bands, 96–98
Energy bands, 84. *See also* Band structure
Energy cutoff, 145
Equilibrium volume, 401
Ergodic hypothesis, 211

F

Fermi surface, 84
  copper, 86
  Cu–Zn alloy, 206
  disordered alloys, 205
  La_2CuO_4, 360, 361
  LaOFeAs, 366, 367
  YBCO, 363, 364
Ferroelectric materials, 339, 346
  $BaTiO_3$, 346
  $Bi_4Ti_3O_{12}$, 347
Fert, A., 369
Feynman, R., 255
First-principles, 1, 233
First-principles approaches and their limitations, 234
First-principles calculation
  atomic clusters, 259
  Fe/Cr, 373
  Fe/Nb, 374
  length and time scales, 235
  surface states, 286
First-principles electronic structure codes, 409
First-principles MD, 209
Formation energy, 391
Friedel oscillations, 283
Functional, 415
Functional derivative, 415

G

Geim, A., 297, 298
GGA (Generalized gradient approximation), 61
Giant magnetoresistance (GMR), 370, 372
Gibbs free energy, 400
Graphene, 297

Green's function, 151
  free electron, 163
  ordered solid, 181
  single impurity, 183
Ground state structure
  hydrogenated Si clusters, 260, 262
  silicon clusters, 260, 261
Grünberg, P., 369
GW method, 63

H

Half-metallic ferromagnets (HMF), 375
Hartree, D.R., 3
  equation, 20
  method, 18, 30
  potential, 21
Hartree–Fock method, 24, 33
Hellmann–Feynman theorem, 223
Herglotz function, 158
Heterojunctions, 290
Heusler alloys, 377
Higgs mechanism, 336
High temperature superconductors, 357
High T_c materials, 357
Hohenberg, P., 4, 41
Hohenberg–Kohn theorems, 49
Homogeneous electron gas, 30, 33
Honeycomb structure, 297
Hubbard model, 110
Hydrogenated amorphous silicon, 248

I

Impurity in an ordered solid, 181
Independent particle approximation, 22
Instantaneous temperature, 212
Integrated density of states, 201
Interface, 281, 290
Interlayer exchange coupling, 370, 373
Inversion symmetry, 82, 335
Iron-based superconductors, 365
Irreducible Brillouin zone (IBZ), 81

J

Jahn–Teller distortion, 387
Jahn–Teller theorem, 387
Jellium model, 1, 30, 256
  atomic cluster, 257
  nanowire, 272

K

Kamerlingh Onnes, H., 357
KKR-CPA, 196
Klein paradox, 307
Kohn, W., 4, 41, 42
Kohn–Sham equations, 52
Koopmans' theorem, 30
Korringa–Kohn–Rostoker (KKR) method, 166
k-point convergence, 145
Kramers' theorem, 82, 83
KS (Kohn–Sham) eigenvalues, 57
KS energies, 54
KS equations. *See* Kohn–Sham equations
KS Hamiltonian, 54
KS orbitals, 54, 57
KS potential, 54

L

Landau levels, 321, 323
  degeneracy, 323
  filling factor, 324
  spacing, 323
LDA. *See* Local density approximation
LDA + U method, 62
Li-ion battery
  cathode, 387
  material related issues, 386
  principle, 385
Linear augmented plane wave (LAPW) method, 174
Linear muffin-tin orbital (LMTO) method, 170
Linear scaling methods, 176
Lippman–Schwinger equation, 162
List of projects, 411
Local density approximation, 55
Local density of states (LDOS), 88
Local spin density approximation (LSDA), 60
Long range order (LRO), 180
Lorentz, H.A., 1, 2

M

Magic clusters, 256, 259
Magnetic multilayers, 370
Mass spectrum, 258, 259
Materials at high pressures, 399, 400
Materials at high temperatures, 403
Materials design, 233, 237
Materials in extreme environments, 399
Mean squared displacement, 212
Melting line of diamond, 405
Metal to insulator transition, 111, 248
Metallic glasses, 241
Metallization of hydrogen, 404
Method of steepest descent (SD), 220
Mobility edge, 248
Molecular dynamics (MD), 209
Molecular hydrogen, 404
Mott, N.F., 111
Mott insulator, 245
Mott transition, 111
Muffin-tin potential, 127
Muffin-tin radius, 127
Müller, K., 357, 358
Multiferroic materials, 339, 351
    $BiFeO_3$, 351
    $GaFeO_3$, 352
Multi-scale approach, 235
Murnaghan's equation, 401
MWNT, 308

N

Nanotubes, 297, 307
Nanowire, 270
    conductance, 276
    metallic, 277
    semiconductors, 277
Nearly free electron (NFE) model, 98
Nearsightedness, 176
Non-periodic systems, 146
Novoselov, K., 297, 298

O

One electron approximation, 22
Oppenheimer, J.R., 12
Orthogonality hole, 119
Orthogonalized plane wave (OPW) method, 112

P

Pair correlation function, 242
Pair distribution function, 242
Pauli spin matrices, 79
Peierls distortion, 271
Perturbation theory using Green's function, 159
Phase diagram
    carbon, 405, 406
    hydrogen, 404
Phase stability, 386, 400
    $LiMnO_2$, 387
    $LiMn_2O_4$, 395
Phase transformation, 400
Phase-shifts, 128
Phonon density of states, 213
Photoabsorption spectra
    hydrogenated Si clusters, 266
    silicon clusters, 266
Plane wave expansion method, 103
Pnictides, 357
Point group symmetries, 80
Polarization, 339, 340
Pressure induced insulator-metal transition, 400
Projected density of states (PDOS), 89
Projector augmented waves (PAW) method, 144
Pseudopotential, 114–121, 125–144
    construction, 133–135
    norm-conserving, 134
    scattering approach, 127
    secular equation, 137
Pseudopotential generation codes, 410
Pseudospin, 304
Pulay force, 229

Q

Quantization of conductance, 272, 276
Quantum confinement, 259
Quantum description of materials, 11
Quantum Hall effect
    fractional, 317, 328
    integer, 317, 325
Quantum phase transition, 334, 335
Quantum spin Hall effect, 317, 328

Quantum spin Hall states, 329
Quantum well structure, 293

R

Radial distribution function (RDF), 242
Random binary alloy, 179, 180
Reduced dimensionality, 255

S

Screening, 46
Self-energy, 63, 186, 195
Self-interaction correction (SIC) method, 63
Short range order (SRO), 180
Sigma and pi bonds, 421
Simulated annealing, 221
Slater, J.C., 4, 24, 168, 172
Sommerfeld, A., 2
Soven, P., 6
sp, sp^2 and sp^3 hybrids, 423
Space group, 82
Spin polarization, 376
Spin–orbit interaction, 76, 329, 330
Spintronic materials, 369
Spintronics, 369
Structural modeling, 243
Structure–property relationship, 234
Supercell, 5, 146
Superlattice, 292
  Fe/Nb, 374
  $SrTiO_3/BiFeO_3$, 293
Surface, 281
  band structure, 286, 287
  electronic structure, 283, 287
  geometry, 282
  Miller index, 282
  reconstruction, 288
  relaxation, 288
  states, 284
  unit cell, 282
SWNT, 308
Symmetry, 80

T

TDDFT. *See* Time dependent density functional theory
Thomas–Fermi theory, 41, 43, 46
Tight-binding method, 104
Time reversal symmetry, 83, 329
Time-dependent density functional theory, 64, 266
t-matrix, 160, 185
Topological insulator, 317, 330
Topological invariant, 330, 332
Transition pressure, 400
Translational symmetry, 69
Tunnel magneto-resistance (TMR), 369

U

Ultrasoft pseudopotential (US-PP) method, 144

V

Velocity autocorrelation function, 213
Verlet algorithm, 210
Virtual crystal approximation (VCA), 185
von Klitzing, K., 317, 318

W

Wannier function, 111
Warren–Cowely short range parameter, 181
Weyl semimetal, 334, 335
Work function, 284
WWW model, 244, 249

Z

Zigzag tubes, 310

For Product Safety Concerns and Information please contact our EU
representative GPSR@taylorandfrancis.com
Taylor & Francis Verlag GmbH, Kaufingerstraße 24, 80331 München, Germany

www.ingramcontent.com/pod-product-compliance
Ingram Content Group UK Ltd.
Pitfield, Milton Keynes, MK11 3LW, UK
UKHW021115180425
457613UK00005B/90